Milestones in Water Reuse

Milestones in Water Reuse

The Best Success Stories

Valentina Lazarova, Takashi Asano, Akiça Bahri
and John Anderson

IWA Publishing
London • New York

Published by **IWA Publishing**
Alliance House
12 Caxton Street
London SW1H 0QS, UK
Telephone: +44 (0)20 7654 5500
Fax: +44 (0)20 7654 5555
Email: publications@iwap.co.uk
Web: www.iwapublishing.com

First published 2013
Reprinted 2013

British Library Cataloguing in Publication Data
A CIP catalogue record for this book is available from the British Library

ISBN: 9781780400075 (Paperback)
ISBN: 9781780400716 (eBook)

Printed and bound in Great Britain by Bell and Bain Ltd., Glasgow

MIX
Paper from responsible sources
FSC® C007785

Contents

List of contributors

EDITORS

Valentina Lazarova*

Technical Advisor and Senior Expert
Suez Environment – CIRSEE
38 rue du president Wilson
78230 Le Pecq, France
valentina.lazarova@suez-env.com

Prof. Takashi Asano

Professor Emeritus
Department of Civil and
Environmental Engineering
University of California at Davis
Davis, CA 95616, USA
tasano@ucdavis.edu

Akiça Bahri

Coordinator
African Water Facility African
Development Bank
13, rue du Ghana, B.P. 323
1002 Tunis Belvédère, Tunisia
A.BAHRI@AFDB.ORG

John Anderson

Afton Water, I Cumbora Circuit
Berowra, NSW 2081
Australia
ja.afton@gmail.com

FOREWORD

Prof. Rafael Mujeriego

Professor of Environmental
Engineering (retired)
President, Spanish Association
for Sustainable Water Reuse (ASERSA)
Universidad Politécnica de Catalunya
Barcelona, Spain
rafael.mujeriego@upc.edu

INTRODUCTORY CHAPTER

Valentina Lazarova*

Technical Advisor and Senior Expert
Suez Environnement – CIRSEE
38 rue du president Wilson
78230 Le Pecq, France
valentina.lazarova@suez-env.com

Prof. Takashi Asano

Professor Emeritus
Department of Civil and
Environmental Engineering
University of California at Davis
Davis, CA 95616, USA
tasano@ucdavis.edu

CHAPTER 1

John Anderson

Afton Water, I Cumbora Circuit
Berowra, NSW 2081
Australia
ja.afton@gmail.com

CHAPTER 2

Joe Walters*

Manager of Business Development &
Regulatory Affairs
West Basin Municipal Water District
17140 South Avalon Boulevard
Suite 210,
Carson, CA 90746-1269, USA
JoeW@westbasin.org

Gregg Oelker

Manager of Water Quality
United Water Services
Edward C Little Water Recycling Facility
1935 South Hughes Way
El Segundo, CA 90245, USA
Gregg.Oelker@UnitedWater.com

---

*Corresponding author

Valentina Lazarova
Technical Advisor and Senior Expert
Suez Environnement – CIRSEE
38 rue du president Wilson
78230 Le Pecq, France
valentina.lazarova@suez-env.com

CHAPTER 3

Lim Mong Hoo*, Harry Seah
Deputy Director
PUB, Water Quality Office,
40 Scotts Road #14-01, Environmental Building
Singapore 228231,
Republic of Singapore
LIM_Mong_Hoo@pub.gov.sg

CHAPTER 4

Lluis Sala
Consorci Costa Brava
Plaça Josep Pla, 4, 3r 1a
17001 Girona, Spain
lsala@ccbgi.org

CHAPTER 5

Iacovos Papaiacovou*, Athina Papatheodoulou
General Manager Sewerage
Board of Limassol Amathus
76, Franklin Roosvelt Block A
PO Box 50622
3608 Limassol, Cyprus
iacovos@sbla.com.cy

CHAPTER 6

Prof. Yu Zhang*, Prof. Min Yang
State Key Lab. of Env. Aquatic Chem.
Research Center for Eco-Env. Sciences
Chinese Academy of Sciences
P.O. Box 2871
Beijing 100085, China
zhangyu@rcees.ac.cn

Fusheng Tang, Dianhai Li, Yuhong Li
Tianjin Water Recycling Co. Ltd., #9-11, B Area
JinDianShiDai Commercial Street Zhujiang Road
Hexi District
Tianjin, 300221, China

Prof. Weiping Chen
State Key Laboratory of Urban and Regional Ecology
Chinese Academy of Sciences
P.O. Box 2871, Beijing 100085, China

CHAPTER 7

Marius Mohr*, Prof. Walter Trösch
Fraunhofer Institute for Interfacial
Engineering and Biotechnology
Nobelstr. 12,
70569 Stuttgart, Germany
marius.mohr@igb.fraunhofer.de

CHAPTER 8

Andrés Deza, Avelino Martínez*
Canal Isabel II - Gestión
Santa Engracia, 125,
28003 Madrid, Spain
amarher@gestioncanal.es
adeza@gestioncanal.es

CHAPTER 9

Scott Edwards*, Fred Layi
Executive VP Communications
Veolia Water North America
200 East Randolph Drive, Suite 7900
Chicago, IL 60601, USA
scott.edwards@veoliawaterna.com

CHAPTER 10

Valentina Lazarova*
Technical Advisor and Senior Expert
Suez Environnement – CIRSEE
38 rue du president Wilson
78230 Le Pecq, France
valentina.lazarova@suez-env.com

Vincent Sturny
SPEA
98713 Papeete-Tahiti
French Polynesia

Gaston Tong Sang
Mayor of Bora Bora
Municipality of Bora Bora, 98730, Vaitape
Bora Bora, French Polynesia

CHAPTER 11

John Anderson
Afton Water, I Cumbora Circuit
Berowra, NSW 2081, Australia
ja.afton@gmail.com

CHAPTER 12

Kiyoaki Kitamura
Tokyo Metropolitan Government
2-8-1 Nishi-shinjuku, Shinjuku-ku
Tokyo, 163-8001, Japan
Kiyoaki_Kitamura@member.metro.tokyo.jp

Kingo Saeki
Executive Director
Japan Sewage Works Association
Uchikanda "Suisui" Building
2-10-12 Uchikanda
Chiyodaku, Tokyo 101-0047, Japan

Prof. Naoyuki Funamizu*
Professor Hokkaido University
Division of Environmental Eng.
Kita-13 Nishi-8,
Sapporo 060-8628, Japan
funamizu@eng.hokudai.ac.jp

CHAPTER 13

Yanjin Liu*, Eugenio Giraldo, Mark LeChevallier
American Water, 3906 Church Road,
Mount Laurel
NJ 08054, USA
Yanjin.Liu@amwater.com

CHAPTER 14

Assoc. Prof. Katsuki Kimura
Hokkaido University, Div. of Environmental Eng.
Kita-13 Nishi-8, Sapporo 060-8628, Japan
kkatsu@eng.hokudai.ac.jp

Prof. Naoyuki Funamizu*
Hokkaido University,
Division of Environmental Eng.
Kita-13 Nishi-8,
Sapporo 060-8628, Japan
funamizu@eng.hokudai.ac.jp

Yusuke Oi
KUBOTA Corporation
1-3, Nihonbashi-Muromachi,
3-chome
Chuo-ku, Tokyo 103-8310,
Japan

CHAPTER 15

Roberto Mazzini*
Operation Division Manager
c/o SIBA SpA
Via Lampedusa 13/D
20141 Milano, Italy
mazzini@sibaspa.it

Luca Pedrazzi
DEGREMONT ITALY
Via Benigno Crespi, 57
20159 Milano, Italy

Valentina Lazarova
Technical Advisor and Senior Expert
Suez Environnement – CIRSEE
38 rue du president Wilson
78230 Le Pecq,
France

CHAPTER 16

Antoine Fazio*
Manager of Urban Water
SCE NANTES
5 Avenue Augustin-Louis CAUCHY
BP 10703, 44307
Nantes Cedex 3,
France
antoine.fazio@sce.fr

Noël Faucher
President
Community of Municipalities
Rue de la Prée au Duc, BP 714
85330 Noirmoutier en l'Ile
cc-iledenoirmoutier@wanadoo.fr

Valentina Lazarova
Technical Advisor and Senior Expert
Suez Environnement – CIRSEE
38 rue du president Wilson
78230 Le Pecq, France

CHAPTER 17

Daryl Stevens

Atura Pty Ltd, Suite 204, 198 Harbour Esplanade
Docklands, Melbourne, Victoria 3008, Australia
daryl@atura.com.au

John Anderson*

Afton Water, I Cumbora Circuit
Berowra, NSW 2081, Australia
ja.afton@gmail.com

CHAPTER 18

Alberto Rojas

Director of Sanitation
State Water Commission
Mariano Otero No. 905
Barrio de Tequisquiapam, SLP
78250 San Luis Potosi, Mexico
saneamientocea@gmail.com

Lucina Equihua*

O&M Director
DEGREMONT MEXICO
Av. Paseo de la Reforma 350 Piso 15
Col. Juárez, 06600 Mexico D.F., Mexico
lucina.equihua@degremont.com

Valentina Lazarova

Technical Advisor and Senior Expert
Suez Environnement – CIRSEE
38 rue du president Wilson
78230 Le Pecq, France

CHAPTER 19

Josef Lahnsteiner*

Director
VA TECH WABAG GmbH,
Dresdner Strasse 87–91
1200 Vienna, Austria
Josef.Lahnsteiner@wabag.com

Srinivasan Goundavarapu

O&M Manager
VATECH WABAG Ltd
Murrays Gate Road, Alwarpat
Chennai 600018, India
g_srinivasan@wabag.in

Patrick Andrade

Head of Industrial Water Group
VATECH WABAG Ltd
Harsh Orchid Bldg., Nagras Road,
New D.P. Road, Aundh
411007, Pune, India
patrick.pune@wabag.in

Rajiv Mittal

MD and CEO
VATECH WABAG Ltd
Murrays Gate Road, Alwarpat
Chennai 600018, India
r_mittal@wabag.in.

Rajkumar Ghosh

Director (Refineries)
Indian Oil Corporation Ltd.
SCOPE Complex, Core -2
7, Institutional Area,
Lodhi Road,
New Delhi - 411 003, India
ghoshrk@iocl.co.in.

CHAPTER 20

Alice Towey

Associate Engineer
East Bay Municipal Utility District
375, 11th St., Oakland, CA 94607, USA
atowey@ebmud.com

Jan Lee

Senior Engineer
East Bay Municipal Utility District
375, 11th St., Oakland
CA 94607, USA
jrlee@ebmud.com

Sanjay Reddy*

Project Director, Black & Veatch
2999 Oak Rd, Suite 490, Walnut Creek
CA 94597, USA
reddysp@bv.com

James Clark

Senior Vice President, Black & Veatch
800 Wilshire Blvd, Suite 600
Los Angeles, CA 90017, USA
clarkjh@bv.com

CHAPTER 21

Prof. Dr.-Ing. Karl-Heinz Rosenwinkel*, Sabrina Kipp
Institute for Sanitary Engineering and
Waste Management, Leibniz University, Hannover
Welfengarten 1, 30167 Hannover, Germany
rosenwinkel@isah.uni-hannover.de

Dr.-Ing. Axel Borchman
Fed. Ministry for Environment
Nature Conserv. & Nuclear Safety
Divsion WA I 1
Robert-Schuman-Platz 3
53175 Bonn, Germany

Dr.-Ing. Markus Engelhart
EnviroChemie
In den Leppsteinswiesen 9
64380 Rossdorf, Germany

Rüdiger Eppers
Volkswagen AG, P. O. Box 1837
38436 Wolfsburg, Germany

Holger Jung
PTS – Papiertechnische Stiftung
Res. Mngt – Waste and Energy
Hessstr. 134,
80797 Munich, Germany

Prof. Dr. Joachim Marzinkowski
Bergische Universität Wuppertal
Gaußstr. 20.
492097 Wuppertal, Germany

CHAPTER 22

Kiyoaki Kitamura
Tokyo Metropolitan Government
2-8-1 Nishi-shinjuku, Shinjuku-ku
Tokyo, 163-8001, Japan
Kiyoaki_Kitamura@member.metro.tokyo.jp

Prof. Naoyuki Funamizu*
Hokkaido University
Div. of Environmental Eng.
Kita-13 Nishi-8
Sapporo 060-8628, Japan
funamizu@eng.hokudai.ac.jp

Prof. Shinichiro Ohgaki
President
National Institute for Environmental
Studies (NIES)
16-2 Onogawa, Tsukuba-City
Ibaraki, 305-8506, Japan
ohgaki@nies.go.jp

Kingo Saeki
Executive Director
Japan Sewage Works Association
Uchikanda "Suisui" Building
2-10-12 Uchikanda
Chiyodaku, Tokyo 101-0047, Japan

CHAPTER 23

Ying-Xue Sun
Associate professor
Beijing Technology and Business University
Beijing 100084, China
syingxue@126.com

Prof. Hong-Ying Hu*
Professor, Env. Simulation and
Pollution Control State Key Joint Lab.
School of Environment
Tsinghua University
Beijing 100084, China
hyhu@tsinghua.edu.cn

Josef Lahnsteiner
Director
VA TECH WABAG GmbH
Dresdner Strasse 87-91
1200 Vienna, Austria

Yu Bai, Yi-Ping Gan
Senior engineer
Beijing Drainage Group, Co. Ltd.
Beijing 100044, China

Ferdinand Klegraf
Senior Process Engineer
VA TECH WABAG GmbH
Dresdner Strasse 87–91
1200 Vienna, Austria

CHAPTER 24

Blanca Jiménez-Cisneros
Investigadora Titular
Instituto de Ingeniería, UNAM
Apdo Postal 70472
04510 COYOCAN DF, MEXICO
BJimenezC@iingen.unam.mx

CHAPTER 25

Bruce Chalmers*
Bruce Chalmers, P.E.
Vice President
CDM Smith
111 Academy Way, Suite 150
Irvine, California 92617, USA
chalmersrb@cdmsmith.com

Mehul V. Patel
GWRS Program Manager
Orange County Water District
18700 Ward Street
Fountain Valley,
CA 92708, USA

CHAPTER 26

Emmanuel Van Houtte*
Geologist, R&D
I.W.V.A.
Doornpannestraat 1,
8670 Koksijde, Belgium
emmanuel.van.houtte@iwva.be

Johan Verbauwhede
Managing Director
I.W.V.A.
Doornpannestraat 1,
8670 Koksijde, Belgium

CHAPTER 27

Robert W. Angelotti*
Director, Technical Services Division
Upper Occoquan Service Authority
14631 Compton Road
Centreville, Virginia
20121-2506, USA
bob.angelotti@uosa.org

Thomas J. Grizzard
Professor
Director of the Occoquan
Watershed Laboratory
Occoquan Watershed
Monitoring Laboratory
9408 Prince William Street Manassas
VA 20110, USA

CHAPTER 28

Troy Walker
Technical Manager
Australia/New Zealand
Veolia Water Australia
PO Box 10819, Adelaide Street
Post Office,
Level 15, 127 Creek Street,
Brisbane, Qld 4000, Australia
troy.walker@veoliawater.com.au

CHAPTER 29

Josef Lahnsteiner*
Director
VA TECH WABAG GmbH
Dresdner Strasse 87-91
1200 Vienna, Austria
Josef.Lahnsteiner@wabag.com

Piet du Pisani
Strategic Executive, City of Windhoek
Dept. of Infrastructure
Water & Waste Management
Box 59, Windhoek, Namibia

Jürgen Menge
Chief Scientist Services,
City of Windhoek Dept. of Infrastructure
Water Technical Services
Box 59, Windhoek, Namibia

John Esterhuizen
Managing Director
Windhoek Goreangab Operating Company
Matshitshi Street, Goreangab Ext. 3,
PO Box 2103, Windhoek, Namibia

Preface

With the growing interest in water reuse, many scientific books and guidelines have been published on this topic. Although there is a long history of successful water reuse projects worldwide, new water reuse projects often face challenges in gaining approval or in implementation and operation. The advantages of water reuse in increasing available water resources remain little-known and not well understood. Furthermore, the economic and environmental benefits of water reuse projects are difficult to assess and demonstrate.

In this context, the IWA Specialist Group on Water Reuse have decided to publish this comprehensive compendium of worldwide practices entitled "The Milestones in Water Reuse: The Best Success Stories", which illustrates the benefits of water reuse in the context of integrated water resources management, its role for the improvement of urban water cycle management, for adapting to climate change and for development of the cities of the future as sustainable and environmentally friendly human habitats. Cornerstone water reuse projects and little-known case studies have been selected from different countries and for different water reuse applications to illustrate the keys to success and lessons learned from their operation.

The main focus and purpose of this book is in unison with our mission: to facilitate the implementation of safe water reuse practices through the development of successful water reuse projects, to highlight innovative water, waste and energy management tools and to share information on state-of-the art water reuse practices via our international knowledge network. This publication is unique and original because it comprehensive overview of water reuse success stories including the drivers, the technical challenges, the regulatory framework, the financial and economic aspects, as well as the political engagement, public attitudes and stakeholders' involvement.

The case studies are categorised and presented in eight parts:

- Introductory Chapter: Milestones in Water Reuse: Main Challenges, Keys to Success and Trends of Development, An Overview — this chapter provides a general overview of current water reuse drivers and practices and defines key milestones in the development of best-in-class water reuse practices.
- Part 1: Role of Water Reuse for Integrated Resource Management — six case studies describe innovative projects in Australia, West Basin (California, USA), Singapore, Costa Brava (Spain), Cyprus and Tianjin (China).
- Part 2: Urban Use of Recycled Water: five case studies present successful water reuse practices in Germany, Madrid (Spain), Honolulu (Hawaii), Bora Bora (French Polynesia) and Australia.
- Part 3: Urban Use: Decentralised Water Recycling Systems: three case studies illustrate the contribution of water reuse to the decentralised urban water management cycles of Tokyo and in high-rise buildings in New York (USA) and Japan.
- Part 4: Agricultural Use of Recycled Water: three case studies discuss the challenges of water reuse in agriculture in Milan (Italy), Noirmoutier (France) and Australia.
- Part 5: Industrial Use of Recycled Water: four studies provide useful information on the success of water recycling of treated municipal effluent as cooling make-up water for power plant in San Luis Potosi (Mexico) and boiler make-up water for power generation facility in California (USA), as well as for industrial wastewater recycling in Panipat (India) and Germany.
- Part 6: Environmental and Recreational Use of Recycled Water: the benefits of environmental water reuse are demonstrated by three case studies in Tokyo (Japan), Beijing (China) and Mexico City (Mexico).
- Part 7: Increasing Drinking Water Supplies: five cornerstone projects are presented: two of aquifer recharge in Orange County (California, USA) and Torreele (Belgium), two of replenishment of surface reservoirs in Virginia (USA) and Queensland (Australia) and the unique direct potable reuse project in Windhoek (Namibia).

We hope that this book will provide water professionals and the public at large — policy makers, elected officials, managers, engineers, planners, operators, recycled water users and community groups — a comprehensive reference with key information that allows to gain a thorough understanding of the best water reuse practices worldwide and will build confidence that recycled water is a safe, economically viable, environmentally friendly product with high social value. We hope also that this publication will motivate more students and young professionals to start working in this challenging field aiming to preserve the beauty of our planet and ensure adequate water supplies for future generations.

Valentina Lazarova

Foreword

Water reclamation and reuse has been an important player in the advancement of water resources management all over the world for more than three decades. Numerous projects and facilities operating in many parts of the world testify the success of this water supply option in improving water resources availability and reliability. By putting together some of its most successful cases, this volume is an emblematic testimony to the contribution that water reclamation and reuse has made to improving water resources management across diverse social, economic and environmental conditions around the world.

The Editors and particularly the authors of each chapter deserve the warmest appreciation for their practical contributions on how water reclamation and reuse concepts have been applied to solve water quality and water availability challenges.

SOME HISTORICAL NOTES

Numerous initiatives at the local, regional and international levels have contributed to the development of water reclamation and reuse throughout the world. The IWA Specialist Group on Water Reuse (Water Reuse Specialist Group, WRSG) has certainly been one of the most active at international level. Under the initiative of Professor Takashi Asano and many other colleagues, the foundations of the WRSG were laid during the IWA biennial Conference in Rio de Janeiro, Brazil in 1986; the Specialist Group was inaugurated during the IWA biennial Conference in Brighton, UK in 1988. WRSG chairs, secretaries, management committees and members at large have played a determining role in the advancement of the group.

The water reclamation and reuse projects presented in this volume have evolved for the most part during the meetings organized by the WRSG over the last 21 years, at a 2 to 3-year interval, since the first Wastewater Reclamation and Reuse Symposium that took place in 1991 in Castell Platja d'Aro, Girona, Spain to the most recent Water Reclamation and Reuse Conference held in Barcelona, Spain, during September 2011. Those and many other worldwide success stories have been instrumental in the advancement of this important water field, by bringing about the subtle perception change from wastewater reuse to water reuse, and by expanding from the initial Symposium in 1991 with 35 oral presentations to the most recent WRSG Conference in 2011 with 125 oral presentations.

CHANGING THEMES OF WATER RECLAMATION AND REUSE

Many things have changed in the water reclamation and reuse field during the last three decades all over the world, as reflected by the success stories described in this volume. One of the most relevant is the recognition of reclaimed (recycled) water as an essential component of integrated water resources management. Reclaimed water has become a new, additional, alternative, reliable water supply source for numerous uses in the diverse environments. As many of the success stories in this volume describe, water supply reliability appears as one of the most valuable feature of reclaimed water, particularly within the prevailing context of "uncertainty" that characterizes all current climate models forecasts.

The water reuse frontier has expanded from agricultural and landscape irrigation and restricted urban uses, to indirect potable reuse and even direct potable reuse at current time. Both traditional and innovative solutions are available to respond to practically any new water demand that may be posed within an integrated water resources management model.

Agricultural and landscape irrigation has expanded from earlier restricted uses to totally unrestricted irrigation of food crops, and particularly to new uses within the urban environment, both under public supervision and private user control.

Health protection, initially centred in microbial quality, has expanded to a wider and more comprehensive view of chemical quality, particularly in association to the commonly known "emerging" contaminants.

Water reclamation technologies have been greatly improved and diversified, in parallel to advances in drinking water research. Reclaimed water and reclamation technologies are being increasingly evaluated taking drinking water quality as a reference, both in developing countries and particularly in developed/industrialized countries.

Success stories included in this volume illustrate the interest that reclaimed water raises among city planners and officials, as an element of urban use development and a local water management option in urban settings, using centralized and decentralized options, providing single or diversified water qualities and uses, and offering always a local and effective response to water shortages.

The success stories described in this volume clearly point out two real challenges: (1) the need to intensify our coverage of "public perception and acceptance" as a frequent limiting factor of the implementation process and (2) the need to recognize the great diversity of water reclamation and reuse solutions available, in contrast to the limited choices we had 30 years ago.

At the same time, the critical conditions brought about by severe droughts and local water scarcity have prompted communities all over the world to turn to indirect and even direct potable reuse as the most reliable and immediate source for water supply. The pressing conditions of "necessity", under no access to conventional water sources, together with the availability of novel technologies, with proven technical and economical capacities, have brought about the "opportunity" for adopting indirect and direct potable reuse schemes. Energy efficiency and greenhouse gas emissions control have also been contributing factors in this new strategy.

IMPORTANCE OF PUBLIC INFORMATION AND PUBLIC PARTICIPATION

Public information and public participation have become an essential component of water resources management, and in particular of reclaimed water management. We still have to involve news media and all means of public communication as our effective partners in those important social issues. Although great progress has been made in developing proactive programs on health and environmental protection, we have to intensify and further develop the scope and reach of our programs.

The last 30 years have witnessed an impressive development of water reclamation and reuse options, as compared to the limited conventional processes available in the 1990's; the success stories in this volume include more sophisticated water reclamation processes and more adaptable treatment lines than those available in the past. Reclamation processes have been developed in close association to drinking water treatment processes, due to the similarity of their ultimate objectives: high water quality, high reliability and high resources and energy efficiency.

Today, water reclamation projects offer a wide spectrum of solutions: from on-site to satellite to centralized facilities; from natural processes to energy intensive processes; from solutions emphasizing basic sanitation to those producing drinking water quality; from solutions addressed to traditional uses (e.g., irrigation) to others addressed to innovative uses, such as indirect and direct potable reuse. Time has also brought about the need to consider additional relevant criteria, like adaptability, reliability, energy efficiency, self-sufficiency, low greenhouse gas emission rates, and in summary, increased sustainability.

RECLAIMED WATER QUALITY GUIDELINES AND STANDARDS

The success stories described in this volume portray different snap shots of an almost universal sequence, as it has been developing in different parts of the world, each with a different starting point and a different rate of progress. Water reuse has been mainly driven by necessity: water scarcity in some areas, serious multiple-year drought episodes in others, and supply unreliability in many others. Reclaimed water users have generally been the main drivers for implementation, with the assistance of water agencies and the support from agencies with less tradition in water reclamation and reuse.

The WRSG has a long history of proposals on the need to develop unified criteria and standards for reclaimed water quality. However, limited progress has been actually achieved, due to differences in understanding of public health, economic, management, and public opinion concerns. As the success stories in this volume present, future solutions to local, regional and international water reclamation and reuse needs will fortunately have a large diversity of regional and national choices, which will have to be selected in accordance with local conditions and international trade regulations.

After decades of intensive debates in most cases, comprehensive reclaimed water quality criteria have been adopted by some countries, marking a turning point in the consideration of reclaimed water as an essential component of integrated management of water resources. The subsequent development of professional associations for the promotion of water reuse has greatly contributed to further advance the objectives and achievements of water agencies.

The Spanish case clearly illustrates this progress pattern. After more than two decades of intensive debates, comprehensive reclaimed water quality criteria and standards were promulgated in 2007, including water rights allocation and reclaimed water cost assignment procedures. The Spanish Association for Sustainable Water Reuse (ASERSA) was established in 2008 with the objective of promoting and documenting water reclamation and reuse and of strengthening international cooperation, in accordance with existing professional associations such as the WateReuse Association in the United States.

SUMMARY

In summary, the main lesson emerging from the reclaimed water success stories presented in this volume is that water reuse is a water resources management issue, strongly supported by the availability of efficient and reliable technologies, but requiring political decision to implement it and adequate management strategies for handling the social, economic and environmental issues involved.

Among the most currently pressing issues, those commonly emphasized are necessity for water reuse, public perception and acceptance, safety and risks of reclaimed water, water rights and economic considerations of reclaimed water. Those success stories clearly illustrate the progress achieved so far in all those areas and offer a practical source of inspiration for the development of effective solutions to the future challenges that integrated management of water resources will be facing in the near future all over the world.

Rafael Mujeriego

INTRODUCTORY CHAPTER

Milestones in water reuse: main challenges, keys to success and trends of development. An overview

Valentina Lazarova and Takashi Asano

Water reuse is increasingly considered a quintessential component of sustainable and integrated water resources management. This chapter provides a general overview of current water reuse drivers and practices and defines key milestones in the development of best water reuse practices.

Why water reuse?

Water is life. It is of fundamental importance for human well-being, quality of life and socio-economic development, as well as for healthy biosphere and ecosystems. Since the birth of humanity, all great civilizations viewed water as a symbol of life and source of life itself.

The human civilization of the 21st century is changing the Earth in ways that threaten its ability to sustain on-going exponential population growth and to maintain existence and survival of other species. A recent review in *Nature* (Barnosky *et al.* 2012) demonstrates that, under the pressure of human activities, a planetary-scale critical transition is taking place leading to drastic changes in the Earth's biosphere, which have the potential to rapidly and irreversibly transform our planet. According to Rockström *et al.* (2009), several planetary boundaries (defined as the safe operation-space preventing unacceptable environmental change), have already been crossed for biodiversity, nutrient cycle and climate change.

While climate change, global warming, energy crisis and population explosion are widely discussed, the alarming extent of water scarcity across the world is not sufficiently understood and taken into account when planning sustainable long-term water supplies for urbanised centres. Water scarcity is not a synonym for water shortage, even though water deficit (shortage) is one of its essential components. Water scarcity includes many other important components such as deterioration of quality of natural water bodies, imbalance between water availability and demand, competition between sectors and even between nations. Water quality degradation is often the major cause of water scarcity and loss of biodiversity, but its impact on global scale has not yet been well assessed. Freshwater bodies have a limited capacity to process the ever increasing pollutant charges from expanding urban, industrial and agricultural water uses. The adverse impact of anthropogenic pollution on marine environment is also on the rise and it is creating dead zones in seas and oceans.

Water scarcity already affects all continents. Water use has been growing at over two times higher rate than population growth in the last century and an increasing number of regions are chronically short of water. Around 1.2 billion people, or almost one-fifth of the world's population, lives in areas of physical scarcity, and over 500 million people are approaching this situation (UN, 2007). Another 1.6 billion people, or almost one quarter of the world's population, faces economic water shortage due to lack funds to build and operate the necessary water supply infrastructure. By 2025, 1.8 billion people will be living in countries or regions with absolute water scarcity, and two-thirds of the world population could be under stress conditions. This situation would be exacerbated in the future as rapidly growing urban areas place heavy pressure on neighbouring water resources.

The daunting problem of water scarcity can be illustrated clearly by the statistics on water availability. The image of our blue planet over 75% which surface is covered with water gives the impression of a global habitat with renewable and inexhaustible water resources. However, the vast majority of the Earth's water resources are saline, with only 2.5% being fresh water. Approximately 70% of the fresh water available on the planet is frozen in the icecaps of Antarctica and Greenland leaving the remaining 30% (equal to only 0.7% of total water resources worldwide) to be available for consumption (IPCC, 2007). Only 0.001% (10,000 to 14,000 km^3) of the planet's total water resources are accessible and renewable freshwater (Figure 1). While the volume of water withdrawn in year 2000 was 4430 km^3, a significant increase in water demand of 5240 km^3 is forecasted by year 2025 (UNEP, 2008). Compared to this raising water demand, the total volume of recycled water remains relatively low, about 11 km^3 in 2009 (30.3 Mm^3/d according to GWI, 2009), which represents only 0.2–0.3% of the total water demand and approximately 5% of the collected and treated wastewater worldwide.

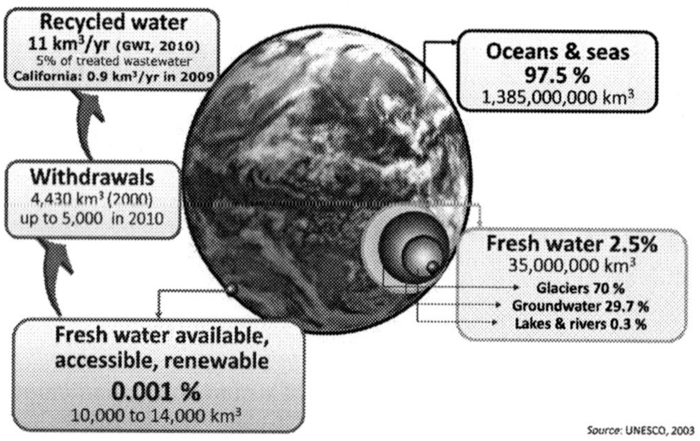

Figure 1 Water statistics – total water demand and volume of recycled water.

The principal driver for water reuse is the increasing water demand caused mainly by rapid population growth and growing consumption which results from improved quality of life and rising per capita income. As illustrated in Figure 2, of the total water withdrawn each year, the volume for agricultural uses represents 70%, that for industry 22% and for municipal consumption of drinking water of 8% only (Shiklomanov, 1999 cited by UNEP, 2008; WWAP, 2009). Approximately one half of the water withdrawn is consumed, mainly in agriculture, while the remaining 50% is returned to water bodies.

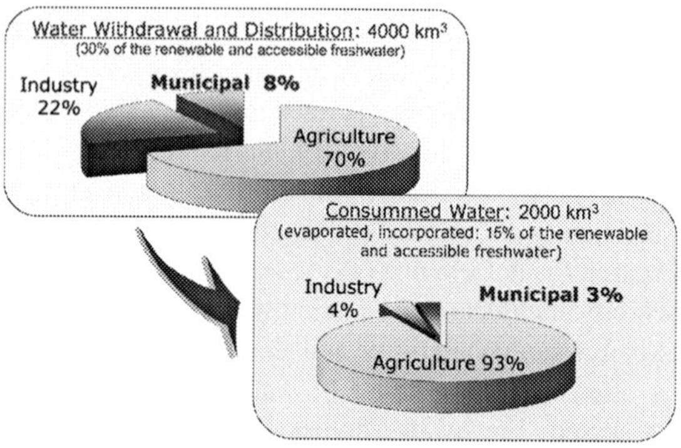

Figure 2 Breakdown by sectors of water withdrawal and water consumption.

The already severe water stress conditions in many regions of the world are expected to escalate as a probable consequence of climate change impacts. As a result, climate change and increased climate variability are likely to be the second major driver for water reuse. The most significant impacts of climate change on water resources are as follows (WWAP, 2012):

- Longer and more severe dry seasons,
- Widespread changes in the distribution of precipitation with more frequent drought and flooding events, leading to overall long-term reduction in river flows and aquifer's recharge rates,
- Increased water use for irrigation,
- Deterioration of the quality of all freshwater sources due to higher temperatures and diminishing flows.

Figure 3 depicts a global map of water stress, expressed by the ratio of the volume of total water withdrawals and total renewable water resources (water availability). This water stress indicator is a criticality important ratio, defined by experts: high water stress is considered to occur when annual water demand is equal to or higher than 40% of the renewable water resources. The map on Figure 3 also shows the regions characterised by another indicator of water stress: the population-water equation – an area is experiencing water stress when annual water supplies drop below 1700 m^3 per person. Finally, few examples of current vulnerabilities of freshwater resources in terms of quantity and quality are also shown on this map, that is the regions in the United States and Canada affected by multi-year droughts; the Murray-Darling basin in Australia with ecosystems damaged by decreasing river flows and polluted aquifers in India.

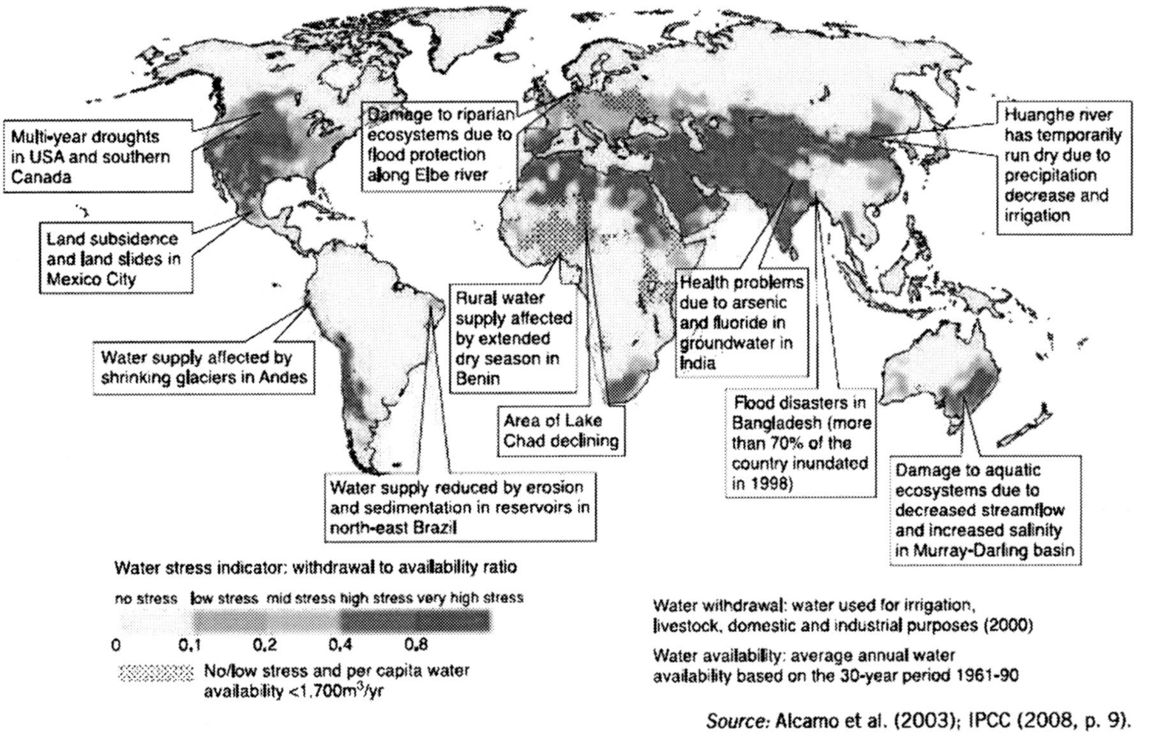

Figure 3 Water stress map with examples of vulnerable freshwater resources and their management (*Source*: WWAP, 2012).

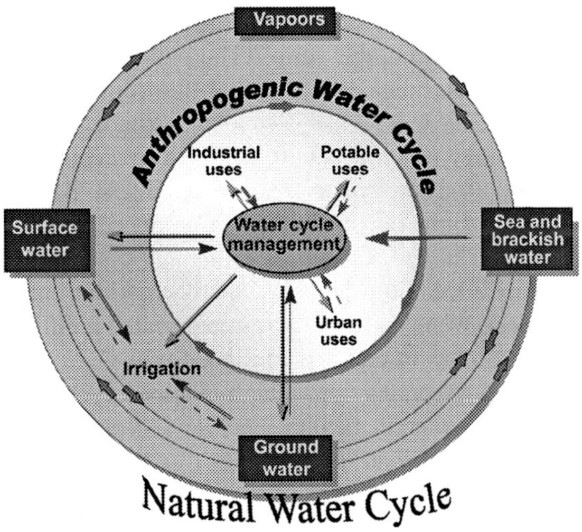

Figure 4 Sustainable water cycle management based on water recycling, energy and nutrient recovery, and efficient environmental protection.

In the context of increasing water stress and scarcity, the most difficult challenge for sustainable long term human habitat planning will be to find creative ways to manage water resources. The conventional linear approach of water management must be converted into an integrated management of systems for water, energy and nutrient recovery and recycling (Figure 4). Adequate sanitation and wastewater management are crucial to human health protection and preservation of freshwater quality and biodiversity. The high investment needed for management of the anthropogenic water cycle (dams; wells; water transportation; treatment and supply; wastewater collection and treatment, etc.) would be easily repaid by the benefits from the recovery of purified water, embodied energy and nutrients (Lazarova *et al.* 2012).

The long-term experience in water reuse — with recent great successes of new water reuse schemes and applications — demonstrates that water recycling provides a unique and viable opportunity to augment traditional water supplies (Asano *et al.* 2007; USEPA, 2012). As a multi-disciplined and important component of water resources management, water reuse can help to close the loop between water supply and wastewater disposal. Effective water reuse requires integration of potable water and reclaimed water supply functions. The successful development of this alternative drought-proof water resource depends upon close coordination of water policies, infrastructure and facilities planning, water quality management and wastewater treatment process selection and reliability.

Over the past three decades, several thousand successful water reuse projects with diverse applications around the world have demonstrated that water recycling is a proven solution to water scarcity, which is an essential tool for mitigation of the impacts of climate change on the diminishing available fresh water resources and is of extreme importance for the protection of the planet's biosphere and for the achievement of the Millennium Development Goals (USEPA, 2012).

The *main objective of this book* is to present 29 selected successful water reuse case studies and illustrate the keys factors to success, the major benefits and the main challenges associated with the development and operation of water reuse projects worldwide.

At present, under the pressure of unprecedented increase in water scarcity and endangered Earth biosphere, the water supply planning paradigm is evolving from reliance on traditional fresh water resources towards building an environmentally sustainable diversified water portfolio where low-cost conventional water sources are balanced with more costly but also more reliable and sustainable water supply alternatives, from which water recycling appears to be the most valuable and a compelling solution for the future preservation of human life and Earth's natural habitats.

Water reuse terminology

The success and public acceptance of water reuse are greatly influenced by the terminology used for outreach to the stakeholders involved in the comprehensive water planning process. For this reason, it is very important that the concepts of water recycling are expressed in simple, understandable, plain English. In general, the term "water reuse" is used synonymously with the terms "water recycling" and "water reclamation". According to the Oxford English Dictionary (6th Edition 2007), these terms are synonyms of "using again" with additional references for recycling to "return to a previous stage of a cyclic process" or "convert waste into a usable form". The public at large is widely engaged in recycling paper, glass, metals, plastics and other waste and clearly understands what the word "recycling" means. For this reason, several new water reuse regulations, such as the California Water Recycling Criteria (2000) for example, have adopted "water recycling" as the most appropriate terminology which is readily accepted by the public at large. Once a water reuse project is presented for review and approval to decision-makers and stakeholders, it is important to use the most suitable terminology which helps them to understand that recycled water is produced by putting wastewater through a recycling process to convert it back into usable water.

An important new concept in water reuse is the "fit-to-purpose" approach, which entails the production of recycled water of quality that meets the needs of the end-users. When water reuse is implemented for different purposes, the most cost effective solution is to use several tertiary treatment trains to produce "designed water" for each type of beneficial use. Advance of wastewater treatment technology is enabling the production of high-quality recycled water equal to or even better in quality than potable water. For this reason and in order to facilitate public acceptance, many new water reuse projects have adopted new terms to describe reclaimed water such as "purified water", "NEWater", "eco-water", and so on.

Water reclamation is the process of treating wastewater and recovering purified water of a quality which is suitable for beneficial use (Table 1). *Water reuse* or *water recycling* is the utilization of this purified water for suitable applications. In addition, water reuse frequently implies the existence of a pipe or other water conveyance facilities for delivering the reclaimed or recycled water to the final users.

Table 1 Glossary of water reuse*.

Term	Definition
Beneficial Use	Use of water directly by people for their economic, social or environmental benefit.
Criteria	Standards rules or tests on which a decision can be based.
Direct Reuse	Use of recycled water delivered directly for beneficial reuse including into a water supply system.
Environmental Buffer	A water body or aquifer which lies between a recycled water discharge and a water supply intake or extraction well. An environmental buffer will often provide additional natural treatment.
Greywater	Wastewater from bathing and washing facilities. Human wastes from toilets and food wastes from the kitchen are excluded.
Guidelines	Recommended or suggested standards, criteria, rules or procedures that are advisory, voluntary and unenforceable.
Indirect Reuse	Use of recycled water delivered into a river, reservoir or groundwater aquifer from which water supply is drawn at a point downstream.
Non-Potable Use	Use of water for purposes that do not require drinking water quality.
Potable Use	Use of water for purposes that require drinking water quality.
Recycled Water	Water recovered by treatment of wastewater, greywater or stormwater runoff to a quality suitable for beneficial use. Synonym of reclaimed water.
Regulations	Standards, criteria, rules or requirements that have been legally adopted and are enforceable by government agencies.
Return Flow	The return of recycled water flows back to the river from which the water supply was drawn.
Standard	An enforceable rule, principle or measure established by a regulating authority for example numerical water quality limits.
Wastewater	Used water discharged from homes, businesses, industry or agriculture.
Water Reclamation	The process of treating wastewater and recovering recycled water of a quality which is suitable for beneficial use.
Water Recycling	Use of recycled water for beneficial purposes.
Water Reuse	Use of recycled water; using water multiple times for beneficial purposes.

*Adapted from Glossary of Water Reuse developed by the IWA Specialist Group on Water Reuse.

The foundation of water reuse is built upon three principles:

1. Providing reliable treatment of wastewater to meet strict water quality requirements for the intended reuse application,
2. Protecting public health,
3. Gaining public acceptance.

Whether water reuse is appropriate for a specific application and project depends upon careful economic considerations, potential uses for recycled water, and the relative stringency of wastewater discharge requirements. Public policies can be implemented that promote water conservation and reuse rather than the costly development of additional water resources with considerable environmental expenditures. Through integrated water resources planning, the use of recycled water may provide sufficient flexibility to allow a water agency to respond to short-term needs as well as to increase the reliability of long-term water supplies (Asano, 2002; USEPA, 2004; Lazarova & Bahri, 2005; Asano *et al.* 2007; USEPA, 2012).

In the planning and implementation of water reuse, the intended water reuse applications govern the degree of wastewater treatment required and the reliability of wastewater treatment processes and operation (WHO, 2006; USEPA, 2012). In principle, wastewater or any marginal quality waters can be used for any purpose as long as adequate treatment is provided to meet the water quality requirements for the intended use.

Just as important as the promotion of new planned water reuse projects, is the acknowledgment of the unplanned reuse of water (Jimenez-Cisneros, 2009; Leverenz *et al.* 2011). This is necessary to: (a) reduce the fears of reusing water and (b) to control undesirable effects, if applicable. Both activities are of interest to developed and developing countries, although for the latter an alternative may be to combine sanitation and reuse goals.

Water reuse applications

The main categories of municipal wastewater reuse applications, related issues and constrains, as well as the most important lessons learned are shown in Table 2. The dominant applications for the use of recycled water include: agricultural irrigation, landscape irrigation, industrial reuse and groundwater recharge. Among them, agricultural and landscape irrigation are widely practiced throughout the world with well-established health protection guidelines and agronomic practices (Lazarova & Bahri, 2005; Asano *et al.* 2007).

Agricultural irrigation was, is and will remain the largest recycled water consumer with recognised benefits and contribution to food security. Urban water recycling, in particular landscape irrigation, is characterised by fast development and will play a crucial role for the sustainability of cities in the future, including energy footprint reduction. Other relevant and cost efficient applications are also emerging such as environmental enhancement, in-building recycling and industrial uses of reclaimed urban wastewater.

Indirect potable reuse, in particular groundwater (aquifer) recharge, after complementary polishing and storage of recycled water in an environmental buffer, has been implemented in many countries as an efficient response to the need to increase water supply. Finally, direct potable reuse, practiced for over 40 years in Namibia, is emerging as a solution to the challenges which some countries will face in the next 20 years (Tchobanoglous *et al.* 2011; Leverenz *et al.* 2011).

As water is used for various domestic, municipal, and industrial applications, its quality changes due to the introduction of various constituents. A conceptual comparison of the extent to which water quality changes through municipal applications is shown schematically on Figure 5. Even when polluted, wastewater contains more than 99.98% of pure water. Today, technically proven water reclamation and purification technologies exist to produce pure water of almost any quality desired including purified water of quality equal to or higher than drinking water.

Worldwide advances in water reuse

Most of the significant developments in water reuse have occurred in arid regions of the world (Figure 6) including Australia, China, Mediterranean countries, Middle East and the United States. For a number of countries where current fresh water reserves are or will be at the point of depletion in the near future, recycled water would be the only significant low-cost alternative resource for agricultural, industrial and urban non-potable water supplies. Even in temperate regions water reuse is characterized by fast development, in particular for industrial purposes, environmental enhancement and urban recycling.

Once the countries implementing water reuse are shown on a world map (Figure 7), it becomes clear that water reclamation is a global trend (Jimenez & Asano, 2008). To better explain the role of water reuse for sustainable development and urban water cycle management, this book presents 29 success stories which are selected from various countries and continents, as well are related to various water reuse applications (see some of the case studies on Figure 7).

At a number of applications, highly treated reclaimed water has been blended with other drinking water sources. In California, the Groundwater Replenishing System (GWRS) in Orange County is the world leader in groundwater recharge using reclaimed water for indirect portable reuse. The health safety and economic feasibility aspects of aquifer recharge with recycled water are demonstrated also in Europe by the project of Torreele, Belgium. The Upper Occoquan Service Authority (UOSA) in Virginia, United States, is a pioneer in indirect potable reuse for replenishment of surface reservoirs. Their long-term operational experience has clearly demonstrated that water quality can be improved by water recycling. In Windhoek, Namibia, because of extreme dry conditions, direct potable reuse was implemented in 1968 and is successfully operated and upgraded, without any adverse health effects, as demonstrated by risk assessment and epidemiological studies. More recently, in Singapore and Australia's Western Corridor water reuse facilities, recycled water has been implemented as a source of raw water to supplement dams and other water supply sources for industries.

Urban growth impacts on infrastructure in developing countries are extremely pressing. In many cities of Asia, Africa, and Latin America, engineered sewage collection systems and wastewater treatment facilities are non-existent or inadequate. For developing countries, particularly in arid areas, wastewater is simply too valuable to waste, as shown by the two case studies in Mexico (San Luis Potosi and Mexico City).

Water reuse will play a crucial role for the urban water cycle management, transforming the cities of the future at water-saving and "leisure-paradise" settlements, as demonstrated in the case studies in China and Japan. For urban applications, in-building water recycling and environmental enhancement are emerging as efficient solutions to reduce water and energy footprints. Even paradisiac islands (for example Bora Bora and Honolulu) and tourist areas (e.g. Costa Brava, Spain) need water reuse to maintain healthy ecosystems and happy visitors.

Table 2 Categories of municipal wastewater reuse applications and related issues or constrains.

Category		Potential application	Issues/ constraints	Lessons learned
Agricultural irrigation	Unrestricted or restricted	Food crops, eaten raw, processed or cooked Pastures for milk production Orchards, vineyards with or without contact with edible fruits Fodder and industrial crops Ornamental plant nurseries	Water quality impacts on soils, crops, and groundwater Runoff and aerosol control Health concerns Farmers acceptance and marketing of crops Buffer zone requirements	Good practices available to mitigate adverse health and agronomic impacts Storage design and irrigation technique are important elements Numerous reported benefits
Landscape irrigation	Unrestricted or restricted	Golf courses and landscape Public parks, school yards, playgrounds, private gardens Roadway medians, roadside plantings, greenbelts, cemeteries	Water quality impacts on ornamental plants Runoff and aerosol control Health concerns Public acceptance	Successful long-term experience Good practices and on-line water quality control can ensure health safety
Non potable urban uses		In-building recycling, Toilet flushing Landscaping (see irrigation) Air conditioning, Fire protection Commercial car/trucks washing Sewer flushing Driveway and tennis court washdown Snow melting	Health concerns Scaling, corrosion, fouling, and biological growth Cross-connection with potable water Pollution of receiving waters	Dual distribution systems require efficient maintenance and cross-connection control No health problems reported even in the case of cross-connections (for tertiary disinfected reclaimed water)
Environmental/ Recreation uses	Unrestricted or restricted	Recreational impoundments Environmental enhancement (freshwater or seawater protection) Wetlands or biodiversity restoration Fisheries Artificial lakes and ponds Snowmaking	Public health concerns Eutrophication (algae growth) due to nutrients Toxicity to aquatic life	Emerging applications with numerous benefits for the cities of the future: improving living environment, human wellbeing, biodiversity, and so on On-line water quality control can ensure health safety
Industrial reuse		Cooling water Boiler feed water Process water Heavy construction (dust control, concrete curing, fill compaction, and clean-up)	Scaling, corrosion, fouling, and biological growth Cooling tower aerosols Blowdown disposal Cross-connection with potable water	Water quality to be adapted to the specific requirements of each industry/process Request for high reliability of operation, cost and energy efficiency
Indirect potable reuse with replenishment of:	Aquifers	Groundwater replenishment by means of infiltration basins or direct recharge by injection wells Barrier against brackish or seawater intrusion (direct recharge) Ground subsidence control	Groundwater contamination Toxicological effects of organic chemicals Salt and mineral build-up Public acceptance	Successful practice since 1970s Multiple barrier treatment ensures safe potable water production Efficient control by means of advanced modelling tools
Indirect potable reuse with replenishment of:	Reservoirs	Surface reservoir augmentation Blending in public water supply reservoirs before further water treatment	Health concerns Public acceptance	Successful practice since 1970s Multiple barrier treatment ensures safe potable water production Improvement of water quality
Direct potable reuse		Pipe-to-pipe blending of purified water and potable water Purified water is a source of drinking water supply blended with source water for further water treatment	Health concerns and issues of unknown chemicals Public acceptance Economically attractive in large scale reuse	Multiple barrier treatment ensures safe potable water production No health problems related to recycled water in Namibia since 1968

Source: Adapted from Asano *et al.* 2007.

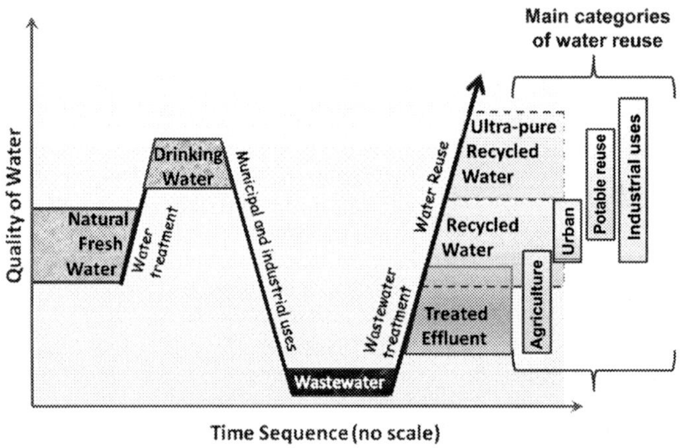

Figure 5 Water quality changes during municipal uses of water in a time sequence (Adapted from Asano, 2002).

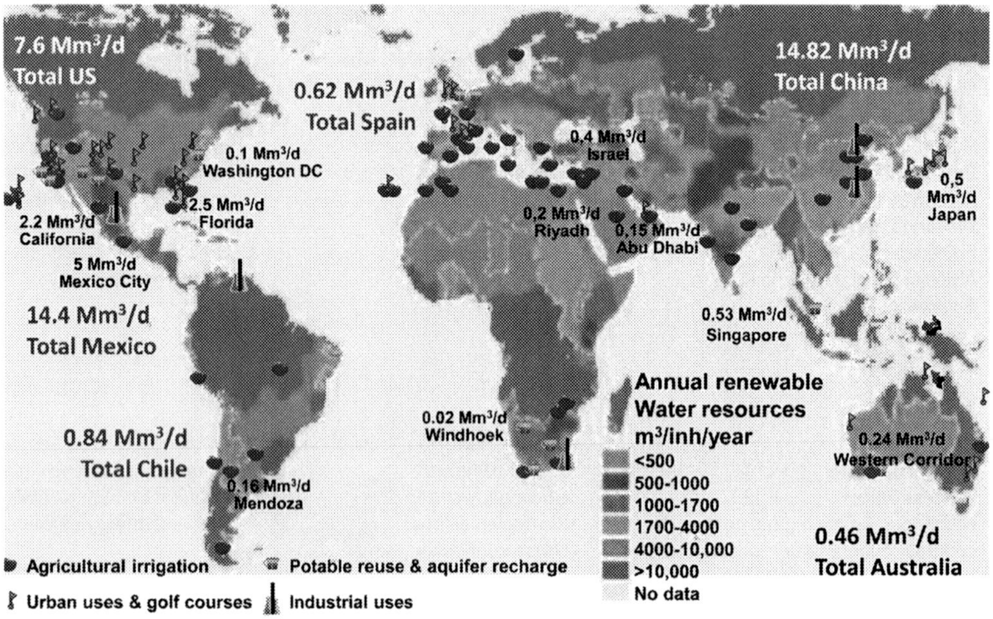

Figure 6 A Global trend towards sustainable water reuse growth at all continents (Adapted from Lazarova *et al.* 2001; Jimenez & Asano, 2008).

Sustainable economic development under the conditions of water scarcity is becoming possible by adopting the use of recycled water, both for agricultural production (examples in Australia, France, Italy and Mexico) and industry (examples in Germany, India, Mexico, Spain, the United States). Very often, water reuse is implemented for different purposes, with the production of "designed water" for each type of use (examples of West Basin, California; San Luis Potosi, Mexico; and Honolulu, Hawaii). In many cities (for example Beijing, Madrid, Tianjin, Tokyo) and countries (for example Australia, Cyprus, Japan, Singapore, Spain, the United States), recycled water is now considered an important component of integrated water resource management making possible to close or accelerate the urban water cycle and to preserve natural water resources and biodiversity.

For example in California, the State Water Code clearly states that "It is the intention of the Legislature that the State undertakes all possible steps to encourage development of water recycling facilities so that recycled water may be made available to help meet the growing water requirements of the State". As a result, water reuse which is legally called "water recycling" in California has been growing steadily since 1970s as shown on Figure 8.

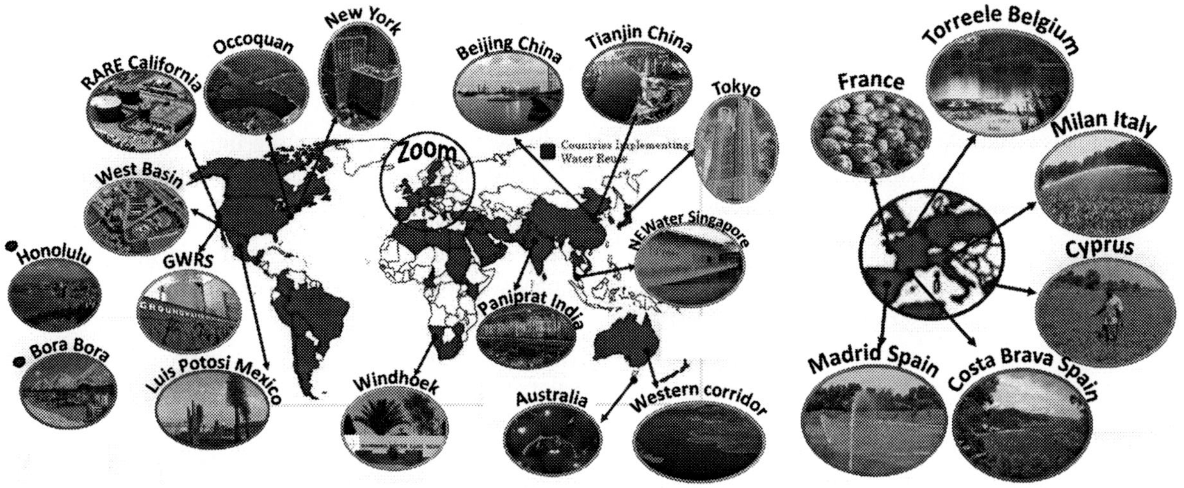

Figure 7 Countries implementing water reuse according to Jimenez and Asano (2008) and some of the selected success stories presented in this book.

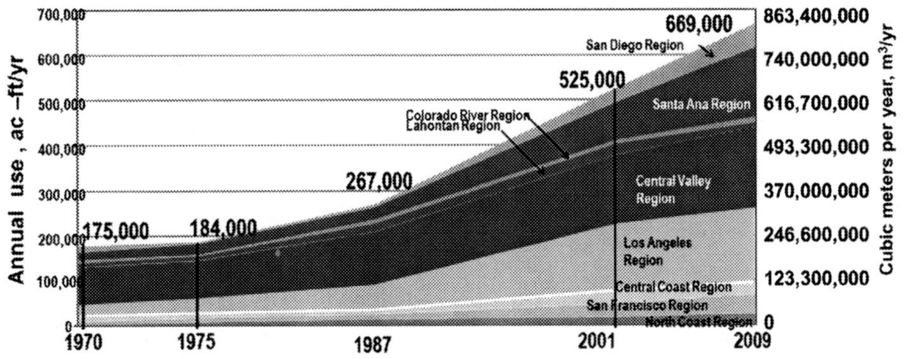

Figure 8 Municipal recycled water use in California, 1970–2009: In 2009, a total of 825 million m^3/yr (669,000 ac-ft/yr) of recycled water was used (*Source*: California Department of Water Resources).

Advanced technologies such as membrane bioreactors, membrane filtration, advanced oxidation, and ultraviolet disinfection are of key importance for the reliable production of high quality reclaimed water. According to a recent market forecast (GWI, 2009), the trend towards an almost exponential increase in the volume of recycled water after tertiary and/or advanced treatment, observed since the 2000s, would continue with a projected growth between 2012 and 2016 of over 13% per year to reach 9,790,000 m^3/d, a tripled volume as compared to 2009. For reference, the volume of secondary effluents reused for agriculture will continue with an almost linear growth of about 6% per year. The countries with the highest projected growth are China, USA, Spain, Mexico, Australia, Peru, India, Saudi Arabia, UAE and Algeria.

Milestones in water reuse for agriculture

The majority of water reuse projects worldwide are implemented for agricultural irrigation and are driven by increasing water scarcity and ever increasing agricultural water demand. The need for alternative water resources has been accelerated over the past few years by the worst droughts, which occur not only in traditionally arid areas in the United States, the Mediterranean region, the Middle East and South Asia, but also in a number of temperate-climate states and countries in Europe and North America. For example, according to a FAO study (FAO, 1999), the drought of 1999 in the Near East resulted in a relative decline of food production of 51% and the economic impacts of the 2003 drought in Europe exceeded €13 billion. In this context, water reuse is becoming more valued, and certain countries are already using a great part of their treated wastewater for irrigation (Argentina, China, Cyprus, Egypt, Israel, Jordan, Kuwait, Libya, Mexico, Saudi Arabia, Spain, Syria, Tunisia and United Arab Emirates).

The basic principle of beneficial water reuse in agriculture is that municipal wastewater may be used for all kinds of irrigation applications as long as the water has been previously treated to appropriate level to meet specific water quality requirements (Ayers & Westcot, 1985; Lazarova & Bahri, 2005). It is worth noting that besides the well-recognized benefits of water reuse, the use of recycled water for irrigation may have adverse impacts on public health and the environment, depending on treatment level, local conditions and irrigation practices. In all cases, the existing scientific knowledge and practical experience can be used to lower the risks associated with water reuse by implementation of sound planning and effective management of irrigation practices with recycled water.

The milestones in development of safe water reuse practices for agricultural irrigation are illustrated on Figure 9. The proposed three categories of water reuse projects are based on the advance in wastewater treatment and scientific knowledge and include:

1. Irrigation of industrial crops, fodders and seed crops, orchards, forests, and so on, irrigated with secondary effluent often after storage and polishing in open lagoons such as maturation ponds. Implementation of large projects in the Unites States and Tunisia in the years 1960s, in Argentina in 1970s, in France and Spain in 1980s.
2. Irrigation of food crops (eaten cooked or processed) with tertiary effluents. In 1980s, two large projects in Florida (e.g. Water Conserv II) and in Israel (Dan Region) have demonstrated the safety and benefits of water reuse by means of extensive scientific studies.
3. Unrestricted irrigation of crops consumed raw with well treated and disinfected recycled water (tertiary filtered and disinfected effluents or ultrafiltration-treated secondary effluents). Long term extensive research completed as a part of the implementation of water reuse projects in California (Monterey) and Australia (Virginia pipeline) has demonstrated the safety of recycled water and has convinced all stakeholders of the benefits of water reuse.

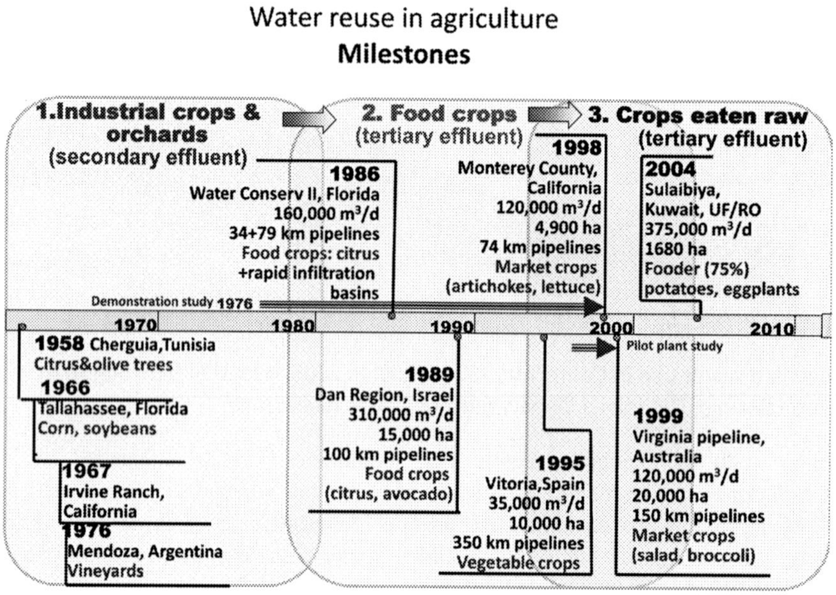

Figure 9 Milestones in water reuse for agriculture with selected cornerstone projects (*Source*: Lazarova, 2011).

Many successful water reuse projects for agricultural irrigation have been developed worldwide. However, the projects presented on Figure 9, and described in more detail in Table 3, are very important for the success of water recycling because they are used as benchmark examples. The safe and beneficial implementation of water reuse schemes could be better guaranteed by the development of appropriate codes of good agronomic and irrigation practices, which are as important for farmers and operators as quality requirements for water reuse (Lazarova & Bahri, 2005).

The *management of health risk* in water reuse for agriculture includes several types of actions that can be classified in three main groups:

- Policy, regulations and institutional initiatives, including set-up of water quality criteria and their enforcement, crops restriction, human exposure control and immunisation of field workers;
- Engineering actions such as wastewater treatment and storage, adequate operation and water quality monitoring, control of reclaimed water application, and in particular, the selection of irrigation method and technology;
- Agronomic practices, including crops selection, control of the timing and frequency of irrigation, leaching and drainage, crop's harvesting measures, and so on.

Table 3 Description of selected cornerstone projects for agricultural irrigation with recycled water.

Project/ Location	Start-up/ Capacity	Type of irrigated crops	Treatment for water reclamation	Storage and distribution	Research studies	Costs and revenue	Lessons learned, benefits and challenges
Water Conserv II, Florida, USA	1986 160,000 m^3/d	1740 ha (max 6000 ha) Food crops (citrus, fruits) and landscape	1) Two recycling facilities, filtration and chlorination (not detected FC/100 mL), filtration before irrigation 2) 65 rapid infiltration basins (40% of recycled water)	34 + 79 km pipes; booster pump station with 4 storage reservoirs; centralised control system	R&D studies on long-term effects and economic viability	Capex $278 million, Opex $4.8 million/yr No charges for growers for the first 20 years	*Main driver*: In 1979, legal proceedings of a citizen group to stop treated wastewater discharge into the river *Benefits*: Recognized better growth and fruit quality, enhanced freeze and drought protection, environmental benefits (protection of surface water, aquifer and biodiversity) *Major findings*: Extensive scientific studies were needed to demonstrate safety and benefits and gain growers' acceptance
Dan Region, Israel	1989 310,000 m^3/d	15,000 ha Food crops (citrus, avocado)	Activated sludge followed by soil-aquifer treatment (SAT) using 4 recharge basins (total area of 80 ha)	100 recovery wells, 87 km pipes, local storage reservoirs (total volume 10 Mm^3)	Several R&D studies and EU projects	Irrigation fees: up to 0.16 €/m^3	*Benefits*: The largest reuse project in Israel with national importance with recognized economic benefits *Major challenges*: Biofouling of the effluent pipelines from the wastewater treatment plant to the SAT and a lack of capacity in the SAT system
Monterey County, California, USA	1978 120,000 m^3/d	4900 ha Food crops and raw-consumed vegetables	Title 22 treatment (coagulation, flocculation, sedimentation, filtration and chlorination), <2.2TC/100 mL, Full monitoring 3–4 times/yr	74 km pipelines, 3 booster pump stations, centralised control system	11-yr R&D studies: food safety, marketing, impact on crop's yield	Capex $78 million, Opex $3.5 million/yr, Total cost 0.24 $/$m^3$, Charge 0.013 $/$m^3$	*Benefits*: Water reuse is considered the key to sustaining the agricultural industry ($3 billion/yr) and tourism ($2 billion/yr) *Major findings*: Buying market was not affected by the type of irrigation water used, as long as the irrigation water met regulatory requirements and no labelling of the produce was required
Virginia Pipeline, Adelaide, Australia	1999 65,000 m^3/d	20,000 ha Food crops and raw-consumed vegetables	Tertiary treatment at the Bolivar WWTP by maturation pond, coag./flocculation, dissolved air flotation, filtration and chlorination, <10 *E.coli*/100 mL	150 km pipes, seasonal storage in aquifer (ASR), 250 users, total recycled volume: 28 Mm^3/yr	3-yr R&D study	Capex AU$23 million Average charge for farmers 0.12 AU$/$m^3$	*Benefits*: (1) Support the AU$120 million vegetable production of Virginia; (2) Protection of sensitive environment and freshwater recharge of brackish aquifers. *Major findings*: Market study, display of recycled water at public meetings, support of Health Authority, change of perception over 3 yrs. *Keys to success*: Extensive research, education and training.
Sulaibiya, Kuwait	375,000 m^3/d	1680 ha	Tertiary treatment by ultrafiltration and reverse osmosis (UF/RO) <2.2TC/100 mL	4 large pipelines (DN 1400), data monitoring system, storage tanks		Capex US$ 442 million, Total cost $1.48/$m^3$, Charge $0.74/$m^3$	*Drivers*: Overload and unfeasible extension of the existing wastewater treatment plant and limited brackish water resources to cover increasing non-potable water demand *Benefits*: Solved problems of environmental pollution and water deficits

Source: Adapted from Crook (2004), Asano *et al.* (2007), Jimenez and Asano (2008) and Alhumoud *et al.* (2010).

The *major findings and lessons learned* from the water reuse projects for unrestricted agricultural irrigation can be summarised as follows:

1. Keys to success: subsidies, efficient and reliable treatment, extensive research, education and training
   - Feedback from operations demonstrated that tertiary disinfected effluent can be safely used for irrigation of food crops, including vegetables consumed raw
   - Extensive scientific studies may be needed to demonstrate safety and benefits and gain farmers' acceptance
2. Major challenges
   - Food safety and public perception are very important issues for farmers (influence of *E. coli* outbreaks such as the fresh spinach issue in the USA in September 2006 resulting in a revenue loss of over $74 million and the "cucumber" crisis in Europe in May 2011 resulting in a revenue loss for farmers of over €600 million)
   - Agronomic aspects: salinity, sodicity and toxic ions management
   - Storage capacity and O&M of irrigation networks (biofouling, bacterial recontamination or regrowth control)
3. Innovation
   - New drip and subsurface irrigation techniques with high water efficiency and centralised control systems
   - Development of salt collecting crops and alternative crops that use less water

Milestones in urban water reuse

Water reuse in urban areas includes a wide variety of applications and schemes with a common characteristic that all these purposes do not require potable water quality. The main categories of urban water reuse are as follows:

- Landscape irrigation, which is the primary use and includes irrigation of public parks, sport fields, green belts, golf courses, as well as private residential areas and gardens.
- Other non-irrigation urban uses such as street cleaning, car washing, fire protection, air conditioning, toilet flushing and some commercial applications.
- In-building recycling which refers mostly to water recycling in high-rise buildings, including office buildings, commercial malls and private residential buildings.
- Environmental enhancement and recreational uses for replenishment of water bodies, lakes and urban streams including those used for swimming (with or no body contact), leisure or fishing purposes.

As a rule, urban water reuse needs an adequate infrastructure, and in particular dual distribution (or reticulation) and dual plumbing systems. Dual distribution and plumbing systems are relatively easy to install in new urban areas or buildings with relatively low initial cost. Because of the high risk of direct contact with recycled water, the water reuse requirements (total disinfection and on-line control), as well as the rules for cross-connection control are very stringent (Asano *et al.* 2007).

Similarly to agricultural irrigation with recycled water, landscape irrigation includes applications of recycled water for irrigation of restricted and unrestricted areas, the latter requiring the highest water quality. Agronomic water requirements are also similar to those for agricultural irrigation and depend on crop's sensitivity. As a rule, irrigation of ornamental plants needs a careful analysis of agronomic water quality parameters, while turfgrasses, for majority of species, are very tolerant to salinity and sodicity (Lazarova & Bahri, 2005). The components of landscape irrigation are, as a rule, more sophisticated than for agricultural irrigation because of the use of water efficient irrigation techniques (sprinklers, micro-sprinklers and drip irrigation) which may require additional filtration after storage, backflow prevention devices, pressure control and automation (Asano *et al.* 2007). Similar devices and efficient operation and cross-connection control practices are required for dual distribution networks.

Relatively few cross connection incidents with backflow from recycled water systems have been reported with no reported illness. The major causes of cross connections between recycled and potable water systems reported in Australia, United States and United Kingdom are:

- Illegal connection of private residences
- Inadequate construction, records and pipe identification
- Higher pressure in recycled water system

Water recycling in urban areas including toilet flushing is becoming an acceptable practice, with confirmed operating experience and recognised economic viability for the following applications:

- At large municipal scale for toilet flushing and irrigation
- In large office buildings (Japan, California-United States)
- In large residential areas (Florida-United States, Australia)

Greywater recycling in residential buildings is under development, but is still at a stage of research and demonstration projects (Canada, France, Germany, UK), very often in combination with rainwater recycling (France, Japan, Germany, UK). The reliability of operation and economic viability of greywater recycling in small buildings and individual homes is not yet demonstrated.

The main categories of urban water reuse are discussed in more detail in parts 2, 3 and 6. Because of the high contribution of this type of reuse for the urban water cycle management and for the cities of the future, the major milestones in urban reuse are classified not by application or technologies, but by the size and scheme of implementation (Figure 10):

1. Recycling in high-rise buildings.
2. Large-scale dual distribution systems.
3. New concepts for eco-cities with low water and energy footprint.

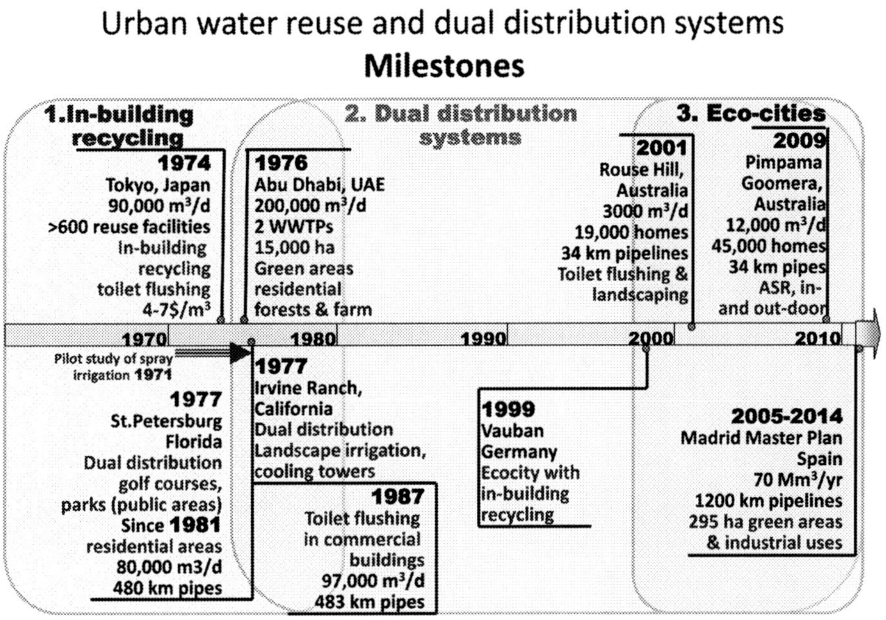

Figure 10 Milestones in urban water reuse with selected cornerstone projects (*Source*: Lazarova, 2011).

The city of St. Petersburg in Florida has implemented one of the largest urban dual water systems of its kind in the world in 1977. Even though the initial objective was to avoid wastewater discharge in surface waters, the water recycling systems have become an essential element of the urban water cycle management (Asano *et al.* 2007). The operational experience and R&D studies contributed to the development of good practices for optimum watering and selection of ornamental plants, as well as for the increase in cost efficiency by using a looped distribution network with pipes of small diameter.

New concepts of sustainable water management and eco-cities with reduced potable water demand by means of dual distribution systems are under development in other countries, and in particular in Australia. On the basis of the operational experience of the first large residential project in Rouse Hill in Sydney, a new project in Pimpama Goomera in Queensland is aiming to cover 45% of the water demand of 45,000 residential homes using recycled water, lowering the drinking water demand to about 16% of its baseline typical level.

Table 4 Description of selected cornerstone projects of urban water reuse (dual distribution systems).

Project / Location	Start-up/ Capacity	Type of uses	Treatment for water reclamation	Storage and distribution	Research studies	Costs and revenue	Lessons learned, benefits and challenges
St. Petersburg Florida, USA	1977 80,000 m^3/d	10,284 customers, 9992 residential lawns, 61 schools, 111 parks, 316 fire hydrants, 6 golf courses, cooling towers	4 recycling facilities with tertiary treatment by coagulation, filtration and disinfection Not detected FC/100 mL 10 deep injection wells in a saltwater aquifer for excess recycled water	Covered storage at each plant (total volume 95,000 m^3), Looped system of 160 km mains plus 306 km small pipes, 9 booster pump stations, Cross-connection control	R&D programs on plant tolerance and optimum watering practice	Total Capex \$135 million, Opex \$5.2 million Resident charge \$295 for connection plus \$13/month; Charge for commercial customers \$0,087/m^3	*Major driver*: Initiative of the city of to stop discharge in surface water, avoid cost associated with nutrient removal and find new water supply *Major benefits*: Postponed development of additional potable water supply *Keys to success*: Extensive research, education and training
Abu Dhabi, United Arab Emirates	1982 300,000 m^3/d	1300 ha green landscape areas, parks and gardens, golf course, 15,000 ha forest land	Two activated sludge plants (Mafraq and Al-Ain) with tertiary treatment by dual media filtration and chlorination <100 TC/100 mL (in 80% of samples)	55 reservoirs with a total volume of 98,400 m^3			*Major driver*: Population growth and water shortage *Major benefits*: Greening of the city, offset pollution, enhanced environment, counteracted desertification, reliable water supply *Keys to success*: Strong institutional support and a *fatwa* issued by religious scholars to support the use of purified water
Rouse Hill, Sydney, Australia	2001 27,000 m^3/d, 19,000 homes (up to 1.7 Mm^3/yr)	Flushing toilets, watering gardens, washing cars and other outdoor uses	Activated sludge followed by tertiary treatment type California "Title 22" coagulation/ flocculation, filtration and chlorination	3 covered storage reservoirs of 2000 m^3 each of wine glass type, 34 km main pipelines		Initial recycled water charge 0.2 AU\$/m^3 (30% of drinking water) to 1.29 AU\$/m^3 (80% of drinking water) in 2009	*Major benefits*: Reduction of drinking water demand by about 40% *Major challenges*: relatively low cost of recycled water. The progression of the Rouse Hill development was based on the principle of full cost recovery, including capital and operating costs.
Pimpama-Goomera, Gold Coast, Australia	12,000 m^3/d, 4400 homes completed (45,000 homes in final)	Dual reticulation system for toilet flushing and garden watering, plus rainwater harvesting	Tertiary treatment of secondary effluent by ultrafiltration, UV disinfection and chlorination, *E.coli* (50 percentile) 100 cfu/100 mL (max150 cfu/ 100 mL)	34 km pipes, seasonal storage in aquifer (ASR) to store 69% of recycled water in winter and supply 100% of summer demand (1 Mm^3/yr)	Studies on public perception and pilot trials	Cost of recycled water 1.34 AU\$/m^3 (60% of drinking water)	*Objective*: To become global leader in sustainable water management with water recycling *Major benefits*: To reduce drinking water supply to about 16% of typical demand and to cover up to 45% of water demand by recycled water (30% reduction yet achieved)

Source: Adapted from Asano *et al.* (2007), Jimenez and Asano (2008).

Other proven concept of integrated urban water cycle management, environmental restoration and water recycling in-building are discussed in the subsequent chapters of this book.

The *major findings and lessons learned* from the projects of urban water reuse can be summarised as follows:

1. Keys to success:
   - Governmental and regional incentives
   - Consistent water quality and supply reliability
   - Community education, communication and market assessment
2. Major challenges
   - **Water quality:** health safety (total disinfection), aesthetic parameters, others (nutrient, dissolved salts, etc.)
   - **Reliability of supply:** interruptible in many cases (fire protection, toilet flushing, cooling) with backup water system and storage capacity for irrigation purposes to meet peak demand
   - **Costs:** very high cost of dual distribution (prohibitive for >15–25 km), dual plumbing viable for new buildings, additional expenses cost for potable water when a substitution is needed
   - **O&M** of dual systems: cross-connection control, management of leaks, corrosion, scaling and bacterial regrowth
3. Innovation
   - Satellite, decentralized and semi-decentralised systems
   - On-line water quality control
   - AMR (automated meter reading)

Milestones in indirect and direct potable water reuse

The success story of potable water reuse started in early 1960s with water augmentation by means of groundwater (aquifer) recharge in Montebello Forebay, California (1962) and direct potable reuse in Windhoek, Namibia (1968). The more than 43-year operational experience direct "pipe-to-pipe" reuse in Windhoek demonstrated the feasibility and lack of adverse health effects of this practice (for more details, see Chapter 29). However, public opposition and concerns of unknown micropollutants were the main constraints for the development of this water reuse practice.

The most common practices of planned indirect potable reuse include aquifer recharge and reservoir replenishment. Indirect potable reuse occurs when some fraction of the raw water used for drinking purposes is of wastewater origin. Unplanned indirect potable reuse is a common situation in which inherent water quality issues are not fully addressed because of the divided responsibilities for wastewater discharge and downstream water supplies.

Because of health concerns, a "multiple barriers" treatment concept is applied in indirect potable reuse projects in order to achieve degree of reliability higher than conventional wastewater treatment and reuse schemes. The feedback from operation and research studies has demonstrated that health safety and high level of treatment and reliability of operation can be achieved using adequate treatment, water quality control and good practices of managed aquifer recharge (Asano *et al.* 2007; Crook, 2010; Kazner *et al.* 2012).

The milestones in development of safe water reuse practices for indirect potable reuse are illustrated in Figure 11. The proposed two categories are based on the advance in wastewater treatment with the technological breakthrough of membrane filtration. Two subcategories, covering both aquifer recharge and reservoir replenishment are dissociated according to the water quality requirement:

1. Surface spreading in aquifers and surface reservoir replenishment
2. Direct injection in aquifers, mostly as seawater intrusion barriers

In fact, the operational experience demonstrated that for all indirect water reuse applications, the advance in membrane efficiency with decreasing cost favoured the implementation of advanced multi-barrier treatment schemes, including the combination of micro- or ultrafiltration, reverse osmosis and advanced oxidation by high UV dose and hydrogen peroxide.

With exception of the oldest planned aquifer recharge project in Montebello Forebay (California), the other cornerstone indirect potable reuse projects are described in detail in the following chapters. The significance and the lessons learned from the aquifer recharge in the unconfined aquifer of Montebello Forebay in the eastern Los Angeles County are of crucial importance for the success of this practice. In fact, the first experience, started in 1962 with the recharge of spreading basins with disinfected secondary effluent from the Whittier Narrows Water Recycling Plant, which demonstrated the economic viability and health safety of this water reuse practice (Crook, 2004). The 5-year health effect epidemiological and toxicological study (1978–1983) and a follow-up epidemiological study (1996–1999) did not demonstrate any measurable adverse effects on groundwater quality or the health of the population drinking this water.

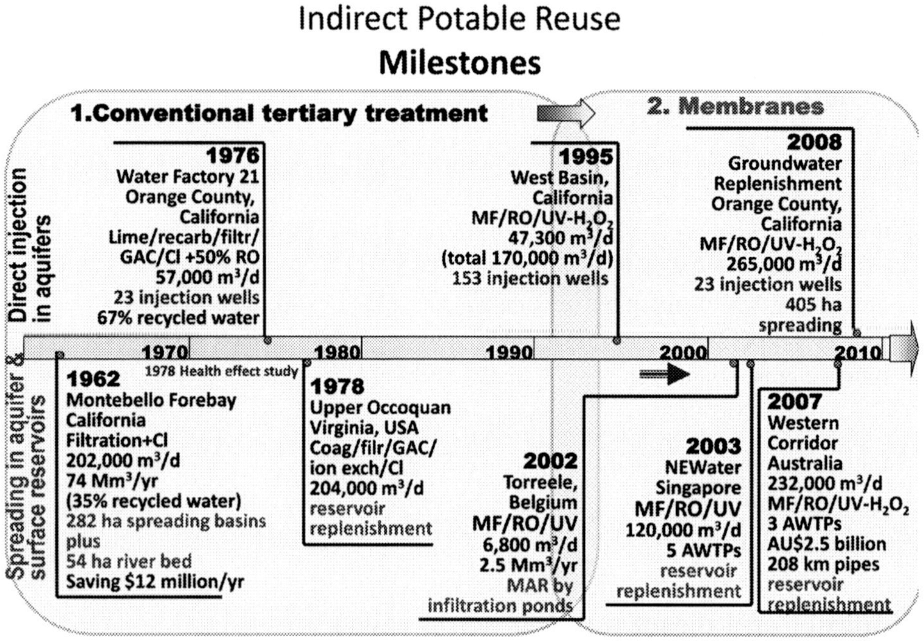

Figure 11 Milestones in indirect potable water reuse with selected cornerstone projects (*Source*: Lazarova, 2011).

The *major findings and lessons learned* from the projects of indirect potable reuse can be summarised as follows:

1. Keys to success:
   - Pilots plant tests and selection of efficient and reliable treatment technologies based on multiple-barriers approach
   - Government and stakeholder collaboration on adopting water reuse projects
   - Continued demonstration of consistent water quality, independent water quality monitoring, on-line process and water quality control and further water quality improvement over project life
   - Effective public outreach, community education, communication and use of positive terminology (e.g. "purified water", NEWater, "eco-water")
   - Governmental grant funding, subsidies, phasing of expansions and public-private partnerships
2. Major challenges
   - Public support and regulatory approval
   - High capital and operation costs and very high monitoring costs for emerging micropollutants
   - Increased emphasis on controlling pollution from urban runoff for both aquifers and surface reservoirs
   - Management of the complex membrane facilities during extreme conditions such as low production rates or flooding
   - Increased salinity and pollution of wastewater, in particular in coastal areas and aged wastewater treatment plants
3. Innovation
   - New membrane development, including low fouling and low energy consumption, nano-membranes
   - More efficient on-line water quality control devices

Milestones in industrial water reuse

One of the greatest potentials for water reuse is to supplement or replace the potable and/or freshwater demands of industries. Industry is the second largest market for water supply after agriculture with around 25% of global world demand. Industrial reuse and internal recycling are becoming current practices in many countries and industries with increasing water demand, closing loop cycles and zero liquid discharge as a long-term goal (see Chapter 18). The inter-sector water reuse, and in particular, the use of recycled urban wastewater for industrial purposes is characterised by fast growth in many countries. The potential for industrial reuse of treated municipal wastewater will increase in the future as raw potable water supplies become more limited, the cost of potable water increases due to more stringent standards and the discharge regulations become more stringent.

The major factors that influence the potential for industrial water reuse include availability of water, the industry's discharge requirements, water quality, volume, economics and reliability.

Although there is a wide range of industrial water uses, the major uses are:

- Cooling system make-up water,
- Boiler feedwater,
- Process water,
- Washdown water,
- Miscellaneous uses, including site irrigation, fire protection, road cleaning, and so on.

The first three categories are of particular interest because they are high-volume and high-quality applications with excellent prospects for using recycled municipal wastewater.

As a rule, the closing of industrial water cycle includes three strategies of water saving and wastewater minimisation:

1. Cascading reuse, involving direct reuse with no or little treatment;
2. Wastewater recycling after appropriate treatment;
3. Source reduction by decreasing the need of water of a given industrial process.

The water quality requirements and fields of application of water recycling in industry differ by type of industry, particular industrial process, as well as their target performance. For this reason, it is not possible to generalise water quality requirements for industrial process water. The greatest concern of the use of recycled water in cooling towers is the risk posed by inhalation of pathogens entrained in the aerosols. Moreover, corrosion, scaling and biofouling problems of equipment and distribution systems are a common issue. As a rule, for water to be suitable to be used as feedwater for boiler or stream generators, it must be of a higher quality than water used for cooling purposes (Asano *et al.* 2007). The quality requirements increase as operating pressure and temperature increase. The control or removal of hardness is required. Insoluble salts of calcium and magnesium are the main contributors to scale formation in boilers and are removed by processes such as ion exchange and reverse osmosis.

Successful water reuse practices are implemented in petroleum industry, oil refineries, thermoelectric power generation plants, pulp and paper facilities, textile industry and even microprocesors, electronic and food industries. Several case studies are presented in detail in part 4.

The milestones in industrial water reuse can be categorised in two groups, similarly to indirect potable reuse (see Figure 10) with the cornerstone being the implementation of membrane technologies, and in particular reverse osmosis. The West Basin Municipal Water District in California is a pioneer in the production of "designer" recycled water from municipal wastewater, from which three qualities are produced for industrial purposes (see Chapter 2). Since the first start-up in 1995 at this plant of the combined MF/RO treatment of municipal wastewater, the long-term operational experience has demonstrated that recycled water provides a reliable source of water for industry when potable water supply is uncertain, and this at lower cost. Several other similar recycling facilities are constructed in Singapore (see Chapter 3), Mexico (Chapter 15), India (Chapter 16), Northern California (Chapter 17) and Hawaii (Chapter 9). The German experience in industrial water recycling (Chapter 18) has demonstrated that there are technical possibilities for water recycling for almost any application. Key factors to the success of industrial water recycling are treatment efficiency and reliability, as well as technical and economic feasibility. As a result, water recycling schemes enabled to reduce wastewater volume by 78–92% in textile, paper and food and beverage industries.

The *major findings and lessons learned* from the projects of industrial reuse can be summarised as follows:

1. Keys to success:
   - Consistent water quality, efficient water quality control and reliable uninterrupted water supply
   - Implementation of proven advanced technologies, in particular membrane bioreactors, membrane filtration (micro- and ultrafiltration, reverse osmosis), ozonation

- Availability of funding, economic success of the clients/investors
- Cost efficiency and heat, energy and resource recovery
- Availability of an adjacent municipal wastewater plant to avoid extensive and expensive reclaimed water distribution network

2. Major challenges
   - Fluctuations of raw water quality and quantity
   - Control of water quality, in particular salinity, silica and hardness
   - Zero liquid discharge target in long-term
   - Disposal and treatment of RO brines and other concentrates
3. Innovation
   - New treatment technologies such as advanced oxidation, more efficient membranes systems, on-line water quality and treatment process control

Challenges for expanding water reuse

As demonstrated by the milestones in water reuse development worldwide, with many regions or communities approaching the limits of their available water supplies, wastewater reclamation has become an attractive option for conserving and extending available water supplies by potentially:

1. Using recycled water for applications that do not require high-quality drinking water,
2. Augmenting water sources and providing an alternative source of supply to assist in meeting both present and future water needs,
3. Protecting aquatic ecosystems by decreasing the diversion of freshwater, reducing the quantity of nutrients and other toxic contaminants entering waterways,
4. Reducing the need for water control structures such as dams and reservoirs,
5. Complying with environmental regulations by better managing water consumption and wastewater discharges.

Producing reclaimed water of a specified quality to fulfil multiple water use objectives is now a reality due to the progressive evolution of water reclamation technologies, regulations, and environmental and health risk protection. However, the ultimate decision to promote water reuse is dependent on economic, regulatory, public policy, and, more importantly, public acceptance factors reflecting the water demand, safety, and need for reliable water supply in local conditions.

Important issues related to the expanding water reuse and some of the foreseeable impediments are summarized below (Asano *et al.* 2007; Lazarova, 2011):

Economic and financial issues

Securing economic viability is an important challenge for the majority of water reuse projects. Unfortunately, water reuse is suffering from the competition with undervalued and/or subsidized conventional water resources. Full-cost recovery is a desirable objective but depends on ability to pay. The cost-benefit analysis of water reuse projects must include other management objectives and socio-environmental criteria, based on a holistic approach and catchment scale.

While water reclamation and reuse is a sustainable approach and can be cost-effective in the long run, the additional treatment of wastewater beyond secondary treatment for reuse and the installation of reclaimed water distribution systems can be costly compared to such water supply alternatives as imported water or groundwater. Similar to the development of any other utilities, the implementation of wastewater facilities generally requires a substantial capital expenditure. In the context of integrated water resources management of the region, government grants or subsidies may be required to implement water reuse. Unfortunately, institutional barriers, as well as varying agency/communities priorities, can make it difficult to implement water reuse projects in some cases.

Several options exist to demonstrate the economic viability of water reuse:

- Implement adequate pricing mechanisms: the price should be a function of delivering water of the required quality in a reliable manner. The competition from undervalued and/or subsidized conventional water resources can be avoided by "full-cost recovery" and "polluter-pays" principles, which must be the basis for establishing water rates.

- Take into account the entire (urban) water cycle and/or catchment area: recent estimation of cost of droughts, environmental impacts of wastewater discharge and cost of replacing lost abstraction in UK demonstrated that water reuse could be a cost competitive solution, even when high-purity recycled water is produced. This fact was confirmed by the cost evaluation of one the most modern and complex advanced water recycling facility for ground water recharge in Orange County, California (see Chapter 25), where the overall annual cost of purified water is 0.39 $\$/m^3$, which is equivalent to 60% of the cost of imported water.
- Need to move from supply-oriented to demand-oriented water supply market and water efficient urban cycle management, as demonstrated in several projects where water reuse is a part of integrated resource management, such as the examples of California, Australia, Singapore and the cities of Beijing and Tianjin in China.

Public and political support and communication

Independent of the type of reuse application and the country, the public's knowledge and understanding of the safety and suitability of recycled water is a key factor to the success of any water reuse programme. Consistent communication and easy to understand messages need to be developed for the public and politicians explaining the benefits of water reuse for the long term water security and sustainable urban water cycle management.

As demonstrated by all the success case studies presented in this book, the public's awareness of sustainable water resources management is essential; thus, planning should evolve through a community value-based decision-making model. Thus, water reuse is placed within the broader context of water resources management and other options in the region to address water supply and water quality problems. Community values and priorities are then identified to guide planning from the beginning in the formulation and selection of alternative solutions.

To date the major emphasis of water reclamation and reuse has been on non-potable applications such as agricultural and landscape irrigation, industrial cooling, and in-building applications such as toilet flushing in large commercial buildings. Understandably, potable reuse raises more public concern. In any case, the value of water reuse is weighed within a context of larger public issues. The water reuse implementation continues to be influenced by diverse debates such as drought and availability of water, growth vs. no growth, urban sprawl, traffic noise and air pollution, perception of reclaimed water safety, and public policy governing sustainable water resources management.

There are universal solutions available to convince the public at large and the project stakeholders regarding the safety and relevance of water reuse. Clearly, the use of a clear and positive terminology and simple explanations on water quality, treatment technology and water reuse benefits are necessary to build-up credibility and trust in water reuse. Existing experience and lessons learned are very important to convince decision makers. Finally, the most important recommendation is to inform and involve the public, politicians and all stakeholders from the beginning of any water reuse project. The increasing media impact and the new communication tools via internet should also be taken into account.

Innovating technology, improving reliability and energy efficiency

The technical challenges facing water reuse are not yet completely resolved. In particular for industrial, urban and potable water reuse applications, it is extremely important to improve performance, efficiency, reliability and cost-effectiveness of treatment technologies. Water recycling facilities are facing tremendous challenges of high variations of raw water quality, peaks in salinity due to salt intrusion in sewers, as well as variation in water quantity due to extreme conditions of lower water demand, flooding or needs of alternative disposal of recycled water.

Energy efficiency, carbon and environmental footprints are becoming important issues (Lazarova *et al.* 2012). The ambitious goals of sustainable development and achieving zero net carbon and pollution emission footprint call for a new holistic approach to the management of the water cycle with an increased role for water reuse. With the further growth of megacities and increasing efforts to optimise energy efficiency, water recycling is of growing interest and will take a leading role in the future of sustainable water management. Decentralized or semi-centralized water distribution systems are more efficient for future cities when water reuse is inevitably considered. Water supply can be tailored to match water demand more closely in centralized water infrastructures, adapting water quality to the given use. With substituting fresh water with appropriately treated recycled water, it is possible to save 30–50% of domestic water demand as demonstrated by the experience of Australia (see Chapter 11).

The energy consumption of water reclamation is significantly lower than that of desalination, and represents only a fraction of the energy intensities for water supply, treatment, and distribution. Nevertheless, energy extensive processes such as reverse osmosis should be limited to high quality purposes whereas alternative solutions (e.g. coagulation/flocculation) might be more energy efficient for irrigation purposes.

More intensive water treatment has a higher environmental impact in terms of carbon footprint. For example the state of the art advanced water reclamation based on high-tech, energy intensive technologies has a carbon footprint five times higher than the conventional water reclamation processes. Although carbon footprint provides a valuable criterion to assess sustainability of industrial products and processes, other environmental impact factors have to be taken into consideration for the assessment of water reuse projects, such as fresh water depletion, reduced pollution, preservation of biodiversity and lower toxicity.

Public water supply from polluted water sources

Due to land use practices and the increasing proportion of treated wastewater discharged into receiving waters, freshwater sources of drinking water are containing many of the same constituents of public health concern that are found in reclaimed water. Much of the research that addresses direct and indirect potable water reuse is becoming equally relevant to *unplanned indirect potable reuse* or *de facto* potable reuse that occurs naturally when drinking water supply is withdrawn from polluted water sources. Because of the research interest and public concerns, emerging pathogens and trace organic constituents including disinfection by-products, pharmaceutically active compounds, and personal care products have been reported extensively. The ramifications of many of these constituents in trace quantity are, unfortunately, not well understood with respect to long-term health effects.

Technological advances and opportunities for potable reuse

In the past, it has been standard practice that whenever additional sources of water supply are necessary but not readily available, non-potable water reuse options have been explored using reclaimed water. However, most of the economically viable non-potable reuse opportunities have been exploited. For example, the typical cost for parallel distribution of tertiary-treated recycled water is 0.3 to 1.7 $\$/m^3$ whereas the typical cost for highly-treated purified water, which could be added directly to the distribution system, is 0.6 to 1.0 $\$/m^3$ (Tchobanoglous *et al.* 2011).

While there has been a clear preference for non-potable and indirect potable reuse applications, a number of factors are making it less feasible to further increase water reuse in these applications. It is inevitable that purified water will be used as a source of potable water supply in the future. Direct potable reuse refers to the introduction of purified municipal wastewater – after extensive advanced treatment beyond conventional secondary and tertiary treatment – directly into a water distribution system after extensive monitoring to assure meeting the strict water quality requirements at all times (see Chapter 29). This implies blending purified water with source water for further water treatment or even pipe-to-pipe blending of purified water and potable water.

Direct potable reuse offers the opportunity to significantly reduce the distance at which reclaimed water would need to be pumped and to significantly reduce the head against which it must be pumped, thereby reducing costs. The other significant advantage of direct potable reuse is that it has the potential to allow for full reuse of available purified water in metropolitan areas, using the existing water distribution infrastructure (Drewes & Khan, 2011; Tchobanoglous *et al.* 2011). Implementation of direct potable reuse will require a confidence in, and reliance on, the applied technology to always produce water that is safe and acceptable to consume.

Towards sustainable water cycle management with water reuse

Each water drop is precious: so the use water again safely and for the right purpose is becoming a worldwide trend. The future of our planet, our ecosystems and our children depends on our ability to shift the paradigm of water resource management by the implementation of sustainable water cycle management.

The main objective of this book is to provide awareness on the tremendous challenges associated with securing reliable water supply and environmental protection in the future worldwide, as well as to give an overview of the solutions and benefits of well designed, integrated and "fit to purpose" water reuse practices. Twenty-nine case studies illustrate the successful implementation of such a holistic approach to water management with water reuse, which represent sustainable and drought-proof alternative water resources.

We hope this book will help you to better understand the challenges of water reuse and to learn more about the currently available, economically viable and environmentally friendly solutions via water reuse.

REFERENCES

Alhumoud J. M., Al-Humaidi H., Al-Ghusain I. N. and Alhumoud A. M. (2010). Cost/benefit evaluation of sulaibiya wastewater treatment plant in Kuwait. *International Business & Economics Research Journal*, **9**(2) 23–32.

Asano T. (2002). Water from wastewater – the dependable water resource. *Wat. Sci. Techn.*, **45**(8) 23–31.

Asano T., Burton F. L., Leverenz H., Tsuchihashi R. and Tchobanoglous G. (2007). *Water Reuse: Issues, Technologies, and Applications*, McGraw–Hill, New York.

Ayers R. S. and Westcot D. W. (1985). Water Quality for Agriculture. FAO Irrigation and Drainage Paper 29 Rev.1, Roma, 174 p.

Barnosky A. D, Hadly E. A., Bascompte J., Berlow E. L., Brown J. H., Fortelius M., Getz W. M., Harte J., Hastings A., Marquet P. A., Martinez N. D., Mooers A., Roopnarine P., Vermeij G., Williams J. W., Gillespie R., Kitzes J., Marshall C., Matzke N., Mindell D. P., Revilla E. and Smith A. B. (2012). Approaching a state shift in earth's biosphere. review. *Nature*, **486**(7 June 2012), 52–58.

California Water Recycling Criteria (2000). California Code of Regulations,. Title 22, Division 4, Chapter 3. California Department of Health Services, Sacramento, California, USA.

Crook J. (2004). *Innovative Applications in Water Reuse: Ten Case Studies*, WateReuse Association, Alexandria, VA, USA.

Crook J. (2010). *Regulatory Aspects of Direct Potable Reuse in California*, NWRI White Paper, National Water Research Institute, Fountain Valley, CA, USA.

Drewes J. E. and Khan S. J. (2011). Water reuse in drinking water augmentation. In. *Water Quality & Treatment: a Handbook on Drinking Water*, J. K. Edzwald (ed.), McGraw–Hill, New York, USA.

FAO, Food and Agriculture Organization (1999). Special Report: Drought Causes Extensive Crop Damage in the Near East Raisong Concerns for Food Supply Difficulties in Some Parts. http://www.fao.org/WAICENT/faoinfo/economic/giews/english/alertes/1999/SRNEA997.htm.

GWI (2009). Municipal Water Reuse Markets 2010. Global Water Inteligence.

IPCC (2007). *Comprehensive Assessment of Water Management in Agriculture*, David Molden (ed.) International Water Management Institute. http://www.fao.org/nr/water/docs/Summary_SynthesisBook.pdf.

Jimenez B. and Asano T. (2008). *Water Reuse: an International Survey of Current Practice, Issues, and Needs*, Scientific and Technical Report No. 20, IWA Publishing, London, UK.

Jimenez-Cisneros B. (2009). Coming to Terms with nature: Water Reuse New Paradigm Towards Integrated Water Resources management. In: *Future Challenges Of Providing High-Quality Water*, **Volume 1**, J. A. Van Wyk, R. Meissner and H. Jacobs (eds), Encyclopedia of Life Support Systems, 195–227.

Kazner C., Wintgens T. and Dillon P. (2012). *Water Reclamation Technologies For Safe Managed Aquifer Recharge*, IWA Publishing, London, UK.

Lazarova V. (2011). Milestones on Water Reuse: Main Challenges, Keys of Success and Trends of Development. Keynote Presentation,. 8th IWA International Conference on Water Reclamation & Reuse, 26–29 September 2011, Barcelona, Spain.

Lazarova V. and Bahri A. (2005). *Water Reuse for Irrigation: Agriculture, Landscapes, and Turf Grass*, CRC Press, Boca Raton, FL.

Lazarova V., Choo K. H. and Cornel P. (2012). *Water-Energy Interactions in Water Reuse*, IWA Publishing, London, UK.

Lazarova V., Cirelli G., Jeffrey P., Salgot M., Icekson N. and Brissaud F. (2001). Enhancement Of Integrated Water Management and Water Reuse in Europe and the Middle East. *Wat. Sci. Techn.*, **42**(1/2), 193–202.

Leverenz H. L., Tchobanoglous G. and Asano T. (2011). Direct potable reuse: a future imperative. *Journal of Water Reuse and Desalination*, **1**(1), 2–10.

Rockström J., Steffen W., Noone K., Persson A., Chapin F. S, III, Lambin E. F., Lenton T. M., Scheffer M., Folke C., Schellnhuber H. J., Nykvist B., de Witt C. A., Hughes T., Van der Leeuw S., Roldhe H., Sörline S., Snyder P. K., Costanza R., Svedin U., Falkenmark M., Karlberg L., Corell R. W., Fabry V. J., Hansen J., Walker B., Liverman D., Richardson K., Crutzen P. and Foley J. A. (2009). A Safe Operating Space for Humanity. *Nature*, **461**(24)472–475.

Tchobanoglous G., Leverenz H., Nellor M. H. and Crook J. (2011). Direct Potable Reuse: a Path Forward. WateReuse Research and WateReuse California, Washington, DC.

USEPA (2004). Guidelines for Water Reuse. U.S. Environmental Protection Agency, EPA/625/R-04/108, Washington, DC, USA.

USEPA (2012). Guidelines for Water Reuse, EPA/600/R-12/618. U.S. Environmental Protection Agency, Office of Wastewater Management & Office of Water, Washington, D.C., USA. http://waterreuseguidelines.org/images/documents/2012epaguidelines.pdf.

UN (2007). Water for Life Decade,. United Nations (UN). http://www.un.org/waterforlifedecade/scarcity.shtml

UNEP (2008). Vital Water Graphics: an Overview of the State of the World's Fresh and Marine Waters – 2nd Edition,. http://www.unep.org/dewa/vitalwater/article42.html.

WWAP (2009). World Water Development Report 3: Water in a Changing World: Facts and Figures. World Water Assessment Programme (WWAP), UNESCO, Paris, France.

WWAP (2012). World Water Development Report 4: Managing Water Under Uncertainty and Risk. Water Assessment Programme (WWAP), UNESCO, Paris, France. http://unesdoc.unesco.org/images/0021/002171/217175e.pdf.

WHO (2006). *WHO Guidelines for the Safe Use of Wastewater, Excreta and Greywater, Vol. II Wastewater Use in Agriculture*, World Health Organization (WHO), Geneva, Switzerland.

View of ornamental plant nurseries in California, irrigated with recycled water.

View of the New Goreangab Plant in Windhoek, Namibia.

PART I

The role of water reuse in integrated water management and cities of the future

Foreword

By The Editors

Traditionally, many cities have relied on the nearest river for public water supplies. River flows are inherently variable and prone to drought, so storage is required to maintain supply during dry periods. In many places, climate change is projected to reduce river flows and groundwater recharge, thereby increasing the frequency and severity of droughts. At the same time, demands are growing to meet the increasing urban water needs and the necessity for more food production. New approaches are needed to bridge the gap between growing demands and declining supplies to reduce the risk of supply failures.

Integrated water management and diversification of supplies

Many water authorities around the world are moving to an integrated water planning approach to maintain the balance between demand and supply, and to reduce the risk of supply failures. Common elements of the integrated water planning approach include water conservation measures to control the growth in demands, and the diversification of supplies to reduce supply risks. Supply diversification often includes the development of non-traditional water sources including decentralised supplies from rain water and stormwater, water reuse (treating used water and recycling it for use again) or desalination of seawater or brackish groundwater.

Using water portfolios to manage risk

Water managers are increasingly building water portfolios to manage supply risks. The concept is similar to that of financial portfolios. Financial managers create investment portfolios that combine high yield but risky investments (e.g. Stocks) with low yield but low risk investments (e.g. bonds and fixed interest) to achieve an acceptable return on investment with a low risk of financial losses. The water manager combines low cost but unreliable supplies (e.g. river supplies) with higher cost but more reliable supplies (e.g. recycled water and desalination) to create a water supply portfolio which is not too costly but is sufficiently reliable and has little risk of supply failure.

Integrated water management case studies

Part 1 of this book "Milestones in Water Reuse: The Best Success Stories" presents a number of case studies showing how water managers have applied integrated water management approaches to meet the future water needs of cities.

Table 1 Highlights and lessons learned from the selected case studies of integrated water resource management with water reuse.

Project/location	Start-up/capacity	Type of uses	Key figures	Drivers and opportunities	Benefits	Challenges	Keys to success
Sydney, Australia	2006, 2010 and 2014 Metropolitan Water Plans	Industry, Agriculture, Toilet flushing, Watering gardens, Car washing	Water savings: 145 Mm^3/yr by 2015. Total volume of treated wastewater: 510 Mm^3/yr. Total volume of recycled water: 70 Mm^3/yr by 2015. Desalination: up to 90 Mm^3/yr during droughts.	Demand reached capacity of the existing supply. Supply restricted due to shortfall in drought security.	Reduced user demands. Increased diversity, reliability and security of water supply. New river water supplies deferred for about 20 years. Improved environmental flows and river water quality.	Impacts of climate change on supplies. Adaptive management to changing conditions. Management of competing sources. Community education.	Government commitment. Direction of the Metropolitan Water Plan by chief executive officers. Major water efficiency, water conservation measures and water recycling initiatives. Additional supplies including groundwater and desalination to increase drought security.
West Basin, California United States	Since 1995 190,000 m^3/d	Industry, Irrigation of a golf course and public access landscape, Toilet flushing, Aquifer recharge for a salt intrusion barrier	Treatment capacity: average 132,500 m^3/d. Production of five types of recycled water. Over 160 km of distribution pipelines. Over 350 customers. Three satellite treatment facilities. Currently fifth phase of expansion to reach 190,000 m^3/d.	Develop Local Water Supply. Develop Reliable Water Supply.	Reduced dependence on interruptible imported water. Avoid loss of industry in region. Recognised economic, environmental and social benefits.	Developing separate facilities and infrastructure. Meeting water quality requirements for industrial customers.	Reliable source of water when potable supplies are uncertain. Match water quality with the intended use. Recycled water provides assured water supply and is less expensive than comparable potable water, even after the users' investment in retrofitting costs. Public outreach and education achieve acceptance.
NEWater, Singapore	Since 2002 531,000 m^3/d	Air conditioning, Cooling and boiler water for industries, Ultra clean water for semiconductor and electronic industries, Augmentation of freshwater water reservoirs (2.5%)	Total volume of treated wastewater: 511 Mm^3/yr. Total volume of recycled water: 194 Mm^3/yr. Recycled water tariffs 1.22 S$/$m^3$. Major benefits: Recycled water meets 30% of total water demand currently (up to 50% in the long term).	Water-stressed region. Limited land to collect rain water. Increasing water demand.	Additional water resource. Attain water supply sustainability with population growth and socioeconomic development. Independent of climate change and extreme weather condition such as droughts. Reduce wastewater effluent discharge. Ultra-pure water quality preferred by semiconductor and wafer fabrication industries.	Initial skepticism expressed by semiconductor and wafer fabrication industries on suitability of NEWater for their process use. Need transmission and distribution infrastructure which is cost intensive. Strict operational control of wastewater treatment plants and NEWater plants. Comprehensive water quality monitoring from source to point of use. Public acceptance for indirect potable use.	Multi-barrier approach to ensure good water quality which includes: source control, high proportion of domestic wastewater, comprehensive secondary wastewater treatment, use of proven advanced technologies, comprehensive water quality monitoring, adhering to strict operating procedures. A single agency managing wastewater and drinking water created synergy of taking a holistic approach on water resource management including water reuse. Strong government support and effective public education and communication.

Costa Brava, Spain	Since 1998 13 facilities	Irrigation of golf courses, parks, public landscape and home gardens, Environmental enhancement (wetlands and biodiversity restoration)	Volume treated wastewater: 30-35 Mm^3/yr. Max volume of recycled water: 6.4 Mm^3/yr (2010). Stringent water reuse standards: <1, <100, <200 *E.coli*/100 mL (private gardens, agricultural and golf course irrigation)	Lack of water resources. High energy consumption of drinking water cycle. Tax increase in drinking water use.	Net increase of resources. Simultaneous water and energy savings. Reduction of discharges. Potential economic savings.	Reliability in quality. Automation for online detection of failures. Pricing. Contractual agreement with municipalities and/or end/users. Refining of present regulations.	Gradual development and improvement of water reuse projects. Dedicated operation and maintenance of facilities. Look for the success of end users. Collaboration with researchers. Active divulgation of projects and their benefits.
Cyprus	Since 2004 Total capacity 150,000 m^3/d	Agriculture, Landscape irrigation, Aquifer recharge, Industrial uses	Total volume of treated wastewater: 19 Mm^3/yr. Total volume of recycled water: 15 Mm^3/yr. Recycled water tariffs: 0.05– 0.21 €/m^3. Water reuse standards: <5 FC/100 mL.	Comply with Directive 91/271/EEC by 2012. Reuse of 100% treated effluents.	Secure good quality water and protect environment. Substantial contribution to the island's water balance. Save freshwater for other uses.	Huge investments needed in a period of economic crisis. Efficient application in appropriate uses and the island's water cycle. Secure adequate quality controls and water quality.	Trust the benefits of water reuse and loyalty in implementation. Systematic work to empower acceptance. Monitoring and dissemination of water reuse outcomes. Ensuring good water quality.
Tianjin, China	2007 260,000 m^3/d	Toilet flushing, Landscape irrigation, Industrial use (cooling water for power plants, construction sites)	Total volume of treated wastewater: 726.4 Mm^3/yr. Total volume of recycled water: 94.9 Mm^3/yr. Total cost: 0.33–0.41 €/m^3. Recycled water tariffs: domestic water 0.28 €/m^3; industrial reuse 0.52 €/m^3. Water reuse standards: <3 *E.coli*/L.	National and local policy & incentives. High water scarcity. Availability of large users (power plant).	Conserving freshwater resources. Saving great amount of high quality water. Improving the environment.	Optimizing the category of reuse. Increasing the ratio of wastewater reclamation. Expanding the scale of reclaimed water and the pipeline construction. Increasing landscape uses.	Water scarcity. Government promotion and decisions. Reclaimed water treatment technologies enabling to consistently achieve the water quality requirements for fit to purpose uses of recycled water.

Chapter 1 describes the "Water for Life" integrated water management plan for Sydney, the largest city in Australia. Limits on the existing sources of water and drought risks have been addressed through an adaptive management strategy in which large savings from water efficiency programs have been combined with new supplies of recycled water and drought contingency supplies from groundwater and desalination to increase drought security. The need for new fresh water supplies has been deferred for two decades while, at the same time, additional water has been allocated for environmental flows to improve river water quality. Recycled water uses include urban and residential use, industrial use, agricultural use and replacement environmental flows to increase reservoir yields.

Chapter 2 describes the West Basin Water Recycling Scheme in California which produces "designer" recycled water tailored to meet customer needs. Recycled water quality is tailored to meet the specific needs for urban landscape irrigation, industrial cooling water, boiler feed water and for aquifer recharge. In a region where normal supplies are uncertain in drought times, the recycled water is a reliable source of water for customers and less expensive than new fresh water supplies. This concept, implemented and successfully operated since 1995, enables successful implementation of integrated resource management in urban and protected areas.

Chapter 3 describes Singapore's "Four Taps" water strategy to diversify its water supply. As an island state, Singapore has insufficient catchment area to supply all of its water needs but does not wish to increase its dependence on imported water. Singapore's NEWater schemes provide purified recycled water for industrial use and for reservoir recharge. NEWater now provides 30% of Singapore's water needs and is a key element in achieving water sustainability in Singapore.

The gradual development of water reuse in Spain's Costa Brava region (Chapter 4) has resulted in beneficial use of recycled water for agriculture, landscape irrigation, urban non-potable uses, environmental uses and for aquifer recharge. In a dry region, gradual and careful development of water recycling has enabled the local water authorities to supply growing water needs.

Chapter 5 describes water recycling in Cyprus, another water-stressed country. The planned implementation of water recycling systems has enabled the community to maximise the beneficial use of available water, minimise discharges and improve environmental water quality. The recycled water is being used for irrigation of permanent crops, urban landscape irrigation, sporting fields and for aquifer recharge.

Water shortages across the whole of northern China have posed major challenges in supplying water for urban and industrial needs without adverse public health and environmental effects. The case study in Chapter 6 describes planned water reuse in the City of Tianjin which has adopted water recycling as its principal strategy to overcome water scarcity. The recycled water is being used for municipal uses, industrial uses and to create scenic lakes to enhance the local environment.

Keys to success

The common themes running through these case studies is that, with integrated water planning, authorities in water-stressed regions with limited catchments have been able to cater for growing demands in the face of declining supplies due to droughts and climate change through water recycling. Keys to success have included strong government support and development of community support through effective public education programs. From a technical point of view, the implementation of adequate recycled water treatment technologies and reliable operation enabling to consistently achieve the water quality requirements for "Fit to Purpose" use of recycled water are critical for the success of water reuse projects.

The major highlights and lessons learned from the selected case studies are summarised in Table 1.

1 Water for life: diversification and water reuse as ingredients in Sydney's integrated water plan

John Anderson

CHAPTER HIGHLIGHTS

Through development of the Metropolitan Water Plan, Sydney has reduced demands, increased drought security, improved environmental flows and river water quality, and deferred the need for new water storages.

KEYS TO SUCCESS

- Government commitment.
- Direction of the Metropolitan Water Plan by chief executive officers.
- Major water efficiency and water conservation measures.
- Major water recycling initiatives.
- Additional supplies including groundwater and desalination to increase drought security.

KEY FIGURES

- Water savings: 145 Mm^3/yr by 2015.
- Total annual volume of treated wastewater: 510 Mm^3/yr.
- Total volume of recycled water: 70 Mm^3/yr by 2015.
- Desalination: up to 90 Mm^3/yr during droughts.

1.1 INTRODUCTION

Australia's largest city, Sydney, has adopted an integrated water management approach to adapt to increasing supply risks and achieve a balance between growing demands and available supplies.

The principal source of supply for the Australian city of Sydney is the Hawkesbury-Nepean river system. The Hawkesbury-Nepean river basin has a catchment area of 22,000 km^2, an average annual rainfall of 890 mm per year, and an average annual discharge of about 3000 Mm^3/yr. Like all Australian catchments, the rainfall and the stream flow are highly variable. The Hawkesbury River enters the Pacific Ocean about 30 km north of Sydney.

To date Sydney has been supplied mainly from a major large storage at Warragamba on the Hawkesbury River, a number of smaller storages on the upper Nepean and Woronora Rivers. The major Sydney water storages command about 44% of the Hawkesbury-Nepean catchment and have average annual inflows of about 1300 Mm^3/yr. In addition, there is a diversion system to pump water from the Shoalhaven River, 160 km south of Sydney.

Since the Nepean River was first tapped for Sydney's water supply in 1886, the population of Sydney has grown from less than 1 million to about 4.5 million people. The population served is projected to grow to about 5.7 to 6.0 million by the year 2036, and potentially could grow to 9.0 million by the year 2100.

The main catalyst for a re-think about meeting future supply needs was a major drought between 2002 and 2007 during which storage levels fell to about 35% of full storage.

Creation of the Metropolitan Water Plan process for Sydney

In 2005, the New South Wales state government initiated a process to develop a Metropolitan Water Plan for the Greater Sydney area. The process brought together the supply authority Sydney Water Corporation, the Sydney Catchment Authority, the NSW Treasury, and the state water resource and environmental regulators. The process was directed by the chief executive officers of the agencies.

The Sydney catchments have a highly variable climate which necessitates a large volume of drought storage. There is a wide range of uncertainty in the projections of climate change impacts on available supplies. It was decided to adopt a flexible, adaptive management approach in the Metropolitan Water Plan. This approach allows the steps that are being implemented to balance supply and demand to be adjusted as better information becomes available.

The initial Metropolitan Water Plan was published in 2006. The process now in place involves annual review of progress against the Metropolitan Plan targets, and updating of the Plan on a four-year cycle. An Independent Review Panel has been appointed to review the adequacy of the Plan and progress against the targets. The second Metropolitan Water Plan was published in 2010 and work has now commenced on the development of the 2014 Metropolitan Water Plan.

1.2 THE 2006 METROPOLITAN WATER PLAN

The 2006 Metropolitan Water Plan (NSW 2006) aimed to keep supply and demand in balance up until 2015 by:

- Water savings of 40% in new detached single residential dwellings and 25% in new medium density dwellings through a mandatory Building Sustainability Index (BASIX) scheme for water efficiency in new dwellings (DIPNR 2004a). The BASIX requirements have also being extended to houses undergoing major extensions or renovations.
- Annual water savings of 120 Mm^3 to 140 Mm^3 by existing consumers through variety of programs including: *Water Efficiency Plans* for large users; the Sydney Water *Leakage Reduction* program; *Water Efficiency Labelling* for new household appliances; a program of *Rebates for Rainwater Tank and Water Efficient Washing Machines*; and a household *Water Audit and Retrofit* program for existing dwellings;
- Up to 70 Mm^3/yr per year of recycled water per year through new water reuse projects for urban, residential, industrial, agricultural and environmental flow supplementation in western Sydney. It is proposed to supply recycled water to 160,000 new houses for residential garden watering and toilet flushing in the new development areas in western Sydney. Initially 18 Mm^3/yr of recycled water will be supplied for environmental flow allocations with possible later expansion to 30 Mm^3/yr or more.
- Reuse of 30 Mm^3/yr of recycled water per year in existing urban areas through development of a recycled water network.
- Significant levels of new supply through rainwater and stormwater harvesting.

The 2006 Metropolitan Water Plan also initiated measures to increase the capacity of the existing water supply system by various measures including:

- *Deep Storage*: Access to the deep storage below the previous minimum operating levels to provide an extra 200 Mm^3 of water in an extreme drought event.
- *Shoalhaven River*: Extra transfers from Shoalhaven River by using the existing pumping system more frequently.
- *Groundwater*: Use of groundwater aquifers as a drought contingency measure.
- *Desalination*: To build a desalination plant as a drought contingency measure if the Sydney storages were likely to fall to about 30% of full storage (desalination readiness strategy).
- Other initiatives and achievements under the 2006 Metropolitan Water Plan have included:
  - Recycling about 33 Mm^3 of recycled water in 2009–2010 for industry, irrigation and agriculture as well as for flushing toilets, watering gardens, washing cars and other outdoor uses. This figure has since increased by 18 Mm^3/yr with the implementation of the Replacement Flows Project at St Marys.
  - Protection of river health by implementing new infrastructure to allow variable environmental flow releases and fish passage from Sydney's dams and weirs in the Upper Nepean River system and from Tallowa Dam on the Shoalhaven River.
  - Roll out of education and training programs, and media campaigns to encourage the community to continue to use water wisely, support implementation of Water Wise Rules and build understanding of greater Sydney's changing water supply system.

1.3 THE 2010 METROPOLITAN WATER PLAN

The 2010 Metropolitan Water Plan has been produced through extensive review and updating of the 2006 Plan. The planning process has also been improved and refined. The key principles in the planning process include:

Water for life

The measures contained in the *Metropolitan Water Plan* (Figure 1.1) provide for a safe, dependable and affordable water supply for greater Sydney now and for the future. Achievements to date – including accessing water deep in dams, recycling, desalination, and reducing demand through water efficiency initiatives – provide water security to at least 2025 while also increasing water for the environment.

Figure 1.1 2010 Metropolitan Water Plan (Photograph courtesy of NSW Office of Water).

However, this security is only achieved with all sectors of the community continuing to play their part by using water wisely and supporting the recycling, river health, and other initiatives within the *Metropolitan Water Plan*. This is why there is a broad community education and engagement program called *Water for Life* underpinning the plan.

Water sharing

The *Metropolitan Water Plan* is designed to ensure sufficient water for people and the environment – with one not more important than the other. Community consultation on the *2006 Metropolitan Water Plan* revealed strong support in the community for maintaining this balance.

The *Metropolitan Water Plan* takes account of the water sharing plans that have been plans are developed under separate processes to ensure that all users who take water directly from rivers or groundwater have equitable access to water.

Water for the environment

The *Metropolitan Water Plan* recognises that water sourced from rivers and aquifers to meet the needs of communities and industry must be balanced against the need to reserve enough water to maintain river health and meet environmental needs. The *Water Management Act 2000* gives priority to water for the environment.

Since 2004, extensive work has been undertaken as part of the *Metropolitan Water Plan*, and through water sharing plans, to make sure rivers and aquifers, and the ecosystems that rely on them, receive adequate *environmental water* and are not overused.

Community planning principles

The community planning principles were developed from the findings of the Phase 1 consultation and validated during Phase 2. These principles will continue to underpin the way in which the 2010 plan is delivered.

- Provide water that is affordable and safe to drink.
- Ensure enough water to meet both environmental and human needs – one not more important than the other.

- Ensure a dependable long-term water supply for current and future generations.
- Maximise water efficiency and recycling, especially capturing stormwater and invest in research and innovation.
- Restore clean healthy waterways and ensure health of catchments by reducing pollution.
- Ensure government and community take joint responsibility for water management.
- Share water – taking into consideration all relevant sectors and regions.

Independent review

The NSW Government established a Metropolitan Water Independent Review Panel to provide expert advice and monitor progress in implementing and adaptively managing the *Metropolitan Water Plan*. The Panel comprises experts in fields such as urban water management, the economics of urban water systems, water conservation, attitudinal research and environmental issues.

- The Panel has provided oversight for the major review of the *2006 Metropolitan Water Plan*, in particular ensuring community input into the development of the *2010 Metropolitan Water Plan*.
- The Panel's advice is that, Sydney residents can depend on their water supply system at least until 2025.
- The Metropolitan Water Independent Review Panel feels that the *2010 Metropolitan Water Plan* is robust, as long diligent adaptive management is applied. It commented, in particular, that the community engagement process has created a healthy, whole-of-community partnership.

Developing the 2010 Metropolitan Water Plan portfolio

Along with the community's input, the review involved complex modelling and analysis to identify a portfolio, or mix of measures, that delivers water security into the future. The water planners took account of a range of factors, including achievements to date, advances in technology, updated population projections, rainfall and dam inflow scenarios, results of climate change research, cost effectiveness analysis, and social and environmental impacts.

The portfolio approach involves analysing different combinations of existing and new water supply and demand measures to identify the mix that provides water security for people and for the environment at the least cost. This challenge includes preparing for extreme drought. Our modelling made sure our water supply system could withstand a drought more than twice as severe as the recent prolonged drought in greater Sydney.

A process summary document has been developed to provide more detail on the portfolio approach to water planning and explain how the security of supply has been established. The document can be found at www.waterforlife.nsw.gov.au/review.

The *key measures* in the 2010 Metropolitan Water Plan portfolio include:

Water efficiency

- The water conservation and water efficiency measures in the Metropolitan Water Plan aim to save 145 Mm^3/yr by 2015 (Figure 1.2).
- Sydney Water is implementing a range of innovative programs to use water more efficiently in homes, businesses and public facilities, including:
  - Almost one million residential water efficiency rebates and offers since the water efficiency program began including WaterFix visits to more than 470,000 homes saving 10 Mm^3/yr and distribution of 201,000 free Do-it-yourself Water Saving Kits saving about 0.8 Mm^3/yr.
  - Water Wise Rules which have saved an estimated 19 Mm^3 of drinking water since they were introduced in June 2009. New homes are being designed to use 40% less water from Sydney's mains water supply.
  - More than 400 business organisations taking part in Sydney Water's Every Drop Counts Business Program, saving about 17 Mm^3/yr of water a year.
  - Sydney Water's leak reduction programs are estimated to be saving around 30 Mm^3/yr through investing almost $100 million and inspecting more than 20,000 km of pipes.
- Thanks to the water-saving efforts of homes and businesses in the greater Sydney region, Sydney is using the same amount of water now as in the early 1970s – even though serving an extra 1.4 million people.

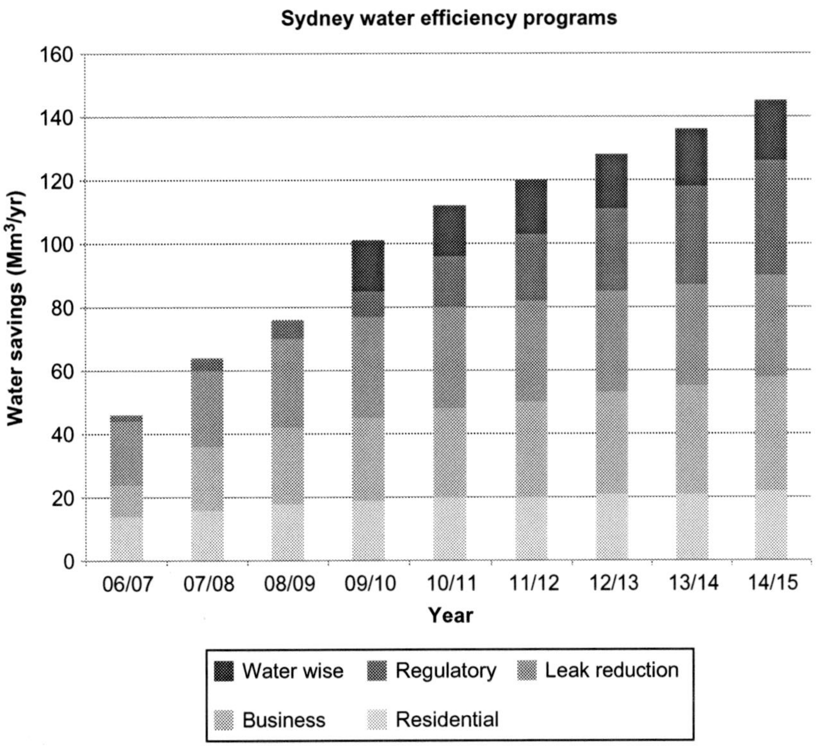

Figure 1.2 Projected water savings in Sydney.

Water recycling

Sydney's rainfall is highly variable and future droughts are likely to be more severe due to climate change of climate change and a growing population, it makes sense to balance supply from the dams with sources of water that do not rely on rain.

- The Metropolitan Water Plan aims to reduce demand for drinking water by around 70 Mm^3/yr by 2015 through water recycling of treated wastewater, greywater and stormwater for non-drinking purposes in homes, industries and agriculture (Figure 1.3).
- Projects completed or underway include:
  - The Replacement Flows Project at St Marys that will provide 18 Mm^3/yr of highly treated recycled water to the Hawkesbury-Nepean River as environmental flows (Figure 1.4). This will save dam water normally released to maintain the river environment downstream.
  - The Wollongong Recycled Water Plant that supplies around 7 Mm^3/yr.
  - Australia's largest dual-pipe residential recycling scheme at Rouse Hill that supplies about 2.2 Mm^3/yr to 19,000 homes for toilet flushing and watering gardens. This project has the capacity to be expanded to supply up to 36,000 homes as the area grows.
  - Sydney Water is now building similar residential schemes at Ropes Crossing and Hoxton Park in south-western Sydney.
  - The Rosehill-Camellia Recycling Scheme now supplies 4.7 Mm^3/yr of recycled water to major industrial and commercial users. It came online in 2011. It is the first project delivered using the third party access regime under the NSW Water Industry Competition Act 2006.
  - Around 70 local scale stormwater reuse projects in the greater Sydney area will save over 2 Mm^3/yr of water a year.
- In 2010, use of recycled water was saving about 33 Mm^3/yr of water a year that might otherwise come from Sydney's drinking water supplies. In addition, the Water Industry Competition Act 2006 is making it easier for the private sector to invest in recycling.

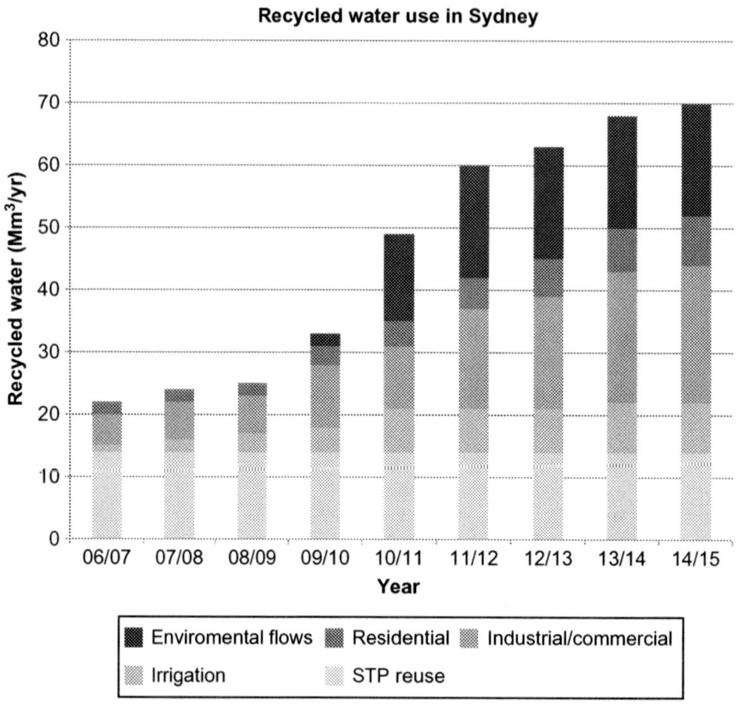

Figure 1.3 Projected use of recycled water in Sydney.

Figure 1.4 St Marys water recycling plant supplies 18 Mm^3/yr for environmental flows (Photograph courtesy of Sydney Water).

Desalination

- In 2007 when the Sydney storages fell to 35% of full storage, the NSW Government initiated the construction of a desalination plant with 90 Mm^3/yr capacity to supply up to 15% of Sydney's current water needs.
- The desalination plant came on line in January 2010 (Figure 1.5). It has run at full capacity during a two year "defects correction period" which ended in June 2012. During this period, the operation of the plant was monitored for water quality, performance and impacts on the supply system. After this period, the following operating rule will apply: the plant will operate at full production capacity and supply desalinated water to Sydney Water's area of operations when the total dam storage level falls below 70% and will continue to do so until the total dam storage level recovers to 80%.

- The desalination plant uses reverse osmosis technology to extract fresh water from seawater. Water from the desalination plant enters the system at Erskineville and is distributed to approximately 1.5 million people across the Sydney CBD, inner west, eastern suburbs, southern Sydney and parts of the Sutherland shire, and at times as far west as Auburn.
- The energy requirements of the plant are 100% offset by *renewable energy* from a 67 turbine wind farm near Bungendore.
- The construction of the desalination plant has made the Sydney supply considerably more secure in prolonged drought sequences.

Figure 1.5 The Sydney desalination plant at Kurnell (Photograph courtesy of Sydney Water).

System upgrading

- Long-term plans are to augment the Shoalhaven water supply transfer system. The timing of this initiative will depend on factors such as future climate predictions, population growth and demand. These factors will be reviewed between now and 2014 with a view to having an augmented system operational around 2025.

Drought contingencies

- A number of measures are available to be implemented in extreme drought to slow the depletion of dams. These include: increasing the amount of water transferred from the Shoalhaven by making more of the stored water at Tallowa Dam available, accessing groundwater, implementing voluntary water usage targets, up scaling the desalination plant, and investigating the optimal rules for reducing environmental flow releases from dams. Pricing is also being examined as a potential tool for managing demand during a drought.

Drought restrictions

- New simpler restrictions, based on the Water Wise Rules, will be implemented should Sydney experience another drought.

The supply-demand balance

- Drinking water use in Sydney peaked at 635 Mm^3/yr in 2002–2003.
- Water use has now been reduced to about 500 Mm^3/yr through the effect of water saving measures and the development of water recycling.
- The capacity of the supply system is currently assessed at around 620 Mm^3/yr after accounting for the severity of the 2002–2007 drought, adding the capacity of the Sydney desalination plan and deducting the increased environmental flow now being released downstream from the storages.

1.4 THE 2014 METROPOLITAN WATER PLAN

Work has now commenced on the preparation of the 2014 Metropolitan Water Plan. The work includes:

Understanding climate variability and climate change

- Balancing greater Sydney's water supply and demand over the long term requires improved understanding of climate cycles and trends, as well as the potential impacts of climate change.
- Climate variability relates to the naturally occurring weather events and patterns that have been tracked over time. Historical records show that Sydney's climate is highly variable, even without climate change impacts. While it is probable there will be severe drought conditions in the future, the exact nature of these droughts—timing, duration and severity—cannot be known. However, we do have a sound understanding of the effects of drought, and the importance of non-rain dependent supplies such as desalination. The *2010 Metropolitan Water Plan* contains a range of measures that can be called on when a prolonged drought hits.
- To better understand the impacts of climate change on greater Sydney's water supply system and future urban water demand, a collaborative study titled "Climate change and its impacts on water supply and demand in Sydney" (known as the Sydney Water Balance Project) was undertaken. The final report, which provides information on how the Sydney supply may be affected over time, was released in November 2010.
- However, climate change modelling is an evolving science and current projections contain considerable uncertainty because of the limitations of the climate data, uncertainties about future greenhouse gas emissions, and limitations of global climate models to estimate climate impacts at the regional level and to model future droughts.
- The second phase of the research will investigates techniques to improve the way extended drought periods are built into the climate modelling.
- Given that information about climate change projections and anticipated impacts will improve over time, it is important to maintain our flexible approach to managing greater Sydney's water supply.

Environmental flows from Warragamba Dam

- Further investigation is underway to determine the optimal environmental flow regime and infrastructure requirements so that a decision on the long-term environmental flow regime downstream from Warragamba Dam can be made in 2014.

Innovation

- Continued investigation and investment into new techniques and technologies, and developing water sources.

1.5 OVERVIEW

The integrated water management approach which has applied to Sydney's water system through the Metropolitan Water Plan process since 2006 has resulted in diversification of the water supply. Sydney has developed a more robust, diversified portfolio of water sources less dependent on rainfall and less prone to drought security risks. Water recycling accounted for 6.6% of the water supply in 2010 and will be doubled by 2015.

The Metropolitan Water Independent Review Panel feels that the *2010 Metropolitan Water Plan* is robust, as long as the adaptive management approach is applied diligently.

The following Table 1.1 summarizes the main drivers for water recycling in Sydney as well as the benefits and challenges.

Table 1.1 Drivers, benefits and challenges in Sydney's integrated water plan.

Drivers/Opportunities	Benefits	Challenges
Demand reached capacity of the existing supply. Supply restricted due to shortfall in drought security.	Reduced user demands. Increased diversity of supply. Increased reliability and security. New river water supplies deferred for about 20 years. Improved environmental flows and river water quality.	Impacts of climate change on supplies. Adaptive management to changing conditions. Management of competing sources. Community education.

REFERENCES AND FURTHER READING

Nicholson C. (2011). Delivering and essential and sustainable water plan for Sydney, Australia, Proc. 8th IWA Specialist Conference on Water Reclamation and Reuse, Barcelona, Sep 2011.

WEB-LINKS FOR FURTHER READING

NSW (2006). *2006 Metropolitan Water Plan: Securing Sydney's Water Supply*. New Sotuth Wales Government, NSW Office of Water, Sydney, Australia. April 2006, www.waterforlife.nsw.gov.au.

NSW (2010). *2010 Metropolitan Water Plan Summary*, New South Wales Government, NSW Office of Water, Sydney, Australia. August 2010, www.waterforlife.nsw.gov.au/mwp

Views of the Sydney Harbour.

View of the Wollongong Sewage Treatment Plant in Sydney (Photograph courtesy of Sydney Water).

View of the Hawkesbury-Nepean River which flow is maintained by the purified water from the St Marys Water Recycling Plant (Photograph courtesy of Sydney Water).

2 Producing designer recycled water tailored to customer needs

Joseph Walters, Gregg Oelker and Valentina Lazarova

CHAPTER HIGHLIGHTS

The Edward C. Little Water Recycling Facility in El Segundo, California is world renown as the only facility worldwide producing five distinct types of "designer" recycled water. Since 1995, these distinct types of treated water are each uniquely suited to specific use, thereby customizing the treatment and cost to the required use: tertiary disinfected water (Title 22 water quality) for irrigation and other urban uses, nitrified water for cooling towers, reverse osmosis followed by advanced oxidation water for direct aquifer recharge in a salt intrusion barrier, and single and double pass reverse osmosis for low and high pressure boiler water. This concept enables successful implementation of integrated resource management in urban and protected areas.

KEYS TO SUCCESS

- Recycled water provides a reliable source of water when potable supplies are uncertain
- Match water quality with the intended use
- Recycled water provides assured water supply and is less expensive than comparable potable water, even after the users' investment in retrofitting costs
- Public outreach and education achieve acceptance

KEY FIGURES

- Treatment capacity: average daily flow of 132,500 m^3/d (35 MGD); able to expand as customer demand dictates
- Production of five types of recycled water
- Over 100 miles (160 kilometres) of distribution pipeline serving over 350 customers
- Three satellite treatment facilities
- Currently fifth phase of expansion to reach 190,000 m^3/d (50 MGD)

2.1 INTRODUCTION

The drought in the late 1980's and early 1990's was the latest in the unpredictable precipitation cycle of southern California which has motivated the local residents to explore innovative methods of bringing water to this arid region. Water recycling at West Basin Municipal Water District (West Basin) came from a very visionary Board of Directors led by Edward C. Little when the West Basin Board made the decision to diversify and localize control of part of its water supply. This tradition of innovation carries on today as West Basin can boast of water supply consisting of desalted groundwater, conservation, recycling and soon, ocean water desalination.

West Basin is the only wastewater recycling facility in the world to produce five different types of "designer" waters including Title 22 irrigation water, cooling tower water, low pressure boiler feedwater; high pressure boiler feed water, and groundwater recharge/ seawater barrier indirect potable recycled water.

Since the mid-1990's West Basin has continually expanded its recycling water program and is currently constructing its fifth expansion to its treatment plant. The recycling plant was recently renamed the Edward C. Little Water Recycling Facility (ECLWRF) in honor of current Board member and past-President Edward C. Little. The ECLWRF recently reached a milestone of over 634 million m^3 (Mm^3) of recycled water produced since the program's inception.

In light of increasing demand and with the uncertainty and limitations of imported water, West Basin is currently implementing a long-term, strategic expansion of its locally-controlled water supplies called Water Reliability 2020, which will reduce the agency's dependence on imported water by 50% by the year 2020. Water Reliability 2020 will double the current water recycling program, double the water efficiency/conservation program and add 75,700 m^3/d (20 million gallons a day, MGD) of ocean water desalination. West Basin was also the first U.S. water agency to actually produce indirect potable water using the new state-of-the-art microfiltration, reverse osmosis and advanced oxidation by ultraviolet light and hydrogen peroxide water purification process.

Brief history of the project development

West Basin Municipal Water District serves an area of southern California with history of western settlement going back to the time of the American Revolutionary War. At that time Franciscan fathers established missions throughout southern California. That was followed by land grants by the King of Spain to his loyal soldiers such as Juan Jose Dominguez who was granted the Rancho San Pedro. While the land was used for grazing cattle and farming, it was limited by the inconsistent rainfall that often left the streambeds dry in the summer heat.

The first well was drilled in 1854 to draw shallow water from areas just below the surface. This worked well enough for ranching, but as the area became more developed, access to water sources limited the growth of the basin. Deep well turbine pumps developed in the late 1800's allowed water extraction from deeper aquifers and enabled the next phase of growth for the region. However, within a few decades after the aquifers were accessed, ground water levels were already declining. Continued pumping reversed the hydraulic gradient and by 1918, seawater intrusion began to affect wells in the coastal cities. With the industrial buildup in the area for World War II, saline intrusion had become a major threat to the groundwater and the lifestyle of many communities.

A citizens committee was formed in 1942 as the West Basin Survey Committee initiating a survey in 1944 that revealed the dire nature of groundwater in West Basin. For over a century and a half, the local water supply had served the needs of the communities. But time had run out. The beach communities were already impacted by seawater intrusion. The inland communities also had groundwater levels below sea level and it was only a matter of time before salt water intrusion impacted the whole region.

West Basin Municipal Water District was formed by a vote of the people in December 1947 to find a new water supply to refill local groundwater basins, which were being rendered brackish from seawater intrusion. At that time, West Basin looked at wastewater recycling and ocean water desalination. Both were determined to be far more expensive than joining the Metropolitan Water District of Southern California (MWD) which imported Colorado River water for its member agencies. West Basin petitioned for, and was granted membership in MWD in 1948.

But the recurring cycle of droughts has challenged the District even with the addition of water from the State Water Project in northern California. With an ever expanding population, and limitations on the available imported water, providing a reliable water supply has been continually challenged. The drought of the late 1980's – early 1990's was the last warning needed for West Basin's Board of Directors to take action. West Basin began diversifying its water supply in the mid-1990's to cushion itself from future water shortages by committing resources to construct a recycled water treatment plant which went into operation in 1995.

West Basin has been rewarded for this leap of faith by seeing the seed they planted as a recycled water treatment plant grow to the world renown facility it is today. West Basin produces approximately 132,500 m^3/d (35 MGD) of recycled water and has plans to more that double that amount in the future to enhance the reliability of its local supply. Much of the success of West Basin's program is due to the development of five types of "designer" water that customizes the water quality to the needs of the customer (described below).

Main drivers for water reuse

The main drivers of the West Basin water recycling are: (1) demand long ago exceeded local water supply, (2) uncertainty and limitations of current imported water supply from 350 miles away, (3) need for assured supply of water for local industry, (4) support of the local authorities to develop diversified water supply.

Milestones of water recycling in West Basin

1991 – West Basin & City of Los Angeles reach agreement to deliver treated wastewater to recycling facility from Hyperion Wastewater Treatment Plant.

1993–1995 – West Basin receives accolades for recycling efforts from WateReuse California, ACWA, U.S. Interior and others.

1993 – West Basin breaks ground on water recycling plant.

1994 – pump station built at Hyperion to bring water to West Basin facilities in El Segundo (about two miles away).

1995 – West Basin recycling facility begins producing water.

1996 through today – West Basin began a series of expansions of its recycling facility and today is designing its fifth expansion to meet the needs of its customers.

Project objectives and incentives

The drought of the late 1980's – early 1990's emphasized the fragile reliability of current water sources and alerted both the local water agencies and the public at large of the precarious nature the local economy and southern California lifestyle without reliable water supplies. West Basin at that time was a small water wholesale agency with minimal staff and less infrastructure. It had never borrowed money to fund a project of this magnitude before. However, projections for potable water rates ten years in the future presented a vision of the future that could support the implementation of a recycled water system. Yet this vision relied on a number of partnerships, timely commitments from potential customers, and successful navigation through the regulatory requirements to sprout from concept to reality. Remarkably, the West Basin Board of Directors had a unified vision of the necessity of the project and voted unanimously to support it.

With the support of the Directors, West Basin General Manager Rich Atwater was the right person to lead this sizable undertaking. In a matter of a few years Mr. Atwater, at the direction of the West Basin Board of Directors, had successfully negotiated an agreement for water supply with the City of Los Angeles, purchased land to site the treatment plant, established a partnership with the U.S. Bureau of Reclamation to share the some of the costs, secured commitment from the first large customer, received voter approval for a Stand-By Charge (tax on land parcels), and began the design and construction of the facilities. In addition, the Metropolitan Water District of Southern California which imports water into southern California for distribution to agencies such as West Basin also provides a Local Resource Program (LRP) whereby it reimburses West Basin $0.2/m^3 ($250 per acre foot) of recycled water it produces.

All of these factors contributed to the West Basin Board approving the significant investment to create a recycled water program. The original plant began operation in 1995.

2.2 TECHNICAL CHALLENGES OF WATER QUALITY CONTROL

Secondary treated wastewater effluent produced at Los Angeles' Hyperion Treatment Plant (HTP) is normally discharged to the Pacific Ocean through a five-mile outfall pipeline. The HTP is the city's oldest and largest wastewater treatment facility and has been operating since 1894. Hyperion treats sewage from 4 million residents, two-thirds of the city of Los Angeles. The plant has been expanded and improved numerous times over the last 100 years. The HTP uses a pure oxygen system and full secondary treatment, biosolids handling, biogas generation. Today, leading edge technological innovations capitalize upon the opportunity to recover wastewater bio-resources that are used for energy generation and agricultural applications. The daily volume of biogas production is 212,400 m^3 (7.5 million cubic feet) which are turned into energy, and 500 tons of biosolids per day are used as fertilizer and soil amendment. A 1.5 m (60-inch) Force Main from Hyperion was constructed from 1992 to 1995 along with a pump station to deliver 15% of the wastewater to the recycling facility.

The ECLWRF produces five types of "designer" water at the main Plant and Satellite Facilities. The average flow is 132,500 m^3/d (35 MGD) with a maximum capacity of 170,300 m^3/d (45 MGD). For comparison, the daily production is enough water conserved to meet the need of 60,000 households for a year.

Recycling facilities and treatment trains

The ECLWRF is the largest water recycling facility of its kind in the world, and was recognized by the National Water Treatment Technologies in 2002 as one of only six National Centers for Water Treatment Technologies. The ECLWRF (Figure 2.1) and the three satellite plants produce five different qualities of "designer" recycled water. The five types include:

- *Tertiary Recycled Water* (Title 22 Product Water) is used for a wide variety of industrial and landscape irrigation uses and has a capacity of 151,400 m^3/d (40 MGD). The process includes conventional tertiary treatment – coagulation/flocculation/filtration/disinfection. Ferric chloride and cationic polymer are used for coagulation, and sodium hypochlorite is used for disinfection. A final minimum chlorine residual of 4.1 mg/L is required to meet the disinfection requirement of a CT of 450 mg-min/L.
- *Barrier Product Water*, or stabilized Reverse Osmosis Water: the process consists of microfiltration (MF), followed by Reverse Osmosis (RO) and Advanced Oxidation Process (AOP – hydrogen peroxide/UV) (Figure 2.2). MF removes suspended solid material, turbidity, and some microbes. RO removes a large percentage of dissolved minerals and

organic material, and most of the remaining microbes. The AOP process provides disinfection, and reduction of Nitrosodimethylamine (NDMA) to below the California Notification Level (NL), 10 ng/L (ppt). The water is stabilized using decarbonation to remove carbon dioxide, and addition of saturated lime solution to add back alkalinity and calcium. The water is used for ground water recharge and the current design capacity is 47,320 m^3/d (12.5 MGD).

- *Pure Reverse Osmosis Water* (MF/RO) is used at refineries for low pressure boilers. Design capacity is 6435 m^3/d (1.7 MGD) at the ECLWRF for use at the Chevron refinery, 18,930 m^3/d (5 MGD) at the Exxon-Mobil satellite facility, and 18,930 m^3/d (5 MGD) at the Carson Regional Plant for use at the BP refinery. This water is very similar to Barrier Product Water, but is not treated with AOP and not stabilized prior to industrial use.
- *Ultra-pure Reverse Osmosis Water* (MF/RO/RO) is produced at the ECLWRF, and used at the Chevron refinery in their high-pressure boilers. The capacity is 9840 m^3/d (2.6 MGD). This water is close to distilled water quality.

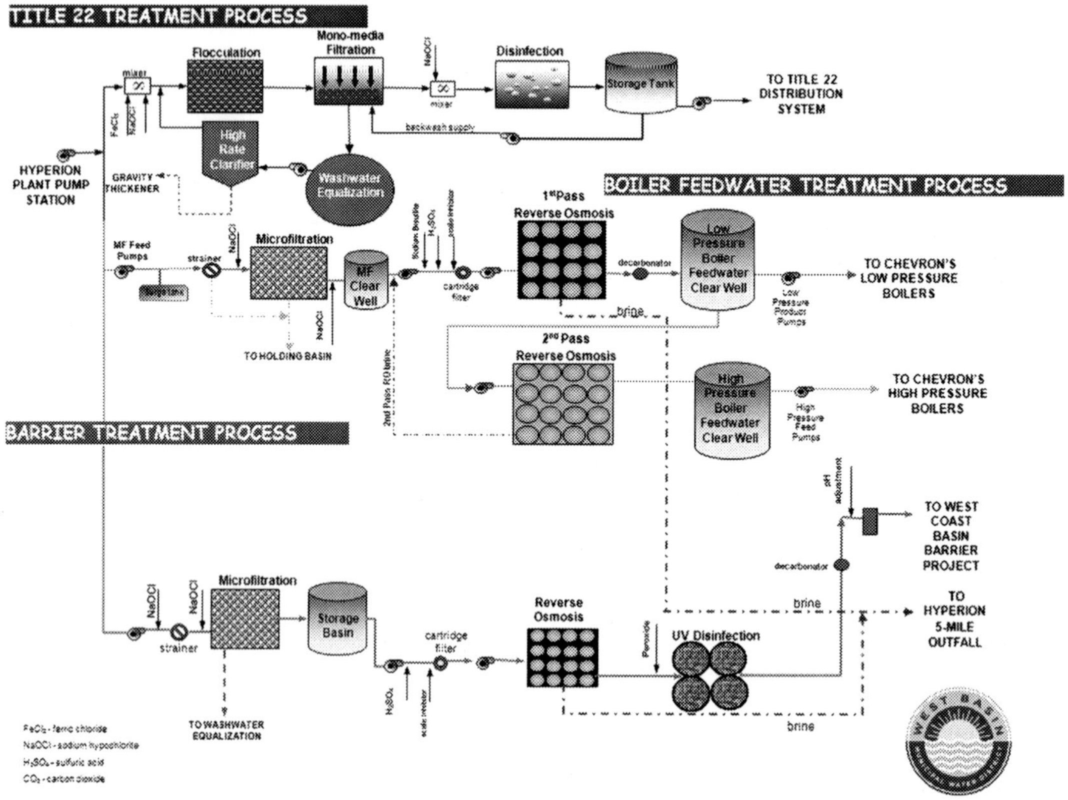

Figure 2.1 Schematics of Edward C. Little Water Recycling Facility.

Figure 2.2 View of reverse osmosis and the advanced oxidation by UV/H_2O_2.

- *Nitrified water* is used for industrial tower cooling towers at three refineries. Ammonia in the Title 22 recycled water is converted to nitrate through a biological fixed film process. Ammonia is problematic for the materials used in the cooling towers, and the conversion to nitrate is necessary. Design capacity at the Chevron satellite plant is 18,930 m^3/d (5 MGD), at Exxon-Mobil is 18,930 m^3/d (5 MGD) and at the Carson Regional Plant is 3785 m^3/d (1 MGD).

The recycling facilities have been constructed in 4 phases as shown in Table 2.1.

Table 2.1 Treatment capacity of recycled water produced by Edward C. Little Water Recycling Facility.

Plant	Year	Design flow m^3/d (MGD)	Application
ECLWRF – Phase 1	1995	56,780 (15.0) 18,930 (5.0)	Title 22 Barrier Water: Lime Clarification-GMF/RO
ECLWRF – Phase 2	1999	56,780 (15.0) 9460 (2.5)	Title 22 Barrier Water: MF/RO
ECLWRF – Phase 3	2003	6430 (1.7) 9840 (2.6)	Industrial water: MF/RO Industrial water: MF/RO/RO
ECLWRF – Phase 4	2006	37,850 (10) 18,930 (5.0)	Title 22 Barrier Water: MF/RO
Chevron	1995 1999	18,930 (5.0) 16,280 (4.3)	Industrial water: Nitrified
ExxonMobil	1998	18,930 (5.0) 12,110 (3.2)	Industrial water: MF/RO Industrial water: Nitrified
Carson	2000	18,930 (5.0) 3785 (1.0)	Industrial water: MF/RO Industrial water: Nitrified

Satellite treatment plants

The three Satellite Plants are as follows:

- Chevron Nitrification Facility (CNF) further treats Title 22 recycled water received from the ECLWRF through a biological nitrification process to remove ammonia for cooling water tower application.
- ExxonMobil Water Recycling Facility (XOM) further treats Title 22 recycled water from the ECLWRF through a biological nitrification and MF/RO processes for use at the ExxonMobil Refinery in Torrance. The water produced is used for cooling tower and boiler feed applications.
- Carson Regional Water Recycling Facility (CRWRF) further treats Title 22 recycled water received from ECLWRF. Water processed through MF/RO and biological nitrification, then blended and sent to the BP Refinery in Carson for industrial use.

Seawater intrusion barrier

The West Coast Seawater Intrusion Barrier consists of 153 injection wells strategically located to prevent seawater intrusion into the West Coast Groundwater Basin. A total of 66,250 m^3/d (17.5 MGD) of water is injected into the Barrier. The recycled water for the Barrier is blended with potable water at the blend station, located in the city of El Segundo. The ECLWRF is permitted for 75% recycled water contribution, or 47,320 m^3/d (12.5 MGD) of the water injected into the Barrier. Currently 34,070 m^3/d (9 MGD) of recycled water, or about 50% of the injected water is recycled water. The permit allows expansion of the facility, and 100% recycled water injection into the Barrier if certain conditions are met. Most have been met, and West Basin is working to meet the last few conditions remaining. Phase V construction has begun to expand the Barrier Product Water production to the full capacity of 66,250 m^3/d (17.5 MGD).

Recycled water distribution system

The Title 22 Distribution system of 160 km (100 miles) conveys tertiary recycled water from the ECLWRF to the Title 22 customers (Figure 2.3), as well as three satellite treatment facilities: Carson Regional Water Recycling Facility (CRWRF), ExxonMobil satellite plant (XOM) and Chevron Nitrification Facility (CNF).

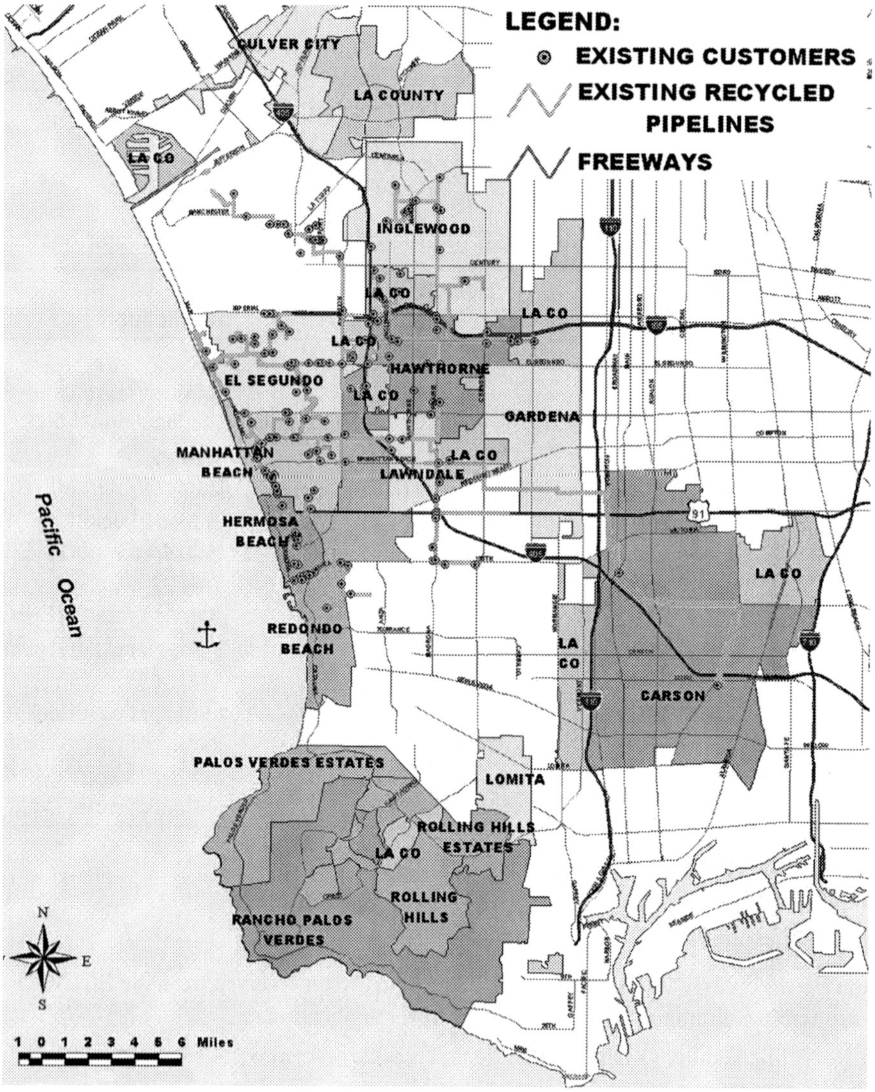

Figure 2.3 Schematics of the distribution system.

There are two chlorine booster stations in the system to meet the needs of customers on the furthest ends of the system. This water is also used for irrigation at parks, golf courses, office buildings, and refineries, for toilet flushing and cooling towers, and for other industrial uses.

The Chevron El Segundo Refinery, located adjacent to ECLWRF, receives four high quality waters which include T-22 water for irrigation, nitrified water for cooling towers, single-pass RO water used in low-pressure boilers, and double-pass RO water used in high-pressure boilers. These four recycled water products are conveyed in four separate distribution system from ECLWRF and the Chevron Nitrification Facility to the refinery. Water use at the Chevron Refinery is more than 95% recycled water.

Recycled water quality

Title-22 product water

The Title-22 Product Water is not drinking water though it meets all drinking water standards but should not be consumed. This recycled water has high levels of ammonia and Total Organic Carbon (TOC). The conventional tertiary process is not designed to remove ammonia, so the product water has the same ammonia level as the source water (35–45 mg/L as N). The TOC is typically 8–12 mg/L. Nutrients (potassium, nitrogen and phosphorus) are beneficial to irrigation customers and fertilizer use can often be reduced (Haering *et al.* 2009).

The source water has a very high chlorine demand, due to the chemistry of wastewater, especially ammonia. To achieve the required 450 mg-min/L CT for disinfection, a much higher dose of chlorine is needed. Currently, chlorine is dosed at over 15 mg/L to meet the demand and maintain the required residual for disinfection. Studies performed at the plant have shown that chlorine decays to zero residual within 3–4 days. Customers at the far reaches of the Title 22 distribution system may see water that is older than 4 days, particularly in the winter when irrigation uses are lower and the water in the system is flowing slower. Sulfate in the water can convert to sulfide when the chlorine is gone, and this can cause the water to have an odour of "rotten eggs". For this reasons, chlorine booster stations have been strategically placed in the distribution system to help alleviate this problem.

Barrier product water

Barrier Product Water is also not drinking water even though it is of equal to or better quality than tap water. Though it could be used for drinking (direct potable reuse), public perception and political roadblocks have stopped this from moving forward in the United States (Tchobanoglous *et al.* 2011). The Barrier water is injected into the ground where it is further treated in the soil. This more accepted practice, indirect potable reuse, is allowed by current regulations. Groundwater models show the injected water has a 20-year residence time in the ground prior to being pumped out in groundwater wells and used as a potable source. The water that was injected into the ground upon plant startup still is in the aquifer.

The membrane processes (MF and RO) are excellent barriers to microbial, organic, inorganic and other contaminants. AOP treatment to reduce NDMA with a UV dose of >115 mJ/cm^2 provides many times the dose required for disinfection. The addition of peroxide in the AOP process is an additional barrier to any unknown compound. Soil-aquifer treatment provides an additional treatment barrier.

Though the Barrier Permit was written earlier, it basically follows the requirements in the Groundwater Recharge Reuse Draft Regulations (CDPH, 2008). Barrier Product water has TOC in the 0.2–0.3 mg/L range, very low in comparison with most drinking waters, and below the permit limit of 0.5 mg/L. Total nitrogen (ammonia + nitrate + nitrite + organic nitrogen) is about 3 mg/L (as N) in the Barrier water, less than the permit limit of 5 mg/L. Coliform results are consistently not detected (daily total and fecal coliforms, and weekly *E. coli*). Trace metals and organics are mostly not detected. Testing of pharmaceuticals and personal care products (PPCP) has been done for many R&D projects and is required in the Barrier Permit. Though many of these compounds are detected in the source water, the only compound detected consistently in Barrier Product Water is caffeine at 10–20 ng/L, or parts per trillion (Snyder *et al.* 2007, 2008; Drewes *et al.* 2008).

Compliance record

The ECLWRF has four main permits for Plant Operation. The Barrier Permit includes required monitoring of the Influent, Barrier Product Water, Barrier Blend (Barrier Product mixed with Potable) and Groundwater Monitoring Wells in the vicinity of the injection system. The Title 22 Permit is focused on monitoring of the Title 22 Product Water as it leaves the ECLWRF and enters the Title 22 Distribution System. The Plant has two NPDES Permits for monitoring of Brine from the RO processes at the ECLWRF and CRWRF. These flows are comingled with secondary effluent discharges to the Pacific Ocean at the Hyperion Plant and LA County Joint Plant respectively (WBMWD, 2002, 2006, 2007).

The ECLWRF has an analytical laboratory (Figure 2.4), the West Basin Water Quality Lab that is certified by the State of California's Environmental Laboratory Accreditation Program (ELAP, 2012). The laboratory is certified to test inorganic and microbiological parameters, TOC, and trace metals in both wastewater and drinking water. The lab utilizes two EPA methods (USEPA, 1993; USEPA, 1994) and all other methods from Standard Methods (SM, 1992, 1995, 1998) for testing. The lab has 6 full time staff, and produces about 2500 analytical results per month. Additional testing of organics, oil & grease, surfactants, asbestos, radioactivity, hexavalent chromium, phenol and pharmaceuticals/EDCs required by the Plant's permits are subcontracted to an outside laboratory. Approximately 70% of all lab testing is done for process control and in support of R&D projects.

In total, the WBMWD reports approximately 40,000 values on compliance reports to the California Regional Water Quality Control Board every year. Over 25,000 of these values come from on-line process meters monitoring flow, pH, chlorine residual, turbidity, conductivity, UV absorbance, and UV intensity. About 15,000 values are reported from lab testing, performed in both the in-house and subcontract labs. In the past fiscal year (July 2010 to June 2011), the ECLWRF has an excellent record of 99.82% compliance (WBMWD, 2010, 2011).

Table 2.2 presents the average annual data for the influent and the final effluent water quality for Title 22 and Barrier treatment trains. The data clearly show the differences in water quality. The low TDS, TOC and turbidity values of the Barrier Water are indicative of the high treatment efficiency of the RO advanced treatment train.

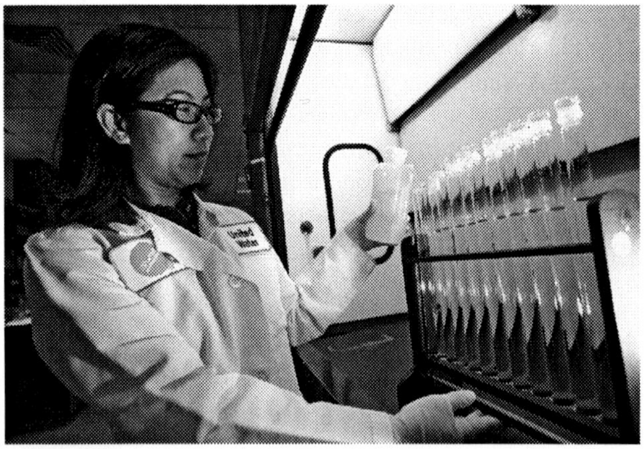

Figure 2.4 Process and water quality control.

Table 2.2 Monthly average water quality data of the Edward C. Little Water Recycling Facility (2009 to 2011).

		Influent		Title 22 Water		Barrier Water	
Parameter	Units	Permit	Measured	Permit	Measured	Permit	Measured
Turbidity	NTU	–	4–30	2 (2.5)	1–2.4	0.5	0.05–0.6 (average 0.092)
pH	pH units	–	6.8–7.2	6.5–8.5	6.5–7.3	6.5–8.5	7.7–8.0
TOC	mg/L	20	8.3–66	20	8.1–14	2	0.14–0.27 (average 0.19)
TSS	mg/L	30	12–43	20	1–17	–	<1
BOD	mg/L	30	5–40	20	<3–5	–	<3
TDS	mg/L	–	–	800	730–1100	300	31–132 (average 83)

The quality of the Title 22 recycled wastewater has been fairly homogenous during the 15 years of operation, without any significant daily or monthly variations, despite high variations in the quality of the source water. The annual average concentrations of the most relevant water quality parameters are <2.2 total coliforms/100 mL, <2 NTU turbidity, and <20 mg/L TOC. The residual chlorine was maintained at a minimum of 4.1 mg/L in order to maintain the required CT of 450 mg-min/L, and even higher to guarantee water quality in distribution systems. The main customers for the Title 22 recycled water are golf courses, parks, schools, cemeteries, commercial centers and private homes for toilet flushing.

Barrier Product Water has also seen homogeneous water quality since plant startup. The membrane processes are excellent barriers to maintain stable, high-quality water. The original plant design included disinfection by sodium hypochlorite and required CT of 450 mg-min/L. The Phase IV plant expansion added the UV/AOP process to destroy Nitrosodimethylamine (NDMA) that had been found in the Product Water. A California Notification Level (NL) of 10 ng/L for NDMA is in place, though exceedance of the limit is not enforceable. UV treatment was nonetheless added to lower the NDMA concentration below the NL. The UV and peroxide addition also ensures an extra 'barrier' to any other compound not known at this time, but could be detected in the future (WBMWD R4-2006-0069).

Table 2.3 presents the annual concentrations for those trace organic compounds that were quantified. It is interesting to note that the compounds that are detected by the conventional base neutral analysis protocol (EPA Methods 525/625) and volatile organic analysis (EPA Method 524.2) are primarily, disinfection byproducts, such as chloroform and bromoform. Other organics are not seen in the Barrier Water.

Industrial water quality

Because it has been treated using flocculation and gravity filtration, many constituents remain in the Title 22 water such as ammonia and salts. The ammonia can be corrosive to cooling towers, so that when it is used for this purpose, the ammonia must be either removed or mitigated using other treatment such as chemicals. For the refineries which use large volumes of recycled water, it is most effective and cost efficient to remove the ammonia by biological process.

Table 2.3 Quarterly average concentrations in μg/L for specific organic compounds.

Constituent	EPA Method	Detection Limit	Title 22 Water		Barrier Water	
			MCL	Range	MCL	Range
Volatile Organics (μg/L)						
Chloroform	524.2	0.021		5.0–7.3		ND-0.9
Dibromochloromethane	524.2	0.034		ND-1.4		ND
Bromodichloromethane	524.2	0.028		1.0–2.2		ND
Bromoform	524.2	0.044		ND-0.86		ND
Total Trihalomethanes		0.127	80	6.0–12	80	ND-0.9
Methylene chloride	524.2	0.03	5	0.97–2.5	5	ND-1.0
1,4–Dichlorobenzene	524.2	0.019	5	ND-0.21	5	ND
Tetrachloroethylene	524.2	0.032	5	ND	5	ND
Acid & Base/Neutral Extractable (μg/L)						
Bis (2–Ethylhexyl) phthalate	525.1	0.47	4	ND-1.9	4	ND-0.58
Phenol	625	0.3	–	ND	–	ND

Source: Adapted from West Basin Municipal Water District Annual Report, 2009, 2010.
ND – not detected above the detection limit, MCL – maximum contaminant level.

High quality ammonia free effluent for cooling water supply is produced by complementary advanced biological nitrification process in aerated bio-filters. Biofor® satellite treatment facilities are installed and produce water at the Chevron, Exxon/Mobil and BP oil refineries. These compact bioreactors are operated at high ammonia loading rates of up to 0.75 kg/d.m^3 and water velocity of 4–5 m/h, which represents a common value for the operation of aerated bio-filters during tertiary nitrification. The nitrification performance of the aerated bio-filters has been excellent until recently. For an average influent ammonia concentration of 40 mgN/L, the residual ammonia concentrations were consistently below the detection limit of 0.1 mgN/L.

Boiler feed water at the refineries requires a much higher level of purity (Table 2.4) and is therefore treated by using reverse osmosis filters with microfiltration as a pre-treatment. This process greatly reduces the mineral content. As an example, TDS is reduced from 800–900 ppm to about 50 ppm. Where even greater purity is required, a second stage of reverse osmosis treatment is performed. This reduces the TDS to below 5 ppm. Special pipes must be used for this mineral free water (comparable to distilled water) to avoid corrosion of the pipes from the water pulling out minerals during distribution.

Table 2.4 Industrial water quality specifications for satellite facilities.

Chevron Low Pressure BF	Chevron High Pressure BF	Carson Industrial RO Water	ExxonMobil Industrial RO Water
Hardness < 0.3 mg/L	Hardness < 0.05 mg/L	Ca 1 mg/L	EC 50 μmho/cm
Silica < 1.5 mg/L	Silica < 0.1 mg/L	Mg 1 mg/L	TOC 0.7 mg/L
TDS < 60 mg/L	TDS < 5 mg/L	Ammonia 4 mg/L	Ammonia 1.9 mg/L
		Silica 1 mg/L	Silica 1.0 mg/L
		TDS 35 mg/L	

Major challenges for operation

The major challenge for the plant operation is the deterioration of inlet water quality over time. Ammonia has increased from 20–25 mg/L (as N) in 1995 to over 40 mg/L in recent years. Chloride has increased from 130 to over 200 mg/L due to changes in water supply in Southern California and conservation of potable water. Allocations of northern California water have been reduced, and Colorado River water is higher in dissolved minerals. TDS has increased from 850 mg/L to sometimes being above 1100 mg/L, also due to potable water supply changes. Processes were designed, and permits were written for water quality of the early 1990s, which was very different. Higher ammonia has posed challenges for operation of the nitrification processes which were designed to convert a lower ammonia concentration. Title 22 Product TDS and chloride limits (800 and 250 mg/L respectively) are routinely exceeded. The addition of ferric chloride as a coagulant in the Title 22 process boosts the chloride above the limit.

In addition to the long-term water quality changes described above, source water quality has been highly variable. Turbidity can vary from 3 to 15 NTU, and has been recorded as high as 130 NTU. There are abnormal organics that sometimes are found in the inlet water. Studies have shown them to be high molecular weight polysaccharides and proteins. These have at times caused microfiltration units to foul rapidly and not recover from standard chemical cleanings. The capacity of the microfiltration units has been reduced over time. Research into processes to stabilize water quality and improve membrane performance has been performed, including ferric addition and diverting side stream flows at the source, ozonation and Dissolved Air Flotation (DAF) addition at the ECLWRF.

The source water is not disinfected, so contains high levels of microbes (bacteria, viruses, protozoa, etc.). Over time, growth of microbial colonies is thought to have caused some nitrification in the influent Force Main. A conversion of some of the ammonia to nitrite and nitrate in transport from the source is noted. A test was performed to inject chemicals at the source pump station to reduce the nitrification, but it was unsuccessful. To resolve this problem, chlorination of the Force Main has been considered.

The high chlorine demand in Title 22 Product water also causes the residual chlorine to dissipate within 3–4 days in the recycled water distribution network. Moreover, sulfide can form from sulfate, and solids can reform under reduced conditions, especially in the far reaches of the system where the water is "older". Consequently, recycled water used for toilet flushing can have suspended solids and rotten egg odour that can cause customer complaints. Irrigation customers sometimes also complain about odour of the water. A flushing permit was used in the early years of plant operation. Water could be flushed from the distribution system and sent to the storm drains nearby. That permit could not be renewed, so chlorine booster stations were placed in strategic locations to increase the chlorine residual in the distribution system and reduce water quality problems.

A challenge for the plant operations personnel is the complexity of the facility. It's not a typical wastewater treatment plant with process A, going to process B, going to process C, then out the outfall. The ECLWRF is an extremely complex plant and there are three satellite facilities located up to 19.2 km (12 miles) away. There are hundreds of alarms to monitor all the time. Though there are operators in the control room at all times, there are different priorities. Customer's needs must be met, and production of the various product waters must be kept up. In addition, there is a diversity of equipment and complex range of processes involved.

It was decided upon conception of the plant to be good environmental stewards, and to keep the waste streams to a minimum. The only waste products from the facilities are RO concentrate that is sent to ocean outfalls, and solids from the solids handling process at the ECLWRF that are disposed through beneficial reuse. Liquid recycle streams from the filtration processes (Title 22 filters, and MF units) all come back to the head of the Title 22 treatment process. This is good for the environment, but poses many challenges for operations. For instance, citric acid is used to clean the MF units. After cleaning, the waste is sent to the pretreatment clarifier in which coagulation and flocculation is done using ferric chloride. Citric acid binds with the iron in a process called chelation (Greenwood, 1997). This prevents the iron from performing its function of coagulation. A connection and permit to discharge the citric waste to the sewer was needed and installed in 2009. Other recycle streams cause issues of turbidity spikes and higher solids loading into the Title 22 process.

The manufacturer of the pretreatment clarifier performed studies on the process, and found that it could run more efficiently with a higher ferric chloride dose. Though the higher dose reduced turbidity coming out of the Title 22 filters, the added ferric also lowered the pH and alkalinity. This meant that the Title 22 Product Water was reaching the lower permit limit of 6.5, and that the alkalinity at the nitrification facilities was significantly reduced. The nitrification process is a biological one in which bacteria convert the ammonia to nitrate while consuming alkalinity. The reduced alkalinity in the Title 22 water was not enough for the bacteria to perform their task. Though sodium hydroxide is added prior to the nitrification process to increase alkalinity, a higher dose had to be added with the increased ferric dose. This higher dose also increased the pH and caused carbonate precipitation within the units. Consequently, ferric dosing to the pretreatment clarifier had to be reduced.

"Squeeze the balloon in one place, it pops up somewhere else". There are competing demands on the same Title 22 distribution system pipe. The refineries have different water quality requirements and each is focused on different parameters. Changes made at the ECLWRF can affect all customers. Early in plant operation, a change in flocculent from ferric chloride to alum was made to satisfy the needs of one refinery. This lowered the iron residual in the system, but caused an outcry from another refinery when the aluminum increased. Requirements of the irrigation customers are also very different than those of the refinery customers. Users for cooling towers and toilet flushing have other desires for water quality.

The Barrier Product permit has a pH limit of 6.5 to 8.5, while allowing from 6.0 to 9.0 for up to 10 minutes per day. At the same time, an agreement with the agency that operates the Barrier Injection System requires maintaining the Langelier Saturation Index (LSI) in a range of −0.5 to +0.5. LSI is a measure of the likeliness to precipitate (positive LSI) or dissolve (negative) calcium carbonate (SM 1992, 1995, 1998). Water in the −0.5 to +0.5 LSI range is considered non-corrosive. To be within that range, pH of the water must be maintained from 7.8 to 8.0 at a minimum. At this pH, it is very difficult to control with lime addition while keeping the pH within the permit limits. Bypass of the decarbonators also reduces the pH and allows for more lime addition to achieve the same pH level. Bypass of the decarbonators has been automated in the system to allow for more stable control.

2.3 WATER REUSE APPLICATIONS

As mentioned previously, West Basin produces five types of "designer" water. It is referred to as "designer" because each type is customized to meet the needs of the customers is serves.

The first type is tertiary (filtered) disinfected recycled water which is approved for use to irrigate turf grass and landscapes for public access area (Figure 2.5). This quality is also approved for a number of other uses such as toilet and urinal flushing, as well as for cooling towers. The main customers are parks, golf courses, cemeteries, university campuses, sports fields and commercial office buildings.

Figure 2.5 View of a golf course and landscape area irrigated with recycled water.

Three oil refineries buy low-pressure boiler feed water. One of them also requires high-pressure boiler feed water and all three use nitrified water in their cooling towers.

Finally, recycled water used in the West Coast (seawater) Barrier is high-purity water which is not only used to create a barrier to prevent seawater intrusion into the aquifers, but also acts as a form of groundwater replenishment.

Evolution of the volume of supplied recycled water

While many smaller sites have converted to using recycled water for irrigation, the majority of growth in recycled water sales has come through four users – three refineries and the West Coast Barrier. Currently these four customers consume about 32–37 Mm^3/yr (26,000 of the 30,000 AFY).

In Fiscal Year 1995–1996, the first full year of production, West Basin produced and distributed 13.4 Mm^3/yr (10,883 AF). This included the West Coast Barrier and Chevron Refinery. As these sites expanded their use and the Exxon Refinery was added, recycled water use exceeded 22.2 Mm^3/yr (18,000 AFY) in a few years. By 2002 the third refinery, BP, was converted and recycled water sales exceeded 33.3 Mm^3/yr (27,000 AFY). Smaller increases throughout the system have created demand for over 37 Mm^3/yr (30,000 AFY) with additional expansion projects being explored.

West Basin recently established a partnership with the Los Angeles Department of Water and Power (LADWP) for West Basin to supply 11.5 Mm^3/yr (9300 AFY) to supply four locations in the LADWP service area immediately adjacent to West Basin. This project is expected to come on-line in early 2014.

Relations and contracts with end-users

As West Basin is a water wholesaler providing water to the local water utilities, a contract is signed for the supply of potable water. When recycled water was added to West Basin's inventory, no additional agreement or contracts were implemented. The end users are able to arrange for tertiary disinfected recycled water from the local water utility in the same manner the site would establish service for potable water service.

This is not the case for water higher water quality requiring additional treatment. Although the water commodity is still sold through the local utility, the end user is asked to contract for repayment of capital facilities used to produce the water plus interest over a period of 20–30 years. This is a separate charge in addition to the commodity charge.

2.4 ECONOMICS OF WATER REUSE

Project funding and costs

The West Basin Recycled Water Program could not have been initiated without the participation of its partners mentioned earlier – Metropolitan Water District of Southern California and U.S. Bureau of Reclamation. Capital facilities for the West Basin Recycling Program have cost nearly $500 million and been financed largely through the issuance of bonds. Grants from the U.S. Bureau of Reclamation have contributed $94 million.

To supplement the product commodity revenue for long term debt payments, West Basin has received revenue in the following manner:

- $95 million from Metropolitan Water District MWD Local Resource Program,
- $96 million: capital payments from large customers requiring higher treatment,
- $187 million: Stand-By Charge.

Pricing strategy of recycled water

Three factors have been key to the expansion of West Basin's recycled water program:

1. Recycled water provides a reliable source of water when potable supplies can be uncertain,
2. Provide water quality suitable for the intended use,
3. Recycled water is generally less expensive than potable water even after the users' investment in retrofitting costs.

West Basin began by pricing recycled water in a manner to attract new users and provide a sufficient price reduction so that costs incurred by the user would be recovered in a relatively short time. This was originally about 40% below the potable rate for irrigation grade recycled water (Title 22). Where potential conversion sites lacked capital funds for the conversion process, West Basin would loan the funds. The user would continue to pay the higher potable rate for recycled water with the differential in rates being applied to repayment.

Annually, West Basin's Board of Directors approves the water rates and charges and the effective date of those rates and charges. Recycled water rates depend on the water quality, and for the Title 22 water on the delivery volumes with declining rate. An example of recycled water rates for 2011 is given in Table 2.5. Compared to 2010, the price increase was from 0 to 16% with a higher increase for Title 22 and nitrified water.

As mentioned earlier in the chapter, expectations were that the price of potable water would rise significantly in the years after the recycled water program went into operation. This would allow recycled water to be priced low enough so that the local water utilities could resell it with a mark up so that they were not negatively impacted financially. As potable water rates rose, recycled water rates would be able to keep pace and eventually the recycled water would become self-sufficient.

However, predicting future potable rates proved unpredictable as they did not escalate as envisioned. As a result, rates for potable water have been established which have worked to subsidize the cost of recycled water. While not part of the original funding plan, this is not inappropriate as everyone benefits from the availability of recycled water which results in a more stable water supply and overall lower water rates.

Table 2.5 Example of prices and percentages of recycled water types produced by West Basin MWD for 2011.

Type of Designer Water	Price \$/m^3 (\$/acre-foot)	Percentage of total production
Title 22 water		11.5
– Distributed inside service area	0.60–0.63 (735–775)*	
– Distributed outside service area	0.56–0.59 (777–817)**	
Barrier water	0.45 (553)	26.0
Nitrified water	0.63 (775)	30.0
Low-pressure boiler feed water	0.74 (914)	25.0
High-pressure boiler feed water	1.1 (1,359)	7.5

*Rate varies with volume delivered per month.
**Rate varies according to service area and volume delivered per month.

Benefits of water recycling

While customers have seen benefit in lower water cost, stability of supply has been greater benefit to the region. The Los Angeles area has been a long standing home to the aerospace industry, especially since World War II. Many large industries have a major presence here because of the location of the US Air Force Base devoted to the management of aerospace contracts including McDonnell-Douglas, Hughes, Raytheon, Boeing, Northup Grumman, and dozens of other firms. The Air Force Base had considered relocating to other locations across the country due to concerns about resources, especially water. When West Basin constructed ECLWTF and thereby assured a stable water supply for decades to come.

As mentioned earlier, refineries make up over half the total demand for recycled water at this time with the potential for that amount to double. It takes about one and a half m^3 of water to produce one m^3 of gas according to refinery sources.

In both these instances, a lack of reliable water supply could have resulted in a loss of key industry to the region. Literally tens of thousands of jobs would have been lost to the region without the addition of recycled water to supplement the supply of potable water.

The benefits of water recycling in West Basin are well recognized by all stakeholders and local population. In addition to be a 'drought proof' water resource for all end-users, enabling the saving of imported water for drinking purposes, the major other benefits of water reuse summarized in Table 2.6.

Table 2.6 Benefits of the water recycling programme for the West Basin region.

Type of benefits	Description
Economic benefits for irrigation	Fertilising value enabling to use less chemical fertilizers Irrigation saves 1.78 kWh of electricity for each m^3 imported water (2200 kWh/AF) Irrigation enables to maintain green public and private areas with positive impacts for the human wellbeing, sport and tourism Increase in land and property value
Economic benefits for industry	Avoided economic losses during drought periods The lack of reliable water supply could have resulted in loss of key industries to the region Cost saving, as for example cooling towers are no longer using potable water and refineries are no longer treating stabilized water Improvement of product quality, which is the case for the textile (carpet) industry
Social benefits	Job saving, as tens of thousands of jobs would have been lost to the region without the addition of recycled water to supplement the supply of potable water
Environmental benefits	Water recycling reduces the amount of treated sewage dumped into Santa Monica Bay by 5 tons/d Reduced pollutant discharge into the bay

The ambitious program of the ECLWRF for sustainable development includes also an energy saving strategy. A solar power generating system of 5574 m^2 (60,000 square foot of solar tiles) with total capacity of 564 kW was installed in 2006

(Figure 2.6) that has reduced emissions of carbon dioxide by over 365 tons per year (equivalent of planting 40 ha (100 acres) of trees or not driving 1,432,316 km (890,000 miles). The solar facility supplies 10% of the peak energy demands at the ECLWRF.

Figure 2.6 Aerial view of the Edward C. Little Water Recycling Facility in West Basin with solar panels.

2.5 HUMAN DIMENSION OF WATER REUSE

Opposition to the recycled water project never appeared to form. As in the centuries since the Los Angeles Basin had first been settled by the Spanish, water supplies were nearing their limits in this desert metropolis. It was once again time to take action to expand the supply of this critical resource in a new way.

Outreach in the form of public meetings began before the plant was constructed. West Basin made every effort to be truthful and forthcoming in presentations to the public about the need for recycled water program, the process it would follow for the project, and the means to finance it (Figure 2.7). City Councils supported the plan as good for the local economy. Endorsement was forthcoming from environmental groups who were delighted to see less waste water discharged into ocean. And the general public voted to help pay for the project with a Stand-By Charge that placed an additional payment on their property taxes.

West Basin has an extensive ongoing public programs involving children, students, residents, business and industries.

Figure 2.7 Visitor center, plant visits and public education materials.

2.6 CONCLUSIONS

West Basin now has a recycled water program that is visited by water professionals from all over the world to see the process of creating designer water. But there were many pieces that fit together to solve the water supply puzzle faced by West Basin in the 1990's. Local leaders supported the idea of more diversity of water resources. Partnerships with federal and local agencies provided a financial foundation to build on. And most critical was a public that understood the need to continue to expand the water supply with locally reliable means, was willing to invest in their future, and put their faith in West Basin to facilitate that solution.

The following Table 2.7 summarizes the main drivers and benefits obtained and some of the challenges for the future.

Table 2.7 Summary of lessons learned.

Drivers/Opportunities	Benefits	Challenges
Develop Local Water Supply	Reduced dependence on interruptible imported water	Developing separate facilities and infrastructure
Develop Reliable Water Supply	Avoid loss of industry in region Recognised economic, environmental and social benefits	Meeting water quality requirements for industrial customers

REFERENCES AND FURTHER READING

California Department of Public Health, CDPH. (2008). Groundwater Recharge Reuse DRAFT Regulation. August 2008, http://www.cdph.ca.gov/certlic/drinkingwater/Documents/Recharge/ DraftRechargeReg2008.pdf (accessed 16 August 2011)

California State Environmental Accreditation Program Branch, ELAP. (2012). Certification of Environmental Accreditation No. 2111 – West Basin Water Quality Laboratory, February 2012. California Department of Public Health, Sacramento, USA.

Drewes J. E., Bellona C. L., Xu P., Amy G. L., Filteau G. and Oelker G. L. (2008). Comparing Nanofiltration and Reverse Osmosis for Treating Recycled Water. AWWA Research Foundation (project 3012), Denver, USA.

Drewes J. E., Reinhard M. and Fox P. (2003). Comparing microfiltration-reverse osmosis and soil-aquifer treatment for indirect potable reuse of water. *Water Research*, **37**(15), 3612–3621.

Faust S. D. and Aly O. M. (1998). Chemistry of Water Treatment, 2nd edn, Lewis Publishers, CRC Press, Boca Raton, Florida, USA.

Greenwood N. N. and Earnshaw A. (1997). Chemistry of the Elements, 2nd edn, Butterworth–Heinemann, Oxford, UK.

Haering K. C., Evanylo G. K., Benham B. and Goatley M. (2009). Water Reuse: Using Reclaimed Water for Irrigation. Virginia Tech, Blacksburg, VA.; http://pubs.ext.vt.edu/452/452-014/452-014.html; (accessed August 24, 2011)

Snyder S. A., Wert E. C., Lei H., Westerhoff P. and Yoon Y. (2007). Removal of EDCs and Pharmaceuticals in Drinking and Reuse Treatment Processes. AWWA Research Foundation (project 2578), Denver, USA.

Snyder S. A., Trenholm R. A., Snyder E. M., Bruce G. M., Bennett T., Pleus R. C. and Hemming J. D. C. (2008). Toxicological Relevance of EDCs and Pharmaceuticals in Drinking Water. AWWA Research Foundation (project 3085), Denver, USA.

Standard Methods for the Examination of Water and Wastewater. (1992). 18th edn, APHA, AWWA and WEF, Washington, USA.

Standard Methods for the Examination of Water and Wastewater. (1995). 19th edn, APHA, AWWA and WEF, Washington, USA.

Standard Methods for the Examination of Water and Wastewater. (1998). 20th edn, APHA, AWWA and WEF, Washington, USA.

Tchobanoglous G., Leverenz H., Nellor M. H. and Crook J. C. (2011). Direct Potable Reuse – A Path Forward. WateReuse Research Foundation, Alexandria, VA, USA.

USEPA. (1994). Method 200.8: Determination of Trace Elements in Waters and Wastes by Inductively Coupled Plasma – Mass Spectrometry, Rev. 5.4. EPA/600/R-94/111, Cincinnati, USA.

USEPA. (1993). Method 300.0: Determination of Inorganic Ions by Ion Chromatography, Rev. 2.1. EPA/600/R-93/100, Cincinnati, USA.

West Basin Municipal Water District. (2002, 2006, 2007). California Regional Water Quality Control Board permits – TITLE 22 Reclaimed Water (RWQCB Order No. R4-2002-0173, Monitoring and Report Program No. 7453; West Coast Basin Barrier Project (RWQCB Order No. R4-2006-0069, Monitoring and Report Program No. 7485); Brine Discharge ECLWRF (RWQCB Order No. R4-2006-0067, Monitoring and Report Program No. 7449); Brine Discharge Carson Regional Water Recycling Plant (RWQCB Order No. R4-2007-0001, Monitoring and Report Program No. 7972).

West Basin Municipal Water District. (2011). Compliance reports to the California Regional Water Quality Control Board – Quarter 3 and 4, 2010, & Quarter 1 and 2, 2011.

View of the ECLWRF facility.

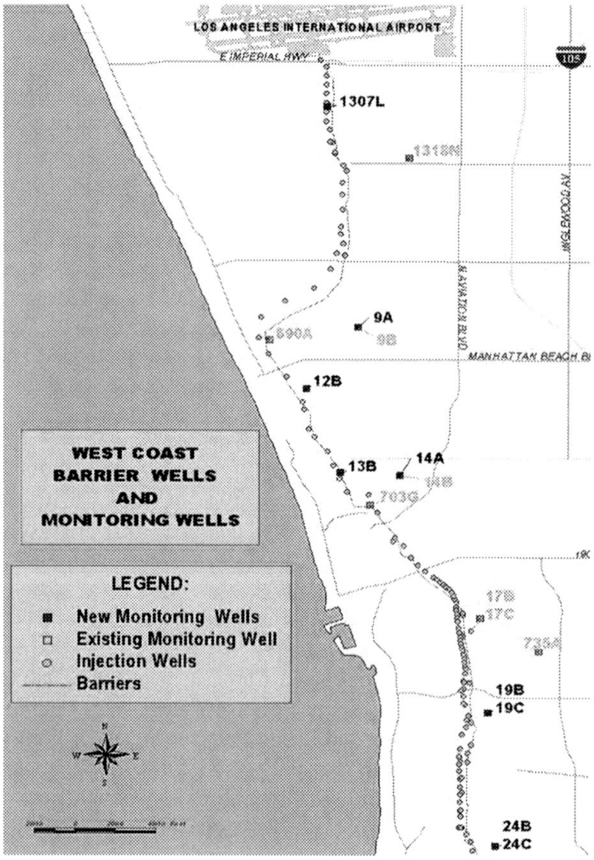

Schematics of the West Coast Barrier wells and monitoring wells.

3 NEWater: A key element of Singapore's water sustainability

Mong-Hoo Lim and Harry Seah

CHAPTER HIGHLIGHTS

Singapore is a water-stressed country with respect to fresh water sources. Although it has abundant rainfall and two third of the country is now water catchment, its limited land area is too small to collect adequate fresh water to meet the densely populated domestic, commercial and industrial water demands. With innovation and use of advanced technologies, Singapore has turned wastewater effluent into a strategic resource to effectively attain its water supply sustainability. Wastewater effluent is treated to high quality for direct non-potable use and indirect potable use. This water reuse programme at municipal scale has provided a substantial additional water source with a multiplying effect on water yield, and is independent of changing climate and extreme weather situations such as droughts.

KEYS TO SUCCESS

- Multi-barrier approach to ensure good water quality which includes: source control, high proportion of domestic wastewater effluent, comprehensive secondary wastewater treatment, use of proven advanced technologies, comprehensive water quality monitoring, adhering to strict operating procedures
- A single agency (PUB) managing wastewater and drinking water created synergy of taking a holistic approach on total water resource management including water reuse
- Fully sewered system enables total collection of wastewater as a water resource for water reuse
- Strong government support and effective public education and communication

KEY FIGURES

- Treatment capacity: 531,000 m^3/d
- Total annual volume of treated wastewater: 511 Mm^3/yr
- Total volume of recycled water: 194 Mm^3/yr
- Recycled water tariffs 1.22 S$/$m^3$
- Major benefits: Recycled water meets 30% of total water demand currently and this percentage will increase to 50% in the long term

3.1 INTRODUCTION

Singapore is a small island city-state, densely populated with about 5 million people living in 710 km^2 of land (Figure 3.1). It has an average annual rainfall of 2400 mm and a daily water consumption of 1.6 million m^3, of which about 55% is for industrial, commercial and other non-potable use and the remainder for domestic use. To achieve a sustainable and robust water supply to meet increasing water demand, Singapore has diversified its water sources, called the "Four National Taps" which consist of:

- Imported water from Johor, Malaysia,
- Local catchment water,
- NEWater,
- Desalinated water.

The third tap, NEWater, is highly purified water produced from treated used water[1] using advanced membrane technology. It was introduced in 2003 in Singapore under the water reuse initiative and plays a very significant role in the overall water supply strategy for Singapore's water sustainability.

[1]Used water and treated used water in Singapore are called wastewater and secondary wastewater effluent, respectively, in other countries.

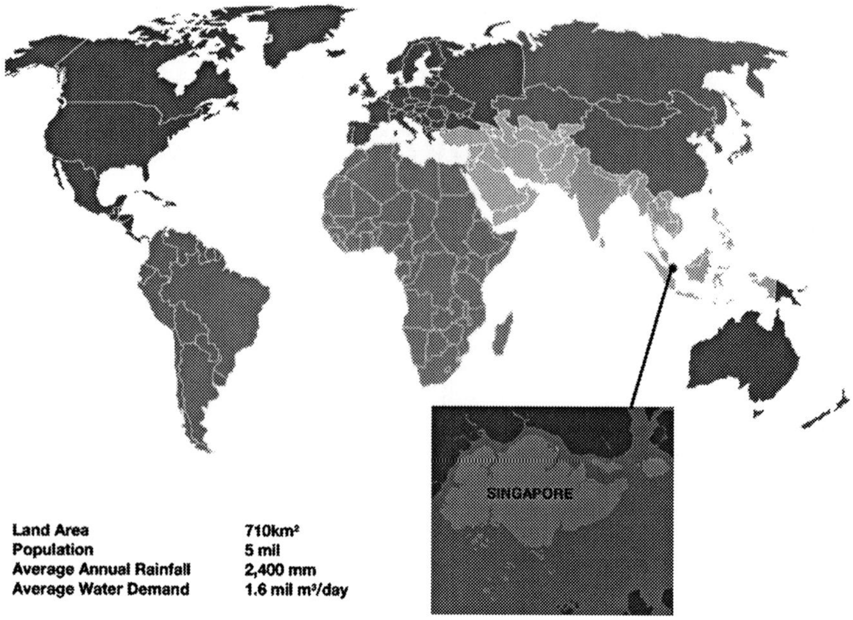

Figure 3.1 Location of Singapore and country information.

Main drivers for water reuse

Singapore is a water-stressed nation. Although it has abundant rainfall, its small land area limits its capability to collect and store the rainfall as substantial land areas are required to cater for socioeconomic developments such as residential housing, commercial and industrial premises, and community facilities. In the 2nd United Nations World Water Development Report (2006), Singapore was ranked 170th among a list of 190 countries in terms of fresh water availability.

With the limited land area, Singapore's water supply needs would not be met with conventional surface water sources. PUB, the Singapore's national water agency which is responsible for the collection, production, distribution and reclamation of water in Singapore, has turned the vulnerabilities of water shortage into opportunities to achieve water sustainability through a holistic integrated water resource management. The fundamentals of its integrated water resource management anchor on 2 aspects: adequate sources to meet demand through source diversification and an effective management of water demand.

Under the water reuse initiatives, treated used water that would otherwise be discharged into the sea, is further treated with advanced membrane technology to produce NEWater. Production of NEWater provides a strategic additional water source. It also marks the closure of the water loop with a faster hydrologic cycle that enables the replacement rate of the total water supply to meet the demand, which is a key step towards Singapore's water sustainability. NEWater also has the advantage of being independent of changing climate and extreme weather situations such as droughts. It augments resilience to Singapore's water supply, achieves the multiplier effect of water yield through recycling, and frees up the potable water previously used for non-potable purposes to meet increasing potable water demand.

The advancement in Singapore's Journey to water sustainability in which NEWater plays a key role, has been in leaps and bounds in the last decade. This did not happen by chance, but through sheer will, careful and holistic planning, and hard work.

Singapore's journey to water reuse

The origin of NEWater could be traced to as early as the 1970s when the idea of reclaiming and purifying used water to high quality water was conceived. In 1974, a pilot plant to produce reasonably good quality water from treated used water was built to experiment the various treatment technologies including reverse osmosis (RO). Although the findings affirmed the feasibility of reclaiming treated used water to supplement the water supply, the costs were astronomical and the membranes operation was unreliable. The idea was then shelved till 1990s when advancements in membrane technology led to higher reliability and cost efficiency. In 1998, PUB initiated the NEWater Study aimed to determine the suitability of NEWater as a source to supplement Singapore's water supply. As part of the study, a dual-membrane demonstration plant of 10,000 m^3/day (2.2 million gallons per day, mgd) was constructed at Bedok in May 2000 (Figure 3.2).

Figure 3.2 Views of the Bedok NEWater demonstration plant.

Box 3.1 NEWater Factories

Name	Year of commissioning (capacity)	Ownership	Comments
Bedok	2002 (32,000 m^3/d, 7 mgd) 2008 (82,000 m^3/d, 18 mgd)	PUB	Expanded from 7 mgd to 18 mgd in 2008
Kranji	2002 (41,000 m^3/d, 9 mgd) 2007 (77,000 m^3/d, 17 mgd)	PUB	Expanded to 17 mgd in 2007
Ulu Pandan	2007 (145,000 m^3/d, 32 mgd)	*DBOO	DBOO by Keppel Seghers for a concession of 20 years
Changi	2011 (227,000 m^3/d, 50 mgd)	*DBOO	DBOO by Sembcorp Utilities for a concession of 25 years

*Design-Build-Owned-Operate

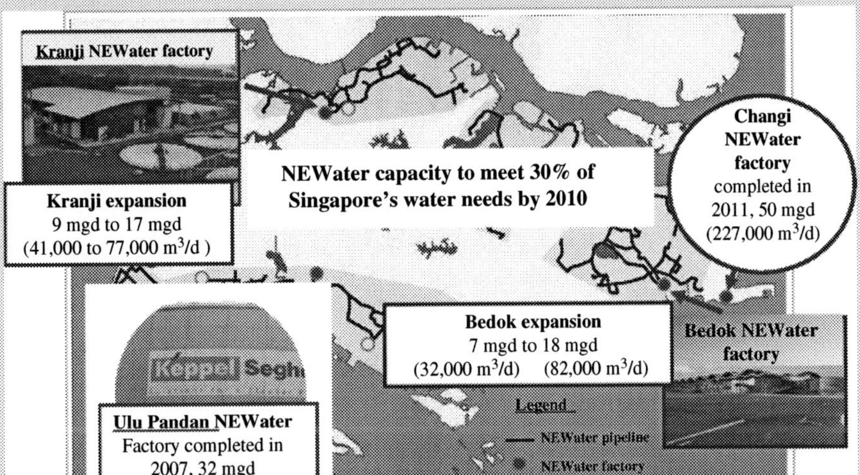

Kranji NEWater factory

Ulu Pandan NEWater factory

Changi NEWater factory

Besides operational issues such as operational reliability and cost, the NEWater study also focused on the public health safety aspect. An extensive two-year water quality study of NEWater was conducted and an independent expert panel comprising the local and foreign experts was convened to evaluate the findings and make recommendations on the suitability of NEWater as a source of supply. The study encompassed more than 20,000 tests carried out on some 190 water quality parameters. The tests showed the NEWater quality to be well within USEPA and WHO Drinking Water Standards.

The expert panel concluded that NEWater was safe for potable use and that Singapore should consider the use of NEWater for indirect potable use (IPU) for the following reasons:

1. Blending with reservoir water would replenish the trace minerals which had been removed in the NEWater treatment process;
2. Reservoir storage would provide an additional safety barrier; and
3. IPU was considered as an appropriate first step to gain public acceptance for NEWater.

Following the positive outcomes of the study, NEWater production plants, also known as NEWater factories, were swiftly constructed and commissioned (Box 3.1). These factories were constructed adjacent to existing water reclamation plants[2] from which the factories receive the treated used water feedstock for NEWater production. This reduces the infrastructural cost and provides synergy between the water reclamation plants and NEWater factories.

While the initial NEWater factories were entirely PUB-owned and -managed, to facilitate private sector participation and cost-effectiveness, the subsequent two NEWater factories were constructed under a Public-Private Partnership (PPP) approach where a private concessionaire, following a successful tender called by PUB, would design, build, own and operate (DBOO) the factory for a stipulated concession period. NEWater produced by the DBOO factories is supplied to PUB for distribution at a tendered price.

In December 2004, the first DBOO tender was successfully awarded under a twenty-year contract to Keppel Seghers. The contract constituted for a 145,000 m^3/d (32 mgd) NEWater factory to be built next to Ulu Pandan Water Reclamation Plant. This Ulu Pandan NEWater factory was completed and commissioned in March 2007. Subsequently in 2008, in response to an increasing demand for NEWater, PUB accelerated the construction of a 227,000 m^3/d (50 mgd) NEWater factory at Changi Water Reclamation Plant through a twenty-five-year DBOO contract awarded to Sembcorp Utilities. This largest NEWater plant, called the Changi NEWater factory, was commissioned in stages from 2009 and reached it full capacity in 2011.

3.2 TECHNICAL CHALLENGES OF WATER QUALITY CONTROL

Multi-barrier approach

In addition to direct non-potable use of NEWater at commercial and industrial premises, a small amount of it (not more than 2.5% of total water demand) is injected into reservoirs for indirect potable use. Although NEWater is well within WHO Drinking Water Guidelines and USEPA Drinking Water Standards, to minimize risks of water safety hazards of indirect potable use, PUB has adopted a multi-barrier approach from source to taps (Figure 3.3) which includes the following:

- Source control
- High proportion of domestic effluent in feedwater (>85%) for NEWater production
- Comprehensive secondary wastewater treatment
- Proven technologies: microfiltration (MF), reverse osmosis (RO) and ultra-violet (UV) disinfection
- Natural attenuation of any residual contaminants in reservoirs
- Water treatment process of coagulation, flocculation, sand filtration, ozonation and disinfection

The principle of the multi-barrier approach is to ensure that potential failure of one barrier will be compensated by effective operation of the remaining barriers, thus preventing contaminants from passing through the entire water treatment system and reaching the customers. The details of the multi-barrier approach and measures taken by PUB are discussed in Seah *et al.* (2008).

[2]Water reclamation plants in Singapore are called wastewater treatment plants in other countries.

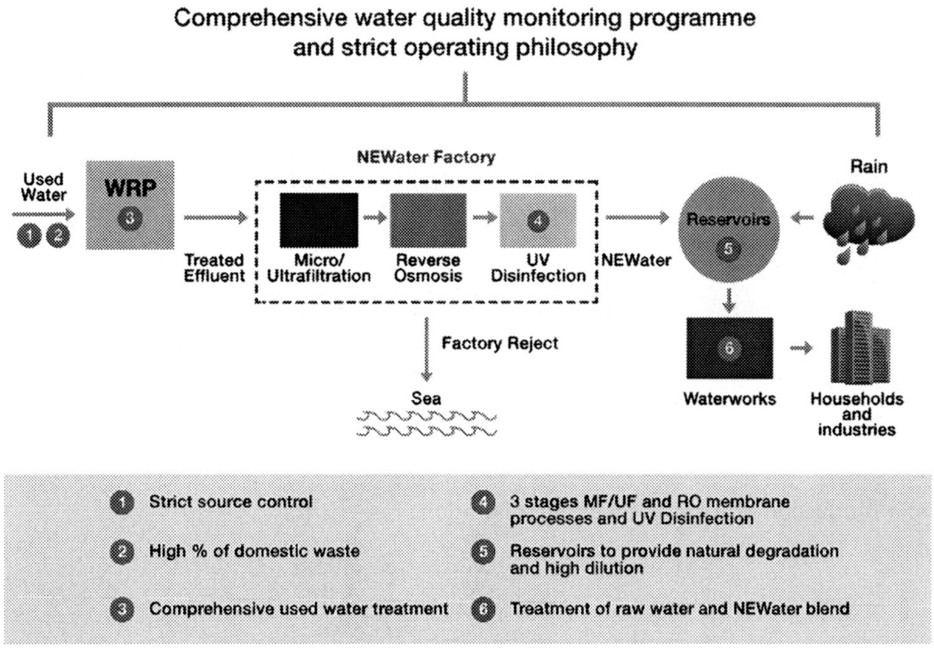

Figure 3.3 Multi-barrier approach for planned indirect potable use in Singapore.

Comprehensive water quality monitoring

The multi-barrier approach adopted for NEWater production is further reinforced with a comprehensive water quality monitoring programme which covers the entire delivery chain of NEWater from source control, wastewater treatment, NEWater production and finally to the taps in households to ensure safety of NEWater for indirect potable use. Some 300 parameters which include physical characteristics, organic and inorganic contaminants, radiological and microbiological quality, as well as contaminants of emerging concern are monitored regularly. The operations and water quality results are reviewed every 6-month by an Internal Audit Panel of local experts, as well as an External Audit Panel comprising local academia and overseas experts of international standing to ensure water safety.

NEWater production process

Used water is first treated in the water reclamation plants through conventional secondary biological treatment processes to meet internationally recognized standards suitable for discharge into rivers. Treated used water is sent to NEWater factories (Figure 3.4), where the first stage of treatment involves micro-filtration (MF) or ultra-filtration (UF). In this process, the treated used water is passed through MF or UF membranes to filter out suspended solids, colloidal particles, bacteria, some viruses and protozoan cysts. The filtered water that goes through the membrane essentially contains only dissolved salts and organic molecules.

The second stage of the NEWater production is the reverse osmosis (RO) process (Figure 3.5). In the RO process, the semi-permeable membrane only allows very small molecules like water to pass through. Consequently, undesirable contaminants such as bacteria, viruses, heavy metals, nitrate, chloride, sulphate, disinfection by-products, aromatic hydrocarbons, pesticides etc, are excluded in the permeate.

The third stage of the NEWater production process acts as a further safety back-up to the RO. In this stage, ultraviolet (UV) disinfection (Figure 3.6) is used to provide an additional barrier for production of clean and safe water. With the addition of some alkaline chemicals to restore the acid-alkali or pH balance, the NEWater is now ready to be piped for its wide range of uses.

Main challenges for operation

The stringent control measures and comprehensive monitoring of water quality reflect that there is no room for complacency with regards to NEWater safety and production reliability. Strict operation procedures were drawn up and regular audits and site checks were conducted to ensure compliance.

Figure 3.4 NEWater production process: view of microfiltration.

Figure 3.5 NEWater production process: view of reverse osmosis.

Figure 3.6 View of UV disinfection and NEWater storage tanks.

The NEWater plants are operated by a pool of competent, highly trained and experienced operators. The focus of operation is to maintain the water quality of the permeate close to the expected baseline readings, which are well within the WHO Drinking Water Guidelines and USEPA Drinking Water Standards. A substantial deviation from the baseline will trigger an alarm and the operator is to investigate immediately and take actions to rectify the faults. In events where deviations are excessive, the NEWater produced will be discarded while investigation and monitoring continue. It is only when the treated water quality returns to baseline readings that the NEWater produced is channelled into the product storage tanks for supply. To ensure the accuracy of the on-line instruments for essential operating water quality parameters, the operators are required to verify the readings with laboratory tests every 8 hours. The operating parameters to be verified and tested include turbidity, pH, conductivity, total and free chlorine, ammonia and oxidation-reduction potential (ORP).

To manage, analyse and evaluate the enormous data on water quality, an Integrated Water Quality Management System is put in place. The system allows for easy review of critical control points and trending which will flash out deviations from baseline readings and breach of control points. This in turn triggers follow-up actions from the operators and managers. In addition, the system enhances both the review process through automated generation of daily and monthly reports, as well as accountability through automated tracking of operators' follow-up actions.

3.3 NEWater APPLICATIONS

NEWater currently constitutes 30% of Singapore's total water demand and is for direct non-potable use. It is supplied to commercial premises for air-conditioner cooling, petrochemical and refinery industries for process cooling, boilers and general washing purposes, and also as ultra clean water to semiconductor and electronic industries. NEWater has proven to be the preferred choice of water supply over potable water for these applications as it is slightly cheaper and has much lower total dissolved solids content than potable water. The projection of NEWater usage is that it will meet 40% of Singapore's total water demand in 2020 and 50% in 2060.

A small amount of NEWater is also used for indirect potable use where it is injected into freshwater reservoirs and the mixed water is treated by water treatment plants for potable use. PUB has initially introduced 9800 m^3/d (3 mgd) of NEWater, which is equivalent to 1% of the daily potable water consumption, into raw water reservoirs. This amount has increased progressively to 2.5% of the total daily potable water consumption in 2011.

3.4 HUMAN DIMENSION OF NEWater

The role of decision makers

Successful implementation of the NEWater programmes which closed the water loop operationally is a crucial element for Singapore's water sustainability. There were several key milestones in the governance of Singapore's water resources which had facilitated the NEWater programme.

In 2001, PUB, which was previously under the purview of the Ministry of Trade and Industry, was transferred to the Ministry of the Environment (ENV). The transfer was the first significant milestone in NEWater development as both PUB and ENV which had been jointly working on water reuse since the 1970s, could then establish a common priority to implement water reuse with good progress.

To streamline the operations following the transfer, used water and drainage functions of ENV were passed on to a reconstituted PUB. This makes PUB the Singapore's national water agency managing the entire water loop from rainwater and used water collection to production of drinking water and NEWater. Traditionally in the world, drinking water and used water are managed by separate agencies. Such an arrangement is not conducive for water recycling initiatives as the barrier and different priorities between the agencies hinder policy changes that can benefit and facilitate water reuse. PUB, as a single agency managing the entire spectrum of the water loop, seized the opportunities and synergy of the integration and made a remarkable progress in water reuse.

Many of the used water management policies implemented in the early days have turned out to be essential factors for the success of NEWater. Singapore's fully sewered system which was implemented a few decades ago to provide proper sanitation for protection of public health, has enabled large-scale collection of used water for water reuse. Also, water reclamation plants which were constructed to treat used water to acceptable standards are now providing reliable effluent of acceptable quality to NEWater factories. The reliable feedstock of acceptable quality avoids disruption of the treatment process and facilitates smooth production of clean NEWater. Used water is now a valuable resource and effective measures, legislation and regulation for source control play an important role in ensuring the reliability of the feedstock for NEWater production.

Public communication strategy

Beside all the measures, technologies and monitoring that have been put in place to ensure good quality NEWater, it is necessary to gain public acceptance of NEWater for indirect potable use. In May 2002, PUB embarked on a public communication plan that not only focused on media and the international experts' positive technical views on NEWater, but also on frequent involvement of political and community leaders to promote NEWater and address the sentiments of Singaporeans.

PUB understood that the media played a key role in conveying information about NEWater to the public. It was therefore essential that they had a clear understanding of the issues related to water reuse. In May 2002, PUB organised a study trip for the journalists to visit water reuse facilities in the United States. The journalists experienced the harnessing and reclaiming of used water for potable uses. In July 2002 when the review on the NEWater study by the External Expert Panel was released in a press conference, the media responded positively to the NEWater programmes.

National campaigns to disseminate the facts of NEWater were led by the concerted efforts of ministers, community and PUB senior officers (Figure 3.7). The information was propagated through platforms such as exhibitions, briefings, advertisements, posters and brochures. Public education campaigns were conducted in schools, community centres and workplaces.

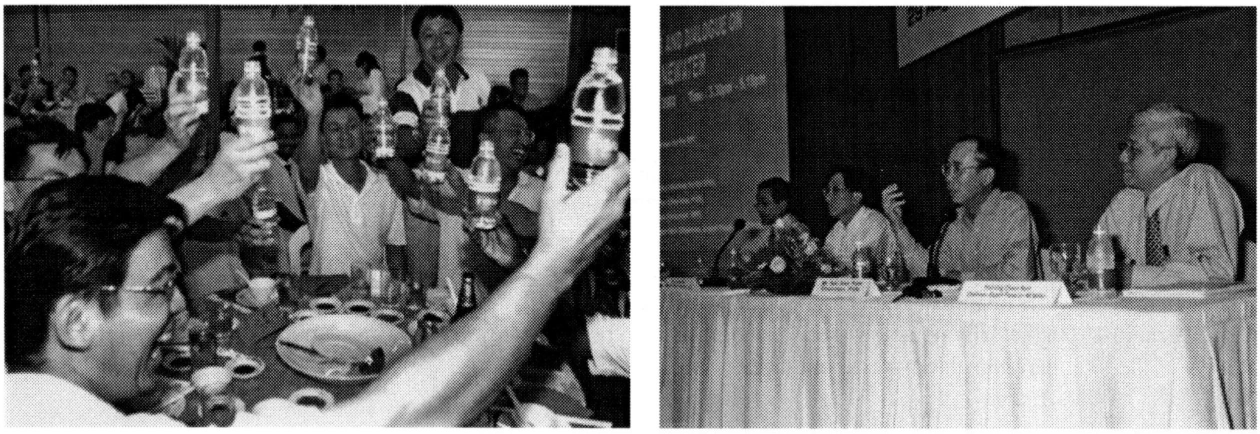

Figure 3.7 Public communication.

Public education: The NEWater Visitor Centre

PUB recognized that public perception and acceptance would be critical to the success of the NEWater indirect potable use programme which, besides breakthrough advanced technologies, also entailed a shift in the mental and emotional perceptions, and correction of misconception on water reuse. To achieve the objectives, PUB constructed the NEWater Visitor Centre which serves as the focal point of the NEWater public education programme. This 2,200 m^2 (24,000 square-foot) facility was designed and constructed in tandem with a new 32,000 m^3/d (7 mgd) NEWater plant, that is the Bedok NEWater Factory.

The mission of the Visitor Centre is to build public awareness, confidence and acceptance of leading-edge technologies that treat used water to produce NEWater which surpasses the World Health Organization Drinking Water Guidelines. PUB knew that it was a great challenge to achieve the mission as historically, indirect potable use has been among the most difficult water initiatives for implementation. A holistic approach with thoughtful planning was taken in the design of the NEWater Visitor Centre which drew on scientific research in psychology, perception, learning and interpretation, with all the senses such as sight, sound, touch and impactful messages incorporated. The objective of integrating the senses and scientific logics is to forge emotional and intellectual connections between water and the advanced technology that leads to confidence and acceptance of the process and product. In addition, architects and interior designs were engaged to amalgamate the elements of engineering technicality with well-planned comfortable "feel-good" spaces to enhance the aesthetic experience to attract visitors. The design also included state-of-the-art interactive computer touch screens, videos and captivating graphic panel and a view of the adjacent actual NEWater plant (Bedok NEWaater Factory) to help sink in the key messages (Figure 3.8).

The NEWater Visitor Centre has brought about broad-based understanding and acceptance of water reuse with respect to NEWater. It also has created public understanding of the preciousness of fresh water, as well as an appreciation of the fact that all water is and has always been "used".

Figure 3.8 NEWater Visitor Centre.

3.5 CONCLUSIONS AND LESSONS LEARNED

Singapore has come a long way to turn the vulnerabilities of water shortage into opportunities to achieve water sustainability through a holistic integrated water resource management. Despite being constrained by the small land area to store the rainfall, the nation has successfully developed diversified sources of water supply, *i.e.* the 'Four National Taps', to attain water sustainability. Development of NEWater, the third tap, is a significant milestone and a key element in Singapore's water sustainability. The Singapore's NEWater journey did not happen by chance, but through sheer will, careful and holistic planning, and hard work.

The following Table 3.1 summarizes the main drivers of water reuse in Singapore as well as the benefits obtained and some of the challenges for the future.

Table 3.1 Summary of lessons learned.

Drivers/Opportunities	Benefits	Challenges
Water-stressed region	Additional water resource	Initial skepticism expressed by semiconductor and wafer fabrication industries on suitability of NEWater for their process use
Limited land to collect rain water	Free up the potable water previously used for non-potable purposes to meet increasing potable water demand	Need transmission and distribution infrastructure which is cost intensive
Increasing water demand due to population growth and socioeconomic development	Attain water supply sustainability with population growth and socioeconomic development	Strict operational control of wastewater treatment plant to supply feed effluent water of acceptable quality to NEWater plants

(Continued)

Table 3.1 Summary of lessons learned (*Continued*).

Drivers/Opportunities	Benefits	Challenges
	Independent of climate change and extreme weather condition such as droughts	Strict operation procedures and control at NEWater plants to produce consistent good quality NEWater
	Reduce wastewater effluent discharge	Comprehensive water quality monitoring from source to point of use
	Ultra-pure water quality preferred by semiconductor and wafer fabrication industries	Public acceptance for indirect potable use of NEWater

REFERENCES AND FURTHER READING

Seah H., Tan T. P., Chong M. L. and Leong J. (2008). NEWater – Multi safety barrier approach for indirect potable use. *Water Science & Technology: Water Supply-WSTWS*, **8**(5), 573–588.

WEB-LINKS FOR FURTHER READING

NEWater. http://www.pub.gov.sg/ABOUT/HISTORYFUTURE/Pages/NEWater.aspx
The Singapore Water Story. http://www.pub.gov.sg/water/Pages/singaporewaterstory.aspx
NEWater, The 3rd National Tap. http://www.pub.gov.sg/WATER/NEWATER/Pages/default.aspx
NEWater Overview. http://www.pub.gov.sg/water/newater/NEWaterOverview/Pages/default.aspx
NEWater Technology. http://www.pub.gov.sg/water/newater/newatertech/Pages/default.aspx
NEWater Quality. http://www.pub.gov.sg/water/newater/quality/Pages/default.aspx
Plans for NEWater. http://www.pub.gov.sg/water/newater/plansfornewater/Pages/default.aspx
NEWater Visitor Centre. http://www.pub.gov.sg/water/newater/visitors/Pages/default.aspx

Views of the NEWater facilities.

4 Integration of water reuse in the management of water resources in Costa Brava

Lluís Sala

CHAPTER HIGHLIGHTS

The gradual development of water reuse in the Costa Brava has led to the beneficial use of reclaimed water for a great variety of uses: agricultural and landscape irrigation, environmental reuse, aquifer recharge and non-potable urban uses. Working together with university researchers, great emphasis has been made to provide a reliable disinfection step in order to protect public health. The measurement of the energy consumption in the water cycle has also been instrumental in order to make steps towards a greater overall sustainability in the area.

KEYS TO SUCCESS

- Gradual development and improvement of water reuse projects
- Dedicated operation and maintenance of facilities
- Look for the success of end users
- Joint work with researchers – mutual benefits
- Active divulgation of projects and their benefits

KEY FIGURES

- Volume of wastewater treated annually: 30–35 Mm^3/yr
- Maximum annual volume of recycled water: 6.4 Mm^3/yr (2010)
- Most frequent water reuse standards: 200 *E. coli*/100 mL (golf course irrigation and municipal non-potable uses); 100 *E. coli*/100 mL (agricultural irrigation); <1 *E. coli*/100 mL (private urban non-potable uses)

4.1 INTRODUCTION

The Costa Brava is the name given to the 200 km-long coastal region of northeastern Catalonia, Spain, in the Girona province. The area is rich in ancient heritage sites, including the renowned Iberian settlement of Ullastret and the Greek colonies of Emporion and Rhode, and the later Roman city of Emporiae. The importance of this coastal area, especially in the north, remained high until the Middle Ages through the era of the County of Empúries, only to fall into oblivion after both the decay of the Catalan-Aragonese power in the Mediterranean and the emergence of Castilian power after the discovery and conquest of America. The area became a quiet backwater until the 1950s and 60s, when it was discovered by central and northern Europeans as a picturesque and charming place to spend summer holidays or even to live all year round. Tourism has completely changed the human geography of the area. With the appearance and later boom of tourism, the region's population has increased seasonally, with each passing year, which has altered every balance in the historically stagnant local societies: from energy supply to urbanisation patterns, from food production to waste collection and from water supply to wastewater treatment. In a short period of time, the villages in the Costa Brava suffered the most radical transformation of their long history and went from a near subsistence economy to a tertiary economy in some 50–60 years.

Currently, in the early 21st century, the Costa Brava area is composed of 22 coastal municipalities, from Portbou in the north, on the border with France, to Blanes in the south, on the border with the Barcelona province. Recent statistics indicate that resident population has doubled up to 250,000 inhabitants in the last 25 years, with a seasonal peak in the summer that is estimated at over 1,000,000 inhabitants. This sudden and huge increase in population has been beneficial for the economic activity of the area, but also has required a quick adaptation of municipal services. In recent decades, both drinking water supply and wastewater treatment have required important investments. Limited and scarce local water resources clearly proved to be insufficient to cope with the water demands, leading to problems such as decline in the groundwater level and seawater intrusion. On the other end of the pipe, non-existent wastewater treatment was causing pollution on the beaches that were so avidly sought after by foreign tourists.

These problems related to water cycle management prompted the creation in 1971 of the Consorci Costa Brava, a supramunicipal organization composed of 27 municipalities (22 coastal plus 5 located inland). The mission of Consorci Costa Brava, together with different higher levels of administration, was to have the required facilities built both in the drinking water field, importing water from richer, nearby areas, and in the wastewater field. Over the years, wastewater reclamation and reuse became the logical next step, in order to help cope with the non-potable demands and also to achieve a reduction of discharges into the environment, further improving the quality of bathing waters.

Drinking water supply

Before the development of touristic activities, Costa Brava municipalities relied mostly on wells tapping from small coastal aquifers and no serious problems or threats were recorded. However, the increase in population made these sources totally insufficient, so water from external ones had to be imported in order to satisfy the new demand, especially that of summer months. Nowadays, roughly 60% of the water supplied in Costa Brava municipalities comes from external sources, which accounts for some 18 million m^3/yr (Mm^3/yr) out of the total volume of around 30 Mm^3/yr, whereas the other 40% comes from local sources all along the coastal strip.

Over the years, three different water transfers have been implemented in order to solve water supply problems in three different areas, respectively called northern, central and southern Costa Brava. This water is sold by the Consorci Costa Brava to the local water companies, which are in charge of the final supply to individual customers. Depending on the specific circumstances, local companies may or may not mix this water with their local water resources before supplying.

Wastewater treatment

The first biological wastewater treatment plant in the Costa Brava area was built in 1972 and was located in the municipality of Blanes. In December 2010, sanitation in Costa Brava was completed, when the last two municipalities with no wastewater treatment were finally connected to a wastewater treatment plant. Wastewater volumes produced in summer or during Easter holidays may be several times larger than those to be treated in winter, which prompts that most of the facilities to require several (up to 3) parallel treatment lines. Starting in spring and until mid autumn, weekend flows are also notably greater than those during the week, adding complexity to the operation and maintenance of these facilities. All the 18 wastewater treatment plants in the Costa Brava use biological treatment (1 rotating biological contactor, 8 conventional activated sludge plants and 9 extended aeration plants).

Conventional activated sludge plants are generally the ones built in the 80′s, whereas the extended aeration ones are those built from the 90's onwards. Despite the fact that all the wastewater treatment plants comply with the limits imposed by the European Urban Wastewater Treatment Directive, these technological differences also produce different effluent quality. Effluents from conventional activated sludge plants have higher COD and nitrogen contents, as well as a higher turbidity and concentration of indicator microorganisms when compared with those of the extended aeration plants.

Treated wastewater is still mostly discharged into the sea through submarine outfalls. However, the growing trend, especially after some serious drought periods, is to reuse treated effluents as much as possible, in order to reduce the net water demand. Despite the fact that percentages of water reuse are still modest, much progress has been made over the last 20 years and several full-scale projects in the area reveal the possibilities of water recycling to improve water resources management.

Water reuse objectives and incentives

Over the last 25 years, water reuse has been slowly growing in Catalonia to recently become an important element of the integrated water cycle management, as a tool for water conservation, territorial balance and environmental protection. Today's flagship water reuse project in the Barcelona Metropolitan Area (Generalitat de Catalunya, 2012) has drawn on previous experiences existing in the region, mostly from Costa Brava, from agricultural irrigation to environmental reuse and aquifer recharge. Not surprisingly, it developed for similar reasons – although on a different scale – to the previous projects that first were implemented in the Costa Brava: insufficient resources to supply the water demand in the Barcelona Metropolitan Area under drought circumstances, plus the evidence that remarkable amounts of water, although scarce, were discharged into the Mediterranean sea once treated. To summarize, the main drivers of water reuse in Catalonia, including those initial projects in Costa Brava are: (1) sharp increase in water consumption due to new water uses, rising population and intensive tourist activity, (2) insufficiency of water resources in the cyclical drought periods of the Mediterranean areas, (3) need for a greater environmental protection by reducing the stress on water sources.

Water reuse projects have the goal of providing new water resources to cope with the part of these demands that do not require drinking water quality (mostly, golf course and landscape irrigation), making the most of the sometimes scarce and increasingly expensive drinking water supplies. This project's practical development also owes to the pragmatic approach by Consorci Costa Brava, accepted and shared first by Junta de Sanejament (former wastewater authority, until year 2000), and later by the Catalan Water Agency until June 30, 2011. During this period, tertiary treatments were located in the wastewater treatment plants and operated and maintained by the same companies taking care of those facilities, which provided some interesting advantages: (1) the O&M of the tertiary treatments benefited from the economy of scale provided by the O&M

until secondary level, mostly by remarkably decreasing the personnel costs; (2) the know-how of such staff, from plant managers to operators and lab technicians, could be directly applied to the tertiary treatment; (3) the decisions made in the operation of the wastewater treatment plants were also made taking into account the later tertiary processes, which improved the consistency of the overall treatment and thus provided a better quality of reclaimed water and a greater reliability in their performance. However, the financial crisis affecting Spain also reached the Catalan Water Agency and starting from July 1st 2012, a new era has begun: all expenses related to tertiary treatments have to be covered directly by users, with no contribution from Catalan Water Agency. This new situation is already putting water reuse to the test from a new financial perspective. If successful, there is no doubt it will be its definitive consolidation; if not, its development may result seriously impaired.

4.2 TECHNICAL CHALLENGES OF WATER QUALITY CONTROL

Treatment trains for water recycling

Because wastewater treatment plants are designed and built to cope with high peak flows, most of the year these facilities work in very favourable conditions, which allow them to produce high quality effluents. The better the secondary effluents, the easier the reclamation treatment needed and the more reliable and consistent the recycled water quality that can be achieved. Since effluents pose a real and short-term risk for human health, Costa Brava reclamation treatments are essentially disinfection-oriented, with or without, preliminary treatment for removal of suspended solids such as coagulation, flocculation, sedimentation and/or filtration (Figure 4.1).

Figure 4.1 View of the tertiary treatment at Tossa de Mar and the gravity storage tank for reclaimed water in El Port de la Selva.

Out of the 30 Mm^3 of wastewater treated annually in the Costa Brava wastewater treatment plants, 6.4 Mm^3 has been reclaimed in 2010 with most having had a beneficial use. Reclaimed water is produced in 13 facilities, which are of the following types: chlorination (1 unit), combined disinfection (UV + chlorination) (1 unit), sand filtration and combined disinfection (1 unit), double-step filtration with chemical addition and combined disinfection (6 units), coagulation, flocculation, sedimentation, filtration and combined disinfection (3 units) and constructed wetlands (1 unit).

Water quality control and monitoring

In order to be able to supply reclaimed water, Spanish regulations (Royal Decree RD 1620/2007) require 90% compliance on a minimum of a 3-month period for quality limits (Table 4.1) that essentially depend on the degree of human exposure. The main parameters chosen for this evaluation are suspended solids and turbidity, as general indicators of the physico-chemical quality of the water and of the likelihood of having an efficient subsequent disinfection, and the concentration of *Escherichia coli*, as indicator of the performance of the disinfection process and a measure of the bacteriological quality of the reclaimed water. The determination of parasitic helminth eggs is also required at an unreasonably high frequency and, because they are removed by sedimentation, their presence or absence owes more to the performance of the biological treatment and the clarifiers than to the disinfection process.

Table 4.1 Summary of the quality criteria for the main categories of water reuse in Spain.

	Quality criteria (percentile 90%ile)[a]			
Type of use	**SS mg/L**	**Turbidity NTU**	***E. coli* cfu/100mL**	**Total nitrogen mg N/L**
Non-potable urban – residential (1.1)	10	2	0	–
Non-potable urban – municipal (1.2)	20	10	200	–
Irrigation of crops to be eaten raw (2.1)	20	10	100	–
Use in cooling towers and evaporative condensers (3.2)	5	1	0	–
Golf course irrigation (4.1)	20	10	200	–
Aquifer recharge by percolation (5.1)	35	–	1,000	<10

[a]Also, parasitic helminth eggs <1/10 L.

Because of the fact that in Spain the wastewater treatment plant performance is evaluated according to monthly arithmetic averages, but the quality of reclaimed water requires a 90% compliance (90% ile), the operation of water recycling facilities represents a great challenge, requiring not only good qualification and equipment, but also an important mentality change. Switching from the perspective of compliance of "one out of two" to that of "nine out of ten" is a necessary change to the operators' method that cannot always be guaranteed to happen quickly and smoothly.

All the reclamation treatments are equipped with online probes for the key parameters, so any water suspected of not complying with the standards is not allowed to leave the facility and reach the user.

The probes and online sensors most commonly installed in Costa Brava water reclamation facilities are:

- Turbidity: This parameter is continuously measured at the inlet of the reclamation system, so any deterioration in quality of the secondary effluent is detected and can be prevented from entering the system. Since the uses supplied until 2010 were either non-potable urban municipal uses (1.2 in Table 4.1), agricultural irrigation (2.1) or golf course irrigation (4.1), all of which have a limit of 10 NTU, a turbidity sensor was not needed at the outlet of the system, but it may be required in the near future when reclaimed water is delivered for uses such as irrigation of private gardens, for which a limit of 2 NTU has to be observed.
- UV output: Any decrease in the intensity of UV light reaching the sensor (i.e. due to lamp aging or burning out, fouling of quartz sleeves, a decrease in water transmittance at 254 nm or a combination of these factors) is detected and recorded. If the intensity falls below a certain level for a certain amount of time, the supply is interrupted as it is no longer possible to guarantee adequate disinfection.
- Redox potential: Because in Costa Brava chlorine is always used as a polishing disinfectant agent, after its addition reclaimed water shows an increase in redox potential values, proportional to the chlorine dose. Once calibrated in each of the reclamation plants according to its local conditions and specificities, online monitoring of redox potential is a powerful and reliable tool for the real-time control of disinfection efficiency. If certain values are not achieved, the control loop automatically increases the dose and, if after a certain amount of time those limits are still not met, supply is interrupted, again due to the lack of guaranteed adequate disinfection.
- Electrical conductivity: In some reclamation plants where water is used for irrigation, or in the presence of risks for high variations and increase of electrical conductivity, sensors are also used to control the agronomic quality of the product delivered to the end user.

Exceptionally, in the case of the Empuriabrava constructed wetland system, where the use of reclaimed water is mainly intended for environmental purposes, the parameter that is continuously monitored at the outlet of the wastewater treatment plant is the concentration of ammonia, which is a key for a proper performance of the wetland system.

Despite the fact that the concentration of *Escherichia coli* is the only parameter required by RD 1620/2007 to measure the efficiency of disinfection, other types of microorganisms are also used to demonstrate the good disinfection efficiency, such as enteroviruses or *Cryptosporidium* oocysts. As proven by detailed studies conducted in two Costa Brava water reclamation plants (Montemayor *et al.* 2008), at the usual doses of the installed equipment, the UV light is more efficient than chlorination in terms of inactivation of somatic coliphages and *Clostridium* spores. The periodic determination of the inactivation values of *Clostridium* spores allows an early detection of a loss of performance of the UV system due to lamp aging. Instead, final chlorine addition enables the better control of bacterial regrowth and biofilms in distribution network and provides a chemical tracer for reliability control.

Table 4.2 Summary of the reclaimed water quality of selected wastewater treatment and reclamation facilities in Costa Brava (data for 2010).

	Water reclamation plant															
Type of facility	**Llançà**		**Port de la Selva**		**Empuriabrava**		**Pals**		**Castell-Platja d'Aro**		**Tossa de Mar**		**Lloret de Mar**		**Blanes**	
Wastewater treatment *(year of start-up)*	Extended aeration *(2003)*		Conventional activated sludge *(1974, upgraded 1996)*		Extended aeration (1995) + chemical phosphorus removal *(2010)*		Extended aeration *(1995)*		Conventional activated sludge *(1983)*		Conventional activated sludge *(1980)*		Conventional activated sludge *(1992, upgraded 2009)*		Extended aeration + chemical phosphorus removal *(1998)*	
Reclamation treatment *(year of start-up)*	Coagulation, flocculation, multi-media filtration, disinfection (UV + chlorine) *(2008)*		Coagulation, flocculation, multi-media filtration, disinfection (UV + chlorine) *(2001)*		Constructed wetlands (7 ha) *(1998)*		Chlorination *(2000)*		Filtration, disinfection (UV + chlorine) *(1998)*		"Title-22"[d] *(2003)*		"Title-22"[d] *(2009)*		"Title-22"[d] *(2002)*	
Treated volume, m^3	49,800		51,900		1,202,400		410,200		798.700		60.800		386.100		2.991.000	
Number of samples	59–69		57–61		15–27		30–36		52–58		49–105		50–96		123–235	
Statistical parameters[a]	**Average**	**90%ile**	**Average**	**90%ile**	**Average**	**90%ile**	**Average**	**90%ile**	**Average**	**90%ile**	**Average**	**90%ile**	**Average**	**90%ile**	**Average**	**90%ile**
SS, mg/L	2.1	4.2	2.6	4.4	3.6	6.2	2.6	3.7	4.4	7.0	2.0	3.2	5.4	8.9	2.5	4.0
Turbidity, NTU	1.2	2.5	2.1	4.0	2.0	3.7	1.3	1.9	2.5	4.1	1.3	1.9	4.0	6.7	1.6	2.5
T254 nm, %	75	64[b]	69	56[b]	64	56[b]	67	60[b]	57	50[b]	69	58[b]	61	55[b]	72	67[b]
Total residual chlorine, mg Cl_2/L	0.4	0.2[b]	1.1	0.3[b]	–[b]	–[c]	0.9	0.4[b]	2.1	0.6[b]	1.2	0.5[b]	4.4	1.3[b]	0.7	0.2[b]
E.coli, cfu/100 mL	2	22	2	17	47	190	<1	<1	<1	4	<1	<1	<1	<1	<1	2
Total nitrogen, mg N/L	7.7	14.2	14.4	28.9	1.1	1.6	4.6	7.4	29.3	44.2	18.9	30.9	25.8	43.1	7.8	10.8
Total phosphorus, mg P/L[c]	3.3	6.4	2.9	5.6	1.9	4.7	3.8	6.6	1.5	2.9	3.0	5.1	2.4	4.1	1.1	1.8

[a]Annual arithmetic mean for all the parameters, except for the concentration of *Escherichia coli*, which is a geometric mean. 90%ile corresponds to the percentile 90 of the annual set of data.
[b]Results correspond to the percentile 10, hence 90% of the results were greater than the value displayed in this table.
[c]Not applicable.
[d]"Title-22" treatment process includes coagulation, flocculation, clarification, filtration and disinfection (UV + Cl_2).

Table 4.2 displays a summary of the performance of eight of the treatment facilities in the Costa Brava area in 2010. This selection shows the diversity of technical configurations of both wastewater treatment and reclamation facilities, and thus better illustrates the results of their performance. Selected facilities have been arranged in geographical order, from the one located furthest north (Llançà) to the one located furthest south (Blanes). Except for the extensive Empuriabrava constructed wetland system, all the facilities are technology-intensive, with different types of mechanical equipment. The water reclamation treatment trains are designed for: (1) an improvement in the physico-chemical quality of secondary effluents (reduction of the concentration of suspended solids and turbidity); (2) an adequate inactivation of both pathogenic and indicator microorganisms; and (3) ensuring a chlorine residual in order to preserve the microbiological quality of the water between the production site and the point of use.

Transmittance values at 254 nm are greater in extended aeration plants than in conventional activated sludge plants. This indicates a superior quality of reclaimed water and the likelihood of an easier and more effective disinfection. As it can be deduced from the example of Pals wastewater treatment plant, a good performance of the biological wastewater treatment is critical for producing reclaimed water of the best possible quality. In this case, a well-designed and well-operated biological treatment followed by simple chlorination outperforms all the other reclamation facilities. Nevertheless, an additional tertiary treatment for removal of suspended solids is greatly recommended to ensure the required high disinfection level in the case of variations in the quality of secondary effluents.

As it can be observed in Table 4.2, in which near worst case values are presented (percentile 90 or percentile 10, depending on the kind of parameter), the reclaimed water produced in the Costa Brava facilities has a consistent microbiological quality that complies with Spanish RD 1620/2007 for the intended uses. In the reclamation treatments that perform better, this consistency is also achieved by the other regulatory parameters, such as suspended solids and turbidity.

Generally speaking for the Costa Brava water reclamation facilities, the residual concentrations of *E.coli* were consistently under the detection limit of <1 cfu/100 mL, thanks to the appropriate equipment and continuous monitoring. Similar results were obtained for the other microorganisms such as sulphite-reducing *Clostridium* spores and somatic coliphages.

The Figure 4.2 illustrates the log inactivation of these indicator microorganisms in the Tossa de Mar reclamation plant. The variations in the monthly average values of inactivation reflect the variations in the concentrations of these microorganisms in secondary effluents. The increasing trend in the inactivation of sulphite-reducing *Clostridium* spores, the bacterial parameter which is the harder to deactivate, proves that the efficiency and consistency of disinfection has been improved over the years. The recent addition of somatic coliphages as an indicator of viral inactivation is also important for the protection of the public health, since the inactivation values are always greater than the maximum concentration of enteroviruses found in secondary effluent (2.3 log pfu/L, unpublished data).

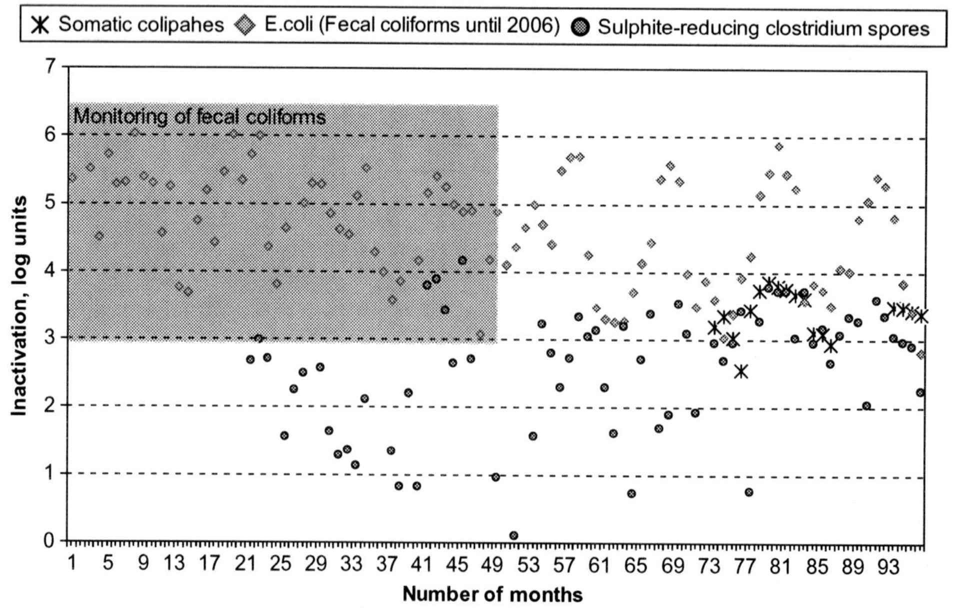

Figure 4.2 Evolution of the monthly average values of inactivation of *E. coli*, sulphite-reducing *Clostridium* spores and somatic coliphages in the Tossa de Mar reclamation plant (January 2003 to December 2010).

4.3 WATER REUSE APPLICATIONS

The evolution of water reuse in the Costa Brava has seen three types of projects, which represent the chronological and conceptual evolution that this activity has undergone during the last 20 years:

1. *Supply to private large users*
   Reclaimed water in the Costa Brava started with the supply of reclaimed water (originally disinfected secondary effluents) for the irrigation of golf courses, pitch & put facilities and some agricultural areas, thanks to the favourable regulations and also because the agreements for the supply of rather large volumes to a single or few users, which management is relatively easy.
2. *Social or environmental beneficial reuse*
   Fostered by EU Cohesion Funds since the mid 90's, Consorci de la Costa Brava and Catalan Water Agency have developed water reuse projects which are beneficial for the sustainable development of the local communities. The two more emblematic projects for environmental enhancement are those of the Empuriabrava constructed wetland system (Figure 4.3), which reclaims water to favour and maintain the biodiversity in a nature reserve, and the Blanes project, in which reclaimed water is used for aquifer recharge in the lower Tordera river.
3. *Urban reuse*
   The most recent water reuse projects happening in some municipalities in the Costa Brava, implemented the installation of reclaimed water distribution pipelines. These pipelines have the long-term intention of delivering water to private users (Figure 4.4, i.e. for the irrigation of home gardens), and represent exactly the opposite situation of the supply to golf courses: small volumes can be supplied to a potentially large number of users and there is no direct connection between the reclamation plant and end-users. Instead, an intermediate operator, usually the municipality, is in charge of recycled water distribution, which often complicates the management of this kind of system.

Figure 4.3 View of Empuriabrava constructed wetland system.

Evolution of the volume of supplied recycled water

Starting in 1989, the production of reclaimed water in the Costa Brava has been steadily growing in volume with every reclamation treatment implemented. In 2010, the overall volume treated beyond secondary level has been of 6.4 Mm^3, most of them with a beneficial reuse. The larger consumer of recycled water is aquifer recharge, which happens mostly in the Blanes reclamation plant with the aquifer recharge of the lower Tordera river (between 2 and 3 Mm^3/yr), followed closely by environmental reuse, a category that includes mainly the Empuriabrava constructed wetland that supplies the Aiguamolls de l'Empordà nature reserve. This category also includes the surplus of reclaimed water produced and not supplied, which is delivered back to the environment with a superior quality (i.e. lower concentration of suspended solids, turbidity and microbial loads) than if it was discharged as secondary effluent.

Depending on the rainfall during a given year, golf course irrigation accounts for approximately 20% to 25% of the reclaimed water production, whereas agricultural irrigation's share ranges between 5% and 10%. Internal and non-potable urban uses presently account for only 2% of the total recycled water production, but the construction and development of reclaimed water distribution systems in some municipalities is likely to lead to a mid-term increase in these figures.

Figure 4.4 Views of the recycled water produced by Llançà reclamation plant and Santa Clotilde gardens, irrigated with reclaimed water in Lloret de Mar.

Relations and contracts with end-users

Since 1989, CCB's supply of reclaimed water to private users has always been made under the coverage of written agreements, in which the conditions of the supply were set. Initially, between 1989 and 1993, CCB's compromise was only to provide the secondary effluent at no cost, whereas the user had to provide the disinfection treatment; if quality was not met, CCB was authorized to stop the supply in order to protect public health, despite no regulations existing at that time. This procedure proved to be not consistent enough, because providing the disinfection treatment was something way beyond the natural capacity of the users (golf course managers and operators). In this context, a new relationship was proposed in 1994, the kind that, with some minor adjustments, has lasted until today.

New contracts had a fundamental change in approach: achieving the quality of reclaimed water was no longer a responsibility of the user, but of the CCB, who would provide the service through the companies operating the wastewater treatment plants. These contracts define the essential issues of what the service shall provide (i.e. volume, required treatment, delivery, quality control and price) and are a useful tool for a fruitful and lasting relationship between CCB and the users.

One of the key points of those contracts is that if the quality cannot be met, the CCB is entitled to stop the supply until the problems are fixed, again to protect public health. Until now, supply cuts have not been frequent and so its impact on golf courses has been small, also because they usually have ponds that can provide the irrigation water when supply is briefly interrupted. The recent development of municipal networks of reclaimed water is turning the continuity of supply into a key challenge, since there will be no storage tanks that could maintain supply during even just a few days of cuts. Operation and maintenance of treatment plants will require fine tuning, but also new management elements will have to be added to the system (i.e. the possibility of occasional introduction of drinking water into the reclaimed water system), so the reclaimed water users are not affected by a varying quality of service.

4.4 ECONOMICS OF WATER REUSE

Project funding and costs

Funding for water reuse projects in Costa Brava has had different sources all over the years. The initial projects of the early 90's, when safe water had to be provided for the irrigation of golf courses, consisted simply of chlorination systems which were funded directly by the users. Equipment was placed in the wastewater treatment plants and, even though it was paid for by the user, operation has always been in the hands of the plant managers and under the responsability of the CCB. The first true water reclamation plants in Costa Brava were installed in the late 90s after CCB's proposal to EU's Cohesion Funds was awarded. Those funds provided 80% of the capital costs and CCB provided the rest, in order to build 7 water reclamation facilities with capacities that ranged from 15 m^3/h to 625 m^3/h (project values) plus a 7-ha constructed wetland system to treat around 1 million m^3/yr of secondary nitrified effluent. The main reason for the funding of those systems was that by recycling effluents that were discharged into the sea a net general benefit could be achieved: for the conventional reclamation treatments, there was a new resource to cope with non-potable demands, thus increasing the guarantee of drinking water supply; and for the constructed wetland system, there would be a water that could be used for environmental uses in a nature reserve.

More recent reclamation treatments were funded at 100% by the Catalan Water Agency and again they were addressed at producing reclaimed water that could be reused and that could have a positive impact in water resources management at a local level. *Title-22* (coagulation, flocculation, sedimentation, filtration and disinfection) tertiary treatments were built in Tossa de Mar, Lloret de Mar and Blanes; also, a combined UV-chlorine disinfection plant was built in Torroella de Montgrí, near the mouth of the Ter river, where it could provide some flows for agricultural irrigation to alleviate the lack of water in dry periods.

Operational costs were fairly easy to calculate in the beginning, when chlorination was the only treatment applied, but became more difficult to be accurately evaluated when other technologies were added. Because in Costa Brava tertiary treatments are located inside the wastewater treatment facilities, economies of scale are happening and resources are shared (i.e. availability of energy, manpower), which makes it not so easy to ascertain which are the real operational costs. In the end, any value given will depend on the assumptions made, like which percentage of personnel costs is assigned to the reclamation facility or what is the payback period.

Pricing strategy of recycled water

The pricing strategy has always mainly been addressed at covering the running costs of the treatment facilities beyond secondary level. Despite slightly different costs in the different plants, due to size or required chlorine doses, from the beginning the strategy was to keep similar prices among users – mainly golf courses-, in order to avoid misinterpretations of the reasons of such differences. However, and despite this intention, differences in price appeared over the years due to specific conditions (i.e. location of the pumping station, whether inside or outside the wastewater treatment plant). Recently, in December 2010, CCB established standardized reclaimed water rates for the different categories of reuse, taking into account annual demand, intended use of water, location of the pumping station and source of electrical power. These rates have been revised and slightly modified in March 2012, but their continuity will be determined not only by the accuracy in the calculation of the operational costs, but also on the intensity and the duration of the economic and financial crisis nowadays affecting Spain.

For municipal reclaimed water distribution systems, negotiations are taking place now in order to establish these prices. Once the prices of water sold to municipalities are established, the prices for the final user could also be established. In this case, the actual price of drinking water in a given community will also be a factor that could possibly influence the final price of reclaimed water. As a mid-term goal, the water reclamation treatment and its supply should be financially self-sustainable, or at least should require the minimum number of subsidies, so it is also less prone to be affected by financial turmoil.

It is also important to mention that recently the Catalan Water Agency has approved a new tax (0.1498 €/m^3) that applies to the use of drinking water used by the municipalities, which until that point did not exist. Because this tax does not apply to reclaimed water, it can result in an additional incentive for the municipalities to develop usable local resources like, wherever feasible, reclaimed water.

Benefits of water recycling

The benefits of water reuse are sometimes hard to measure. A way to try to better understand them is to imagine and assess what the situation would be if these efforts (technical and economical) had not been done for the last 20 years. First of all, and more evident, the Costa Brava would have 6 million m^3/yr less water, which means that the effects of periodic droughts would be more intense. It is very likely that some activities, such as the irrigation of golf courses would not be possible under those circumstances, since the priority would be to guarantee the supply to the municipalities. In a situation of reduced rainfall and a decrease in the level of the drinking water reservoirs, this would also mean that coastal aquifers would be overdrafted and possibly affected by salinization.

In those cases where drinking water production and supply has greater energy consumption than the production and distribution of reclaimed water, there would also be an increase in the overall energy consumption in the municipality. Wastewater treatment plants would not need to perform to the level required for a subsequent reclamation step, which means that more treated wastewater would be discharged into the environment and, despite compliance of treatment standards, possibly with greater concentrations at least of BOD, suspended solids and turbidity, without the existing positive impact on water quality level.

And, finally, staffs from wastewater treatment plants that produce and supply reclaimed water necessarily need to deal with disinfection and health-related water microbiology issues, something that is not so critical in conventional wastewater treatment plants that just discharge treated effluents. This represents a higher degree of technical skills in operators and a higher commitment to the pursuit of sustainability, making their daily jobs more interesting and rewarding – as seen in our plants.

4.5 HUMAN DIMENSION OF WATER REUSE

Water reuse in Costa Brava has seen no opposition by citizens. Initially, when the first golf courses were forced to use reclaimed water there was some reluctance to its use, but the drought conditions aided acceptance of the situation. Management of reclaimed water, which meant not only irrigation of turf but also fertilizer contribution, required a couple of seasons to be mastered, but later quickly became standard practice, just like the management of storage ponds. Even though no active informative campaign has been made, since 2004 data on treated volumes and quality have been regularly published on CCB's website, making this information easily available to anyone interested on the topic. CCB's website also contains many technical articles and presentations, publications and videos, which are also another rich source of information to citizens.

Another unplanned key to success can be observed in Tossa de Mar, where water reuse started in a very humble way. The first use for chlorinated secondary effluent in the area was to help turn an uncontrolled landfill into a recreational park, called Parc de Sa Riera, which quickly became a town favourite. Because of that, Tossa de Mar residents were immediately aware of the potential of reclaimed water and the Catalan Water Agency and the city council decided to investment in the reclamation treatment and the distribution network system, respectively- to improve the treatment and promote water reuse in the community. In Lloret de Mar, the successful irrigation of the picturesque, Italian Renaissance style Santa Clotilde gardens since 2007 (Figure 4.4) has also been important to public acceptance of reclaimed water in Costa Brava.

4.6 CONCLUSIONS AND LESSONS LEARNED

After a certain degree of quality and consistency is reached, reclaimed water can be effectively included in water resources planning. Promoting water reuse is very interesting in coastal areas, especially those exposed to water shortages or with insufficient water resources. In fact, water reuse represents not only a net increase in water availability, but in many cases it contributes to a lower water footprint for the community. Water reuse in the Costa Brava started as a supply activity for certain type of users, mainly golf courses and pitch & putt facilities, but over time it has evolved towards becoming a new resource to be taken in account by municipalities.

Despite the fact that currently most of the treated wastewater is still discharged into the sea, over the last two decades some perceptions have deeply changed in the Costa Brava. Mentality has changed from the necessary disposal of something still considered a waste product to the dominant perception that effluents are resources that need to be reclaimed and reused for a greater benefit for the community, and they only have to be discharged when there is no possible beneficial use for reclaimed water. In people's minds water is not to be wasted if it can be recycled, a perception that has also been favoured by the frequent droughts of the last decade. As a proof of this, the most recent projects in northern Costa Brava, the area with the lowest average rainfall, are proposing a maximum beneficial reuse and a minimum discharge into the sea, based on the direct use of reclaimed water for all kinds of non-potable uses in summer, mostly irrigation, and the indirect recharge of local aquifers via soil percolation from October to May in case of drought. Practical experience gained with the aquifer recharge project of the lower Tordera River is providing the quality criteria, technical skills and confidence to make the most of water that otherwise would be discharged into the sea.

Another important goal for the next few years will be the development of supply schemes to private users, in collaboration with local municipalities. Reliability of the supply, implementation of safety measures, financial arrangements, water pricing and user perception will be essential issues to be tackled in a cooperative way by the Catalan Water Agency, the Consorci Costa Brava and the municipalities that have already installed a reclaimed water distribution system, or that plan to do so in the near future.

The following Table 4.3 summarizes the main drivers of the water reuse in Costa Brava as well as the benefits obtained and some of the challenges for the future.

Table 4.3 Summary of lessons learned.

Drivers/Opportunities	Benefits	Challenges
Lack of water resources High energy consumption of drinking water cycle Tax increase in drinking water use	Net increase of resources Simultaneous water and energy savings Reduction of discharges Potential economic savings	Reliability in quality Automation for online detection of failures Pricing Contractual agreement with municipalities and/or end/users Refining of present regulations

REFERENCES AND FURTHER READING

Agència Catalana de l'Aigua. (2008). Proposta de gestió de la ITAM Tordera i de les extraccions de l'aqüífer als horitzons 2009 i 2025. (ITAM Tordera proposal for management and withdrawals from the aquifer at horizons 2009 and 2025).

Borràs G., Soler M. and Sala L. (2007). Summary of data concerning the quality of the reclaimed water produced at the Blanes Reclamation Plant (Costa Brava, Girona, Catalonia). Proceedings of the 6th Conference on Wastewater Reclamation and Reuse for Sustainability, October 9–12, 2007, Antwerp, Belgium. http://www.ccbgi.org/docs/antwerp_2007/article_antwerp_2007_managed_aquifer_recharge_028.pdf (accessed October 2011).

Generalitat de Catalunya. (2011). Water Reuse: The Catalan Experience. Case Studies. Departament de Territori i Sostenibilitat. http://waterbcn2011.org/download/E-04-112-11%20Water%20Reuse%20Cataleg%20DEF.pdf (accessed 18 April 2012).

Kampf R. and Claassen T. (2004). *The Use of Treated Wastewater for Nature: The Waterharmonica, a Sustainable Solution as an Alternative for Separate Drainage and Treatment*. IWA-Leading-Edge Technology, LET2004, WW5, 3 June 2004, 9:00, Prague, Czech Republic. http://waterharmonica.nl/publikaties/2004_praag/Paper_Kampf_et_al__Let2004_07.pdf (accessed October 2011).

Montemayor M., Costan A., Lucena F., Jofre J., Muñoz J., Dalmau E., Mujeriego R. and Sala L. (2008). The combined performance of UV light and chlorine during reclaimed water disinfection. *Water Science and Technology*, **57**(6), 935–940.

Sala L., Sala J., Ordeix M., Boix D., Couso J. and Serra M. (2007). Les rieres de la Costa Brava: evolució històrica recent, estat actual i perspectives de futur. *SCIENTIA gerundensis*, **28**, 47–61.

WEB-LINKS FOR FURTHER READING

http://www.ccbgi.org/reutilitzacio.php
http://www.ccbgi.org/activitats.php
http://www.ccbgi.org/publicacions.php
http://www.ccbgi.org/videos.php?video=1&ln=en

View of the constructed wetland in Empuriabrava.

5 Integration of water reuse for the sustainable management of water resources in Cyprus

Iacovos Papaiacovou and Athina Papatheodoulou

CHAPTER HIGHLIGHTS

Herewith the story of Cyprus, a water stressed country, is presented which has successfully implemented water recycle scheme, overcoming various technical and social constrains. By recycling, 91 millions m^3 of water have been beneficially reused since 2004. Both the numbers and social acceptance are fruitfully increasing.

KEYS TO SUCCESS

- Trust the benefits of water reuse and loyalty in implementation
- Systematic work to empower acceptance
- Monitoring and dissemination of water reuse outcomes
- Ensuring good water quality

KEY FIGURES

- Treatment capacity : 150.000 m^3/d
- Total annual volume of treated wastewater: 19 Mm^3/yr
- Total volume of recycled water: 15 Mm^3/yr
- Recycled water tariffs: 0.05–0,21 €/m^3
- Water reuse standards: <5 fecal coliforms/100 mL Suspended solids <10 mg/L, BOD_5 <10 mg/L

5.1 INTRODUCTION

In the Mediterranean region, water resources are limited and drought incidents occur frequently. This led to the need for full utilization of water resources, while water demand is increasing, as a result of population growth, tourism development, and increasing standards of living. The population of the Mediterranean region is expected to rise from 454 million people in the year 2005 to around 520 million in 2020, while tourist arrivals are due to increase from around 200 million in the year 2000 to 300 million in 2020 (MediTerra, 2008). These developments are anticipated to put additional pressure on the availability of water in the region.

Cyprus, the third largest island in the Mediterranean region, has an area of 9.251 km^2. The area over which the Republic exerts effective control and to which the information presented refers to, is about 5.800 km^2. The population is 803.200 inhabitants according to 2009 demographic estimations with 0.9% annual growth in 2010 (Statistical Service, 2011).

Cyprus has been officially given – by the EU – the status of a "country with scarce water resources" (MANRE, 2005). Prolonged drought periods combined with the increasing demand, emphasize the need to improve the efficiency of water use, to implement water demand management practices and to augment the existing sources of water with more sustainable alternatives.

The reuse of treated wastewater in Cyprus has been steadily growing and is now widely practised. Several reuse schemes are operational and many more are under study or construction. Water reclamation and reuse has become substantially important for environmental, economic and social reasons and its application could be considered as an example of sustainable development.

Main drivers for water reuse

The rapid development of the island, the changes in traditional agriculture practices and land use, the over-abstraction of groundwater (causing intrusion of sea water into the groundwater tables) in combination with climate change, have led to a negative balance of supply and demand of water resources in Cyprus. The reuse of water for non potable applications or potable water substitution has been proven internationally in water stressed regions to be a drought proof source and one of the most effective water scarcity solutions (Durham *et al.* 2005). In Article 12 of the EU Wastewater Directive 91/271/EEC it is stated that "treated wastewater shall be reused wherever appropriate". The adoption of the Directive into National legislation was an additional incentive to take the initiative forward.

Brief history of the project development

Following the events of 1974, the mass influx of refugees and the turn towards tourism industry saw record levels of groundwater and marine deterioration. It was then that the authorities urged for the construction of Wastewater Treatment Plants (WWTPs) and the sustainable use of their effluent. The operation of sewerage networks started as early as 1980. Nowadays, 30 years later, the island is being served by more than 220 plants, the majority of which are privately owned plants of hotels and tourist villages (Table 5.1).

Table 5.1 WWTPs and effluent applications in Cyprus.

Category	Number	Effluent application
Hotels, Apartments, Tourist villages	178	N/A
Industries and Private Hospitals	10	N/A
Refugee housing	3	Landscape irrigation
Hospitals	4	Landscape irrigation
Military Bases	9	Landscape irrigation
Rural (<2000 p.e.)	4	Agriculture, Landscape irrigation
Rural (>2000 p.e.)	6	Agriculture, Landscape irrigation
Urban	7	Agriculture, Aquifer Recharge

*p.e.: people equivalent.

In the pursuit of full compliance with the European Directive 91/271/EEC, the country's implementation plan requires the full connection to a sewerage network and subsequent treatment plants of 7 urban agglomerations and 50 rural agglomerations (total 860.000 p.e.) by the 31st of December 2012. At present, 66% of urban areas are in compliance with the Directive's requirements, compared with only 8% of rural areas. It is anticipated that by the end of 2012, 30 WWTPs will be serving the areas with population greater than 2000 (ED & WDD, 2010). All new plants have provisions for tertiary treatment, in order to provide an alternative water resource on the island. The water recycling scheme is managed and operated by the Water Development Department (WDD) of the Ministry of Agriculture Natural Resources and Environment (MANRE).

The vast majority of the island's coastal waters are being used for bathing and recreational activities. The results of a monitoring program shows that the coastal waters are in conformity with the guideline values set by the Bathing Water Directive 2006/7/EC. Cyprus has been recognised as "star performer", with 100% of its bathing water sites in compliance to the Directive for 2010 (EEA, 2011), the installation of central sewerage networks, WWTPs and reuse programs has considerably contributed to this achievement.

Project objectives, incentives and water reuse applications

Until recently the use of reclaimed water as an essential part of integrated management of the water resources of the island, was somewhat overlooked. The frequent occurrence of droughts, the decreasing availability of the resource and the simultaneous increase of competition over water among the economic sectors, led to the intensification of the application of treated effluent.

Within this context, the allocation of treated effluent to farmers, landscape irrigation and groundwater recharge has been an essential step towards good water governance. This approach included guidelines on conservation measures and treated effluent application methods. A pricing policy was set by the authorities to favour the use of reclaimed water. Currently, about 15 million m^3 (Mm^3) of recycled water are being distributed; it is however expected that this amount will be multiplied in the near future. The overall volume of water to be recycled for beneficial purposes is expected to be 65 Mm^3 by 2015 and 85 Mm^3 by 2025 (Karavokiris *et al.* 2010). As shown on Table 5.2, the major part of the reclaimed water is used for irrigation purposes.

Table 5.2 Quantities of treated effluent, reused in 2010.

WWTP	Treated effluent produced, m^3	Irrigation, m^3	**Groundwater, m^3
*Urban	14,918,181	11,042,163	2,341,268
Rural (>2000PE)	587,650	587,650	
Rural (<2000PE)	204,400	204,400	
Refugees housing	323,025	323,025	
Hospitals	540,200	540,200	
Military Bases	249,660	249,660	

*Excluding Mia Milia bicommunal WWTP, for which there is no available data on the reuse scheme.
**Applies only to Paphos WWTP. (Source: Water Development Department)

The case of the Sewerage Board of Limassol-Amathus

The Sewerage Board of Limassol – Amathus (SBLA) is a public utility organization with the primary mission of the construction, operation and maintenance of the central sewerage system for the collection and treatment of municipal wastewater of the Greater Limassol area (Figure 5.1), as well as the construction and maintenance of the main infrastructure of the storm water drainage system. The sewerage and drainage project is constructed in phases and covers the entire urban Greater Limassol Area (Papaiacovou, 2001). The construction works of the first phase of the sewerage and drainage system began in 1992 and were completed in 1995, while the works of phase two, which began in 1999, were completed in 2004.

Figure 5.1 View of Limassol coastline (Source: Limassol municipality). The installation of its sewerage network has played an essential role for the award of 15 blue flags to its beaches.

Currently, the final phase is being implemented and construction works continue to be carried out with the main objective being the expansion and completion of the works up to 2013. The works of the current phase began in August of 2006 with the extension works of the WWTP in Moni area which were completed in September 2008. The plant has a treatment capacity of 40.000 m^3/d. Since 1992, when the staged construction of the sewerage and drainage system began and up until 2011, over 250 million € have been invested in the project. The works covered about 70% of the SBLA area and about 20,000 premises (30,000 house connections) have been connected to the sewerage system, providing service to about 130,000 people.

A new WWTP is now tendered out for construction and operation until 2013, in order to satisfy the increasing needs of the city. In the meantime, the sewerage network is being expanded in order to cover the remaining areas of SBLA area. The total additional investment is expected to be over €150 million. Apart from these works, the SBLA has the responsibility for the construction of the flood prevention and storm water drainage works valued at over €100 million.

5.2 TECHNICAL CHALLENGES IN WATER QUALITY CONTROL

Treatment trains for water recycling

In order to ensure the high quality of reused effluent, the appropriate technology is being employed by each Sewerage authority. In the case of the SBLA, sand filtration has been chosen to further treat the secondary effluent which is subsequently transferred to a meandering contact tank for chlorination (Figure 5.2). The disinfected effluent is pumped to the WDD's storage tank from where it will be distributed to the end users. Even though sand filtration is the most common method applied in the island for tertiary treatment, a membrane biological reactor (MBR) has been recently installed in Anthoupoli WWTP and will be also constructed in the extension of Larnaca WWTP expected by late 2013, as well as at the new SBLA treatment plant expected to be operating by 2014.

Figure 5.2 (a) A general view of the biological WWTP of Limassol (b) the tertiary treatment stage of sand filtration and (c) chlorination (the storage tank can be also seen at the back).

Water quality control and monitoring

Due to the nature of the initial source there are strict quality guidelines in order to minimize health and environmental risks. The criteria related to the use of treated wastewater for irrigation purposes have been established in June 2005 (Decree 269/2005) and are stricter than the World Health Organisation guidelines, taking into account the conditions specific to Cyprus. The criteria are followed by the Code of Good Agricultural Practice (Decree 263/2007), in order to ensure the best possible application of water for irrigation.

The Department of Environment is responsible for setting the quality characteristics of effluent in all the plants serving areas with population greater than 2000. For plants serving smaller agglomerations; the WDD is the responsible authority for monitoring the quality of effluent. Quality criteria are set through the Disposal Permit, issued to the different Sewerage Boards every four years. Some examples of the quality parameters as expected for the outflow as well as the desired sampling frequency are shown in Table 5.3. Indicative of the treatment efficiency and the quality of effluent are the analyses results for the treated effluent produced by the SBLA in 2010, also presented in Table 5.3.

Main challenges for operation

The results of the European Surveillance of Antimicrobial Consumption project revealed that Cyprus registers the second highest consumption of antibiotics among 34 participating countries. Recent studies on the island have shown the presence of pharmaceutical compounds in treated effluent, some of which are known or suspected endocrine disruptors (Fatta *et al.* 2010; Makris & Snyder, 2010). No study has however, proven the ability of these compounds to enter the trophic chain, thus posing a threat to human health. In the light of this realization, research projects have been launched in this field, with the active contribution of some Sewerage Boards.

To safeguard the good quality of effluent and the operability of WWTP, not only the operators of the plant are responsible for conducting regular analyses according to the Disposal Permit, but also the WDD is taking monthly samples. To ensure the validity of results, samples are analyzed in accredited laboratories.

Table 5.3 Qualitative characteristics of treated effluent as expected by the Disposal Permit of the Sewerage Board of Limassol and the mean results obtained in 2010 analyses.

Parameter	Frequency of control*	Disposal permit limits	Mean results for 2010
BOD_5, mg/L	15 days	10	4
COD, mg/L	15 days	70	33
Suspended solids, mg/L	15 days	10	4.2
Total nitrogen, mgN/L	15 days	15	6.9
Total phosphorus, mgP/L	15 days	10	2
Total coliforms/100 mL	15 days	5	0

*The maximum number of samples that can differ with respect to the number of samples taken during any year is three.

Another important challenge for the operators of WWTPs of the island is the elevated salinity levels either due to the intrusion of sea water to the sewerage network or due to the disposal of the desalination plants brine in the sewer network. A study of the WDD, suggests the use of reverse osmosis technology to further treat an amount of sewage, so that its reuse would be acceptable for irrigation in sensitive soils and plants. However, financial and technical constraints appear towards this suggestion. The use of MBR is also being discussed (Karavokiris *et al.* 2010). Existing control measures regarding the application of the provisions of Code of Good Agricultural Practice should also be highlighted and reviewed, as it comprises an issue that should be dealt by the Department of Agriculture to the end users level.

5.3 WATER REUSE APPLICATIONS

The excellent quality of reclaimed water enables its broad application. The prevailing effluent disposal methods are mainly irrigation of permanent crops (Figure 5.3) as well as watering of landscape areas, football fields, hotel gardens, and so on. In the case of Paphos – the fourth biggest city located in southwest of Cyprus – the entire quantity of treated effluent produced is being used for Ezousa aquifer recharge, which is subsequently pumped for irrigation through diversion in an irrigation channel. Irrigation with reclaimed water is regulated by the Code of Good Agricultural Practice. The treated effluent can be applied to all kind of crops excluding leafy vegetables, bulbs and corms eaten raw.

The application of the treated effluent from the WWTPs of the island is mainly irrigation of:

- Citrus trees, olive trees, fodder crops (clover, sorghum-sudangrass, perennial ryegrass), industrial crops, vegetables (potatoes, beetroots), cereals (corn, barley).
- Landscape.
- Football fields.

Figure 5.3 Irrigation with reclaimed water in Limassol (a) olive trees (b) clover.

Evolution of the volume of supplied recycled water

The recycled water resource was initially met with scepticism. However, as can be seen in Figure 5.4, this was soon overcome as the quantities of reclaimed water increased significantly through the years. During winter months, where the demand for water in agriculture decreases, treated wastewater in the case of Limassol is pumped to Polemidia Dam (irrigation dam) and the excess – due to dimensional restrictions of the conveyor-is overflowing to the sea. Disposal to the sea is also applied in the case of the Sewerage Board of Larnaca, only in the cases where the storage tanks are being maintained or when there is no demand.

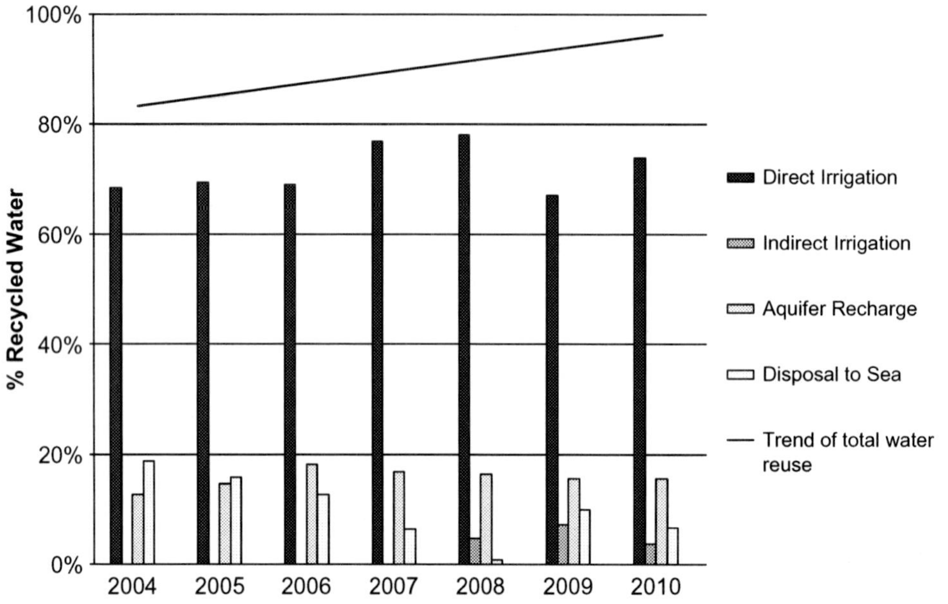

Figure 5.4 Percentage and applications of reused water in the period of 2004–2010. The line shows the trend of total water reuse in the aforementioned period (disposal to the sea is not taken into consideration).

The Government's long-term plans and water policy is to fully utilize reclaimed water and incorporate it to the island's water balance. In 2011, 4.8 Mm^3 of recycled water of the Limassol Project were added to the water balance through the Southern Conveyor Water Works. The volume is expected to increase (Karavokiris *et al.* 2010).

5.4 ECONOMICS OF WATER REUSE

Project funding and costs

It is estimated that the full compliance of Cyprus to the Urban Wastewater Directive will have a total cost of over €1.5 billion, out of which over €750 million will be spent on urban areas. To meet the regulatory obligations, by 2011, the Urban Sewerage Boards had invested over €620 million whereas rural agglomerations €33 million, in constructing expenses.

The Government does not participate to the capital investment or operational costs of the collection and primary and secondary treatment facilities in the urban areas. However, in order to encourage wastewater reuse throughout the country, the Government's policy is to cover the construction and operational costs of the tertiary treatment facilities. In addition, in order to promote the development of non-urban areas, the Government contributes by providing substantial subsidies on the construction cost of sewerage systems of smaller rural communities. Remaining expenditure such as operational and maintenance costs of the plants are contributed by the communities or the Sewerage Boards, through the imposition of charges to the citizens.

The Government through the WDD has the overall responsibility for the management of wastewater reuse system throughout the Government controlled areas of Cyprus. The treated effluent produced by the urban and rural Sewerage Boards is delivered to the WDD which undertakes its distribution to the various users. The tertiary treatment cost is being reimbursed by the Government to the producers of the treated effluent that is, the Sewerage Boards. The users are being charged on the basis of volumetric water consumption (Table 5.4).

Table 5.4 Treatment cost per cubic meter of the sewerage system of Limassol Amathus.

Cost category	2008 €/m³	2009 €/m³	2010 €/m³	2011 €/m³	2012 €/m³
Energy Cost	0.20	0.22	0.23	0.24	0.24
Total Operating Cost	0.80	0.852	0.85	0.88	0.87
Administrative Cost	0.45	0.51	0.51	0.50	0.47
Depreciation	0.74	0.71	0.64	1.10	1.58
Financing Cost	0.27	0.56	0.86	1.10	1.16
Total Cost	2.26	2.82	3.09	3.82	4.32

Pricing strategy of recycled water

In compliance with Article 9 of the Water Framework Directive (2000/60/EC), a new pricing policy was launched, the aim of which was to recover the cost of water services. In the case of recycled water, the main components of financial cost are (a) capital cost of the tertiary treatment, (b) capital cost of the distribution system, (c) operational and maintenance costs and (d) administrative cost.

As a motivation for its use, treated effluent is supplied to the consumers even without full cost recovery. Therefore the recycled water is sold to the farmers at 25% lower price in comparison to the fresh irrigation water (Table 5.5). The cost recovery reaches 88% of the actual cost. The SBLA is charging WDD the amount of 0,152 €/m³ for the tertiary treatment cost recovery. This amount covers the operational, maintenance and administrative cost of the tertiary treatment facilities (MANRE, 2010).

Table 5.5 Proposed prices for the different applications of reclaimed water in comparison to the prices of the fresh irrigation water (Adjusted by MANRE, 2010).

Beneficiaries	Proposed charge, €/m³	Proposed charge for the fresh unfiltered irrigation water, €/m³
Agricultural organisations for agricultural production	0.05	0.15
Individuals for agricultural production	0.07	0.17
Sport activities	0.15	0.34
Landscape and hotel gardens irrigation	0.15	0.34
Golf courses irrigation	0.21	0.34
Abstraction from aquifers recharged by reclaimed water	0.08	–
Community parks and landscapes in rural areas, where the WWTP is within their boundaries	Free	–

5.5 HUMAN DIMENSION OF WATER REUSE

Public education and communication strategy

The public education component of the treated effluent reuse scheme is one of the most essential parts of government's strategy. The Department of Agriculture is responsible for the education of farmers (as end users) in the aspects related to agricultural production, including the use of treated wastewater. The selection of crops and irrigation systems to be used, as well as the preparation of irrigation schedules are amongst the responsibilities of the Department. The follow up of the guidelines of the Code of Good Agricultural Practices is also the responsibility of the Department of Agriculture. In addition to the above, the WDD and Sewerage Boards are also contributing towards that direction. For example, the SBLA has been undertaking educational campaigns in the form of presentations, leaflets, trips to wastewater application systems abroad, and so on.

Public acceptance and involvement

Public perceptions and acceptance of water reuse are recognised as part of the main ingredients of success for any reuse project. The acceptance of treated wastewater reuse in the island, especially for the farmers as end users was initially slow. Farmers appeared to be hesitant over the public health risks of water reuse, quality control concepts and effects of its application on the cultivations. However, about 15 years since its first application, things have changed; the acceptance of treated effluent is very high. Contributing factors to this change were the success stories of water reuse schemes, the recent drought periods and the reduction of available surface and ground water as well as the irrigation water supply cuts. The pricing policy applied also affected the acceptance of the resource.

5.6 CONCLUSIONS AND LESSONS LEARNED

The reuse of treated wastewater in Cyprus has proven to be a success story, as the demand rises and its beneficial effects have become obvious. It is expected that the amount of wastewater produced and allocated for reuse will be steadily increasing as more agglomerations will get connected to the sewerage network. The coming challenges for the competent authority and Sewerage Boards are the treatment methods for safeguarding public health by the application of treated water, its storage and reuse. It could be considered appropriate to reproduce the policy claim *"Not a drop of water to the sea"* coined for maximum capture of run-off by dam construction- in the case of treated wastewater reuse. In an officially declared water stress country we cannot afford disposing water without fully exploiting its beneficial properties.

The following Table 5.6 summarizes the main drivers, benefits and challenges of the water reuse in Cyprus.

Table 5.6 Summary of lessons learned.

Drivers/Opportunities	Benefits	Challenges
Comply with Directive 91/271/EEC by 2012	Secure good quality water and protect environment	Huge investments needed in a period of world economic crisis
Increase volume of Recycled water	Substantial Contribution to the island's water balance	Efficient application in appropriate uses and the island's water cycle
Reuse of 100% of Recycled water	Save fresh water for other uses	Secure adequate quality controls and water quality

REFERENCES AND FURTHER READING

Council Directive 91/271/EEC. (1991). Urban waste-water treatment. *Official Journal L*, **135**(30/05), 40–52.

Decree 269/2005 on Water pollution control (General guidelines on the disposal of wastes from urban wastewater treatment plants).

Decree 263/2007 on Code of good agricultural practice for the limitation of nitrate pollution.

Durham B., Angelakis N. A., Wintgens T., Thoeye C. and Sala L. (2005). Water recycling and reuse: a water scarcity best practice solution. Proceedings of the ARID Cluster International Conference, Coping with drought and water deficiency: From research to policy making. Limassol-Cyprus.

ED and WDD- Environment Department and Water Development Department. (2010). Report on Urban wastewater treatment during 2007–2008. Article 16- Directive 91/271/EEC. Ministry of Agriculture Natural Resources and Environment.

EEA- European Environment Agency. (2011). European water quality in 2010. EEA Report No.1 /2011. Copenhagen.

Fatta-Kassinos D., Hapeshi E., Achilleos A., Meric S., Petrovic M. and Barcelo D. (2010). Existence of pharmaceutical compounds in tertiary treated urban wastewater that is utilized for reuse applications. *Water Research and Management*, **25**, 1183–7793.

European Parliament and Council Directive 2000/60/EC. (2000) Establishing a framework for community action in the field of water policy. *Official Journal L*, **327**(22/12), 1–73.

Karavokiris and Associates Engineers Consultants and Kaimaki P. S. (2010). Preliminary report of water policy- Report 3. Water Development Department, Nicosia.

Makris K. and Snyder S. A. (2010) Screening of pharmaceuticals and endocrine disrupting compounds in water sapplies of Cyprus. *Water Science and Technology*, **62**(11), 2720–2728.

MANRE- Ministry of Agriculture Natural Resources and Environment. (2010). Water Framework Directive EU- Reporting sheets on economics. Nicosia, Republic of Cyprus.

MANRE- Ministry of Agriculture Natural Resources and Environment. (2005). Water Framework Directive EU- Summary report. Articles 5 and 6. Nicosia, Republic of Cyprus.

MediTerra. (2008). *The Future of Agriculture and Food in Mediterranean Countries*. International Centre for Advanced Mediterranean Agronomic Studies, Paris.

Papaiacovou I. (2001). Case study – Wastewater reuse in Limassol as an alternative water source. *Desalination*, **138**, 55–59.

Statistical Service. (2011). Demographic Report 2009, Population Statistics, Series II, Report No 47. Nicosia, Republic of Cyprus.

6 Role of water reuse for Tianjin, a megacity suffering from serious water shortage

Yu Zhang, Fusheng Tang, Dianhai Li, Yuhong Li, Weiping Chen and Min Yang

CHAPTER HIGHLIGHTS

Municipal wastewater reclamation is the most active and largest water conservation measure in Tianjin to create a water-saving city. It proves to be an effective way to solve the water scarcity in Tianjin. This case study is aiming to review the development of water reuse in Tianjin and challenges of the construction of water recycling plants, treatment technology, reuse category and water supply services.

KEYS TO SUCCESS

- Water scarcity
- Government promotion and decisions
- Reclaimed water treatment technologies enabling to consistently achieve the water quality requirements for fit to purpose uses of recycled water

KEY FIGURES

- Start up 2007
- Treatment capacity: 260,000 m^3/d
- Total annual volume of treated wastewater: 726.4 Mm^3/yr
- Total volume of recycled water: 94.9 Mm^3/yr
- Capital and operation costs: 2.6–3.2 Yuan RMB/m^3 (0.33–0.41 €/m^3)
- Recycled water tariffs (2012): domestic water 2.2 Yuan RMB/m^3 (0.28 €/m^3); industrial reuse 4.0 Yuan RMB/m^3 (0.52 €/m^3)
- Water reuse standards: <3 *E.coli*/L

6.1 INTRODUCTION

The elderly always say that Tianjin lies on the tip of the Nine River, which is the general name for the many basins of Haihe. Just like the people of Beijing often call the Yongding River the "mother river" of Beijing, Tianjin residents have been grateful for their mother river, Haihe, for generations (Figure 6.1a). It is the historical natural river that has nourished the life of Tianjin. The Haihe basin begins with over 300 streams, and these streams finally converge into a 72 km mainstream at the Sancha estuary that pours into the ocean. This particular river section is the Haihe River today; it is the river with the shortest mainstream in the world, and it is this mainstream next to which Tianjin was formed (Figure 6.1b).

The formations of the Haihe river trial and the Haihe plain have only 4000 years of history; it began as a "continental marsh" left behind after the Bohai River retreated. Over thousands of years, the Haihe plain converged with many other rivers because of its low terrain, forming countless lake swamps, and the plain area has experienced sights similar to a galaxy of glitter swamps.

During the Sui dynasty (581AD–618AD), the Sui Yang Emperor demanded the Tongji drainage and the Yongji drainage to be connected so that this 1400 km canal could communicate the Haihe River with the Yellow River and the Yangtze River. The opening of this canal established Tianjin as the shipping center because it was "located at the Nine River and connected the transportation from 7 provinces," which made Tianjin a water city that "gathered thousands of streams and connected shipping from all directions." Since Beijing became the capital during the Jin dynasty (1115AD–1234AD) and before modern transportation existed, Tianjin was the only pier that shipped grains to Beijing. From the Jin dynasty to the Qing dynasty (1616AD–1911AD), grains from southern China could all be shipped to Beijing via the Haihe River.

However, through its history, Tianjin has become a city lacking in water resources because of its population growth, city expansion, and exploitation of an upstream water source. The total water resource capacity of the city is 181.6 million m^3 (Mm^3), and the average surface water capacity is 105.5 Mm^3. The per capita water capacity is only 160 m^3/cap, 1/15 of the nationwide per capita and 1/60 of the world per capita water availability, which makes it a severe water scarce region (Table 6.1). During recent years, although with the water conveyance from the Luan River and the Yellow River, the per capita water availability is still only 380 m^3/cap. This scarcity of water is still a significant factor that constrains the economy and the social sustainable developments of Tianjin. Particularly, Tianjin is one of the most important industrial cities in China, which relies heavily on water, and the percentage of industrial water usage is 80% of the total water utilization in Tianjin.

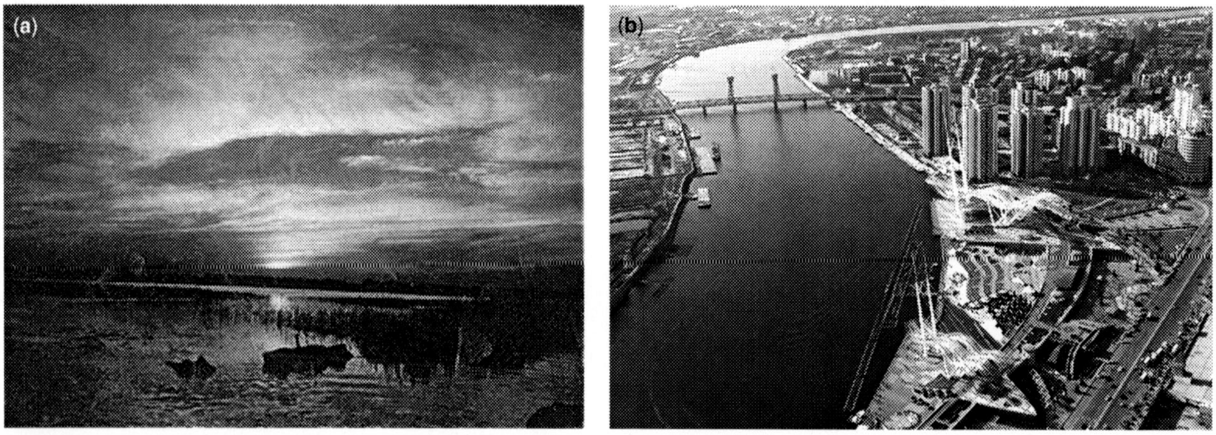

Figure 6.1 Views of the Haihe River in ancient time (a) and the Tianjin city today around the Haihe River.

Table 6.1 Total water resources in Tianjin in million m^3 (Mm^3).

Area	Surface water	Ground water	The amount of water resources
Ji Yunhe mountain area	190	69	259
Bei Sihe plain	468	552	949
South Dian Dong-Qing plain	397	211	608
Total	1055	832	1816

In order to solve the water scarcity issue, water conveyance from south China, desalination and water reclamation have become the major approaches. Compared to desalination and water conveyance, wastewater reclamation possesses obvious advantages in project investments and costs, so Tianjin has made water reclamation its main approach in solving water scarcity. The water reclamation sources in Tianjin are very rich in terms of available wastewater. As shown in Table 6.2, four wastewater treatment plants are in operation with additional two plants under construction. As for the wastewater management ability, they can all become reclaimed water sources.

Table 6.2 Municipal wastewater treatment and recycling plants in the urban area of Tianjin.

Wastewater treatment and recycling plants	Wastewater treatment capacity, m^3/d	Status	Reclaimed water treatment capacity, m^3/d
Ji Zhuangzi	540,000	In service	70,000
Xian Yanglu	450,000		50,000
Bei Cang	100,000		20,000
Dong Jiao	400,000		50,000
Shuang Lin	200,000	Under design	10,000
Zhang Guizhuang	300,000		60,000

Brief history of the project development

Tianjin has started research on the wastewater reclamation since the 1980s and has entered the comprehensive development stage today. Tianjin was one of the first cities in China to begin a water reclamation planning program in the year 2000. In 2004, the city government included the "Tianjin Center City Water Reclamation Plan" in the "General Planning of the City of Tianjin".

By the end of 2002, the Ji Zhuangzi wastewater reclamation facility completed its construction and began its operation. This reclamation facility was officially selected as part of the demonstration programs that was launched by the central government in five water scarce cities. Currently, the central city of Tianjin has completed the construction of four water reclamation facilities, with a total treatment capacity of 190,000 m^3/d (see Table 6.2).

In 2001, with the approval of the city government, the city council established the Tianjin Water Reclaimed Co., Ltd. (Water Co.) in charge of the management, financing and construction of water reclamation facilities, as well as the production, supply, and marketing of reclaimed water.

Project objectives, incentives and water reuse applications

Water reuse is an important element of the sustainable urban water cycle management of the city of Tianjin along with other measures such as water conservation, climate change mitigation and reduction of water pollution.

Meanwhile, a series of water reclamation regulations and policies were introduced including the Tianjin Water Saving Act, the Tianjin City Drainage and Reclaimed Water Application and Management Ordinance, the Tianjin Housing and Public Building Reclaimed Water Supply System Construction and Management Act, the Tianjin Central City Reclaimed Water Utilization Planning Act, the Notification on Further Developments of the City's Infrastructure Construction and Management Act, and the Notification on the Fee Standard of Reclaimed Water Pipeline Network Construction Act. These regulations formed the governance mode with water supply investment, government support on the construction of the core networks, local developers' investments on building pipeline networks, and industrial factories with self-supporting reclaimed water pipeline infrastructure. Tianjin has constructed 427 km of core pipeline network for recycled water and 263 km of community pipeline network. The main materials of the pipelines include ductile iron, polyethylene (PE), etc., and the diameter range is DN300-DN1600.

Currently, the main functions of reclaimed water in Tianjin are urban miscellaneous uses including toilet flushing, landscape irrigation, construction sites and as cooling water for power plants. Therefore, the Water Co. strictly follows the general national standards and monitoring frequency requirements stated in the Urban Wastewater Reuse Water Quality Standard for Urban Miscellaneous Water(GB/T 18920-2002), the Urban Wastewater Reuse Water Quality Standard for Scenic Environment Water(GB/T 18921-2002), and the Urban Wastewater Reuse Water Quality Standard for Industrial Water (GB/T 19923-2005).

6.2 TECHNICAL CHALLENGES OF WATER QUALITY CONTROL

Treatment trains for water recycling

The water source of the first Ji Zhuangzi water reclamation plant, constructed in 2002, is the Ji Zhuangzi wastewater treatment plant that has an initial total production capacity of 50,000 m^3/d. Two different treatment trains are used. The first one produces 30,000 m^3/d of recycled water for industrial uses by means of coagulation, sand filtration, and disinfection. The 20,000 m^3/d of recycled water for the residential areas (urban miscellaneous uses and landscaping) is purified by coagulation, continuous external pressure microfiltration (CMF) and ozonation.

Over the past ten years, these technologies have been proven to be efficient, feasible and reliable in operation. However, because of the higher requirements in water quality stability, desalination and consumer demands, the Ji Zhuangzi water reclamation plant was upgraded and extended. The initial quartz sand filter was transformed into a submerged microfiltration (SMF) unit with a treatment capacity of 40,000 m^3/d, which greatly reduced operational energy consumption.

As situated in a coastal area, the wastewater of Tianjin contains high concentration of salt, which requires the use of reverse osmosis (RO) technology to control recycled water salinity. Nowadays, the four water reclamation plants in Tianjin mainly apply the core treatment train shown in Figure 6.2 including coagulation, micro- or ultrafiltration (MF or UF), reverse osmosis (RO), ozonation and final chlorination.

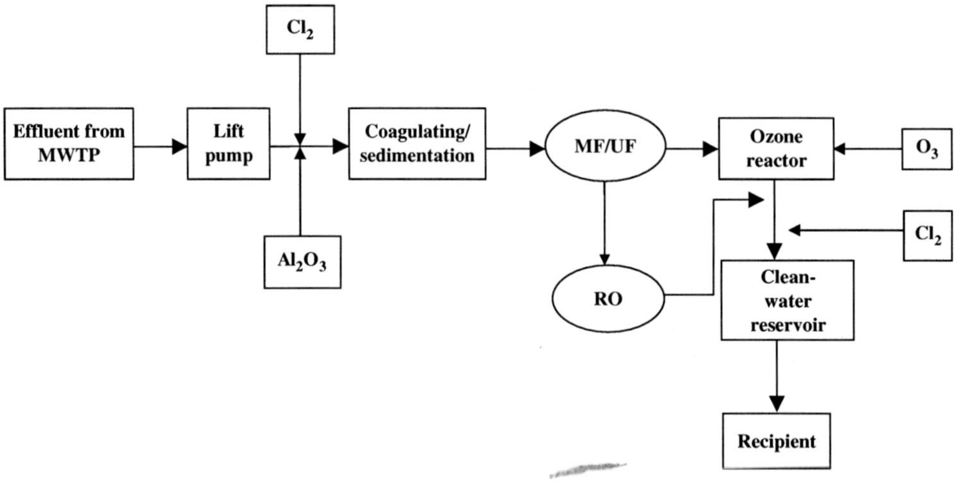

Figure 6.2 The main treatment train used in the reclaimed water treatment plants in Tianjin.

Water quality control and monitoring

According to the actual reclaimed water utilization status of the central city of Tianjin, reclaimed water is mainly applied for urban miscellaneous purposes, industrial uses and landscape irrigation. The water quality monitoring follows the national standards, given in Table 6.3. As for the not regulated industrial uses, the water quality monitoring is according to the standard requirements of the corresponding industries.

Table 6.3 Reclaimed water quality for residential and industrial reuse according to the Chinese national standards.

		Residential reuse		Industrial reuse	
Water constituent	Unit	National standard	Recycled water quality	National standard	Recycled water quality
pH	/	6.0~9.0	6.5~7.5	6.5~9.0	6.5~7.5
Colour	Degree	≤30	10~25	≤30	10~25
Odour	/	Odorless	Odorless	Odorless	Odorless
Turbidity	NTU	≤5	0.1~2.0	≤5	0.1~2.0
Chemical oxygen demand (COD_{Cr})	mg/L	–	–	≤60	10~25
Biochemical oxygen demand (BOD_5)	mg/L	≤10	0.5~4.0	≤10	0.5~4.0
Total coliform	coli/L	<3	<3	≤3	≤3
Chlorine residual	mg/L	0.5~1.0 (after 30 min)	0.5~1.5	0.5~1.0 (after 30 min)	0.5~1.5
Total dissolved solids	mg/L	1000	600~950	1000	600~950
Ammonia	mg/L	5(10)	0.1~8.0	10(5)	0.1~8.0
Anionic surfactant	mg/L	0.5	0.05~0.5	0.5	0.05~0.5
Iron	mg/L	0.3	<0.3	0.3	<0.3
Manganese	mg/L	0.1	<0.1	0.1	<0.1
Dissolved oxygen	mg/L	1.5	1.5~8.0	1.5	1.5~8.0
Total phosphorus	mg/L	–	–	1.0	0.01~0.5
Petroleum	mg/L	–	–	1.0	<0.2
Total hardness ($CaCO_3$)≤	mg/L	–	–	450	200~430
Total alkalinity ($CaCO_3$)≤	mg/L	–	–	350	150~280
Sulfate	mg/L	–	–	250	150~220
Silicon dioxide	mg/L	–	–	50	10~20
Chloridion	mg/L	–	–	250	180~240

The test parameters and sampling frequency are divided into four groups:

- Once a day: pH, colour, odour, chemical oxygen demand (COD), suspended substance, total phosphorus, fecal coliforms, total dissolved solids (TDS), ammonia, manganese.
- Twice a day: turbidity and chlorine residual.
- Once a week: biochemical oxygen demand (BOD), anionic surfactants, iron, dissolved oxygen, total nitrogen, hydrocarbons, total hardness, total alkalinity, sulphate, silicon dioxide, chloridion.
- Three times a week: total coliforms.

Meanwhile, certified laboratories are consigned to monitor regularly the reclaimed water quality in order to ensure their compliance with regulations.

Main challenges for operation

Source water quality issues

Tianjin is an important coastal industrial city. The municipal wastewater plants accept a certain amount of industrial wastewater, and as a consequence, the quality and quantity of the raw wastewater are unstable. Further optimisation of the operation and management of the wastewater plants are needed to improve the quality of the produced secondary effluents.

The main problems in water quality of secondary effluents, used as the source water for water reclamation plants, are the high salinity, pathogenic microorganisms, colour, odour and toxic micropollutants that may cause ecological and health risks. For this reason, the membrane technology was selected to improve recycled water quality, as well as the ozonation, which helps to remove colour and odour in addition to disinfection.

Lifetime and stability of membrane operation

The combination of micro- or ultrafiltration membranes with reverse osmosis is selected as the best available technology enabling to ensure the stability of the reclaimed water quality. Several pre-treatment processes (chlorination, coagulation/sedimentation and pre-filtration) are used to reduce membrane fouling. To improve the reliability of operation, equipment operation management systems and technical operation manuals were established. Since 2002, the membrane systems have steadily been used for nearly a decade, which is far beyond the design lifetime of 5 years.

Application of ozone in wastewater reclamation processes

The purpose of ozonation is mainly to decolorize, eliminate odour and disinfect the recycled water. After several pilot studies to compare various technologies, the Ji Zhuangzi wastewater reclamation project was the first in China to implement ozone technology as a core process for wastewater reclamation. Currently, ozone technology has been widely applied to other water reclamation plants in Tianjin, as well as to some water reclamation projects in Beijing.

6.3 WATER REUSE APPLICATIONS

The repartition of the volumes of recycled water used for the major water reuse applications in the central city of Tianjin is shown in Table 6.4.

Table 6.4 Structure of reclaimed water reuse in Tianjin (2010 and 2020).

Year	Industrial reuse (%)	Municipal reuse (%)	Scenic environment reuse (%)	Others (%)
2007	20	65	15	–
2010	68.5	18.9 (toilet flushing 8.9%)	10.2	2.4
2020	28.0	14.0	52.0	6.0

In 2007, 20% of the recycled water was used for industrial uses, 15% for landscaping (scenic environment) and 65% for urban miscellaneous uses. In 2010 the industrial water reuse reached nearly 69%, while landscape water went down to 10%. According to the city planning, the recycled water demand of the city will reach 797 Mm^3 in 2020, from which 52%

will be used for scenic environment (Figure 6.3a), 28% for industrial use (Figure 6.3c), 10% for municipal miscellaneous purposes, 4% for residential miscellaneous purposes (Figure 6.3b) and 6% for other purposes.

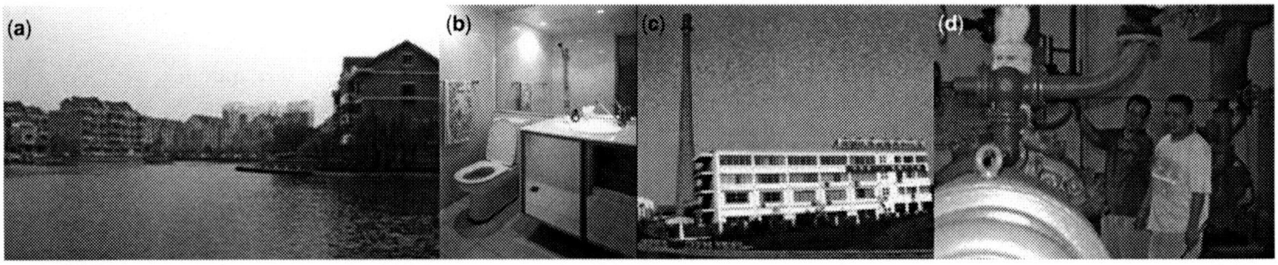

Figure 6.3 Illustrations of water reuse applications in Tianjin: a) reuse for scenic environment, b) toilet flushing, c) industrial use for heating and d) cooling water.

The first end-users of the first wastewater reclamation demonstration project of Ji Zhuangzi, launched in 2002, were the Tianjin Meijiang ecological living community and the surrounding residential housing. The main uses of the recycled water were toilet flushing, landscape irrigation and water ponds. These non-potable applications were well accepted and approved by the local residents because of the good and stable water quality and secure water supply. Since the recycled water pipeline network of the Ji Zhuangzi water reclamation plant was built and extended, more communities and parks have started to use reclaimed water for landscaping, green space irrigation and street cleaning. Because the community landscapes and irrigation systems require large amounts of water, the utilization of reclaimed water enabled a significant reduction in the cost of landscape irrigation, as well as a reduction in water pollution; increase in water use efficiency, and finally accomplishment of the goal of potable water saving.

In 2008, the four artificial lakes surrounding one of the major landscape projects of the Tianjin Olympic Stadium, the European Icicle and the Water Drop, a large fountain project, and the nearby green spaces of these projects, have all been using recycled water as the main source of water. Using "water" as the design theme has reflected the purpose of the 2008 Olympic Games – "green Olympics, technical Olympics, and humanitarian Olympics," with the Water Drop representing "the source of life".

The Tianjin Road is an "ecological landscape road" that represents the ecological and environmental protection ideologies. It is the major road that connects the central city of Tianjin and the new coastal city. The green spaces on the side and at the center buffer zone along the road are all using recycled water as water source.

In 2009, recycled water was applied to the Chen Tang Zhuang thermal power plant and the Dong Bei Jiao thermal power plant as circulating cooling water, which marked the end of the first stage of recycled water development: the extension of water reuse to industrial end-users.

In September 2010, the Annual Meeting of Tianjin Summer Davos Forum for New Leaders was hosted in the Meijiang Convention Center. Fifteen-hundred representatives from over 90 countries and regions gave the world an impressive meeting of ideas and economics. In order to express our nation's environmental protection ideology and to achieve sustainable social development, recycled water was used as the water source for miscellaneous purposes, such as green space irrigation and toilet flushing at the convention center, providing thus a secure water source and ecological friendly environment for the convention.

Evolution of the volume of supplied recycled water

Tianjin's water reclamation has experienced 10 years of developments, with a steady increase in recycled water volumes (Table 6.5). Since the end of 2009, the major proportion of recycled water usage was not for urban miscellaneous purposes, but for industrial cooling which has accounted for over 50% of the total recycled water volume. The most important increase in water reuse was observed in 2010 (+333% compared to 2009).

Table 6.5 The amount of reclaimed water in Tianjin urban area (2006–2010).

Year	2006	2007	2008	2009	2010	2011
Recycled water volume, m^3	1,809,000	3,280,000	3,526,000	3,720,000	12,373,000	14,270,000
As percentage of treated wastewater (%)	1.40	2.54	2.73	2.88	9.85	11.06

Relations and contracts with end-users

To satisfy consumer demands, the Water Co. formed a management system including water supply planning, pipeline network construction, market research, client development, water supply contract signing and specific client services. The water supply contracts play the most important part in the water supply relationships. Currently, the recycled water supply contracts can be divided into residential contracts, industrial (thermal power plants and heating) contracts, and other contracts including city administration contracts. The residential contract is the only one signed with individual residents through a third party (property companies), while other contracts are signed directly by the client and the water company. The signing process is completed during the recycled water connection process.

6.4 ECONOMICS OF WATER REUSE

Project funding and costs

The Water Co. is responsible for the construction of the Tianjin central city water reclamation projects. The funding mainly comes from the Water Co. funding and loans, with some government investment supports.

For example, the total investment for building a membrane + ozone + RO water reclamation plant with 50,000 m^3/d water production capacity is approximately 110–125 million Yuan (€14.0–15.9 million): 26% for construction expenses, 51% for infrastructure investments and approximately 23% for other expenses. The cost of construction is approximately 2200–2500 Yuan/m^3 (279–317 €/m^3).

The specific cost of recycled water is 2.5–2.6 Yuan/m^3 (0.32–0.33 €/m^3), including energy consumption cost of 0.55–0.65 Yuan/m^3 (0.07–0.08 €/m^3) and the chemical consumption costs of 0.4–0.8 Yuan/m^3 (0.05–0.10 €/m^3).

Pricing strategy of recycled water

The city government established the cost of recycled water from a promotional point of view. In 2003, the price department of Tianjin approved the sales price of recycled water as shown in Table 6.6.

Table 6.6 Recycled water rates in 2003.

Water reuse categories	Price of recycled water, €/m^3 (Yuan RMB/m^3)	Price of tap water €/m^3 (Yuan RMB/m^3)
Domestic water	0.14 (1.1)	0.37 (2.9)
Water for schools, government, hospitals and nursery schools	0.15 (1.2)	0.56 (4.4)
Process water, cooling water in industry	0.17 (1.3)	0.59 (4.6)
Reclaimed water for municipal reuse, public building, gardens, scenic environment, road spraying, etc.	0.19 (1.5)	0.59 (4.6)
Reclaimed water for vehicle cleaning and construction	0.23 (1.8)	2.29 (18.0)

The sales price was adjusted in April 2009 and the recycled water rates were categorized depending on the purpose of use, which can be divided into four categories including residential water, power plant water, special industry and industrial water, administrational water and water for business services (Table 6.7). Currently, the Tianjin government is attempting to lower the cost of recycled water by tax promotions and subsidy policies, along with increasing the price of potable water from traditional water sources in order to make reclaimed water more competitive.

Table 6.7 Recycled water rates in 2009.

Water reuse categories	Price of recycled water, €/m^3 (Yuan RMB/m^3)	Price of tap water €/m^3 (Yuan RMB/m^3)
Domestic water	0.14 (1.1)	0.50 (3.9)
Reclaimed water used in heat-engine plant	0.19 (1.5)	0.85 (6.7)
Reclaimed water used in industry, administration services, service industry	0.38 (3.0)	0.85 (6.7)
Reclaimed water used in special trade (vehicle cleaning, temporary use)	0.51 (4.0)	2.69 (21.1)

Benefits of water recycling

The economic analysis shows that water recycling does not have high economic benefits for reclaimed water suppliers. However, there are certain benefits to industrial clients such as owners of power plants, heating stations, etc. since they are using cheaper recycled water. In addition, for the city of Tianjin as a whole, water recycling reduces the cost and avoids new expenses for water conveyance and seawater desalination, providing thus financial benefits to the city.

In addition, water reclamation has very high non-economic benefits. Water reuse not only relieves water scarcity, but also reduces pollution and improves the water environment. In particular, the reduction of the amount of water conveyance has high ecological impacts upstream. The implementation of wastewater reclamation also promotes more stringent requirements for plant's operation and enhances resident consciousness toward environmental protection and water conservation.

6.5 HUMAN DIMENSION OF WATER REUSE

Public education and communication strategies

To educate the public, the Water Co. launched a series of activities to improve communication with clients, including periodic client meetings. The Water Co. is now becoming the public education foundation for environmental protection.

To improve the communication in water recycling, the company has built a work service hotline and the "12319" Tianjin city development hotline to provide customers with continuous services 24 hours a day, 365 days a year. The service provides information on insurance, emergencies, reports, complaints, suggestions, complements, consulting, etc., which also represents an efficient tool for complaint monitoring.

The role of decision makers

The city government of Tianjin is very concerned about transforming urban wastewater into a beneficial alternative water source. Water recycling is defined as an important, publicly beneficial, environmental friendly, and sustainable plan. Important accomplishments include the following:

1. Building a water reclamation business – the Tianjin Reclaimed Water Co., Ltd.
2. Completing a national pilot project – the Ji Zhuangzi wastewater reclamation project.
3. Establishing a specific city master plan – the Tianjin Central City Reclaimed Water Application Plan.
4. Introducing a regional regulation policy – the Tianjin City Drainage and Reclaimed Water Application and Management Ordinance.
5. Constructing a network of reclaimed water treatment plants (RWTPs) – the Xian Yang Lu RWTP, the Dong Jiao RWTP, the Bei Ceng RWTP, and the Zhang Gui Zhuang RWTP.
6. Approving an investment mode for water reclamation infrastructure construction – plants are invested and constructed by the water companies, while main pipelines are developed with government supports, pipeline networks are built by developers, and other supporting infrastructures are self-managed by the water recycling plants.
7. Establishing and implementing a series of regulations regarding reclaimed water development – Government Supporting Regulations for Pipeline Network Construction; Government Reclaimed Water Pricing Standards; Electricity Price Reduction for Water Reclamation Plants; Water Source Fee Exemption for Reclaimed Water Businesses; Double Pipeline for New Residential Housing. Meanwhile, the social acceptance toward reclaimed water has increased.

Public acceptance and involvement

Effective public involvement begins at the earliest planning stage and continues through implementation and beyond. The city of Tianjin's reclaimed water project included citizen surveys, open houses, program websites, media relations, briefings for government officials, plant or project tours, etc. Recycled water is currently being used for toilet flushing, indicating that water reuse has been accepted by local residents.

6.6 CONCLUSIONS AND LESSONS LEARNED

The main lessons learned can be summarised as follows:

1. As a drought-proof alternative water source, recycled water should be widely used in megacities with severe water scarcity such as Tianjin, since it has few impacts on the environment and lower costs compared to long-distance water conveyance and seawater desalination.

2 Tianjin selected the reclaimed water treatment technologies depending on the water quality requirements and the purpose of recycled water uses.

3 Government promotion and decisions were extremely important.

4 The main factors that constrain water reclamation include: (1) pipe networks are not in place, (2) large clients, such as industries are using low water volumes, (3) low public acceptance of residential and municipal reuse.

The following Table 6.8 summarizes the main drivers, benefits and challenges of water reuse in Tianjin.

Table 6.8 Summary of lessons learned.

Drivers/Opportunities	Benefits	Challenges
National and local promotion on utilizing the municipal wastewater as a water resource Serious water scarcity Big reclaimed water user (power plant)	Conserves freshwater resources Saving great amount of high quality water Improve the environmental quality	Optimizing the category of reuse and continue to increase the ratio of wastewater reclamation Expanding the scale of reclaimed water and the pipeline construction Augmentation of landscape use

REFERENCES AND FURTHER READING

Chen C., Gao G., Liu W. Y. and Zhang Y. Q. (2009). Advanced treatment of municipal wastewater by modified microfiltration membrane. *China Water & Wastewater*, **13**, 61–63. (In Chinese).

Li C. J., Liu J. H. and Liu W. Y. (2007). Understand and explain 'the city of Tianjin regenerates water design specifications'. *Tianjin Construction Science and Technology*, **2**, 9–11. (In Chinese).

Liu W. Y., Tang F. S., Jiang W. and Zhang L. (2010). Interpretation on regeneration of urban water operations, maintenance and safety technical regulations. *Tianjin Construction Science and Technology*, **1**, 6–8. (In Chinese).

Lv B. X., Liu W. Y. and Li D. H. (2007). Introduction of Ji Zhuangzi reclaimed wastewater project. *Water Technology*, **1**(3), 58–61. (In Chinese).

Tang F. S. (2009). Wastewater recycling and water quality control in Tianjin. *Construction Science and Technology*, **11**, 56–57. (In Chinese).

Zhang Y., Yan Z. M., Yuan H. Y., Tian Z., Tang F. S., Li D. H. and Li Y. H. (2010). Urban wastewater reuse and water security in Tianjin. *Construction Science and Technology*, **23**, 41–43. (In Chinese).

WEB-LINKS FOR FURTHER READING

http://www.tj.gov.cn/
http://www.tjhb.gov.cn/
http://www.tjcac.gov.cn/
http://www.tstc.gov.cn/
http://www.tjcep.com/index22.aspx

View of landscape irrigation with recycled water in Tianjin.

View of the Ji Zhuangzi water reclamation plant.

View of the reverse osmosis (RO) technology used to control recycled water salinity in Tianjin.

PART II

Urban use of recycled water

Foreword

By The Editors

Historical development of urban use

For more than eighty years, second reticulation systems have been used in water short regions to supply non-potable water for various purposes to conserve limited drinking water supplies. In the USA reclaimed water was used in a dual water system at Grand Canyon National Park in 1926 for lawn watering, toilet flushing, cooling water and boiler water. The City of Pomona initiated a system for irrigation of lawns and gardens in 1929. A system to water lawns and supply ornamental lakes in San Francisco's Golden Gate Park started in 1932.

In response to drought and water shortages in the 1970s, a number of cities in the USA developed large recycled water reticulation systems to supply water for irrigation of urban landscaping and municipal uses. In some cases these system also supplied water for irrigation of residential lawns and gardens and various industrial uses. The Pomona virus study published by the Los Angeles County Sanitation District in 1977 demonstrated that the key to successful use of recycled water in these schemes was treatment of reclaimed water by filtration and disinfection to produce high quality recycled water which could be used safely by consumers to irrigate public spaces. The California Department of Health Services issued the California Wastewater Reclamation Criteria (Title 22 regulations) in 1978. Examples of major urban reuse schemes include the systems in Los Angeles County, Irvine Ranch Water District, the Las Virgenes Water District and the St Petersburg system in Florida.

There are numerous examples of urban water recycling systems that supply water for toilet flushing in office buildings, shopping centres, sporting venues and airports. Around the world, there has been widespread use of recycled water for municipal purposes including street cleaning, sewer flushing and washing down water in treatment plants. Recycled water has also been used for washing municipal buses and metropolitan train fleets. The economics of urban water recycling schemes have been improved by supply of recycled water to large commercial, municipal and industrial water users.

Urban reuse case studies

Part 2 of this book "Milestones in Water Reuse: The Best Success Stories" presents a number of case studies showing how water managers have used recycled water to meet the urban water needs which do not require drinking water quality. Systems have varied in scale from large centralised dual reticulation schemes down to decentralised schemes at a local housing subdivision scale or in some cases for a single large public building or residential building scale. The case studies of decentralised systems are presented in Part 3.

Chapter 7 presents a demonstration study of a semi-centralised (local) water recycling project in Knittlingen in Germany to supply recycled water in an urban subdivision of 105 houses. The main objective is to demonstrate the performance, feasibility and cost efficiency of a small-scale anaerobic membrane bioreactor for energy recovery and nutrient valorisation as fertilisers for agricultural irrigation.

Chapter 8 describes how recycled water is being used to meet the water needs of population growth in the autonomous communities of Madrid in Spain. Water reuse was included in the 2005 Master Plan "Madrid Dpura Plan" with the

Table 1 Highlights and lessons learned from the selected case studies of urban use of recycled water.

Project / Location	Start-up/ Capacity	Type of uses	Key Figures	Drivers and Opportunities	Benefits	Challenges	Keys to Success
Demonstration project DEUS 21, Knittlingen, Germany	2004 MBR for 175 inhabitants (semi-centralised systems)	Evaluation of the quality/ cost for use in agricultural irrigation	Development area with 105 plots MBR performance: At 12.8–26.5°C Average COD in effluent: 135 mg/L with 85% removal Volatile suspended solids: 20–25 g/L Sustainable flux: 13–14 $L/m^2 \cdot h$	Implement semi-centralized systems that operates economically and more sustainably than other concepts	Optimisation of the anaerobic membrane bioreactors at small scale and low temperature. Potential application in warm climate. Potential benefits from energy recovery (biogas) and nutrients (irrigation).	Combination of anaerobic biological treatment and membrane filtration (AnMBR) at low T°C. Start-up problems due to lack of adequate inoculum. Re-contamination of the effluent.	Process choice of an Anaerobic Membrane Bioreactor with pre-treatment by rotating disk filter. Vacuum sewer system reduces amount of wastewater. Rainwater is collected separately, treated and used in-door. Early involvement of stakeholders.
Madrid, Spain	2005 "Dpura Plan" 10 existing plus 20 new reclamation facilities Tertiary treatment capacity: 267,000 m^3/d		Total volume of treated wastewater: 563 Mm^3/yr Total volume of recycled water 2010: 6.82 Mm^3/yr Investments in Madrid Dpura Plan: €200 million Water reuse standards (urban uses): *<200 E.coli/100 mL <1 helm. egg/10 L <10 NTU*	Increase in water consumption. Climate change with severe droughts. Political and institutional commitment.	Environmental benefits such as increasing water reserves in the region's reservoirs. Cheaper price than drinking water. Efficient communication and involvement of stakeholders.	Supply reclaimed water to 52 municipalities in the Autonomous Community of Madrid. Construction and operation of the dual distribution system.	Citizens and municipalities have become aware of the advantages of using recycled water Awareness and commitment of the regional authorities led to the decisions for the necessary investments. Municipalities had not borne any cost for the construction of infrastructures. Cost efficiency and environmental benefits.
Honolulu, Hawaii	2000 Treatment capacity: 45,400 m^3/d		Total annual volume of treated wastewater: 12.6 Mm^3/yr Total volume of recycled water: 11 Mm^3/yr Capital and operation costs: approximately \$3.3 million/year Water reuse standards: *<2.2 Fecal coliform/100mL*	Conservation of potable resources. Energy efficiency. Examining newer technologies to increase effluent efficiency.	Use of recycled to lower demand and pressure on water resources. Lower operating costs, beneficial to client and customers. Increase overall flow efficiency of facility to deliver more reuse water.	Meeting demand as growth of reuse water increases. Controlling operational costs with unpredictable energy prices. Capital costs and footprint required for new equipment.	Customized local solution that is demand-based to match customer needs. Connected technology and design with local industrial, commercial, residential, environmental and recreational needs. Minimizes impacts of periodic droughts. Superior quality ultra-pure process water for industry.

Bora Bora, French Polynesia	2005 300 m^3/d, 2008 500 m^3/d	Landscape irrigation of hotels, stadium, sport fields, nursery, Boat washing, Filling fire protection boats, Filling waterfalls and ponds, Cleaning and industrial uses	Volume of recycled water consumption: ~70,000 m^3/yr O&M costs: 0.68 €/m^3 (38% of the annualised life cost) Recycled water charge: 0.67 to 2.51 €/m^3 plus fixed annual charge depending on volumes Water quality standards: 0 *E.coli*/100 mL	Water stress and increasing water demand due to tourism development. Municipal policy of sustainable development. Pollution of the lagoon with risk for the fragile ecosystem.	Saving of high quality freshwater water for potable water supply (10% of the demand). Safeguard of the lagoon and its biodiversity. Cost saving for large end users. Prevention of revenue loss of building and tourist companies during droughts. Fast and easier implementation than new freshwater supply.	Ensure high reliability of operation of membrane filtration and efficient preventive maintenance. Manage reliable operation through flood conditions and recycled water supply without interruptions. Define the most adequate pricing strategy of water services. Overcome an initial poor perception of recycled water by luxury hotels.	Adequate choice of the treatment technology and the ability to provide high-quality recycled water without any interruption. Strong political engagement and commitment of the elected officers. Implementation of an adequate public communication and education program. Good collaboration with the local health authorities to overcome negative risk perception of urban spray irrigation. Public-private partnership. Economic viability with ability to cover O&M costs of tertiary treatment and distribution.
Australia	Since 2001 Residential reuse in 250,000 houses by 2030	Flushing toilets, Watering gardens, Washing cars, Clothes washing, Industrial use	Projected future volume of residential recycled water use: about 37 Mm3/yr by 2030. Recycled water tariffs: typically 80% of drinking water price. Water reuse standards: fit for purpose based on risk assessment: *<1 E.coli/100 mL and virus-free*	Limits on new drinking water sources. Higher standards for discharges to the environment.	Annual drinking water use reduced by 40% or more. Peak demands on drinking water systems reduced by 60% or more.	Recycled water quality monitoring. Quality controls to prevent cross-connections. Community education	Limits on new drinking water sources. Development of clear guidelines for urban and residential use. Government support and Ministerial endorsement. Support from health authorities based on quantitative health risk assessment. Adaption of U.S. experience to local conditions.

objective to use the recycled water produced by 10 existing plus 20 new reclamation facilities by means of a large dual distribution networks for urban and industrial uses.

Chapter 9. In Hawaii, an island surrounded by ocean but located in a drought-affected area, the Honolulu facility utilizes sophisticated technology, distinct process chains and cooperation between government and the private sector to produce adequate water products that meet the growing needs of industry, commerce and the public. The major specificity of this urban water recycling scheme is the distribution of two grades of recycled water, one for irrigation of landscaping, and a second higher grade for industrial processes.

Chapter 10 presents an outstanding example of a successful water reuse project in conditions of very restrictive water reuse regulations that had been overcome thanks to the strong political engagement of elected officers, trust of large end-users and support by local population. On the island of Bora Bora, a coral atoll in French Polynesia, fresh water supplies are distinctly finite. An urban reuse scheme has been developed to supply recycled water for landscape irrigation and other urban non potable uses with on-going projects for aquifer recharge. The reliable supply of high quality recycled water without interruptions and a suitable pricing of water services ensured the technical and economic viability of the project enabling to compensate for the water stress and to protect the lagoon, the most precious heritage of the Island.

Chapter 11. There has been widespread development of urban and residential reuse schemes in Australia. Recycled water is supplied for irrigation of household lawns and gardens, and connected indoors for toilet flushing. In some schemes, where recycled water quality is suitable, a recycled water tap is installed in the laundry to supply water for household clothes washing.

Keys to success

The common themes running through these case studies are that urban reuse has offset pressures from growing demands and water shortages due to drought and climate change by producing large savings in drinking water needs. There have been worthwhile economic and environmental benefits. Keys to success have included strong government and regulatory support, stakeholder involvement, development of community support through effective public education programs, effective treatment, careful monitoring and quality control, and recycled water priced lower than drinking water.

The major highlights and lessons learned from the selected case studies are summarised in Table 1.

7 Semi-centralised urban water management as prerequisite for water reuse

Marius Mohr and Walter Trösch

CHAPTER HIGHLIGHTS

This chapter presents a demonstration study of the concept of semi-centralised urban infrastructure systems. It has been shown that the treated wastewater could be used for agricultural reuse and that a reuse of the anaerobically treated wastewater would be the most sustainable solution.

KEYS TO SUCCESS

- Process: Anaerobic Membrane Bioreactor
- Filtration unit: Rotating disk filter
- Vacuum sewer system reduces amount of wastewater
- Rainwater is collected separately
- Early involvement of stakeholders

KEY FIGURES

- Development area with 105 plots
- Temperature in bioreactor: 12.8–26.5°C
- Average COD in effluent: 135 mg/L, 85% elimination
- Volatile suspended solids in bioreactor: 20–25 g/L
- Sustainable flux: 13 – 14 L/m^2.h

7.1 INTRODUCTION

Worldwide, urban growth and increasing consumption have tremendous effects on the infrastructure of supply, treatment, and disposal of water, wastewater, and solid waste. To ensure the sustainable development of urban areas, a relatively new and very relevant and economically viable concept of decentralised or semi-centralised systems is gaining increasing interest. The main principle is to treat wastewater as a resource and to recycle as much as possible the purified water and the other resources it contains. The semi-centralized approach, where depending on the characteristics of the settlements between 1000 and 10,000 inhabitants are treated as a unit, reduces the transport distances of the water and the recovered substances and therefore promotes closed cycles on a regional basis.

The concept of semi-centralised urban infrastructure systems was evaluated by the demonstration project DEUS 21 in Knittlingen, Germany. The major objective is to present and optimize processes developed in lab scale in a technical scale under realistic conditions. The next step is the implementation of this concept under conditions where it operates economically and more sustainably than other concepts. The demonstration study was implemented not too far away from the wastewater treatment facilities of Fraunhofer IGB in Stuttgart, as regular visits on the plant for sampling and maintenance were necessary. As the demand for water reuse in this region is not very high, the treated wastewater has not been reused in the context of this project. However, it has been shown that the treated wastewater could be used for agricultural reuse, though, and that a reuse of the anaerobically treated wastewater would be the most sustainable solution.

Main drivers for water reuse

Water scarcity is a problem in many regions of the earth. Especially densely populated areas with arid climate suffer from lack of water. Most water is used in agriculture: in 2000, 70% of the withdrawn water has been used for agricultural irrigation, in some regions even more than 80% (UNESCO, 2009). In future, the demand for water in agriculture is expected to grow even more, due to:

- Population growth worldwide.
- Changing diet: in newly industrializing countries, the consumption of meat and dairy products is increasing.
- Production of energy crops, which might lead to a doubling of water demand in agriculture until 2050 (Beringer *et al.* 2011).

- Climate change, leading to less regular rainfalls.

The demand for agricultural products is highest close to urban settlements, where at the same time the highest amount of wastewater accumulates. To minimize distances for transport of water and crops, a regional economy with small-scale agriculture in the surrounding of cities is a promising alternative. Semi-centralised urban water management concepts with treatment processes that allow the reuse of water and nutrients in agriculture would complement this approach.

Brief history of the project development

In 2004, construction works for this demonstration study started in Knittlingen, "Am Römerweg". An area with 105 plots was developed based on the concept DEUS 21 in the context of a research project. In 2005, the service building for the treatment facilities, called "Wasserhaus", was constructed inside the living area (Figure 7.1). In December 2005, the first inhabitants moved in, and the vacuum system for wastewater collection started to operate. The first pilot plants started running in 2006 inside the "Wasserhaus". The houses were not constructed all at the same time, so the population in the development area increased little by little. In 2010, about half of the plots were occupied. To be able to run it with reasonable loads, the demonstration plant for the treatment of wastewater started operation in early 2009, the bioreactor being designed for about 175 inhabitants (half of the estimated final population).

Figure 7.1 Photographs of the "Wasserhaus" in Knittlingen and the vacuum station in the basement.

Project objectives and incentives

The main objective of the project was to demonstrate and optimize the processes needed for the concept of semi-centralised urban management DEUS 21 in technical scale. The research part of the project was funded by the German Federal Ministry for Education and Research BMBF and by German industry interested in utilizing the results of the project. The town of Knittlingen financed the development of the area, which was less expensive than the conventional development would have been due to the small diameters of the vacuum sewer pipes.

The concept DEUS 21 is illustrated in Figure 7.2. Wastewater is collected separately from rainwater with the vacuum system. Vacuum toilets and kitchen waste macerators can be utilized in the households, leading to a higher concentrated wastewater (less water consumed and higher organic load). Due to extra costs, only about 20% of the inhabitants chose to install vacuum toilets and about 25% are using macerators in their kitchens. For wastewater treatment, an anaerobic process has been chosen, because it transforms the organic load into biogas, and the effluent still contains the nutrients, so they can be utilized as well. The rainwater is collected by gravity. In Knittlingen, a separation between rain off from roofs and from streets was not realized. The rainwater is stored in underground cisterns and treated with the objective to use it in the households for flushing toilets, taking showers, dish washing, and washing machines. The vacuum station as well as the treatment plants for wastewater and rainwater is located inside the "Wasserhaus". The treated wastewater can be used for agricultural irrigation. The addition of collected rainwater to recycled wastewater is possible, too, and reduces the effort for its treatment compared to utilization in households due to lower quality standards to be met.

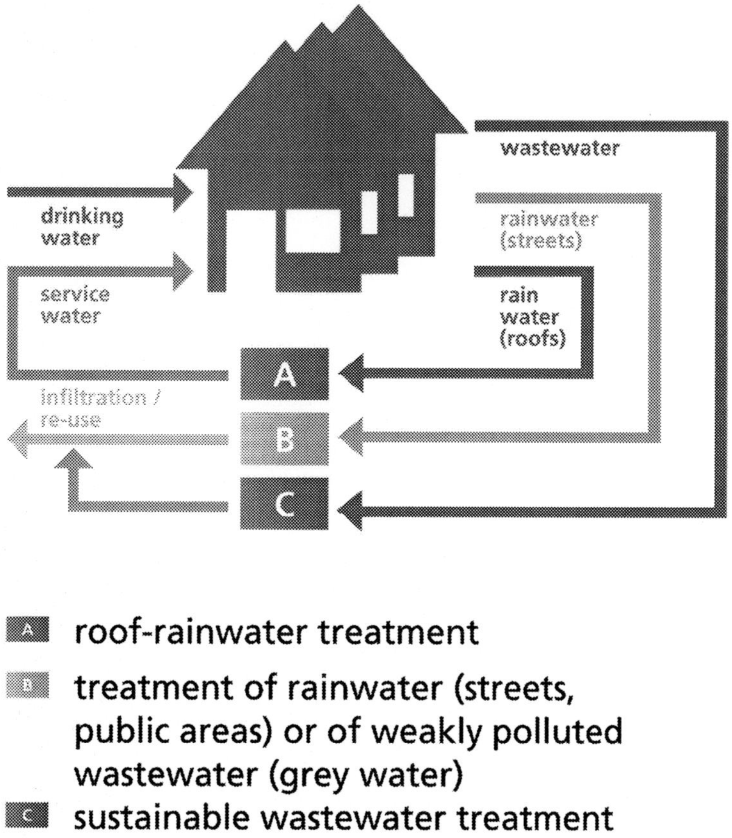

Figure 7.2 The concept DEUS 21.

7.2 TECHNICAL CHALLENGES

The main challenge of the wastewater treatment process was the combination of anaerobic biological treatment and membrane filtration (anaerobic membrane bioreactor process (AnMBR), especially at relatively low temperatures. Typically, large scale anaerobic wastewater treatment plants are designed for influent temperatures above 18°C (Lettinga *et al.* 2001). Although currently no large scale plants for treatment of sewage according to the anaerobic membrane bioreactor process could be identified, this process has been characterized as promising by scientific reviews (Sutton *et al.* 2004; Liao *et al.* 2006).

Treatment train for water recycling

The sewage is collected in the vacuum station (Figure 7.3). From there, it is pumped into the equalization tank. This tank is designed for the equalization of the quantity of sewage over the day, so that the effluent of the plant can be kept steady. This is as important as the separation of the rainwater, because the membrane area, is designed according to the maximum hydraulic load.

The next stage is a sedimentation tank (Figure 7.4a), where the solids are settled for separate treatment. The solids, which are up to 1–2% of the influent quantity, are treated in a high-load digester tempered to 37°C with a retention time of five to ten days. To make the treatment more efficient, water is removed from the digestion process by a rotating disk filter.

The water with non-settleable solids flows from the sedimentation tank into a non-tempered, fully mixed anaerobic bioreactor. The mixed sludge is circulated over an external rotating disk filter (Figure 7.4b), which is continuously withdrawing the effluent from the bioreactor.

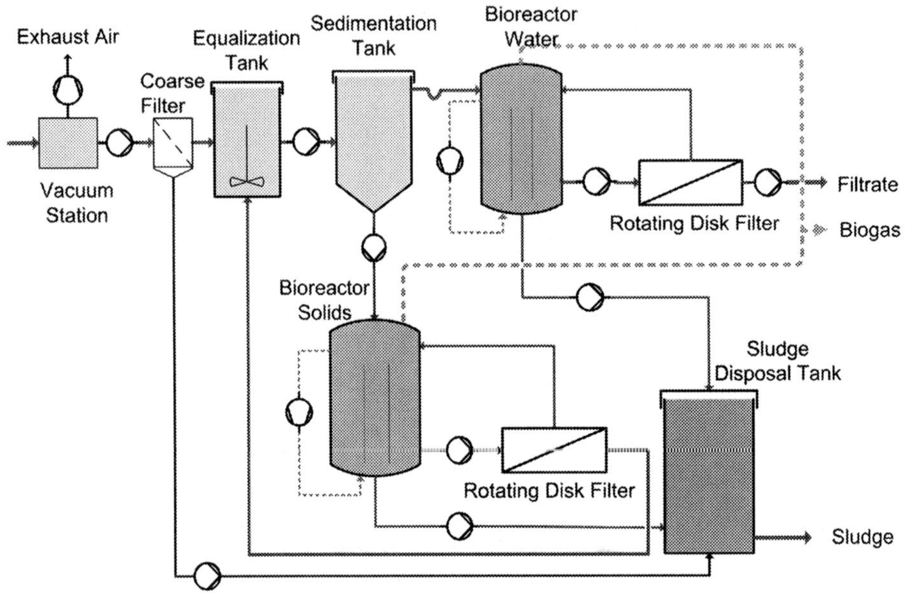

Figure 7.3 Scheme of the anaerobic MBR process in Knittlingen.

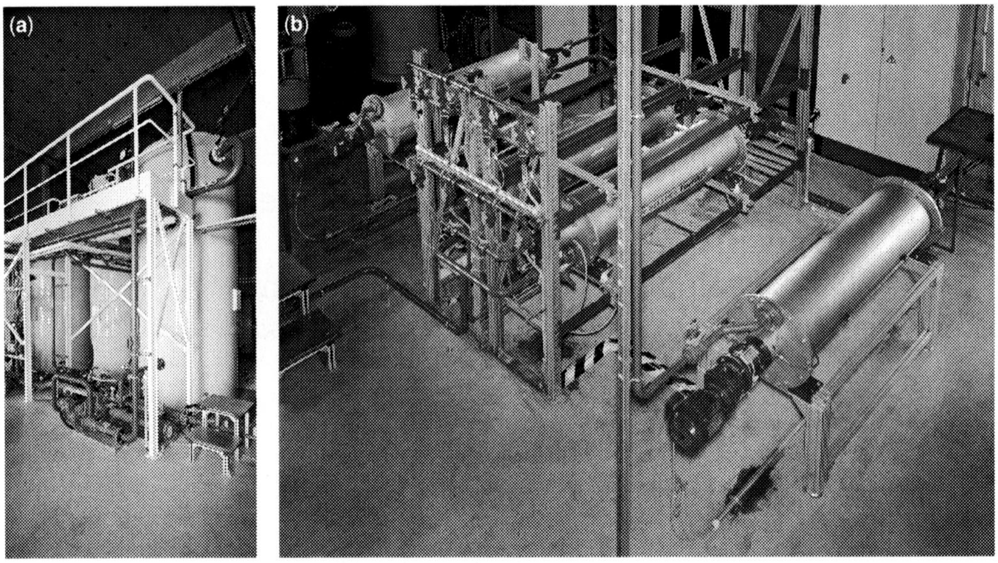

Figure 7.4 View of the wastewater treatment: left photographs, from left to right: equalization tank, sedimentation tank, bioreactor (water) and rotating disk filters at the right photographs.

The rotating disk filters are equipped with ceramic microfiltration membranes with a nominal pore diameter of 0.2 μm. Fouling control is achieved by rotating the shaft, on which the filtration disks are fixed, creating a centrifugal force which causes the solids to flow off.

The treatment process splits up the wastewater into the solids-free filtrate, the biogas from the two anaerobic bioreactors, and the stabilized solids after anaerobic digestion.

The bioreactors were initially inoculated with anaerobic sludge from a sewage sludge digestion plant. Since for anaerobic treatment of wastewater a different composition of the anaerobic biocenosis is needed than for the digestion of sludge, the bioreactor for the wastewater treatment had to be started up slowly. In November 2009, pellet sludge from a plant for the treatment of wastewater from the production of fruit juice was added, leading to a more stable operation immediately due to the higher concentration of methanogenic microorganisms.

Water quality control and monitoring

Samples of the influent of the bioreactor and the effluent of the filters were taken about twice a week during the plant operation. The measured concentrations of COD (chemical oxygen demand) and nutrients are presented in Table 7.1 for the period from November 2009 until July 2010. The temperature in the bioreactor ranged from 12.8 to 26.5°C, due to seasonal changes. High COD values above 155 mg/L occurred only in winter when the temperature in the bioreactor was below 14°C. This can be explained by a reduced activity of the methanogens at low temperatures. The nutrients nitrogen and phosphorous are not eliminated by the anaerobic treatment process and leave the system dissolved in the effluent, mainly as ammonium and phosphate.

Samples for the analysis of microbiological parameters were taken on three different days during operation of the plant, in the effluent of each of the four filter modules. Average values for total coliforms and *E. coli* are presented in Table 7.1. It can be seen that most bacteria are retained by the membrane, although re-contamination of the effluent is an issue, due to the relatively high concentrations of nutrients and COD. Experiments showed that the real cut-off is far below the nominal pore size of 0.2 μm, which makes it likely that most viruses, which tend to attach to solids anyway, are retained by the membrane as well.

Table 7.1 Water quality of influent and effluent of the demonstration plant between November 2009 and July 2010.

Parameter	Influent*	Effluent*
COD, mg/L	914 (425–1215)	135 (63.8–280)
N_{tot}, mg/L	109 (77.5–151)	110 (76.0–135)
NH_4-N, mg/L	88.0 (55.8–127)	94.4 (70.3–116)
PO_4-P, mg/L	13.1 (9.87–17.0)	12.6 (8.62–13.8)
Total Coliforms, CFU/mL	$1.5 \cdot 10^7$	5.8 (0–17)
E. Coli, CFU/mL	$2.9 \cdot 10^6$	0

*Average value and limit of variations.

In the period from January 2010 until July 2010, the average influent flow of the bioreactor for wastewater treatment was 5.7 m^3/d, leading to an average hydraulic retention time of 43 hours (minimum 26 hours) and an average organic loading rate of 0.51 g COD/L·d (maximum 0.84 g COD/L.d). The average elimination rate of COD was 85%. The eliminated COD was transformed into biogas, most of it leaving the reactor over the gas meter, while about 10% of the produced biogas could be found dissolved in the effluent. By air stripping, it could be removed from the effluent and added to incineration.

Main challenges for operation

After securing the effluent quality, the achievable flux of the membrane filtration is critical for the sustainability of the process. The sustainable flux according to Bacchin *et al.* (2006) depends mainly on the characteristics of the sludge. The higher the solids concentration of the sludge, the lower is the sustainable flux. For the biological elimination of the COD, on the other hand, a certain concentration of microorganisms is required. Temperature of the sludge is important as well. With decreasing temperature, the activity of the microorganisms decreases, requiring a higher concentration of microorganisms. At the same time, the dynamic viscosity of the sludge increases with decreasing temperatures, which has a negative impact on the attainable fluxes.

Measures to increase the attainable flux are the rotation of the filter disks, regular back-flushing with filtrate and chemical maintenance cleaning. It could be shown that with higher rotation velocities the sustainable flux increases, at least up to rotation velocities between 300 and 350 rpm (Mohr, 2011). Back-flushing with filtrate should be automatically carried out every few hours for some minutes, while the interval for chemical maintenance cleaning of the rotating disk filter with ceramic membranes is about a year (Zech *et al.* 2011).

The demonstration plant in Knittlingen was operating with a VSS (volatile suspended solids) of 20 to 25 g/L. With a rotation velocity of 300 rpm and a temperature between 20 and 25°C, sustainable fluxes of 13 to 14 $L/m^2 \cdot h$ have been determined. At temperatures between 10 and 15°C, the sustainable flux is likely to be around 12 $L/m^2 \cdot h$. There are many options to optimize the sustainable flux, though. If, for example, the ratio of methanogens – being responsible for the limiting step in the degradation chain – in the sludge can be increased, the VSS of the sludge can be decreased, leading to a higher sustainable flux. Other options are optimizing the automatic back-flushing and increasing the rotation velocity even more.

A major challenge in the operation of the plant in Knittlingen was the start-up, as there was no adequate inoculum available. With the sludge that has grown in the bioreactor during operation, a new plant could be started up in less than a month. Even over a period of several months without operation, the sludge does not lose its activity, as reported by Lettinga *et al.* (2001).

If the nutrients cannot be recycled together with the purified water for irrigation, they must be eliminated or, better, regained out of the effluent. Depending on the climate conditions, outside the growing periods or during rainy seasons, there might be months during which the water cannot be reused in agriculture. Phosphate can be eliminated by precipitation with iron or aluminum salts. This leads to compounds where phosphorous is bound in a way that a further utilization is not economical any more. If phosphate can be precipitated as magnesium-ammonium-phosphate (MAP, struvite) from the effluent, it can be utilized as a fertilizer.

Even if most of the phosphate is precipitated as MAP, only a fraction of the ammonium of the effluent is eliminated this way (see Table 7.1). For the elimination of nitrogen by a nitrification/denitrification process, as it is usually applied in aerobic sewage treatment, there is not enough carbon left in the effluent of the plant. A solution could be a by-pass of a part of the influent. This way, the nitrogen can be eliminated, but will not be recycled. Recycling of the nitrogen as ammonium sulphate is possible by applying an process with zeolite as ion exchanger and ammonia stripping (Aiyuk *et al.* 2004), which has been successfully tested with the effluent of the plant in Knittlingen.

7.3 POTENTIAL WATER REUSE APPLICATIONS

The reuse of the AnMBR effluent in agriculture enables to valorise both water and nutrients. It also saves the effort of operating an additional process for the removal of the nutrients. As the effluent passes a microfiltration membrane, it is free of helminth eggs and parasites. Even though the effluent could contain bacteria due to regrowth, bacteria originating from the wastewater are retained by the membrane as well, preventing the spreading of diseases through the wastewater. If operating with a microfiltration membrane, it is not sure that all viruses are retained, as they are smaller than the pore diameter. Anyway, the secondary layer on the membrane and the fact that viruses tend to adsorb to particles make it probable that most viruses are held back as well. To increase the chance to keep the effluent free of viruses, ultrafiltration membranes with smaller pore diameters can be used instead.

Compared to the effluents of modern aerobic wastewater treatment plants, the COD of the AnMBR process is relatively high. This can cause increased microbial re-growth in distribution networks. If the effluent is reused, blocking of irrigation pipes by biofilms has to be prevented by regular flushing. When the effluent has to be stored for more than few days, odour has to be controlled or prevented by aeration. If the demand for irrigation should be covered only by the effluent of the depicted process, the risk of over-fertilization would be very high due to the high concentrations of nitrogen. As the amount of the effluent that can be provided to an area is limited by the maximum nitrogen demand of the plants, the demands for water and phosphorous could not be covered. This means that water and phosphorous have to be added to plants irrigated with an AnMBR effluent.

Another potential disadvantage of irrigation with wastewater, independently of the treatment process, is the high concentration of dissolved salts that can cause damages to sensitive plants, the soil, and the groundwater. The amount of salts in the water can be reduced by supplying a relatively soft drinking water, but a certain amount of salts will be added to the water by usage anyway. To prevent evaporation, which would increase the risk of salinization even more, subsurface irrigation systems should be used. The addition of rainwater collected separately from the effluent of the wastewater treatment plant from the roofs of the houses would decrease the concentration of salts and of nitrogen, decreasing the risks of salinization and over-fertilization.

Risks by heavy metals and organic trace contaminants cannot be excluded when reusing wastewater, although substances adsorbing to particles are retained by the membrane in the depicted process. In the semi-centralised approach, the addition of industrial wastewaters carrying higher concentrations of these substances can be prevented more easily and inhabitants can be informed about the importance of not adding certain substances to the sewage.

7.4 ECONOMICS OF WATER REUSE

Project funding and costs

The costs for investment and operation of the research plant cannot be compared to those of plants operating under economic conditions, because the objectives are totally different. Still, factors influencing the costs of the studied process can be identified. As for any MBR-process, the specific amount of water to be treated should be kept small, because costs associated with the membranes rise linearly with the amount of water. This means, intelligent water management concepts should keep the rainwater away from the wastewater and help minimize the per capita water consumption.

Concerning the investment costs, a single semi-centralised plant will certainly be specifically more expensive than a centralised plant due to scale effects. The effect of scale can be realized for semi-centralised plants if a large number of them are constructed in a region. Currently, the rotating disk filters used in the research plant are still prototypes. Further development is going on, with the objective of having an economically feasible filter ready in 2012.

The optimization of the wastewater treatment process can also have a share in cost reduction. Increasing the sustainable flux to values around $20\ L/m^2 \cdot h$ seems realistic. This has again an effect on the specific membrane area, influencing both capital and operation costs.

The geographic location also influences the economics of the DEUS 21 concept. If there is a demand for heat, the energy that can be won from the biogas can be utilized to a large extent, resulting in savings of other energy sources. The digestion of the solids in the separate reactor achieved degradation rates of 60 to 70% of the VSS. Assuming that all households add their kitchen wastes to the wastewater, a biogas production of at least 80 liter per inhabitant and day (80 L/cap·d) can be expected from the digestion of the liquid and the solid organics together. This amount of biogas has an energy potential of about 200 kWh per inhabitant and year. The same accounts to the utilization of the water and the nutrients phosphorus and nitrogen. If there is a demand close to the treatment plant, these savings can be used to decrease the operating costs of the process. Moreover, a process to eliminate the nutrients becomes unnecessary, if a utilization throughout the year (e.g. with greenhouses) or the storage of the effluents is possible.

Because of the low growth rates of the anaerobic microorganisms, the amount of excess sludge in the AnMBR process is only 10 to 20% of those of secondary sludge produced by the aerobic activated sludge process (Mohr, 2011). This saves costs for the treatment, transport and disposal of sludge. Labour costs are low in the DEUS 21 concept, because the wastewater treatment plant is fully automated and can be remote controlled from any computer that has connection to the internet. Therefore, few experts are able to control a large number of automated, semi-centralised treatment plants, while once or twice a week a trained person has to physically visit the plants to check and if necessary maintain the aggregates, like pumps and filters.

Pricing strategy of recycled water and benefits of water recycling

As water is a basic need, it is in most cases directly or indirectly subsidized. For recycled water that can be used in agriculture, the price should be adapted to the possibilities of the users. For the effluent of the depicted treatment process, the price includes not only the water but also the nutrients dissolved in the water that can be used to substitute fertilizers. The costs of the transport of the water and the treatment process can be covered by fees for wastewater purification paid by the original users of the water, the revenues from biogas utilization, and the price paid by the users of the treated water. Depending on the economic situation, additional subsidies might be necessary to cover the costs of the treatment process. For a sustainable operation, it would be valuable if at least the costs for operation could be covered by the users, making the operation independent from decisions of external sponsors.

Irrespective of these considerations, the benefits of such a system for the overall economy are immense: Most important, the wastewater is treated and pathogenic microorganisms that can be spread by the wastewater are retained, being a crucial prerequisite for health care. Additionally, this concept helps to build up local small scale economies by utilizing the water for growing fruit and vegetables in the vicinity of settlements. Inhabitants can benefit from a healthier diet and a better supply with vitamins as well as by employment and an extra income by selling the products on local markets. This semi-centralised approach can lead to a system of almost closed cycles of water, nutrients, and energy.

7.5 HUMAN DIMENSION OF WATER REUSE

Essential precondition for the successful realization of water reuse according to the concept DEUS 21 is the early involvement of the affected stakeholders: inhabitants, regional and local authorities, infrastructure planners, construction enterprises, land owners, and future users of the treated water. The early determination of tariffs and prices is important to ensure acceptance by the users and to prevent annoyance. In Knittlingen, several meetings with the town's administration, the town council, the inhabitants, and the involved planners and constructors have been organized, before and during the implementation of the project. During the project, the acceptance of the new infrastructure system has been checked by interviewing the inhabitants of the development area, by visiting them, and by questionnaires. The public attitudes were predominantly positive.

Especially for the implementation of a water reuse scheme, the communication between the users of the recycled water, that is, the farmers, the authorities, and the operators of the wastewater treatment plant is important to ensure a safe utilization of the recycled water. Potential problems in operation have to be communicated to the end users, if the quality of the effluent could be affected. It is also important to keep in mind that a sound irrigation and fertilization management is necessary to guarantee that the water is used in the right way. This helps prevent problems with over-fertilization and salinization. The initial users of the water, the inhabitants, have to be informed about substances that must not be added to the sewage like leftovers of pharmaceuticals and of cleaning agents, and other chemicals used in households. The awareness to add as little chemicals as possible to the sewage will be higher when the inhabitants are involved in the utilization of their treated sewage and profit from their correct behaviour.

If a concept like DEUS 21 is realized in a certain region for the first time, a research institution located in the region with specialists in social and agricultural sciences should be involved to adapt the concept to the local culture and biogeographic conditions like climate, soil characteristics, and local plant species.

7.6 CONCLUSIONS

In the research project DEUS 21 in Knittlingen, a semi-centralised water management concept has been demonstrated. The effluent of the wastewater treatment plant, operated according to the AnMBR process, is characterized by a relatively low concentration of organic substances (COD), no particles, almost no pathogenic microorganisms, and a relatively high concentration of the nutrients ammonium and phosphate. This quality, which has not been achieved before with an anaerobic process at low temperatures and a comparable influent quality, makes the recycling of the water and the nutrients in agriculture a feasible option. This could lead to a change in wastewater management and an increase in reuse applications.

As the concentrations of ammonium are relatively high, ammonium-free water, for example, rain water, and phosphorous have to be added and the irrigation and fertilization has to be realized according to a management scheme, taking into account the different growth stages of the plants.

The following aspects are relevant for a sustainable implementation of the concept:

- The specific water consumption should be low and rainwater should be collected separately, as the necessary membrane area correlates with the amount of water to be treated and has a significant impact on investment costs, operation costs and energy demand.
- High temperatures lead to higher growth rates of the anaerobic microorganisms and therefore allow the operation with lower solid concentrations, and increase the filterability of the sludge. This results in higher fluxes and a reduction of the necessary membrane area as well.
- A demand for the produced biogas, for example, for cooking, or the waste heat of electricity production close to the treatment plant saves fossil energy resources. The utilization of the effluent in agriculture saves water and nutrients and substitutes a process to eliminate the nutrients, which would be necessary for a discharge into receiving waters.
- A further optimization of the treatment process should aim at generating sludge with a higher ratio of methanogens, allowing a lower concentration of solids. The operation of the filtration should be optimized to achieve a higher sustainable flux. Both measures would reduce the necessary membrane area. The rotating disk filter itself is currently refined, reducing its investment and operational costs as well as the energy demand.

All in all, with minor optimization of the treatment process, the concept DEUS 21 is suited best for the implementation in regions with warm climate and a demand for irrigation water, in which no wastewater treatment exists yet. This semi-centralised approach with water reuse can help to build up sustainable water infrastructures and local markets for fruit and vegetables and decrease the dependency of regions on the world markets.

REFERENCES AND FURTHER READING

Aiyuk S., Xu H., Van Haandel A. and Verstraate W. (2004). Removal of ammonium nitrogen from pretreated domestic sewage using a natural ion exchanger. *Environmental Technology*, **25**, 1321–1330.

Bacchin P., Aimar P. and Field R. W. (2006). Critical and sustainable fluxes: theory, experiments and applications. *Journal of Membrane Science*, **281**, 42–69.

Beringer T., Lucht W. and Schaphoff S. (2011). Bioenergy production potential of global biomass plantations under environmental and agricultural constraints. *GCB Bioenergy*, **3**(4), 299–312.

Lettinga G., Rebac S. and Zeeman G. (2001). Challenge of psychrophilic anaerobic wastewater treatment. *Trends in Biotechnology*, **19**(9), 363–370.

Liao B. Q., Kraemer J. T. and Bagley D. M. (2006). Anaerobic membrane bioreactors: applications and research directions. *Critical Reviews in Environmental Science and Technology*, **36**, 489–530.

Mohr M. (2011). Betrieb eines anaeroben Membranbioreaktors vor dem Hintergrund der Zielstellung des vollständigen Recyclings kommunalen Abwassers und seiner Inhaltsstoffe (Operation of an Anaerobic Membrane Bioreactor in the Context of the Objective of Complete Recycling of Municipal Wastewater and its Contents). Ph.D. thesis. In: Berichte aus Forschung und Entwicklung, Vol. **40**, Fraunhofer Verlag, Stuttgart.

Sutton P. M., Bérubé P. and Hall E. R. (2004). *Membrane Bioreactors for Anaerobic Treatment of Wastewaters.* Water Environmental Research Foundation, University of British Columbia, Canada.

UNESCO (2009). *The 3rd United Nations World Water Development Report: Water in a Changing World (WWDR-3).* United Nations Educational, Scientific and Cultural Organization, UNESCO Publishing, Paris, France. http://www.unesco.org/new/en/natural-sciences/environment/water/wwap/wwdr/wwdr3-2009/downloads-wwdr3/.

Zech T., Mohr M., Sternad W. and Trösch W. (2011). Comparison between aerobic and anaerobic wastewater treatment with a ceramic membrane filtration system. In: Proceedings to 6th IWA Specialist Conference on Membrane Technology for Water and Wastewater Treatment, Aachen, Germany.

8 The exciting challenge of water reuse in Madrid

Andrés Deza and Avelino Martínez

CHAPTER HIGHLIGHTS

The population of the Autonomous Community of Madrid has increased by 30% over the last two decades during which time the frequent periods of drought have become increasingly severe and long-lasting. These circumstances have given rise to the need for an alternative source of water, Recycled Water, in order to cope with the supply to customers, whether they are domestic, industrial or municipal. Political authorities have become aware of this need and have adopted the appropriate decisions to invest in the new infrastructures necessary, in order to turn the Reuse Water Plan for the region Madrid into a reality.

KEYS TO SUCCESS

- Citizens and municipalities have become aware of the advantages of using recycled water
- Awareness and commitment of the regional authorities led to the decisions for the necessary investments being taken
- Municipalities had not borne any cost for the construction of infrastructures
- Recycled water prices are cheaper than drinking water and reduce water bills
- Environmental benefits such as greater water resources in the region's reservoirs

KEY FIGURES

- Total Tertiary Treatment capacity: 267,000 m^3/d
- Total annual volume of treated wastewater: 563 million m^3/yr
- Total volume of recycled water 2010: 6.817 million m^3/yr
- Investments in Madrid Dpura Plan: €200 million
- Water reuse standards (urban uses):
  - *E.coli* 200 cfu/100 mL
  - Intestinal nematodes 1 egg/10 L
  - Turbidity 10 NTU
  - Suspended solids 20 mg/L

8.1 INTRODUCTION

The Autonomous Community of Madrid is a region made up of 179 municipalities, including the capital city of Spain, with a surface area of 8000 km^2 and a population of 6.45 million inhabitants. Canal de Isabel II is the publicly-owned company in charge of the region's comprehensive water cycle management. Its origins, which date back to 1851, are inextricably linked to the region's history. Canal de Isabel II is responsible for 14 reservoirs, 81 groundwater abstraction wells, 22 large water holding tanks and 271 smaller water distribution tanks, 18 pumping stations, 13 drinking water treatment plants, plus 16,000 km of water mains and distribution networks, 150 wastewater treatment plants and 6800 km of sewerage networks. It is also the largest producer of renewable energy in the Autonomous Community of Madrid.

Over the last two decades, the population of the Autonomous Community of Madrid has seen a 30% increase, linked to a growing economic development together with an increase in living standards. Changes in habits and new uses of water (recreational, sporting, industrial and domestic) have pushed up water consumption as a result. In addition, climate change, with severe droughts periods which occur increasingly often, have led to the need for taking important decisions to efficiently preserve and protect this essential and scarce resource.

Main drivers for water reuse

In recent decades, the Autonomous Community of Madrid has experienced considerable demographic growth, with its population increasing from 4.97 million in 1991 to 6.5 million in 2010, representing 30% growth in twenty years. Furthermore, current forecasts indicate that the population is expected to climb to between 8.6 and 9.2 million people by 2025. This demographic growth is one of the most relevant causes of the registered increase in water consumption.

In addition to the demographic factor, the last four decades have seen the climatology of the river basin in which the Madrid region is situated affected by a dryer climatic cycle, which, according to data from the Water Authorities, resulted in a 30% reduction in water inflows when compared to the immediately preceding wet cycle. The effects of this climatological phenomenon will foresee ably be accentuated by those of climate change, which could have a serious impact on the particularly vulnerable fresh water resources, resulting in a range of consequences for the human settlements and ecosystems.

Two main conclusions have been formulated: firstly, essential water resources are scarce and very limited in the region of Madrid; and secondly, any efforts must focus on the optimum development and use of the various available water sources, as well as on managing demand.

The need to guarantee citizens' access to an adequate water supply within the context of the limitations described above has therefore led to the adoption of new water policies, among which the use of alternative water sources, such as the water reuse, is particularly relevant.

Brief history of water reuse development

Aware of this situation, in 2005, the Autonomous Community of Madrid Government presented one of its largest and most ambitious plans for the region's environmental consolidation. The plan was called "Plan Madrid Dpura" (Madrid Wastewater Treatment Plan) and the tasks of drafting, executing and operating the plan were entrusted to Canal de Isabel II as was, even more crucially, its complete financing. Investment in the Madrid Dpura plan amounted to €700 million, of which more than €200 million were allocated exclusively to the regional strategy for water reclamation and reuse.

The "Water Reuse Plan" for urban uses was expounded by the Autonomous Community of Madrid, with an initial development stage that would last until 2014. Actions to date have been designed to improve and extend wastewater collection and treatment facilities to cover all the municipalities in the Madrid region. The plan includes the construction of 20 tertiary treatment plants to supplement the 10 existing ones, in order to make these resources available and incorporate them into the new distribution system as reclaimed water (Figure 8.1). These facilities are currently being developed.

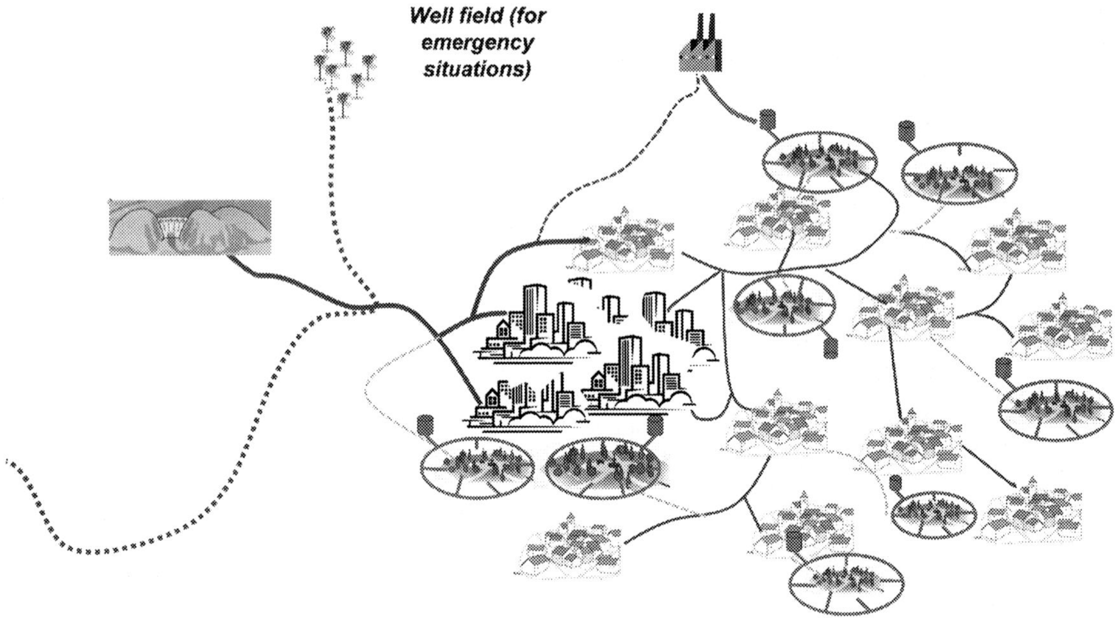

Figure 8.1 Schematics of the water reuse network in Madrid.

The success of this plan would depend on two main pillars: the sizeable investment that was required in order to implement it and the commitment of customers to use recycled water. With this in mind, in 2005, Canal de Isabel II developed master plans covering a total of 51 municipalities of the Autonomous Community with a view to signing agreements with the town councils of the municipalities involved for the execution of the projects that would be drafted.

The plan developed by the public company Canal de Isabel II is the largest in Europe at this time and constitutes one of its most significant investments which, when combined with other measures that have been adopted (wells, water transfers and other alternatives), will make it possible to guarantee the future water supply to the population of the Autonomous Community of Madrid, particularly in times of drought.

These actions are supplemented by a reclaimed water ring over 140 km long which transports the reclaimed water produced in the tertiary treatments of Madrid's main wastewater plants for the irrigation of 295 ha of the city's green areas.

Project objectives, incentives and water reuse applications

With the implementation of this Plan, it is expected that 70 million m^3 (Mm^3) of reclaimed water will be produced each year, of which around 40% will be used in the city of Madrid alone. The remaining 60% will be used to supply 51 municipalities of the Autonomous Community of Madrid, benefiting 2.5 million of the region's inhabitants. The reclaimed water supply will irrigate 6500 ha of public green areas, the equivalent of 9300 football fields, year round, regardless of water reserve levels. It may also be allocated to industrial uses, street flushing and the irrigation of golf courses. All this will permit 9 million m^3 of drinking water to be saved per year. In order to distribute such a significant volume of reclaimed water, the Madrid Dpura Plan also foresees the construction of a water distribution network consisting of 1200 km of pipelines that will convey recycled water to users requiring this service.

8.2 TECHNICAL CHALLENGES OF WATER QUALITY CONTROL

Water quality standards

Water reuse in Spain must meet the regulations of the Royal Decree 1620/2007 issued on 7 December 2007, which establishes the legal requirements for water reuse. The targeted water quality parameters required for the major uses of recycled water in Madrid are given in Table 8.1.

Table 8.1 Water quality requirements for water reuse in Spain according to the Royal Decree 1620/2007.

Use of purified wastewater	Intestinal nematodes	*E.coli*	Suspended solids	Turbidity
Industrial uses				
a) Processing and cleaning waters except in the food industry	–	10,000 cfu/100 mL	35 mg/L	15 NTU
b) Other industrial uses				
c) Processing and cleaning waters for use in the food industry	1 egg/10 L	1000 cfu/100 mL	35 mg/L	–
d) Refrigeration towers and evaporative condensers	1 egg/10 L	Absent	5 mg/L	1 NTU
Recreational uses				
a) Irrigation of golf courses	1 egg/10 L	200 cfu/100 mL	20 mg/L	10 NTU
Urban uses				
a) Irrigation of urban green areas (parks, sports fields...)	1 egg/10 L	200 cfu/100 mL	20 mg/L	10 NTU
b) Street flushing				
c) Fire prevention systems				
d) Industrial vehicle washing				

According to the intended uses and the related quality requirements, Canal de Isabel II has selected the most appropriate tertiary treatments described in the following sections.

Treatment trains for water recycling

The tertiary treatment processes currently in operation in Madrid follow the conventional scheme including physical-chemical treatment and settling or air flotation units, filtration and disinfection. In some specific cases, more advanced technologies such as ultrafiltration and reverse osmosis have been adopted to obtain higher quality recycled water that can be used in urban areas such as for the irrigation of parks, green areas, golf courses or for industrial processes.

For the *physical-chemical treatment*, typical processes include static mixers (although some of them also have mixing and flocculation chambers), followed by lamellar settling tanks (Figure 8.2a) or air flotation units, which enable the removal or reduction of parasitic eggs or cysts, detergents, remaining suspended solids and associated organic matter. The chemicals used in this process are usually inorganic coagulants (iron or aluminium salts) and organic polymer-based flocculants (polyelectrolytes), which are sometimes combined.

Figure 8.2 Views of tertiary pre-treatment processes: from left to right lamella settling (a), ring filters (b) and sand filters (c).

Filtration is the next step after physical-chemical treatment (Figure 8.2a and b), and its purpose is to retain any remaining suspended particles by means of porous media. A variety of filtration systems are implemented such as pressurised sand filters (the most commonly used so far, over 40% of facilities), moving bed continuous washing sand filters (21%), ring filters (13%), open sand filters (13%), cloth filters (4%), micro-sieves (almost 5 %).

The purpose of *disinfection* is to remove pathogenic microorganisms through the addition of chemicals (*hypochlorite, chlorine gas, ozone)* or physical devices *(UV light).* The target pathogenic microorganisms include those of viral origin (poliomyelitis, hepatitis, gastroenteritis), bacterial origin (cholera, typhus, dysentery, tuberculosis) and parasite origin. All the tertiary treatment facilities in Madrid are equipped with UV lamps installed in closed reactors (Figure 8.3a), except for one facility, where the UV system is installed in channels.

Figure 8.3 Views of UV disinfection (a) and ultrafiltration modules (b).

The *advanced treatment* technologies in operation include various processes that are classified according to pore size and the *driving force* responsible for permeate flux (Figure 8.3b): Microfiltration (MF), Ultrafiltration (UF), Reverse Osmosis (RO).

An *ultrafiltration* system was implemented in one of the facilities dedicated to golf course irrigation. A new advanced recycling plant is under construction using ultrafiltration followed by reverse osmosis and disinfection to supply process water to a paper mill, which manufactures recycled paper.

In summary, the conventional tertiary treatment included in the "Water Reuse Plan" of Madrid aims to produce reclaimed water for irrigation of green areas and golf courses, while the advanced treatments are dedicated to producing water for industry.

Water quality control and monitoring

The quality control of reclaimed water is also regulated in Royal Decree 1620/2007 as illustrated by Table 8.2.

Table 8.2 Monitoring frequency of recycled water according to the Royal Decree 1620/2007.

Parameter	Minimum frequency range
Intestinal Nematodes	Weekly-fortnightly
Escherichia coli	2 times/week-weekly
Suspended solids	Daily-weekly
Turbidity	Daily-weekly
Other contaminants[1]	Weekly-monthly

[1] *Legionella*.: 3 times/week, only for industrial use.

Frequency range goes from twice-weekly up to monthly depending on water uses (urban, agricultural, industrial and recreational).

According to Spanish regulations, the reclaimed water quality control programme determines the sampling location, as well as the minimum frequency, analytical methods to be used and limit concentration values for each parameter. Finally, it also defines the guidelines for assessing the reclaimed water quality and the contingency plan in the case of non-compliance.

Sampling is performed at various points throughout the reclaimed water production and distribution system. Essential points are therefore the tertiary treatment effluent, holding tank outlets, automatic sampling stations distributed throughout the network and the points of water use.

In addition to the controls required by the regulations, Canal de Isabel II has a series of Official Sampling Stations (O.S.S.) located throughout the reclaimed water network to provide data regarding water quality so that the appropriate measures can be taken in the event of deviation from the established standard. The parameters measured automatically are total chlorine and turbidity.

In the event of non-compliance, service is suspended until the cause is identified and resolved. During the suspension period, water from the urban drinking water network can be used as an alternative supply to ensure the continuity of service. For this reason, all reclaimed water tanks in the distribution system have an alternative drinking water supply.

In order to avoid cross contamination between the reclaimed water and drinking water supplied to a customer, it is absolutely essential that a cross-connection control is performed before the start-up and periodically during operation of the dual distribution system.

8.3 WATER REUSE APPLICATIONS

The main uses of reclaimed water in the Autonomous Community of Madrid are irrigation of public parks and gardens (Figure 8.4), golf courses, street flushing and industrial uses.

Figure 8.4 Views of landscape irrigation in Madrid.

In public parks it is essential that the existing irrigation networks (previously fed with drinking water) be disconnected from the drinking water network for fountains, before recycled water is fed into them. The development of the Water Reuse Plan has meant that since mid-2011 almost 60 public parks, covering an area of more than 275 ha, can be irrigated using reclaimed water.

Similarly, a number of points throughout the urban areas have been furnished with hydrants in order to allow filling of reclaimed water tankers used to flush town streets.

Particularly notable is the case of the supply of reclaimed water into the manufacturing process in a paper mill, whose production amounts 470,000 tonnes a year of 100% recycled paper, with an annual water consumption of 4.3 Mm^3. The significant volume of drinking water consumed by this paper mill, combined with the increasingly frequent periods of drought threatening the Autonomous Community of Madrid, made it apparent that alternative water sources were needed in order to ensure optimum operation of the plant and guarantee its supply at all times and under any circumstances. In this context, an agreement was signed between Canal de Isabel II and the paper mill to fully replace the drinking water supply with recycled water. The use of reclaimed water in the factory's industrial processes requires water with a low salt content and no risk of contamination from bacteria, protozoa or viruses and of particular importance, no trace of endotoxins. This last aspect is critical in this particular case due to the generation of aerosols in some of the manufacturing processes.

Based on these requirements, the recycling facility for the paper mill was designed with a flow capacity of 12,400 m^3/day, with a physical-chemical treatment, sand and activated carbon filtration, ultrafiltration using hollow fibre membranes, two-stage reverse osmosis and final remineralisation of the treated water. The total investment for this advanced tertiary treatment is €19 million.

Evolution of the volumes of supplied recycled water

Ten water reclamation facilities are currently in operation in Madrid, with the capacity to supply 3.4 million m^3/year (Mm^3/yr) for irrigation in municipal green areas, three golf courses, an entertainment park and the so-called "Culebro axis", which is a municipal complex located in the south-east of the Autonomous Community of Madrid, formed by the municipalities of Pinto, Parla, Humanes, Getafe, Fuenlabrada and Alcorcón.

In addition, Canal de Isabel II plans to supply 51 municipalities, 14 golf courses and one industrial unit by means of 30 water reclamation plants as part of their existing respective waste water treatment plants. The annual production of reclaimed water will amount to 40 Mm^3/yr, which correspond to 24 Mm^3/yr for municipal green areas; 12 Mm^3/yr for golf courses and 4 Mm^3/yr for industrial use. Another 30 Mm^3/yr should be added as production in the facilities located in the city of Madrid, making a total annual volume of 70 Mm^3/yr.

8.4 ECONOMICS OF WATER REUSE

Project funding and costs (capital and operation)

The sourcing of the funds necessary to finance this Plan is included in Canal de Isabel II's Investment Plan. As a publicly-owned company dependent on the Autonomous Community of Madrid, Canal de Isabel II has also established a tariff policy in order to recover all costs deriving from the infrastructure construction work and the operation and maintenance of the implemented water reuse system.

Pricing strategy of recycled water

Pricing strategy of recycled water is established from two basic points:

- To make the users aware of the need to behave responsibly in efficient water consumption
- To draw up a tariff system which is fair and equitable

In order to achieve this, tariffs have been structured into different levels where price is established according to water consumption and volume contracted by user. The closer are the values between water consumption and volume contracted, the lower the price of recycled water will be. As both values diverge the price of recycled water will increase.

The tariff consists of two different services (Table 8.3):

- Production of recycled water service. It includes all essential physical-chemical treatments of wastewater to produce high quality recycled water. The cost of this service is changeable according of water consumption by users
- Recycled water transport service. It includes transport from the treatment plants to users and is a fixed part which takes into account the investment made by Canal in all recycled water facilities

Table 8.3 The pricing of tertiary treated recycled water in Madrid (data for 2011).

Recycled water consumption as percentage of the contracted volume	Tariffs for production of recycled water, €/m^3	Tariffs for recycled water transportation, €/m^3
Less than 25% of the contracted volume	0.2840	0.0541
Between ≥25% to ≤75% of the contracted volume	0.2073	0.0396
More than 75% of the contracted volume	0.1306	0.0249

Benefits of water recycling

The use of reclaimed water offers excellent economic benefits for users as recycled water is cheaper, which translates into significant savings for large water consumers.

Furthermore, the use of recycled water for certain purposes facilitates improvement in energy efficiency, particularly in years of drought. In a dry year, the reclamation of 70 Mm^3 of recycled water can constitute, as an average, a reduction of 41 GWh of the annual energy consumption in the drinking water supply of the region, equivalent to the consumption of 13,700 homes or 41,800 inhabitants.

The use of recycled water has also significant non-monetary benefits as it allows the allocation of drinking water quality for the applications where the highest quality is demanded.

The Reuse Plan, in turn, contributes significantly to the protection and enhancement of the cultural and natural heritage of the Autonomous Community of Madrid as it allows the watering of historic gardens in periods of water restraints due to scarcities and droughts.

8.5 HUMAN DIMENSION OF WATER REUSE

Public education and communication strategy

Public education by means of the distribution of information on the use of recycled water has been an important aspect of the Reuse Plan. For this purpose, informative leaflets have been prepared and distributed to create a general awareness of the initiative. These leaflets include a description of good practices to ensure the appropriate use of recycled water.

Suitable signposting has also been installed in parks where, following a study of every specific park, the appropriate panels have been installed in the most strategic points in order that the people are accurately informed about the use of recycled water (Figure 8.5).

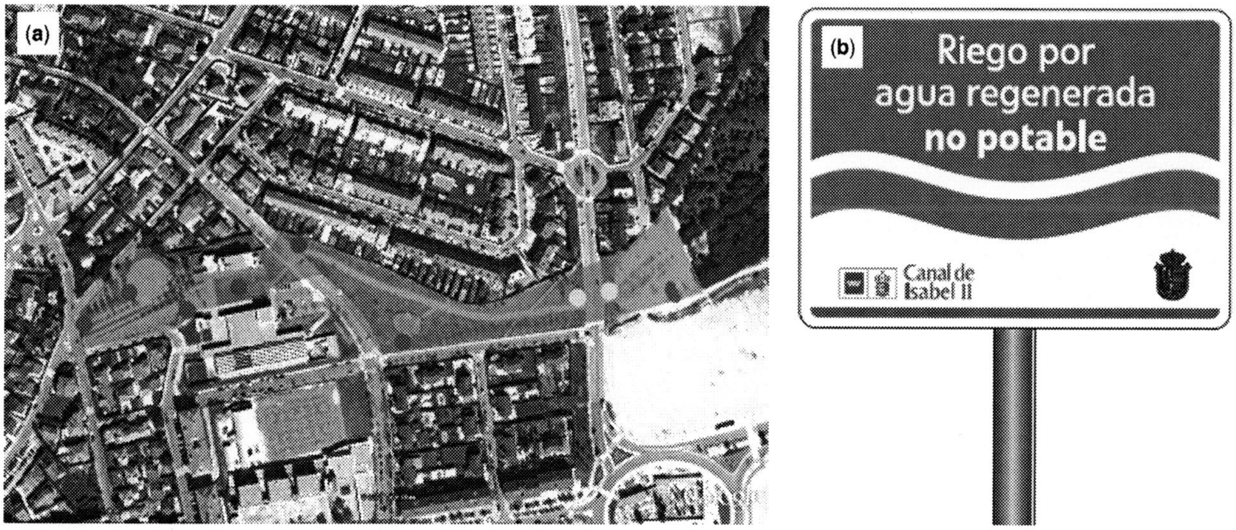

Figure 8.5 Aerial views of the irrigated areas (a) with the location of the warning signs (b).

The role of decision makers

In order to understand the success of the water reuse plan in the Autonomous Community of Madrid, attention should be drawn to the political and institutional commitment to its successful implementation. The regional authorities would obtain a clearer awareness of the necessity of such a plan if the scarce water resources of the Autonomous Community of Madrid were to be optimised, and if decisions in terms of economic investments were adopted accordingly.

Therefore, Canal de Isabel II received the mandate to develop the plan, with the necessary investment to be charged to its own budgets, meaning no cost to the municipalities that were to benefit from the service.

Furthermore, with a view to fostering and increasing the use of reclaimed water in the region, authorities belonging to the River Basin Authority (Hydrographic Confederation of river Tajo) and the Autonomous Community Government Authority (Department for the Environment) decided not to authorise any new golf courses unless they are irrigated with recycled water, a policy which is also extended to public green areas, which are more likely to be authorised if drinking water is substituted for reclaimed water for irrigation.

Public acceptance and involvement

The environmental benefits of the Reuse Plan, together with its economic benefits, with lower prices and related savings in water bills, have led to widespread acceptance of water reuse by the three main types of users: municipal councils, golf courses and industry.

The environmental benefits, which include increased water reserves in the region's reservoirs – seriously affected by the increasingly frequent periods of drought – will also undoubtedly lead to favourable opinion among the citizens of Madrid as they learn of the initiative.

8.6 CONCLUSSIONS AND LESSONS LEARNED

The reuse of water in the Autonomous Community of Madrid is becoming an essential aspect of the comprehensive management of water resources, favouring a significant net increase in water resources. The security and reliability of water supply is highly important in a region such as Madrid, where periods of drought are increasingly common and severe.

The ambitious Reuse Plan, with its first milestone in 2014, is a ground-breaking project which constitutes a benchmark at national level as the supply of recycled water is set to exceed 10% of total water consumption, providing a quality product that will substitute the consumption of drinking water for uses such as irrigation of green and sporting areas and industrial processes.

To do so, significant economic efforts have been made in order to allocate sufficient resources to implement not only conventional processes but also advanced processes such as ultrafiltration or reverse osmosis.

It is important to stress that it would be not possible to implement this Plan without the existence of a solid legal and regulatory framework and a clear political commitment.

Lastly, the environmental commitment with society and the effective dissemination of information about the Reuse Plan have resulted in a very good acceptance by municipal councils, industry and private clients (in the case of golf courses), partly because of the increasing awareness of the scarcity of water resources, and partly due to a tarification policy that favours the use of recycled water over that of drinking water.

The clear and decided commitment to incorporating recycled water into the whole water cycle of the Autonomous Community of Madrid has thus constituted an exciting challenge.

The following table summarizes the main drivers of the water reuse in Madrid as well as the benefits obtained and some of the challenges for the future.

Table 8.4 Summary of lessons learned.

Drivers/Opportunities	Benefits	Challenges
Important population increase, growing economic development and an increase in living standards Change in habits and new uses of water, with an increase in water consumption Climate change with severe droughts periods Political and institutional commitment by the regional authorities adopting decisions in terms of economic investments	Environmental benefits such as increasing water reserves in the region's reservoirs Prices are much cheaper than drinking water and savings in water bills Citizens have been aware of the advantages to use reclaimed water and acceptance by municipal councils, industry and private clients Municipalities have not borne any cost for the construction of facilities	Supply reclaimed water to 52 municipalities in the Autonomous Community of Madrid

View of a park irrigated with recycled water.

View of the advanced reclamation facility using ultrafiltration hollow fiber membranes and two-stage reverse osmosis to produce high quality recycled water for the Holmen Paper industry.

View of the water reuse holding tank and pumping station in Parla, Madrid.

Panoramic view of one of the tertiary treatment process in the Community of Madrid.

9 A double dose of water reuse in the middle of the Pacific Ocean – how Honolulu is supplying a growing population and industry

Scott Edwards and Fred Layi

CHAPTER HIGHLIGHTS

Surrounded by ocean but located in a drought-affected area, the Honouliuli Water Recycling Facility was designed to far exceed EPA consent decree requirements. The facility utilizes sophisticated technology, distinct process chains and cooperation between government and the private sector to produce two distinct recycle water products that meet the growing needs of industry, commerce and the public.

KEYS TO SUCCESS

- Customized local solution that is demand-based to match customer needs
- Connected technology and design with local industrial, commercial, residential, environmental and recreational needs
- Minimizes impacts of periodic droughts
- Superior quality ultra-pure process water for industry

KEY FIGURES

- Treatment capacity: 45,400 m^3/d
- Total annual volume of treated wastewater: 12.6 Mm^3/yr
- Total volume of recycled water: 11 Mm^3/yr
- Capital and operation costs: approximately $3.3 million/year
- Water reuse standards: <2.2 Fecal coliform/100 mL

9.1 INTRODUCTION

It is counterintuitive to think that the Hawaiian island of Oahu, surrounded by a vast blue Pacific Ocean, is in need of water resources. Yet, water reuse in the Honolulu area is both a reality and symbol of increasing pressure on water resources globally. With a growing population, periodic droughts and the U.S. Environmental Protection Agency mandating that a portion of wastewater effluent be recycled from the City and County of Honolulu's wastewater treatment facility, government officials became convinced of the wisdom of using treated wastewater. Contemplating water availability on the island and the environmental and economic impact of producing new sources of potable water, Honolulu Board of Water Supply (BWS) approved a design-build-operate contract for the development of the 45,400 m^3/d (12 million gal/day) Honouliuli Wastewater Reclamation Facility (WRF).

Main drivers for water reuse

Oahu is a study in contrasts – microclimates range from desert-like conditions to lush tropical forests (Figure 9.1). Impacted by trade winds that sweep up mountains and generate high levels of precipitation, the southwestern end of the island suffers from periodic and sometimes prolonged droughts. To that end, as noted by Hartley and Chen (2010), rainfall accumulation varies considerably in the summer months with precipitation collected during May-July 2002 and May-September 2003 at Manoa Lyon showing an accumulation of 259.59 mm versus a weather station at Waianae recording just 10.16 mm. Only an average of 460.0 mm (18.11 in) of precipitation falls each year at Ewa Beach, near the Honouliuli reuse facility.

During the 1970s, the City & County of Honolulu received a NPDES discharge waiver from the United States Environmental Protection Agency allowing discharge of primary effluent into an ocean outfall. In 1995, the City and County of Honolulu entered into a consent decree with EPA requiring the city to recycle 37,850 m^3/d (10 million gal/d) by the end of 2001 as part of Honolulu's request to extend the NPDES discharge waiver.

Figure 9.1 A view of Honolulu from Diamond Head on Oahu *(Photograph courtesy of Veolia Water and Bryan Spear).*

Human population and tourism continued to expand. Honolulu's population grew 8.8% from 2000 to 2010, expanding to 953,207 people according to the last U.S. Census. With limited groundwater supplies and increasing levels of tourists and locals enjoying golf at a multitude of area courses, along with a substantial cost for locating and developing new water supplies required to supply those golf courses, there was a growing need to ensure a sustainable potable water supply by offsetting use with recycle water.

Brief history of the project development

Given the consent decree, City & County officials considered their options and estimated development costs for new potable water sources at approximately 1600–1800 U$/$m^3$ (U$6–7 million per one million gallons potable water).

Veolia Water was selected by the City and County of Honolulu in a competitive procurement in October 1998 to design, finance, build and operate the Honouliuli WRF, which is located less than 16 km (10 miles) west of Pearl Harbor in Ewa Beach.

In 2000, construction was finalized and operations commenced. In 2003, the Honolulu Board of Water Supply (BWS), a semi-autonomous city agency, purchased the reclamation facility at a cost of approximately US$48.1 million based on analysis by outside consultants and a review of the company's costs related to the design and construction of the facility. As part of the agreement, Veolia Water was selected to operate and guarantee the facility's performance under a 20-year partnership designed to maximize savings for BWS and help ensure rate stability for public users. Under the new partnership, BWS owned all treatment and distribution facilities and was the public authority (replacing the City and County of Honolulu) to sell all produced water, just as it did throughout the island of Oahu.

At the time, as the only public water agency and water purveyor on Oahu, BWS held some concern over the potential for privatization of municipal water systems. However, Veolia Water's technical and operational expertise, along with its willingness to partner with BWS, led to a successful project. The resulting agreement established project and rate stability and ensured that BWS had access to technical experience it did not possess, allowing the agency to concentrate on potable water supply production, treatment and delivery.

Project objectives and water reuse applications

There were multiple program objectives for the reuse project. The overriding principle was to meet with EPA compliance requirements and deadlines, but planners had more foresight into potential long-term benefits. These included:

- Improved water quality for irrigation users, primarily area golf courses.
- Dependable, high-quality water for local industrial users.
- Savings of potable water to ensure local supply and future growth of the community.

- Rate stability for customers.
- Minimization of the potential impact of periodic droughts.
- Cost savings through innovation, private-sector involvement and the offset of needs to locate new groundwater or surface water supplies.

During the project's initial phase, the City and County of Honolulu also sought to benefit from the private-sector company financing the project, preserving city funds.

9.2 TECHNICAL CHALLENGES OF WATER QUALITY CONTROL

The Honouliuli Water Reclamation Facility is designed to produce a "double dose" of water reuse – specifically, two grades (R1 and RO, Figure 9.2) of beneficial water reuse from effluent supplied by the city's adjacent Honouliuli Wastewater Treatment Plant. Two separate process trains are required to produce the two grades of water product, introducing complexity and risk for both BWS and Veolia Water-R1 for landscaping and irrigation, and R0 water for industrial use by refineries and power production facilities.

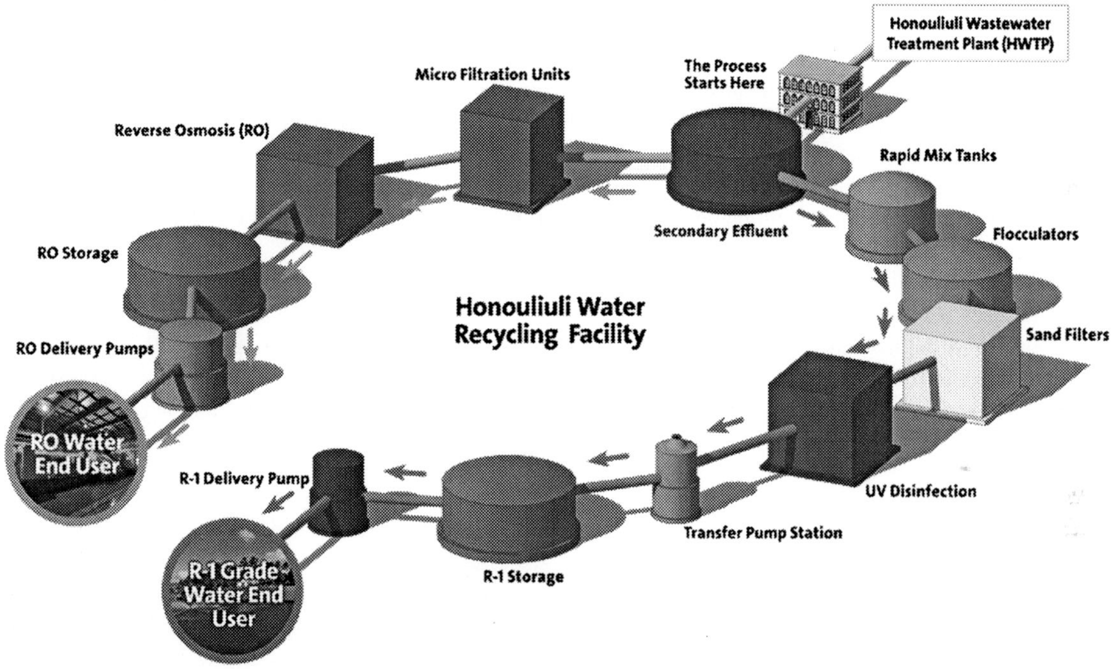

Figure 9.2 The process trains for the Honouliuli Water Reclamation Facility-R1 for landscaping and irrigation, and RO water for industrial use by refineries and power production facilities.

A second and fundamental challenge was transporting the water to multiple users in several locations. Veolia Water designed, financed and built a 24 km (15-mile) distribution system to deliver the two water products to end users. Because industrial manufacturers – especially power and oil refining companies – were customers, financial risks associated with potential operational shutdowns were considered.

Reverse osmosis (RO) demineralized recycled water, described further below, is produced and distributed to industrial customers who use it as a high-quality, ultrapure process water for boiler feed. Most of these customers use potable water as a backup in case of RO water distribution disruptions. The RO water system includes two storage tanks at different locations that have the capability to pump to its customers which gives the system redundancy. Veolia Water has a contractual obligation to produce specified RO water to these users. While these contractual requirements have always been met, several temporary disruptions have occurred, including a pipeline failure due to unrelated construction activities and an island-wide energy disruption due to an earthquake.

A unique on-island challenge is the continuing increase in the cost of fuel that directly impacts the electrical and transportation costs. According to a news report in early September, less than 1% of the continental U.S. electricity is generated by burning oil. According to the Hawaii State Department of Business, Economic Development & Tourism, 75% of the island's electricity is generated by burning oil – driving the costs per kWh nearly 4 times that of the Western U.S. coast. To help mitigate these high costs, several operational practices and engineering controls were implemented including: maintaining high tank levels in the R1 & RO distribution systems which allows the pumps to run at optimum pump efficiency; looping the R1 distribution system decreased the pumping pressure in half (from 90 psi to 45 psi) and maintaining daily preventative maintenance procedures with the UV system prevents unnecessary power consumption.

The other impact of high fuel costs is the increasing cost of transportation. Much of the parts, materials, and chemicals needed to operate the facility are shipped in. The facility utilizes both maintenance and operations programs and databases that help maintain a predictive maintenance practice. Since economical transportation options usually take several days (and in some cases, weeks), inventory of both chemicals and spare parts are closely monitored.

Treatment train for water recycling

Treated to the highest level as regulated by the Hawaii State Department of Health, the R1 water is supplied to the city and to eight area golf courses for landscaping and irrigation purposes. Treated effluent from the Honouliuli Wastewater Treatment Plant enters rapid mix tanks, is chemically treated and flocculated. Sand filtration is utilized to meet the maximum 2.0 NTU turbidity standards. Effluent is then treated using disinfection via Trojan Technologies' UV4000 ultraviolet irradiation (UV).

While the R1 water does not meet drinking water standards, it slightly exceeds standard EPA reuse requirements. Ultraviolet light emits electromagnetic radiation that destroys bacteria and viruses by causing a rearrangement in their genetic code that inhibits the bacteria and virus from reproducing. The process eliminates the handling of potentially dangerous substances and adds no toxic compounds to the effluent.

The RO water represents ultra-pure demineralized water, requiring a rigorous process chain due to the strict process requirements of power and refining companies located in a nearby industrial park. Two oil refineries, four power production facilities and one SNG (Synthetic Natural Gas) facility use the RO product.

Effluent from the wastewater treatment facility is first treated via Siemen's CMF® continuous microfiltration process (Figure 9.3b) that uses a membrane filter to remove particles greater than 0.2 μm (bacteria is typically larger than 1 μm). The water is then run through six reverse osmosis (RO) skids arranged in parallel, and equipped with a high-pressure feed pump and a cartridge filter with 5-μm cartridges. In terms of the log scale in micrometers, ultra-pure demineralized water product following reverse osmosis reaches the ionic range, eliminating a multitude of particles and leaving particles that are observable only with scanning tunneling microscopes or scanning electron microscopes.

Figure 9.3 Views of the the Honouliuli Water Reclamation Facility and the membrane microfiltration unit *(Photograph courtesy of Veolia Water and Bryan Spear).*

M... of the industrial users make steam for power generation from the RO water. The potable water supplied in the area is ...ica content and if boiled causes scaling – an expensive proposition for industry that requires on-site demineralizers. ...r producers have reported significant reduction in chemicals, maintenance and regeneration costs associated with ...of its demineralizers due to the Honouliuli WRF's RO process chain.

Water quality control and monitoring

The effluent from the Honouliuli Wastewater Treatment Plant essentially represents the reclamation facility influent. The monitoring battery consists of routine sampling for a variety of constituents as demonstrated in Table 9.1.

Table 9.1 The Honouliuli WRF influent monitoring programme.

Constituent	Units	Type of Sample	Sampling Frequency
Total Daily Flow	Mgd	Continuous	Daily
Total Dissolved Solids	mg/L	24-hour composite	Weekly
BOD_5	mg/L	24-hour composite	Weekly
Total Suspended Solids	mg/L	24-hour composite	Twice Weekly
Turbidity	NTU	On-line	Continuous
Silica	mg/L	24-hour composite	Monthly
Calcium	mg/L	24-hour composite	Monthly
Magnesium	mg/L	24-hour composite	Monthly
Sodium	mg/L	24-hour composite	Monthly
Potassium	mg/L	24-hour composite	Monthly
Ammonium	mg/L	24-hour composite	Monthly
Barium	mg/L	24-hour composite	Monthly
Strontium	mg/L	24-hour composite	Monthly
Bicarbonate	mg/L	24-hour composite	Monthly
Carbonate	mg/L	24-hour composite	Monthly
Sulfate	mg/L	24-hour composite	Monthly
Chloride	mg/L	24-hour composite	Monthly
Fluoride	mg/L	24-hour composite	Monthly
Nitrate	mg/L	24-hour composite	Monthly
pH	Standard Unit	On-line	Continuous

Certified by a third-party laboratory, more than 3200 tests are conducted per year regarding the R1 and RO water products. Table 9.2 shows the specifications for the two types of recycled water, R1 reclaimed water for various public users, including golf courses, and high-purity RO water for industrial users with more stringent standards.

Main challenges for operation

To set the stage for an effective partnership, formal partnering meetings were conducted involving 20–30 participants at any one time, including consultants, regulatory entities, public works officials from the Honolulu Board of Water Supply, the City and County of Honolulu and Veolia Water. These meetings were used to establishment and align the goals and objectives, project philosophy and a unified approach to managing the project.

The Veolia Water project team has worked closely with BWS to make adjustments to operations and reduce costs. For instance, variable frequency drives (VFDs) and SCADA (supervisory control and data acquisition) systems have been integrated to ensure greater system reliability and reduce energy costs. As the distribution system has expanded to 54.7 km (34 miles), telemetering has been installed to ensure accuracy with more connected users.

Regarding the RO process train, the latest advanced RO membranes have been installed to treat the secondary effluent, which has improved the plant performance. The facility has also performed a Bioassay which will lower the UV dose requirements from 140 mJ/cm^2 to 100 mJ/cm^2 on the UV system and increase the hydraulic profile from 37,850 m^3/d (10 mgd) to 60,570 m^3/d (16 mgd).

The microfiltration units have to continuously operate for production of water but also to keep the RO membranes from fouling due to the biological foulant behavior of the influent water (effluent from the city's wastewater treatment plant). The demand for RO water is constant throughout the year and the average daily production is approximately 5700 m^3/d (1.5 mgd). Since the overall demand is about 75% of the total capacity, it is essential that all microfiltration and reverse osmosis equipment be maintained in full operational condition.

Table 9.2 Water quality specifications for the R1 reclaimed water and for the high-purity RO water.

R1 reclaimed water			High-purity RO water		
Constituent	**Units**	**Range**	**Constituent**	**Units**	**Range**
Calcium (Ca)	rng/L as $CaCO_3$	0–79	Calcium (Ca)	mg/L as $CaCO_3$	0–1
Magnesium (Mg)	mg/L	0–119	Magnesium (Mg)	mg/L	0–1
Sodium (Na)	mg/L	0–174	Sodium (Na)	mg/L	0–3
Potassium (K)	mg/L	0–15	Potassium (K)	mg/L	0–1
Bicarbonate (HCO_3)	mg/L as $CaCO_3$	0–245	Bicarbonate (HCO_3)	mg/L as $CaCO_3$	0–1
Sulfate (SO_4)	mg/L	0–46	Sulfate (SO_4)	mg/L	0–1
Chloride (Cl)	mg/L	0–300	Chloride (Cl)	mg/L	0–2
Silica (SiO_2)	mg/L	0–84.5	Silica (SiO_2)	mg/L	0–1.5
Strontium (Sr)	mg/L	0–0.33	Electrical Conductivity	μS/cm	0–30
Fluoride (F)	mg/L	0–5	pH	Standard Unit	5–8
Nitrate (NO_3)	mg/L	0–1200	Carbon Dioxide (CO_2)	mg/L	0–37
pH	Standard Unit	6–9	Temperature	°F	Ambient
Temperature	°F	Ambient			
BOD_5	mg/L	0–30			
Total Suspended Solids	mg/L	0–30			
Settleable Solids	m1/L	0–0.1			
Dissolved Oxygen	mg/L	2–6			
Total Dissolved Solids	mg/L	0–800			
Fecal Coliforms	no./100 mL	0– 2.2			

The Honouliuli facility is a demand-based process that only produces water to match the customer needs. During the summer months, the demand for R1 water is greater than in the winter months due to more rainfall and wetter conditions.

Because moderate pressure is required in the system, a plan has been developed to supplement the facility's 19,000 m^3/d (5 mgd) storage tank with a larger elevated tank off site to ensure constant pressure on the R1 distribution system and eliminate the necessity to pump water to customers (Figure 9.4).

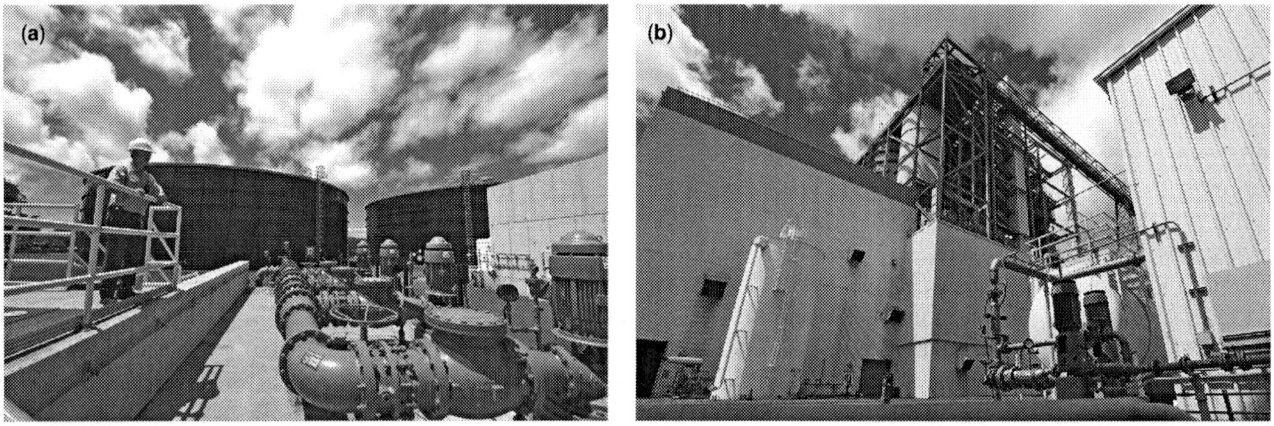

Figure 9.4 Views of the pumps and storage tanks for the R1 recycled water (a) and the recycled water pumps at the largest power plant in Honolulu (b) *(Photograph courtesy of Veolia Water and Bryan Spear).*

ditionally, to ensure long-term performance of the on-site equipment and maintain efficiencies, the site has implemented lity of all assets to prioritize their work and spend maintenance dollars wisely. An on-going condition assessment of the assets is then used to establish the Capital Replacement Plan that helps provide a planning tool for replacement of aging or degrading equipment at the appropriate time.

Finally, accidents and injuries at complex facilities such as this one are not uncommon. A strong safety program and culture have been built based on the company's successful corporate program. Not a single lost-time accident has occurred since the project began full operations.

9.3 WATER REUSE APPLICATIONS

Two different classes of water users are served by the Honouliuli Water Reclamation Facility (Figure 9.5). Approximately eight different golf courses, the City of Kapolei and the City and County of Honolulu facilities are provided with R1 water, all which use it for landscaping and irrigation.

Figure 9.5 Views of a golf course and a power plant supplied with recycled water in Honolulu *(Photograph courtesy of Veolia Water and Bryan Spear).*

The second class represents a broader swath of the island's economy, including refiners, manufacturers and power companies which are largely located in the same industrial park. These organizations are provided with RO water, a high-quality, ultrapure process water. Customers include Chevron, Tesoro, Kalaeloa Partners, AES, Hawaiian Electric Co. (HECO) Peaking Plant, HECO Kahe Power Plant and Citizens Gas.

Evolution of the volume of supplied recycled water

The project's commencement history includes the delivery of R1 water to two substantial golf development properties in September 2000, with Kalaeloa Partners, an energy provider, becoming the first RO water customer in November 2000. A year later, two energy customers were added, along with two additional golf courses, and most notably, a Chevron refinery was added in late 2002. Product volumes increased with additional RO and R1 customers, culminating in a new golf course using R1 in 2008 and two additional power plants purchasing RO water as regular customers in 2009.

The total volumes of R1 and RO recycled waters produced during the first, fourth, seventh, and eleventh years are noted in Table 9.3. In 2012 a total volume of 11.5 milion m^3 (Mm^3) was produced, from which 83% of R1 recycled water for irrigation and 17% high purity water for industrial purposes.

Table 9.3 Evolution of total volumes of R1 and RO waters produced.

Year	R1 Water, m^3	RO Water, m^3	Total, m^3
2000	1,211,331.20	75,708.20	1,287,039.40
2003	8,888,142.68	1,525,520.23	10,413,662.91
2006	9,240,185.81	1,885,134.18	11,125,319.99
2010	9,573,301.89	1,922,988.28	11,496,290.17

Relations and contracts with end-users

During the initial project phase, Veolia Water maintained customer relations as the facility owner. This role was assumed by BWS when the agency purchased the facility assets.

BWS provides customer support services, training and compliance monitoring for its recycled water customers. Due to the agency's status as a public agency mandated by City Charter to provide water supply, BWS represents long-term stability for customers beyond the facility's operating contract term. On a go-forward basis, BWS has established an objective to ensure the recycled water program is self-sustaining by initiating a rate structure that will recover the operating, maintenance and capital replacement over the life of the infrastructure assets. Through education and their own experiences, customers understand the agency's recycled water is resource based and conserves potable water and that the agency establishes recycled water rates that will be less than potable water rates to ensure customers' continued use. In addition, these commercial and industrial developers accept that they must connect to the recycled water system, where available, because BWS, by rule, will not allocate potable water for irrigation or industrial process water.

BWS is currently in the process of renegotiating existing R1 and RO contracts, and while customers prefer the lowest rate possible, they understand the need for higher rates to ensure the recycled water system is repaired and replaced over time to maintain current levels of reliable service.

9.4 ECONOMICS OF WATER REUSE

Project funding and costs

City officials initially estimated that a traditional procurement – as opposed to one bundling all phases and components of the project including design engineering, construction and operation – would have resulted in subsidizing up to 0.4 US$/m^3 ($1.50 per thousand gallons) due to development and production expenses.

Initially funded by Veolia Water, the treatment facility was purchased by BWS for approximately US$48 million based on a third-party estimate of approximately US$60 million. BWS invested another US$12 million in additional pipelines that connected new users and looped the system for reliability. As part of the agreement, Veolia Water provided a guaranteed cost for operations and maintenance (O&M) on an ongoing basis.

Pricing strategy of recycled water

As pricing models evolve, arguments are being developed for charging the full cost of water service to encourage efficient water use. Such pricing would include not just O&M cost and capital costs but all related direct costs, scarcity costs and environmental costs.

BWS believes recycled water should be less costly than potable water to ensure buyers. R1 water pricing is affordable, reflecting the lower cost of tertiary filtration, disinfection and low-pressure distribution. The R1 rate recovers O&M and future capital replacement costs. Initial capital funding to purchase the treatment facility and system was financed through bonds based in part on the avoided costs of potable water development, meeting the previous EPA consent decree to recycle 37,850 m^3/d (10 mgd), and the opportunity benefits of additional revenue streams from the future sale of recycled water to a new and expanded customer base.

The foresight to construct the RO water system in addition to the R1 water process provided a larger revenue stream that enhanced the economic feasibility of the project. Industrial water users pay approximately 1.3 US$/m^3 (US$5 per 1000 gallons) for the RO water, which represents a substantial reduction in costs for users whose costs to demineralize potable water amounts to approximately 2.1–2.6 US$/m^3 (US$8–10 per 1000 gallon), which represented a substantial savings to industry. BWS reports that the movement toward renewable energy, bio-fuels, photovoltaics and waste-to-energy power plants continues to provide positive growth potential for RO water.

R1 water ranges in price from 0.15 US$/m^3 to 0.53 US$/m^3 (US$0.55 per 1000 gallons for golf courses to US$2.00 per 1000 gallons for other irrigation users). This price disparity exists due to negotiated prices with historical customers to meet the previous EPA consent decree and due to market pricing strategies to attract new users of R1 water. For comparison, the cost of potable water is currently 0.854 US$/m^3 (US$3.13 per 1000 gallons) for non-residential users and may increase as high as approximately 1.364 US$/m^3 (US$5.00 per 1000 gallons) if BWS is successful in adopting a proposed rate increase.

Benefits of water recycling

Considerable financial and non-financial benefits have been established including all project objectives stated previously.

A dependable supply of ultra-pure process water (RO water) exists to help stimulate economic development. R1 water is being used for irrigation and recreational purposes. Likewise, reclaimed water has reduced the need to rely solely on potable water supplies and deliver financial and environmental benefits.

The overriding principle, however, was compliance with EPA consent decree deadlines. Additionally, cost savings have helped support stable water rates for BWS customers.

During the project's initial phase, the City and County of Honolulu also benefitted from the private-sector company financing the project, preserving city funds. Reclaimed water rates have not significantly increased since the water reclamation facility was constructed. Plant operations have been award winning, compliant (passing every annual inspection) and safe (Veolia Water has maintained operations for approximately 11 years without a single lost-time accident).

9.5 HUMAN DIMENSION OF WATER REUSE

Public education and communication strategy

As an agency accustomed to regular interface with the public through public meetings and routine reporting by local news media, BWS and City and County of Honolulu officials knew the importance of both public education and customer education. Officials, along with the private-sector partner, began an outreach campaign to potential commercial and industrial users to secure commitments to purchase the two water recycling products. Securing industrial users, developers knew, would help reduce overall development expenses.

The partnership developed multiple communications tactics to ensure transparency, promote reuse, and reduce the possibility of incorrect information about recycling, which would have a potential negative impact on the sale of future water product. A web site was established, and a newsletter and brochure were developed for distribution to the community and to customers. Prior to full operations, an opening ceremony with numerous public officials was held featuring traditional "O'o" sticks and an untying of a *lei* of *maile* (pronounced my-lay) leaves.

All communications featured the benefits of water reuse – conservation of drinking water supplies, economical rates for customers, and a safe and sustainable supply that is less impacted by drought. Especially during the project's initial operational phase, a number of news reports occurred outlining the merits of the project.

Tours and public speaking at local schools are regularly conducted by BWS and Veolia Water staff. Because of the water supply challenges on the island, program goals were also integrated into a BWS demand-side management program to increase public awareness, ultimately saving millions of liters of potable water supply. From 1990 to 2010, potable water production on Oahu has stabilized at approximately 570,000 m^3/d (150 mgd) despite major growth in the Ewa area and Central Oahu.

The role of decision makers

The Honolulu Board of Water Supply is a semi-autonomous agency that manages O'ahu's municipal water resources and distribution system. The agency has a seven-member Board of Directors, five of which are appointed by the Mayor and approved by the City Council. Two additional directors are the Director of Hawaii's State Department of Transportation and the Chief Engineer of the Honolulu City Department of Facility Maintenance. This Board appoints the BWS Manager and Chief Engineer to supervise overall operations.

The private-sector partner, Veolia Water, reports to the BWS Manager on a daily and weekly basis, with formal written documentation provided on a monthly and annual basis. The relationship is governed by a contract, noting water quality metrics and contractual performance requirements. Specific production reports can be provided to all users on a daily, monthly and annual basis.

Public acceptance and involvement

Perhaps due to the sensitivity of water resource management issues, wide acceptance has occurred. There is general public awareness of the famous purple pipes denoting water reuse. BWS continues to promote the benefits of reclaimed water and conduct public tours. Low-flow plumbing fixtures are now common on the island and a part of daily life.

Industrial users who are accustomed to production capacity affecting profit margins are largely demonstrating a reduction in chemicals and maintenance versus previous reliance on potable water use. Similarly, golf course managers who use the R1 water agree with claims that the water is "nutrient rich."

BWS foresees and has established long-term plans to convert existing brackish water supplies to water reclamation technologies.

9.6 LESSONS LEARNED AND MAIN KEYS TO SUCCESS

BWS entered into the agreement with some initial concern about perceived competition by Veolia Water. Ultimately, the flexibility and expertise demonstrated by the private-sector partner and the willingness of BWS to partner with a private-sector company lead to a successful project.

The BWS purchase provided immediate financial stability to the project and ensured larger combined resources, drawing on Veolia Water's technical and O&M expertise. The partnership meant that BWS did not have to build new capacity or operate the complex facility, allowing the agency to concentrate on potable water supply production, treatment and delivery. Similarly, the addition of the RO product line provided a larger revenue stream that enhanced the project's economic feasibility.

Common customer interests and a wide range of benefits were critical to the project's overall success, highlighting the success of sustainability as applied to this project. From an economic vantage point, customers are assured of an affordable supply of high-quality reuse water with multiple benefits and positive characteristics. From an environmental perspective, there is a reduction in the need for surface and groundwater supplies and a reduction in wastewater effluent discharge. Society benefits from a much-needed supply, along with the inherent environmental and economic benefits.

A true partnership among public and private-sector officials working with common goals established the framework for immediate and long-term success. Identifying all the potential benefits of potential water customers in light of their other options also helped establish the project's success. For instance, industrial users, such as petroleum refiners, have multiple options for process water. Turning to a third party can involve risk, especially when considering that the source is a wastewater treatment facility. Real trust of the product and the partnership has been well established.

The following Table 9.10 summarizes the main drivers of water reuse in Honolulu as well as the benefits obtained and some of the challenges for the future.

Table 9.10 Summary of lessons learned.

Drivers/Opportunities	Benefits	Challenges
Conservation of potable resources. Energy efficiency. Examining newer technologies to increase effluent efficiency.	Use of recycled to lower demand and pressure on resources. Lower operating costs beneficial to client and customers. Increase overall flow efficiency of facility to deliver more reuse water.	Meeting demand as growth of reuse water increases. Controlling operational costs with unpredictable energy prices. Capital costs and footprint required for new equipment.

REFERENCES AND FURTHER READING

Edwards S. (2001). The big partnership in the big pond. *Water Environment & Technology*, **13**(1), 30–34.

Hartley T. M. and Chen Y. L. (2010). *Characteristics of Summer Trade Wind Rainfall Over Oahu*. Department Meteorology, University of Hawaii, Honolulu, HI 96822 (revised 2010).

Usagawa B. (2012). *Personal Communication and Collaboration with the Honolulu Board of Water Supply*, contributed to this chapter.

View of a large power plant in Honolulu.

10 The keys to success of water reuse in tourist areas – the case of Bora Bora

Valentina Lazarova, Vincent Sturny and Gaston Tong Sang

CHAPTER HIGHLIGHTS

The relatively small water reuse system in Bora Bora is an outstanding example of a successful water reuse project in conditions of very restrictive water reuse regulations that were overcome thanks to the strong political engagement of elected officers, trust of large end-users and support by local population. The reliable supply of high quality recycled water without interruptions and a suitable pricing of water services ensured the technical and economic viability of the project enabling to compensate for the water stress and to protect the lagoon, the most precious heritage of the Island.

KEYS TO SUCCESS

- Adequate choice of the treatment technology and the ability to provide high-quality recycled water without any interruption
- Strong political engagement and commitment of the elected officers
- Implementation of an adequate public communication and education program
- Good collaboration with the local health authorities to overcome negative risk perception of urban spray irrigation
- Public-private partnership
- Economic viability with ability to cover O&M costs of tertiary treatment and distribution

KEY FIGURES

- Start up in 2005 with a capacity of 300 m^3/d, extended to 500 m^3/d in 2008
- Volume of recycled water consumption: ~70,000 m^3/yr
- O&M costs: 0.68 €/m^3 (38% of the annualised life cost)
- Recycled water charge: 0.67 €/m^3, 1.68 €/m^3 and 2.51 €/m^3 plus fixed annual charge depending on the volume used
- Water quality requirements: 0 *E.coli*/100 mL

10.1 INTRODUCTION

The economic development of Bora Bora, known as the "Pearl of the Pacific" and globally renowned for the beauty of its lagoon, is carried out through a municipal policy of sustainable development. This is supported by an intense willingness of its elected representatives and based on a meticulous respect of the environment. The preservation of the island's charm and the design of buildings adapted to the landscape is ensured to prevent rejection by the local population. A very important element in this policy is the management of water resources and water reuse is playing a crucial role.

Main drivers for water reuse

Since 1998, the island of Bora Bora has regularly faced water shortages, due primarily to decreasing rainfall and frequent droughts combined with hotel development and population growth. As a result, severe water shortage has been observed with periodic interruptions of water supply. In order to secure water supply for the local and tourist population, seawater desalination and water recycling have been implemented as alternative resources.

In addition to water stress and increasing water demand, another important driving factor for water reuse has been the motivation of the municipality to protect water resources.

Brief history of the project development

The freshwater resources of the island are not sufficient to provide for the resident population (8000 inhabitants) and 200,000 tourists per year. Following a growth in water demand and increasingly more severe dry seasons, the production capacity of drinking water was due to be increased several times, with the introduction of alternative resources such as desalination of seawater (3 desalination plants constructed in 2000, 2005 and 2007) and water reuse.

The raw sewage of the island is collected and transported by a pressurized network by means of 70 pumping stations, then treated by two wastewater treatment plants, one on the north side and one on the south side of the island. The first water reuse project for irrigation was implemented in the mid 1990s using polished secondary effluents in maturation ponds. Recurrent odour problems and bacteria regrowth in the distribution network associated to the relatively high cost of recycled water, led the consumers, in particular the luxury hotels, to limit their water reuse demand. To avoid these problems and produce high quality recycled water for unrestricted urban uses, an advanced tertiary membrane treatment was designed in 2003 and implemented in the wastewater treatment plant of Povai. The first stage was commissioned in April 2005 with a treatment capacity of 300 m^3/d and was extended to 500 m^3/d in 2008. The recycled water distribution network has been extended to completely cover the demand of non-potable water of all luxury hotels (Figure 10.1). In 2010, a new storage reservoir of 450 m^3 was built, as well as a new pumping station in order to increase the pressure to feed 20 hydrants for fire protection. The operation of the water reuse system was also completely automated. Water recycling scheme is operated by SPEA (Société Polynésienne de l'Eau et d'Assainissement, Water and Wastewater Treatment Agency), including wastewater treatment, tertiary ultrafiltration and distribution network.

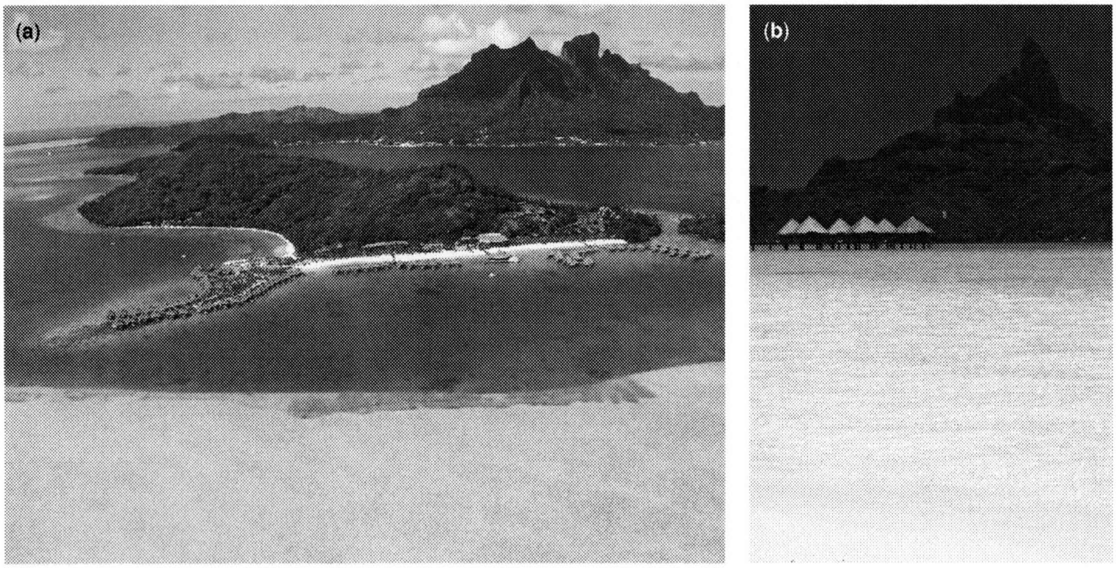

Figure 10.1 View of the Island Bora Bora (French Polynesia): aerial view (a) and water villas of luxury hotels (b).

Project objectives and incentives

The water reuse program in Bora Bora is an element of the global strategy of decision makers for integrated resource management. This approach also includes drinking water conservation measures and implementation of stringent leakage control in drinking and recycled water distribution networks. The needed funds and financial arrangement for water recycling scheme have been found by means of a public-private partnership, governmental subsidies and loans. An adequate tariff policy has been implemented in favour of the local population and enables the recovery of operating and maintenance costs.

During the first stage of implementation of the water reuse program in Bora Bora, only non-potable reuse applications were implemented. Currently, new satellite recycling facilities are under construction for golf course irrigation and aquifer recharge for indirect potable reuse.

10.2 ROLE OF WATER QUALITY AND TREATMENT TECHNOLOGY FOR THE TRUST IN WATER REUSE

The adequate choice of the treatment technology, the tertiary ultrafiltration which allows almost total disinfection and removal of suspended solids, is the first of keys factors that ensured the fast growth of recycled water demand. In addition, the ability to provide high-quality recycled water without any interruption provided the required trust in water reuse from large end users such as the luxury hotels.

Treatment train for water recycling

The ultrafiltration hollow fibbers submerged membranes (ZeeWeed 500) have been chosen to polish a part of the secondary effluent and implemented in the existing wastewater treatment plant of Povai (Figure 10.2). The ultrafiltration membranes have small pore size of 0.035 μm, which represents an effective physical barrier for all microorganisms and pathogens, including protozoa, cysts, bacteria and viruses. The initial treatment capacity was 300 m^3/d and was extended to 500 m^3/d in 2008. Recycled water is stored in two covered reservoirs and pumped into the industrial (non-potable) water distribution network after chlorination in order to maintain 0.5 mg/L chlorine residual.

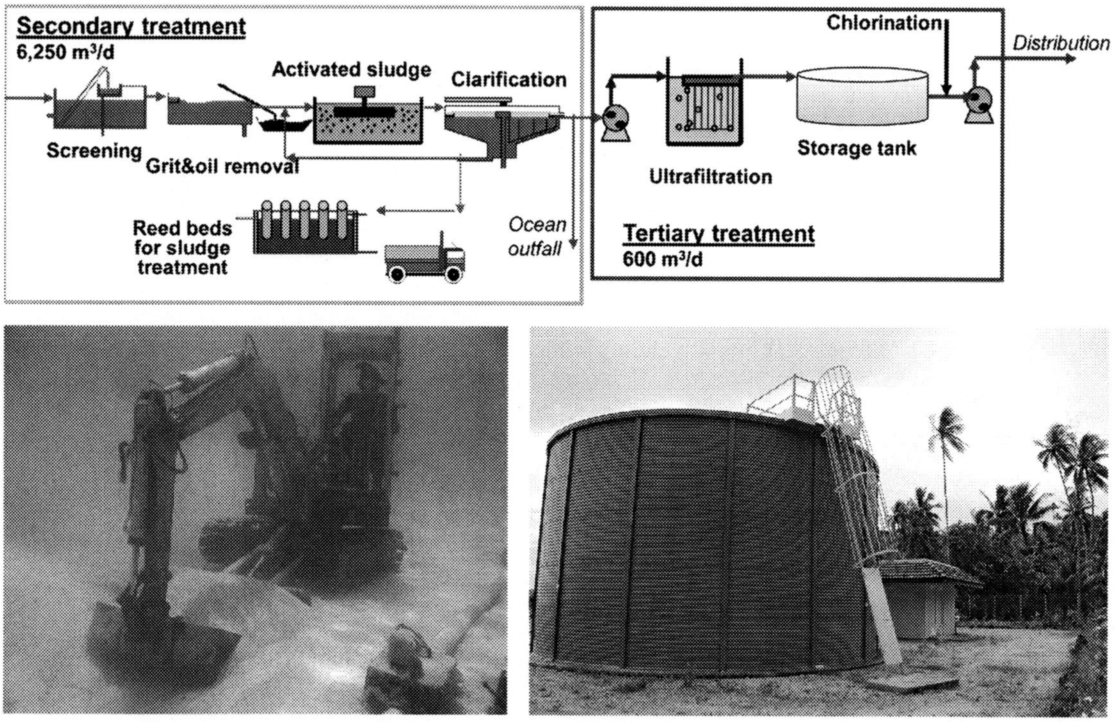

Figure 10.2 Schematic flow diagram of wastewater treatment and recycling plant and views of the construction of the dual distribution system and the new storage tank.

Water quality control and monitoring

Table 10.1 illustrates the wastewater quality (raw sewage, secondary effluent and recycled water) for the last three years.

Table 10.1 Water quality characteristics of the water recycling facility in Bora Bora.

		Secondary effluent		Recycled water (UF permeate)	
Parameter	Raw sewage*	Measured*	Consent	Measured	Guide value
COD mg/L	595 (270–837)	31 (21–65)	90	15 (4–34)	40
BOD_5 mg/L	349 (200–540)	7 (<5–22)	25	4 (1–6)	20
TSS mg/L	238 (125–275)	9.5 (4–19)	35	<5	20
N_{tot} mg/L	47 (30–70)	8.3 (2–18)	20	7.3 (2–17)	20
P_{tot} mg/L	6.8 (4.1–8.1)	2.5 (1.0–5.8)	–	1.9 (0.45–5.8)	–
E.coli / 100 mL	not analysed	10^5–10^7	–	not detected	0/100 mL
Streptococci/100 mL	not analysed	–	–	not detected	0/100 mL

*Average value and limit of variations (32 monthly composite samples, excluding *E.coli* and *Enterococci* that were monitored monthly in grab samples as colony forming units).

Despite the high variations of raw sewage characteristics, tertiary ultrafiltration consistently produced an effluent with a very good quality, free of suspended solids and with a very low content of organic carbon (COD and BOD below the detection limits). During the 5-year operation of the submerged UF membranes, three cases of contamination of the permeate with fecal coliforms have been detected due to valve leakage and overflow of unfiltered secondary effluents (up to 200 cfu/100 mL). *Enterococci* (56/100 mL) and *Clostridium* spores (2/20 mL), which, as a rule are not present in the membrane permeate, have been also detected during this plant failure. To improve the reliability of operation of the ultrafiltration, the treatment plant was upgraded to not allow any by-pass of unfiltered water or external contamination of the permeate. An on-line turbidity measurement was recommended as an efficient measure to control any valve leakages or breaks of membrane fibbers. An additional barrier for efficient disinfection and control of biofilm growth in the distribution system is ensured and maintained by a final chlorination after storage tanks.

Main challenges for operation

By applying the recommended maintenance and cleaning procedures, a good recovery of membrane permeability was observed, 202 to 247 $L/h \cdot m^2$ bar at 20°C compared to the initial value of 250 $L/h \cdot m^2$ bar. It is important to underline that despite some minor operational failures and a major damage of the WWTP in September 2005 due to tsunami with seawater intrusion and loss of activated sludge from the clarifier, an excellent membrane flux was consistently maintained at 24 ± 2 $L/h \cdot m^2$ (20°C).

Preventive measures against membrane fouling are the crucial factor avoiding excessive loss of membrane permeability (on-line cleaning using scouring air, backpulses, membrane relax, automated maintenance cleaning). The frequency of chemical recovery cleaning varied from 2 to 6 months depending on the wastewater quality. By safety, recovery cleaning was carried out before reaching the fixed minimal value of 100 $L/h \cdot m^2 \cdot$ bar. Compared to other membrane systems, more frequent recovery cleaning was required, especially during winter periods.

10.3 WATER REUSE APPLICATIONS

The main water reuse application is for landscape irrigation, mostly for luxury hotels (Figure 10.3), but the production of high quality recycled water enabled the development of the following other urban uses:

- Boat washing.
- Filling the water reservoirs of all fire protection boats.
- Supply of 20 hydrants for fire protection.
- Cleaning and landscape irrigation in all water and wastewater plants, as well as in the 70 pumping stations.
- Cleaning and industrial uses in the composting facility.
- Washing of construction engines and preparation and tests of concrete at 3 to 7 building sites, depending on the year.
- Municipal nursery for the production of flowers, shrubs and other plants.
- Municipal workshops (construction equipment, school buses).
- Municipal stadium Teriimaevarua and the associated sport facilities.
- Environmental enhancement by filling waterfalls and ponds.

Figure 10.3 Views of the hotel landscape in Bora Bora irrigated with recycled water.

Evolution of the volume of supplied recycled water

The adequate choice of the treatment technology and the ability to provide high-quality recycled water without any interruption were the two key factors that ensured the fast growth in recycled water demand (Figure 10.4). A twofold increase in the recycled water consumption was observed shortly after the start-up of the new recycling facility, with an average annual volume of about 70,000 m^3/yr and a daily demand up to 600 m^3/d. The number of end users after the start-up of the new membrane recycling facility tripled compared to that in 2004.

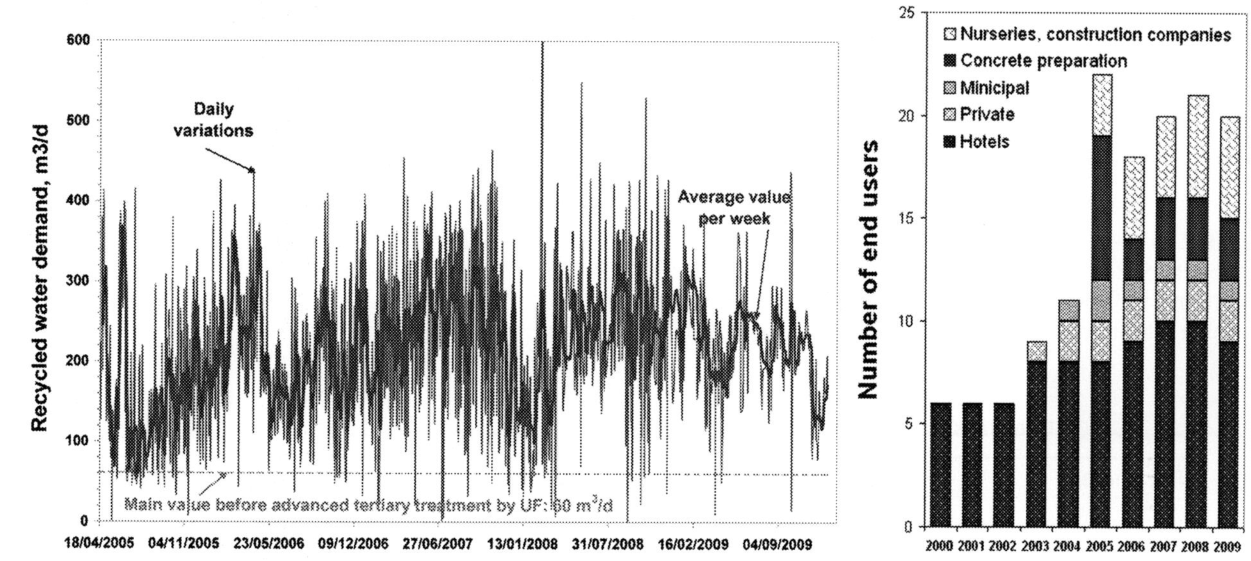

Figure 10.4 Evolution of the recycled water demand and the number of end users.

10.4 ECONOMICS OF WATER REUSE

Project funding and costs

The capital cost of the water reuse program was covered by the international, territorial and French funding. The main challenge of the municipality was to minimise the risk of failure of tertiary treatment at acceptable operation and maintenance costs below the cost of seawater desalination. For the first 3 years of operation of the membrane facility, the operating and maintenance cost was about 0.68 €/m^3, which is equivalent to 38% of the annualised life cost.

The operating costs of the recycling facility in Bora Bora include the operation and maintenance not only of the tertiary membrane treatment, but also of the distribution system, which is predominantly underwater (submerged in the lagoon). The main expenses comprise fixed costs for labour, repairs and maintenance, membrane replacement and water quality monitoring, as well as variable costs for chemicals and energy consumption (Figure 10.5).

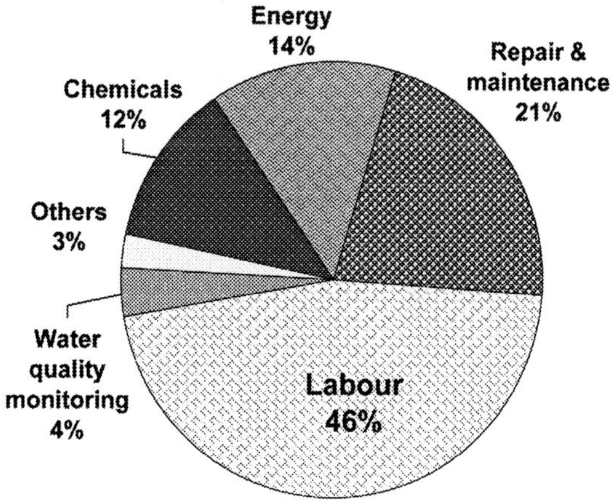

Figure 10.5 Distribution of operation costs (tertiary treatment and distribution network).

The main component of operation costs is labour, accounting for 46% of total operation costs. Despite the high price of electricity in this isolated tourist island (0.18 €/kWh), the contribution of energy costs is only 14%, which is relatively low compared to typical values of tertiary MF/RO treatments of 26–32%. Chemical costs for membrane cleaning and final chlorination contribute to another 12% of operation costs. A new important part of operation costs is repair and maintenance, which rises to 21%, including membrane replacement. It is important to underline that the cost of spare parts and scouring equipment is also higher because the remoteness of this island, and thus, the high transportation costs. Water quality monitoring remains relatively low, 4% of operation costs, but a great part of the expenses are included in the labour costs. The average energy consumption at nominal flow is estimated at 0.36 kWh/m^3, but a twofold increase to 0.62 kWh/m^3 is observed when hydraulic load drops to 50%. During operation with flow rates below 30%, the energy consumption raises to 1.0–1.6 kWh/m^3 due to the high fixed energy needs of membrane treatment.

Pricing strategy of recycled water

After dialogue with the stakeholders and local communities, the municipality of Bora Bora decided to implement a tariff policy in favour of the local population: a two-part tariff with an ascending rate structure (Table 10.2). The social signal of this decision was 'The less you use the less you pay'. The progressive rate with the increasing volume of recycled water supply is a reminder of the lack of water and the necessity to save the resource. At same time, for a given category of end-users, a declining rate structure is implemented to encourage the use of recycled water.

Table 10.2 User fees for recycled water in Bora Bora.

Parameter	Criteria	First block	Second block	Third block
Volume for large users, m^3/month	>350 m^3	<550	550 to 800	>800
Recycled water charge, €/m^3		2.35	2.18	1.65
Volume for medium users, m^3/month	<350 m^3	<110	110 to 200	>200
Recycled water charge, €/m^3		1.16	1.08	0.88
Volume for small users, m^3/month	<30 m^3	<5	5 to 10	>10
Recycled water charge, €/m^3		0.76	0.71	0.67

According to this new recycling water pricing, established since November 2005, recycled water charges vary from 30 to 100% of potable water rate, depending on user category (water demand). In addition to these consumption-based rates, a fixed annual charge of 187 € is required for each connection. For comparison, the previous recycling water charge for polished secondary effluent (maturation pond effluents are not allowed for spray irrigation and other urban uses) was fixed at 0.67 €/m^3 regardless of water consumption. It is important to emphasize here that the use of recycled water (industrial non-potable water) and the subscription to this service are strictly reserved for professional needs of the population.

With the increasing number of end-users, a simplified pricing of recycled water was under discussion to provide additional financial incentives for water reuse (Table 10.3). In fact, it was observed that during rainy periods, a strong decrease in the demand for recycled water occurs. In such a context, the simplified pricing of recycled water that is expected to be implemented in 2011 will favour water reuse. The discount rate of recycled water will be from 26 to 48% of the drinking water rates for large users with a monthly consumption above 75 m^3.

Table 10.3 New user fees for recycled water in Bora Bora under validation.

		Volume of recycled water, m^3/month		
Category of end-users	Fixed annual charge, €/yr	0–30 m^3	31–75 m^3	>75 m^3
Small users DN 25	209	0.67 €/m^3	1.68 €/m^3	2.51 €/m^3
Midium users DN 32/40	293			
Large users DN > 40	419			

High drinking water rates for large consumers due to desalination (200 to 700% of the rate for small consumers, i.e. local population), as well as increasing water demand of luxury hotels, and in particular the requirement for landscape irrigation and other non-potable urban uses, are important factors for the acceptance of water reuse. Consequently, the main challenge of plant operation was to minimize risk of failure of tertiary treatment at acceptable operation and maintenance costs below the cost of seawater desalination.

Benefits of water recycling

Large users such as luxury hotels were the first to recognise the economic benefits of water reuse as the cost of high-quality recycled water is, in fact, 2.5 to 3 times less expensive than potable water. Consequently, recycled water demand increased, encouraged also by the declining rate.

Another economic benefit that can be easily estimated is the prevention of revenue loss of building and tourist companies. In fact, because of severe drought, at the end of 2005 the use of drinking water for non-potable purposes was proscribed and four important building sites and their landscaping were interrupted because of the lack of water for concrete preparation and tests. It would require the construction of one desalination plant for 3 hotel extensions, each with 100 luxury suites of 100 m^2 each. The economic damage that could be caused by the delay of construction has been estimated at €2 to 3 million, and without taking into account the potential loss of revenue during the peak tourist season that would have been over €50 million. The economic and social benefits of this new recycling scheme were also recognised by the local and regional authorities, as the start-up of the membrane facility enabled them to postpone the construction of an additional desalination plant and to limit water supply interruptions for local residents that were frequent and unavoidable during drought periods.

In addition to the increasing reliability of water supply, the local population underlined another very important environmental benefit of the reuse of well-treated wastewater: the safeguard of the lagoon and its biodiversity. The fragile lagoon system of Bora Bora can be rapidly deteriorated by human activities such as accidental discharge of wastewater and/or wastes.

For the local decision makers and elected officers, the most important benefit of water reuse was the protection of local natural freshwater resources, for example, saving of 10% of drinking water for domestic and potable uses. The protection of the lagoon was also a crucial benefit. Consequently, Bora Bora has become the only municipality of French Polynesia, which has been awarded the "Blue Flag of Europe". This label is highly prized by foreign visitors, mostly from northern Europe, who are used to select their vacation based on environmental criteria.

10.5 HUMAN DIMENSION OF WATER REUSE

Generally, technical issues are ranked as less important in the surveys of public perception of water reuse by individual consumers. However, these issues are crucial for the trust and acceptance of water reuse by large end-users such as industries, farmers and tourist facilities, as demonstrated by the experience of Bora Bora. In this case, water quality compliance, availability and reliability of the supply of recycled water were very important factors for the acceptance to supplement potable water supply with recycled water in tourist resorts with a high risk for direct human contact. The success of the water recycling scheme in Bora Bora is almost the only water reuse project with sprinkler irrigation approved by French Health Authorities since the mid 2000s. An important specificity of the new French regulation on water reuse for irrigation, published in August 2010, is the existence of several restrictions on spray irrigation in urban areas and golf courses, despite requirements on water quality and minimal distances. Health concerns and the precautionary principle were the main factors for some opposition of regulatory bodies for the implementation of landscape irrigation with recycled water (Lazarova & Brissaud, 2007).

In addition to technical factors, other economic and environmental factors greatly influenced the Bora Bora community acceptance of water reuse.

Public education and communication strategy

The public education and involvement program was an essential part of the water reuse project with strong involvement of the local media. As a results, a numbers of brochures, flyers, articles in the newspapers, TV and radio speeches have been published to convey information to a large number of people.

Community consultation and information were performed by the organisation of forums (public meetings) with broad interest groups including local populations, local entities, municipalities, water utilities, legislative officers from the earliest stage of development of water reuse programs. Several workshops have been organised with interest groups to validate proposed technologies, their performances, risks associated, costs and benefits to catchment scale water conservation and sustainable development.

The sustainable development policy was the key issue of public forums and education programs. Water reuse is included as an important element of this strategy as a complementary and cost effective alternative resource to desalination. The highlights from the discussions on the perception of water reuse during the public forums and workshops are as follows:

- Strong support of the local population for the protection of the environment, and in particular the lagoon from wastewater discharge,
- Recognition of environmental benefits of water reuse by all stakeholders not only for the protection of the environment, but also for the valorisation of desalted water by its recycling for non-potable purposes,
- Understanding of the relevance of the holistic approach for sustainable development including water, waste and energy management with social and economic advantages for the local population.

The role of decision makers

One of the most important keys to the success of the water reuse scheme in Bora Bora was the strong political engagement and motivation of the elected representatives, in particular, the leading role of the mayor Gaston Tong Sang. Periodically, public meetings had been organised to present and explain the policy of integrated resource management respecting the natural environment and integrating the spirit of the unique Polynesian traditional way of life (Figure 10.6). The Water Supply Master Plan for the next ten years includes water reuse as an important element of the integrated resource management strategy.

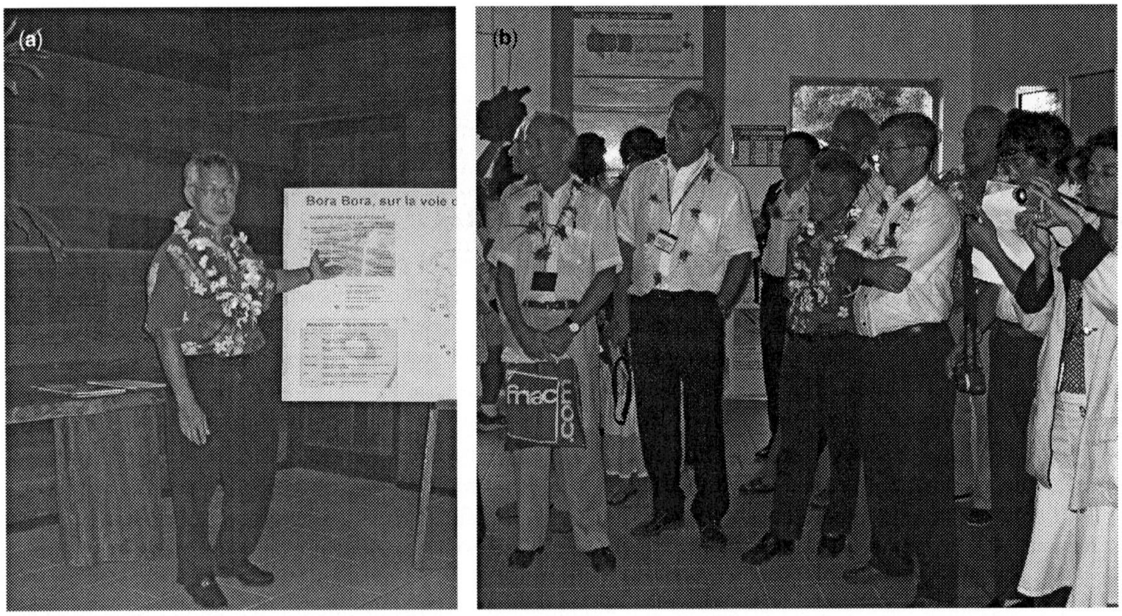

Figure 10.6 View from a scoping meeting on integrated water resource management and water reuse by the mayor of Bora Bora (a) and participant in a water reuse workshop (b).

Health authorities have been included in the public consultations since the beginning of the project and actively participated in the scoping meeting and workshops. The management of water quality and the reliability of operation were the critical factors for the trust of local authorities in the effective risk management. The positive feedback from operation and from large end-users allowed obtaining permits non only for the landscape irrigation, but also for various other non-potable urban uses.

Public acceptance and involvement

The good acceptance of water reuse and the diversification of water reuse applications were favoured by the trust of stakeholders in the safety of recycled water and the advantages of water recycling. The water reuse scheme and advanced tertiary treatment have been presented and discussed during public and scoping meetings, 3 to 6 times a year.

In particular, public forums organised in the treatment plant were very successful (Figure 10.7). During these meetings, not only major stakeholders were invited, but also different interest groups and the local population.

Figure 10.7 Public forums organised in the water recycling plant.

10.6 CONCLUSIONS AND LESSONS LEARNED

To better preserve public health and overcome all constraints of public perception, a membrane tertiary treatment has been implemented in the tourist Island of Bora Bora for the production of high-quality recycled water. The advanced wastewater treatment using ultrafiltration helped to develop trust in water reuse. Consequently, the demand for recycled water steadily increased during the last few years with a wide diversification of urban uses, not only for irrigation, but also for other urban uses such as cleaning, industrial and commercial uses and fire protection.

The main keys to success for this water reuse scheme were the strong commitment of elected officers and the implementation of adequate public communication and education programmes. The leading role of politicians and the good governance, with the suitable pricing of water services, enabled the achievement of good cost efficiency of water reuse.

The close involvement of local authorities, water professionals, all stakeholders and local population have made it possible to recognize the economic viability of water reuse in Bora Bora and to clearly identify the numerous advantages and benefits. This includes the preservation of the outstanding environment and the release of an equivalent volume of potable water for domestic purposes, local economic development, and a reliable water supply to the local population in case of drought.

The presence of luxury tourism on the Island is not such a source of environmental degradation, as might be expected. On the contrary, thanks to the suitable pricing of water services, which calls for a greater contribution to the hotels, tourism provides funding for high quality public services. This includes highly efficient desalination facilities and water recycling that compensate for the water stress and protect the lagoon, which is the most precious heritage of the Island. The strong effort of the community of Bora Bora for integrated resource management was granted in 2000 with the label "Blue Flag of Europe", which is the symbol of an exemplary environmental quality.

The main outcomes have been the trust in recycled water and the recognition of the economic and environmental benefits of water reuse with perspectives for new water reuse projects. Following the great success of urban water reuse in Bora Bora, several new water reuse projects with the extension of the existing recycling facility are now under consideration, including the extension of the fire protection network and construction of a new fire reservoir and the construction of a new membrane recycling facility with the production of multi-quality recycled water for golf course irrigation (150 ha) and aquifer recharge (UF/RO treatment facility) for indirect potable reuse.

The main drivers, benefits and challenges of water reuse in Bora Bora are summarised in the following Table 10.4.

Table 10.4 Summary of lessons learned.

Main drivers/Opportunities	Benefits	Challenges
Water stress and increasing water demand due to tourism development. Municipal policy of sustainable development. Pollution of the lagoon with risk for the fragile ecosystem.	Saving of high quality freshwater water for potable water supply (10% of the water demand). Safeguard of the lagoon and its biodiversity. Cost saving for large end users (charge for recycled water 2.5–3.0 times cheaper). Prevention of revenue loss of building and tourist companies due to potable water use restrictions during droughts (estimated up to €3 to 50 million). Fast and easier implementation than new freshwater supply.	Ensure high reliability of operation of membrane filtration and efficient preventive maintenance. Manage reliable operation through flood conditions and recycled water supply without interruptions. Define the most adequate pricing strategy of water services. Overcome an initial poor perception of recycled water by luxury hotels.

REFERENCES AND FURTHER READING

Anderson J., Baggett S., Jeffrey P., McPherson L., Marks J. and Rosenblum E. (2008). Public acceptance of water reuse. In: Water Reuse, An International Survey of Current Practice, Issues and Needs, B. Jiménez and T. Asano (eds), IWA Publishing, London, UK, pp. 332–350.

Asano T., Burton F. J., Leverenz H. L., Tsuchihashi R. and Tchobanoglous G. (2007). Water Reuse: Issues, Technology, and Applications. McGraw–Hill Professional Publishing, New York, USA.

Jiménez B. and Asano T. eds. (2008). Water Reuse, An International Survey of Current Practice, Issues and Needs. IWA Publishing, London, UK.

Jiménez B. and Asano T. (2008). Water reclamation and reuse around the world. In: Water Reuse, An International Survey of Current Practice, Issues and Needs, B. Jiménez and T. Asano (eds), IWA Publishing, London, UK, pp. 3–26.

Hatton MacDonald D. and Proctor W. (2008). The economic dilemmas of water management and reuse. In: Water Reuse, An International Survey of Current Practice, Issues and Needs, B. Jiménez and T. Asano (eds), IWA Publishing, London, UK, pp. 299–315.

Lazarova V. and Brissaud F. (2007). Intérêt, bénéfices et contraintes de la réutilisation des eaux usées en France. *L'eau, l'industrie, les nuisances*, **299**, 43–53.

Lazarova V., Carle H. and Sturny V. (2007). Economic and environmental benefits of urban water reuse in tourist areas, Proc. of the 6th IWA Spec. Conf. on Wastewater Reclamation and Reuse for Sustainability "Guiding the Growth of Water Reuse", October 9–10, Antwerp, Belgium.

Lazarova V., Rougé P. and Sturny V. (2006). Evaluation of economic viability and benefits of urban water reuse and its contribution to sustainable development, Proc. IWA Water Congress, Beijing.

Morris J., Lazarova V. and Tyrrel S. (2005). Economics of water recycling for irrigation. In: Irrigation with Recycled Water: Agriculture, Turfgrass and Landscape, V. Lazarova and A. Bahri (eds), CRC Press, Boca Raton, pp. 266–283.

View of the landscape of the luxury hotels in Bora Bora.

11 Australia's urban and residential water reuse schemes

John Anderson

CHAPTER HIGHLIGHTS

In Australia, the drinking water needs of some new residential housing areas are being reduced by 40% or more through use of recycled water for garden watering, toilet flushing and other non-potable purposes. Recycled water is supplied to residential houses through a second reticulation and clearly marked recycled water services lines in each property. Current plans provide for recycled water supply to about 250,000 houses by 2030.

KEYS TO SUCCESS

- Limits on new drinking water sources
- Development of clear guidelines for urban and residential use
- Government support and Ministerial endorsement
- Support from health authorities based on quantitative health risk assessment
- Adaption of U.S. experience to local conditions

KEY FIGURES

- Projected future volume of residential recycled water use: about 37 Mm^3/yr by 2030
- Recycled water tariffs: typically 80% of drinking water price
- Water reuse standards: fit for purpose based on risk assessment: Typically <1 *E.coli*/100 mL and virus-free

11.1 INTRODUCTION

The drivers for water reuse in Australia

Australia is a dry country. Mean annual runoff is about 50 mm, a total of 400,000 Mm^3/yr. About 70% of this runoff occurs as floods. About half of the 120,000 Mm^3/yr of divertible water occurs north of the Tropic of Capricorn and Tasmania, remote from the main centres of population. Of the remainder, about 24,000 Mm^3/yr is harvested for agricultural, industrial and urban use.

Most water is used on a "once-through" basis before return to the natural water cycle by discharge or evaporation. About 18% is used for urban and industrial uses, 74% for irrigation and 8% for rural stock and domestic use. Over 50% of this use occurs within the Murray-Darling Basin where the available surface water resources are almost fully committed to irrigation. Urban use, including commercial and industrial use, averages about 3,500 Mm^3/yr. Over half this amount, about 2,000 Mm^3/yr, is returned to the environment as wastewater flows after various forms of treatment.

In the past, Australia has been extremely liberal in its approach to water rights and water use. The available water is now heavily committed in a number of river basins. Some are showing signs of environmental stress, manifested through declining water quality. The rivers of the Murray-Darling Basin have deteriorated markedly since European settlement, principally because of agricultural land use practices and diversion of water for irrigation, but also because of nutrients in urban wastewater and stormwater discharges. Similar deterioration has occurred in the coastal rivers that supply the major cities, caused principally by urban diversions and urban runoff. These occurrences highlight the impacts which occur as water use approaches or exceeds the limits of sustainability in individual catchments and the constraints imposed on the use of water resources by the combined impact of water diversions and urban runoff.

Parts of Australia were affected by severe droughts from 1978 to 1983, in the 1990s and again from 2002 to 2007. The recognition that water resources are finite, limited and affected by drought has generated new approaches to the management of water systems including pursuit of water savings and the development of water recycling.

New South Wales residential reuse initiative – historical background

In the 1970s, the NSW Public Works Department helped local government authorities to build 20 dual reticulation water systems in western New South Wales. A second filtered water reticulation was constructed for indoor use and the original unfiltered river water supply was retained for outdoor use. This approach allowed for cost-effective filtration of water supplies as a health measure while providing ample water for garden irrigation in inland areas where annual evaporation is about 3 m.

Between 1978 and 1983, parts of south-east Australia experienced their worst drought since the commencement of rainfall records. In response to this event, the New South Wales Minister for Public Works, Mr Ferguson, established the NSW Recycled Water Coordination Committee to promote reuse and to advise the Government on guidelines and technology. The committee helped establish new guidelines for use of recycled water for agricultural irrigation in 1987.

The NSW dual reticulation experience led to recognition that there was potential to reduce demands on limited drinking water supplies by supplying recycled water for garden irrigation and other needs through in a second reticulation (Anderson, 1986).

In 1989, the NSW Recycled Water Coordination Committee, with assistance from Shoalhaven City Council and the NSW Public Works Department, organised a pilot project at Shoalhaven Heads. A second reticulation was constructed to deliver filtered and disinfected recycled water from the local water reclamation plant to 19 residential homes in Shoalhaven Heads for garden watering and other outdoor uses for an 18- month trial period.

In 1991 the NSW Recycled Water Coordination Committee prepared draft guidelines for residential use of recycled water using the operating experience gained during the Shoalhaven Heads pilot project (NSW RWCC, 1993). The NSW draft guidelines also drew on US experience including the risk management work of Rose and Gerba (1990), the California and Florida recycled water systems and the California Title 22 regulations.

After endorsement by the health and environmental agencies, the NSW guidelines for residential use of recycled water were published in May 1993. A consultation between Health department officials and Professor Joan Rose during her visit to Australia in May 1992 greatly facilitated the approval process.

The guidelines allowed residential use of filtered and disinfected recycled water for toilet flushing, garden watering, washing of cars and other outdoor uses, but not for uses with high ingestion risks such as filling of home swimming pools. The guidelines also set out some basic requirements for water quality testing, recycled water reticulation systems, cross-connection control and system management.

Australian confidence in the viability of residential reuse systems was enhanced when an Australian firm developed a cost-effective modular system for microfiltration of recycled water in 1993. The use of microfiltration as a disinfection step provides greater certainty that multi-barrier water reclamation systems will remove pathogens.

Development of guidelines

Subsequent to the development of the New South Wales guidelines for urban and residential use of recycled water in 1993, the other Australian states developed similar guidelines providing for production of Class A or Class A+ recycled water suitable for urban and residential uses.

Australia has now developed national water recycling guidelines (EPHC, 2005, 2006, 2008a, 2008b, 2009a, 2009b). Unlike the earlier state guidelines which were prescriptive in terms of recycled water quality and treatment processes, the national guidelines follow an approach which requires the production of recycled water quality that is fit for purpose based on comprehensive health risk assessments for the planned application.

The national water recycling guidelines series (Figure 11.1) now includes guidelines for stormwater reuse; augmenting drinking water supplies (EPHC, 2008b) and managed aquifer recharge (EPHC, 2009).

11.2 CASE STUDIES: THE PIONEERING PROJECTS

Rouse Hill (New South Wales)

Soon after the publication of the NSW guidelines for urban and residential use of reclaimed water, Sydney Water announced its intention to develop a major dual reticulation scheme at Rouse Hill in Sydney's north-west. A major driver for this scheme was the need to limit treated wastewater and nutrient discharges from large new housing developments to small local streams that flow into the Hawkesbury River. By recycling water, environmental impacts and constraints on growth were reduced. The scheme commenced operation in 2001. On average the Rouse Hill scheme has reduced demand for drinking water by about 40%.

Sydney Olympic Park (New South Wales)

As part of the "green" credentials of the 2000 Olympic Games in Sydney, the Sydney Olympic Park Authority developed the Homebush Bay Water Reclamation and Management Scheme (WRAMS) at the Olympic Park site (Figure 11.2).

Figure 11.1 Australian guidelines for water recycling (Photograph courtesy of National Health and Medical Research Council).

Figure 11.2 Homebush Bay Water Recycling Plant (Photograph courtesy of Sydney Olympic Park Authority).

Stormwater from the site was collected in a disused brickpit and recycled for irrigation of sporting fields and landscaping in the Olympic Park precinct, and toilet flushing in the sporting venues. Also wastewater from local catchments was diverted from a local pumping station, treated in a 2200 m^3/d wastewater treatment plant. The recycled water and stormwater are further treated in recycled water treatment plant which incorporates continuous microfiltration plant with 7500 m^3/d capacity, reverse osmosis units of 2000 m^3/d to reduce the salinity of the recycled water, and chlorination before delivery to the residential reuse area and the sporting venues (Chapman, 2005).

Subsequent to the Olympic Games, the athletes' village at Newington was converted to 2000 residential homes which are supplied with recycled water through the dual reticulation network for garden watering, toilet flushing and clothes washing.

11.3 CASE STUDIES: RESIDENTIAL REUSE AUSTRALIA-WIDE

New South Wales

Western Sydney

The current planning schemes in Sydney provide for up to 300,000 new homes in western Sydney, 140,000 in the north-western sector and 160,000 in the south-western sector. The building sustainability requirements for new dwellings in Sydney require new houses to achieve water savings of 40% relative to 2001 levels. Residential reuse is one option for achieving the level of water savings required.

Rouse Hill, Sydney

Sydney Water is continuing the expansion of Australia's largest residential water recycling scheme in the Rouse Hill area in Sydney's north-west (Figure 11.3). The scheme started in 2001, and over 19,000 homes are now using up to 1.7 Mm^3/yr of recycled water for flushing toilets, watering gardens, washing cars and other outdoor uses. On average the Rouse Hill scheme has reduced demand for drinking water by about 40%. Eventually the Rouse Hill scheme will serve 36,000 homes. The recycled water supply area includes parts of Acacia Gardens, Beaumont Hills, Castle Hill, Glenwood, Kellyville, Kellyville Ridge, Parklea, Quakers Hill, Stanhope Gardens, The Ponds and Rouse Hill. The Rouse Hill Water Recycling Plant treats about 4.7 Mm^3/yr of wastewater to residential use standards. The capacity of the Rouse Hill Recycled Water Plant has recently being doubled to 27,000 m^3/d (Sydney Water 2011a).

Figure 11.3 Rouse Hill Water Recycling Plant (Photograph courtesy of Sydney Water).

Hoxton Park, Sydney

The new Hoxton Park Recycled Water Scheme will be one of the largest residential water recycling projects in Australia. Sydney Water is building the scheme in the Liverpool and Campbelltown local government areas of south-western Sydney. It will supply recycled water to about 14,000 future homes as well as industrial developments in areas including Edmondson Park, Middleton Grange, Ingleburn Gardens, Panorama Estate and Yarrunga Industrial Area.

The Hoxton Park Recycled Water Scheme will be delivered in two stages. Stage 1 will be ready from 2013 and Stage 2 in about 2017, depending on the progress of development in the serviced areas. Recycled water will be used for non-drinking uses like watering gardens, flushing toilets, washing cars and in some factory processes. It will be treated to a very high standard at a new water recycling plant at Glenfield which will be completed in 2013. Stage 1 of the scheme is expected to supply about 0.9 Mm^3/yr of recycled water to about 7,000 homes and businesses by 2015 (Sydney Water, 2011b).

Ropes Creek, Sydney

At Ropes Creek, construction of dual reticulation pipelines is continuing as the development near St Marys expands. The Ropes Crossing Recycled Water Scheme will provide recycled water to new homes and for irrigation of local playing fields. When completed, it will provide 1550 homes with up to 0.5 Mm^3/yr of high quality recycled water.

Pitt Town, Sydney

A private developer is proposing to develop a dual reticulation scheme for 1000 houses at Pitt Town in Sydney's north-west. This scheme will be the first such scheme developed under the NSW Water Competition Act which allows for competitive provision of water services.

Hunter

Hunter Water serves the large urban area around Newcastle, 160 km north of Sydney. Hunter Water is planning to supply recycled water to new residential development areas at Gillieston Heights, North Cooranbong and Thornton North. The dual reticulation schemes currently planned will serve about 10,200 homes and will use about 1.2 Mm^3/yr of recycled water.

Ballina

Ballina Shire is a fast growing coastal community in northern New South Wales, about 800 km north of Sydney. Ballina Shire has made dual reticulation water recycling mandatory in all new residential subdivisions. The recycled water supply will be commissioned in 2012. It is anticipated that up to 7200 residential properties will be connected to the recycled water system over the next 20 to 30 years.

Victoria

Yarra Valley Water, Melbourne

Yarra Valley Water has embarked on a major program to supply Class A recycled water to new residential developments in north-eastern Melbourne. Yarra Valley Water has committed to supplying recycled water to around 100,000 homes or approximately 200,000 people in the area. Recycled water is approved for laundry water use in the Yarra Valley area and a recycled water tap is being installed in the laundry of new houses for connection to the washing machine.

Kalkallo Project

Yarra Valley's innovative Kalkallo Stormwater Harvesting and Reuse Project is one of international significance, with the ultimate possibility of supplementing the Kalkallo region's local drinking water supply with treated stormwater. It will involve capturing and treating around 365,000 m^3/yr of stormwater harvested from a commercial development 38 km north of Melbourne's central business district (Yarra Valley Water, 2011a).

Doncaster Hill principal activity centre

An innovative scheme developed by Yarra Valley Water will provide recycled water to developments in Doncaster Hill by the end of 2013. While recycled water schemes are becoming common in greenfield urban developments, Doncaster Hill is believed to be Australia's first high density urban redevelopment to incorporate a dual pipe system to deliver recycled water to residents. Yarra Valley Water is working together with Manningham City Council on the project, which will see recycled water delivered to around 4000 new residential dwellings in Doncaster Hill developments by the end of 2013 (Yarra Valley Water, 2011b).

City West Water, Melbourne

Melbourne Water is working with City West Water and government departments to supply recycled water to residential developments in the City of Wyndham. One of these, a master plan for Werribee Fields, is a new 2000-home 'green' suburb displaying market-leading sustainable water and energy use including a dual reticulation recycled water system (City West Water, 2011).

Western Water, Melbourne

Eynesbury

Homes in the new township of Eynesbury are using Class A recycled water from the Melton Recycled Water Plant for toilet flushing and outdoor use. The first residential connection occurred in December 2008. When fully developed, the Eynesbury area will have 2900 houses. In 2009/10, 226,000 m^3 of recycled water were used at Eynesbury. Residents used 6500 m^3 of drinking water and 9,000 m^3 of recycled water. An additional 217,000 m^3 of recycled water was used for irrigating recreational areas, including the golf course at Eynesbury (Western Water, 2010).

Toolern

Melton is a particular focus for Western Water's integrated water cycle management strategy. It is one of the lowest-rainfall areas in Victoria. It is also one of the fastest-growing urban areas in Australia, with the new suburb of Toolern expected to absorb an extra 55,000 residents by 2030. Toolern's completed Precinct Structure Plan requires a minimum 50% reduction in average household drinking water consumption. Just like in the Eynesbury development, all 24,000 new homes at Toolern will be supplied by Western Water with Class A recycled water as well as drinking water, through a dual pipe system. Class A recycled water is for garden use, car washing and toilet flushing. At Toolern, the integrated management of stormwater involves stormwater capture, storage in wetlands and treatment for appropriate use, while environmental flows to rivers and streams will be prioritised. Western Water's aim is to make Toolern an IWCM showcase to achieve 100% net reduction in drinking water use, making it Australia's first water-neutral suburb (Western Water, 2011).

South East Water, Melbourne

Class A Recycled water is now a reality for a growing number of customers across South East Water's region. More than 7000 customers are now connected to recycled water infrastructure across an increasing number of estates. With the recent drought resulting in reduced water levels in our catchments, much is being done to look at sustainable and innovative water supply solutions. The proposed upgrade of the Pakenham treatment plant and the Eastern treatment plant in 2012 will provide a major boost to the supply of Class A recycled water and allow further growth of this valuable resource.

The dual reticulation scheme at Hunt Club Estate in Cranbourne East was recently opened (Figure 11.4). Around 1200 homes at the estate will eventually be connected to recycled water, saving about 200,000 m^3 of drinking water each year. These are the first of up to 43,000 lots to be supplied with recycled water from the Pakenham and Eastern treatment plants Customers with recycled water supply receive a range of great benefits.

Figure 11.4 Dual meter installation (Photograph courtesy of South East Water).

Barwon Water, Geelong

Barwon Water, based in Geelong, Victoria, will invest around AU$82 million (66 million €) over the next 10 years in water, sewage and recycled water infrastructure to the new Armstrong Creek residential development south of Geelong. With a land area of 2500 ha, Armstrong Creek, between Grovedale and Mt Duneed, is currently the largest residential growth area in Victoria. The "super-suburb" is expected to provide for 22,000 homes. Armstrong Creek residents will have access to recycled water on tap.

All Armstrong Creek residents will benefit from a dedicated "purple pipe" delivering Class A recycled water for flushing their toilets, washing their cars and watering their gardens. Recycled water will also be used for irrigating public spaces and sporting grounds. The use of recycled water at Armstrong Creek is expected to save more 2.4 Mm^3 of drinking water a year. It will also reduce the amount of water discharged to Bass Strait from the Black Rock Water Reclamation Plant.

Queensland

Pimpama-Coomera Waterfuture Project, Gold Coast

Pimpana-Coomera Waterfuture Master Plan which provides for water sensitive urban design, rainwater tanks, reduced infiltration gravity sewers and dual reticulation recycled water networks with the 7700 ha Pimpama-Coomera urban development area in the Gold Coast hinterland south of Brisbane. The area is expected to cater for a future population of about 120,000 people.

The Pimpama-Coomera Master Plan requires new housing in the areas to have a dual reticulation recycled water supply to garden watering and toilet flushing and rainwater tanks to supply laundry water and one outside tap. The Plan is likely to reduce drinking water supply needs to about 16% of typical demand in conventional residential developments. The Pimpama water recycling plant is designed to provide Class A+ recycled water using a reclamation process that includes ultrafiltration and UV disinfection followed by chlorine disinfection to maintain a residual greater than 0.5 mg/L.

About 4400 homes were completed prior to the commissioning of the water recycling plant and initially were initially supplied with drinking water in the recycled water pipes. The recycled water system commenced operation in December 2009.

South Australia

Mawson Lakes, Adelaide

Mawson Lakes is a suburb about 11 kilometres north of Adelaide and a major feature of the Mawson Lakes development is the innovative AU$16 million (13 million €) water recycling system which complements the normal mains water supply (Figure 11.5). Recycled water is water derived from sewerage systems and treated to a standard which is satisfactory for its intended use.

Figure 11.5 View of the landscape areas in Mawson Lakes irrigated with recycled water (Photograph courtesy of SA Water, Adelaide).

The system distributes a mixture of highly treated wastewater from SA Water's Bolivar Wastewater Treatment Plant and stormwater harvested in Salisbury that has been cleansed and treated through a series of engineered wetlands. Residents can use recycled water for toilet flushing, watering the garden and washing the car. The recycled water is also being used for irrigation of public parks and reserves. Recycled water is delivered via clearly distinguishable purple coloured pipes, mains, meters and taps (SA Water 2011a). Recycled water is delivered to about 4000 homes in the area and can save about 800,000 m^3 of drinking water being drawn from the River Murray each year.

Southern Urban Reuse Project, Adelaide

The AU$62.6 million (50.8 million €) Southern Urban Reuse Project is jointly funded by the South Australian Government and the Federal Government. The project was completed in June 2011 (SA Water, 2011b).

Each year up to 1.6 Mm^3 of treated wastewater will be transferred from the Christies Beach Wastewater Treatment Plant to the Aldinga Wastewater Treatment Plant where it will be stored. The wastewater will then be further treated before being provided to up to 8000 new homes in the south (Seaford Meadows in the first instance) for use in dual reticulation systems (Figure 11.6). This is similar to Mawson Lakes where treated wastewater is used for domestic purposes including toilet flushing and garden watering.

Figure 11.6 Views of dual service lines in a house under construction (a) and dual service water meters for irrigation (b) (Photograph courtesy of SA Water, Adelaide).

Western Australia

Perth

A new dual reticulation scheme will supply groundwater to 1,000 homes at Brighton, a northern suburb of Perth in Western Australia. The non-drinking water can be used for toilet flushing, watering lawns and gardens, clothes washing and car washing.

11.4 MAIN CHALLENGES AND LESSONS LEARNED

System management

Experience has shown that supply authorities need to put sound system management and monitoring system in place for residential reuse schemes to ensure that the recycled water is of high quality, is used as intended and that the integrity of the drinking water system is not compromised.

There have been a number of cross-connection incidents since the first residential reuse schemes commenced operation in the year 2000. The Rouse Hill scheme has had five cross-connection incidents caused by incorrect household plumbing.

A cross-connection incident marred the start-up of the recycled water supply in the Pimpama-Coomera area in Queensland, Australia. The incident occurred shortly after the recycled water supply was commissioned in December 2009. Recycled water was detected in the drinking water supply in an area of about 630 homes. The incident was traced an incorrect service line connection made by a sub-contractor. The defect was not detected in pre-commissioning tests (Health Stream, 2010).

There was a single house incident in the Sydney Olympic Park scheme where a householder connected an ice-making machine to a recycled water pipe within his home.

The major incidents have been quickly detected and corrected. No serious health effects have been reported from these incidents.

The adopted system management and monitoring measures include:

- Implementation of ISO 14001 quality management plan for recycled water systems
- The use of purple pipe and marking tapes for reticulation pipes and household service lines.
- The installation of backflow prevention valves on potable water connections.
- Inspection and testing of household meter and valve installations at house completion and when properties change ownership.
- Education of plumbing contractors in the correct installation of dual systems.
- Education of householders in correct use of recycled water.
- On-line monitoring of key recycled water quality parameters.
- Mandatory reporting of any recycled water quality events and incidents to health authorities.

South East Water requires householders that receive recycled water to conduct a simple self-check of the household plumbing annually to test for any cross connections (South East Water, 2011).

Health studies

The safety of recycled water use at Rouse Hill was tested in an epidemiological study conducted by the Hawkesbury-Hills Division of General Practice in association with Monash University. More than 35,800 patient records collected from 11 general practitioners across the areas serviced by the dual-reticulation water system and from areas not serviced by the system. The records were checked for the incidences of gastroenteritis, respiratory and skin complaints, urinary tract infections and muscular-skeletal diseases.

No increased incidence of disease was found in people served by the dual-reticulation system.

Costs and pricing

The construction of a second reticulation network for recycled water in new greenfield housing areas involves additional capital costs. There may be, however, some capital cost offsets:

- If fire hydrants are on the recycled water supply, there are savings on the drinking water network because peak flows are lower.
- Drinking water treatment costs are lower and there may also be savings in water supply headworks costs.
- Cost for discharge of reclaimed water may be lower.

The most common practice in Australia is to charge for recycled water in residential schemes at a price which is about 80% of the drinking water price. The additional capital costs of the recycled water system are most commonly built into the price of new housing lots, reflecting the environmental costs associated with providing new housing lots when new drinking water sources are limited and there are environmental constraints on new discharges in local rivers or estuaries.

Public education and acceptance

The water authorities responsible for new residential reuse schemes conduct public education programs for residents. Community meetings are held and educational information is supplied to all households. Education programs are also conducted for local plumbers and contractors likely to work in the recycled water supply area. The Sydney Water educational materials include recycled water factsheets "Using recycled water around your home" and "Gardening with recycled water" (www.sydneywater.com.au/Water4Life).

Sydney Water has also established a water recycling education centre at the St Marys water recycling plant. Students can tour a working water recycling plant and see how water recycling technology works. They can also learn about the current and future role of recycled water in the Sydney water supply.

There have been high levels of public acceptance and government support for the residential water recycling schemes in Australia, as evidenced by the widespread adoption for new housing areas and the rapidly growing number of houses served by recycled water.

11.5 OVERVIEW

Following the success of the early residential reuse projects in New South Wales, Australian water authorities have adopted residential use of recycled water for both large and small systems as a means of catering for growth, diversifying supplies and reducing the impact on limited local water resources.

At the end of 2011 there were about 30,000 homes connected to residential reuse schemes in Australia. Full development of current areas, together with proposed schemes, could see another 290,000 homes connected over the next 20 years.

Residential reuse schemes are most attractive in the southern parts of Australia where annual rainfalls are around 500 mm and summers are usually hot and dry. The benefits are not as great in the temperate areas along the east coast where annual rainfalls are 900 mm or more and there is usually some summer rainfall.

For the initial residential reuse schemes in Sydney, the combination of residential reuse and irrigation of public open space areas in new development areas reduces annual demands on the drinking water supply by about 40%. Larger water savings in drinking water use are projected for the new residential reuse schemes in Melbourne and Adelaide which have drier climates and higher garden water needs in summer.

In Australia, garden watering needs make up the majority of peak water demands in dry periods. The transfer of garden irrigation to the recycled water reticulation reduces peak demands on drinking water systems by 60% or more.

The following Table 11.1 summarizes the main drivers for urban residential water reuse in Australia as well as the benefits obtained and some of the challenges for the future.

Table 11.1 Summary of lessons learned.

Drivers/Opportunities	Benefits	Challenges
Limits on new drinking water sources. Higher standards for discharges to the environment.	Annual drinking water use reduced by 40% or more. Peak demands on drinking water systems reduced by 60% or more.	Recycled water quality monitoring. Quality controls to prevent cross-connections. Community education.

REFERENCES AND FURTHER READING

Allconnex. (2011). Pimpama-Coomera Water Future Project, www.allconnex.com.au/MYHOME/RECYCLEDWATER/Pages/PCWPack.aspx

Chapman H. (2005). WRAMS, Sustainable water recycling. In: *Proc Integrated Concepts in Water Recycling*, S. J. Khan, M. Muston and A. I. Schafer (eds), University of Wollongong, Australia 2005.

City West Water. (2011). *West Werribee Dual Supply Project*, www.citywestwater.com.au/our_assets/west_werribee.aspx

De Rooy E. and Engelbrecht E. (2003). Experience with residential water recycling at Rouse Hill. Proc Water Recycling Australia, Aust Water Assn., Brisbane 2003.

EPHC, NRMMC & AHMC. (2005). Australian Guidelines for Water Recycling: Managing Health and Environmental Risks – Impact Assessment. Environment Protection and Heritage Council, Natural Resource Management Ministerial Council & Australian Health Ministers Conference. Sep 2005.

EPHC, NRMMC & AHMC. (2006). Australian Guidelines for Water Recycling: Managing Health and Environmental Risks. Environment Protection and Heritage Council, Natural Resource Management Ministerial Council & Australian Health Ministers Conference. Nov 2006.

EPHC, NRMMC & AHMC. (2008a). Overview Document – Australian Guidelines for Water Recycling: Managing Health and Environmental Risks. Environment Protection and Heritage Council, Natural Resource Management Ministerial Council & Australian Health Ministers Conference. Mar 2008.

EPHC, NRMMC & AHMC. (2008b). Australian Guidelines for Water Recycling: Augmentation of Drinking Water Supplies. Environment Protection and Heritage Council, Natural Resource Management Ministerial Council & Australian Health Ministers Conference. May 2008.

EPHC, NRMMC & AHMC. (2009). Australian Guidelines for Water Recycling: Managed Aquifer Recharge. Environment Protection and Heritage Council, Natural Resource Management Ministerial Council & Australian Health Ministers Conference. Jul 2009.

Health Stream. (2010). The Pimpama-Coomera Cross-Connection Incident. Health Stream n57 March 2010, http://www.wqra.com.au.

NSW RWCC. (1993). NSW Guidelines for Urban and Residential Use of Reclaimed Water. NSW Recycled Water Coordination Committee. May 1993.

SA Water. (2011a). Mawson Lakes Recycled Water System, www.sawater.com.au/SAWater/WhatsNew/MajorProjects/mawson_lakes.htm

SA Water. (2011b). Southern Urban Reuse Project, www.sawater.com.au/SAWater/WhatsNew/MajorProjects/SthUrbanReuse.htm

South East Water. (2011). Have you done your plumbing check? *Recycled Water Newsletter*, Apr 2011(2), p. 2. www.southeastwater.com.au/SiteCollectionDocuments/Recycled_Water/Recycled_Newsletter_Edition2.pdf

Sydney Water. (2011a). Rouse Hill Recycled Water Scheme, www.sydneywater.com.au/Water4Life/RecyclingandReuse/RecyclingAndReuseInAction/RouseHill.cfm

Sydney Water. (2011b). Hoxton Park Recycled Water Scheme, www.sydneywater.com.au/MajorProjects/SouthWest/HoxtonPark/index.cfm

Western Water. (2010). Eynesbury update, Western Water Recycled Water News Autumn 2010, Mar 2010.

Western Water. (2011). Toolern development: A showcase for Integrated Water Cycle Management, Western Water Recycled Water News, April 2011.

Yarra Valley Water. (2011a). Kalkallo Stormwater Harvesting and Reuse Project, www.yvw.com.au/Home/Aboutus/Ourprojects/Currentprojects/Kalkallostormwaterharvesting/index.htm

Yarra Valley Water. (2011b). Doncaster Hill Principal Activity Centre, www.yvw.com.au/Home/Aboutus/Ourprojects/Currentprojects/DoncasterHill/index.htm

PART III

Urban water reuse: decentralised water recycling systems

Foreword

By The Editors

Historical development

The history of recycled water use in individual buildings in Japan dates back to 1965 and earlier, although the number of installations was small. A severe drought in 1978 prompted the use of recycled water in buildings in Fukuoka in southern Japan. There have been about 100 new installations annually in Japan since 1980, with more than 1475 individual office buildings and apartment complexes equipped with on-site water recycling systems in 1997. A factor in the installations in public buildings in Tokyo is their function as emergency service centres following natural disasters. In addition, large dual distribution systems for toilet flushing and landscape irrigation have been constructed in many other cities (Fukuoka, Kobe, Nagoya, etc.).

In 1991 the Irvine Ranch Water District in California adopted a specific policy and implemented the use of recycled water in six high-rise commercial buildings in Irvine for toilet flushing, followed in 2002 by the first high-rise building equipped with water reuse system for the air conditioning cooling towers. Many other projects for in-building recycling, including greywater recycling, have been implemented in other states (California, Florida, Texas, etc.).

Water reuse and on-site recycling in urban buildings: case studies

Part 3 of this book "Milestones in Water Reuse: The Best Success Stories" presents a number of case studies showing how decentralised water recycling systems have been used to meet non-potable water needs in public and residential buildings.

Chapter 12 presents the first milestone in Japan of water reuse in large urban areas in Tokyo's Shinjuku area. Few years later, the water reuse program was extended to five other large areas, mainly for toilet flushing. One of the first exhibition spaces on water recycling in the world was opened in 1986 in Shinjuku area which greatly contributed to promote and explain water reuse to large public and decision makers. The lessons learned demonstrated how to resolve the technical and economic challenges, maximising the benefits and establishing win-win relationship with building owners.

Chapter 13 describes the first "green" high-rise apartment building in North America, the Solaire in Manhattan, New York. The reuse system is located in the building basement and the recycled water is used for toilet flushing, cooling water, and landscape irrigation. The system has been able consistently to offset potable water use through recycling 100% of the treated water.

Chapter 14. More than 2500 high rise buildings in Japan now have on-site water recycling. This chapter points out the main technical challenges for the selection, design and operation of on-site treatment systems. Four standardized treatment process trains for on-site water recycling are approved in Japan, including the promising membrane bioreactor technology.

Keys to success

The water recycling systems presented in the case studies are typically halving the drinking water needs of high-rise buildings. Benefits include savings for the building owners and savings for the supply authorities by reducing system expansion costs to

Table 1 Highlights and lessons learned from the selected case studies of decentralised water recycling in urban areas and high-rise buildings.

Project/Location	Start-up/Capacity	Type of uses	Key Figures	Drivers and Opportunities	Benefits	Challenges	Keys to Success
Shinjuku area and five other large urban areas in Tokyo, Japan	Since 1984: Shinjuku area (30 buildings) Since 1990s 5 other areas Total treatment capacity: 450,000 m^3/d	Toilet flushing, Garden watering, Recreational use	Total volume of treated wastewater: 138.6 Mm^3/yr Total volume of recycled water: 1.1 Mm^3/yr Recycled water tariffs: 3.41 US$/$m^3$ Water reuse standards: $<$10 total coliforms/mL for toilet flushing	Willingness of the Tokyo Metropolitan Government to develop an effective water cycle. Good collaboration of stakeholders.	Saving of potable water which avoided the expansion of the capacity of potable water supply system in congested urban areas. Cost saving for building owners.	Reliable recycled water supply despite the high variations of hourly water demand of high-rise buildings. Good aesthetic quality of recycled water to avoid complaints on colour or odour.	Win-win relationship between the building owners and the Tokyo metropolitan government. Reliable water supply in terms of quality and quantity with lower cost. Reduced cost of reclaimed water transportation and distribution due to the low transportation distances. Relatively low initial cost for the dual pipe distribution due to early planning since the building design.
The Solaire building, Battery Park City, New York, United States	2006 Treatment capacity: 95 m^3/d	Toilet flushing, Landscape irrigation, Cooling water make-up, Cleaning	Total volume of treated and recycled water: 35,000 m^3/yr. Capital costs: US $560,000. Recycled water tariffs: included in rental fee Water reuse standards: $<$1 fecal coliform/ 100 mL	Regulatory requirements. Reduction of sewer overflow. Population growth and increasing water demand.	A show case of "green" building. Reduction of water demand and water fee. Reduction of discharges of stormwater and wastewater. Avoided cost to upgrade and expand the City's existing water and sewer systems	Limited space in the building basement for the reuse system. Relatively high energy demand for the membrane bioreactors.	Regulatory mandate for sustainable development by the Battery Park City Authority. Public support through grants and incentives. Utilization of an emerging technology at that time – membrane bioreactor. A team of people including the building manager, engineers and system operators who continue to optimize the system.
On-site recycling in high-rise buildings, Japan	Start-up before 1965 with accelerated growth since 1980s Number of installations about 2500	Toilet flushing (water quality requirement: not detected *E.coli*/ 100 mL)	Total floor area of an example building: 563,800 m^2 Day-time population: 20,000 Treatment capacity of the on-site system: 680 m^3/d Footprint of a treatment system: 600 m^2 Number of operators: 2 Frequency of monitoring: twice a week	Water shortages caused by droughts. Local regulations encouraging water reuse. Favourable taxes.	Reduction in volume of public water supply. Reduction in loading of sewer system. Maintenance of public/economic activities in disaster situations.	Combination with rainwater harvesting A water balance: the amount of lightly contaminated grey water cannot cover the amount needed for toilet flushing Treatment of heavily contaminated grey water containing grease and fats.	Local regulations and favourable taxes. Availability of a design manual for on-site wastewater reclamation. Water reuse regulations. Wastewater is divided into several fractions and is separately collected. Technological know-how of design and operation of wastewater reclamation system.

cater for high density urban areas. Incorporating this type of water recycling into new buildings has the potential to ensure that the drinking water needs of the world's rapidly growing cities can be met with less impacts on natural water resources.

Keys to success have included identifiable economic and environmental benefits, effective treatment capable of providing good quality recycled water while handling large variations in hourly demand, clear separation and marking of household pipework and good quality control.

The major highlights and lessons learned from the selected case studies are summarised in Table 1.

12 Semi-decentralized water recycling in megacities: the example of Tokyo Shinjuku Area

Kiyoaki Kitamura, Kingo Saeki and Naoyuki Funamizu

CHAPTER HIGHLIGHTS

Ochiai wastewater treatment plant (currently, Ochiai wastewater reclamation plant) started to supply 1400 m^3/d of reclaimed wastewater polished by tertiary treatment via rapid sand filtration followed by chlorination to high-rise buildings in Shinjuku area for toilet flushing in 1984. This project is the first milestone in Japan of water reuse in large urban areas.

KEYS TO SUCCESS

- The win-win relationship between the building owners in the project area and the Tokyo metropolitan government.
- The buildings are able to receive stable water supply in terms of quality and quantity with cheaper cost.
- Since the Shinjuku area is very high density of high-rise buildings, costs of reclaimed water transportation and distribution was reduced due to the low transportation distances.
- The relatively low initial cost for the dual pipe distribution system in the building due to early planning since the building design.

KEY FIGURES

- Treatment capacity : 450,000 m^3/d
- Total annual volume of treated wastewater: 138.6 Mm^3/yr
- Total volume of recycled water: 1.1 Mm^3/yr
- Recycled water tariffs: 273 JPY/m^3 (3.41 US$/$m^3$)
- Water reuse standards: Not detected *E.coli* for toilet flushing <50 total coliforms/100 mL for garden watering and recreational uses

12.1 INTRODUCTION

General description of water reuse in Tokyo

The population in the Tokyo Metropolitan District is approximately 33 million and their water is supplied from the dam lakes built in three watersheds, the Tone River, the Ara River, and the Tama River. All domestic wastewater from this region is treated and discharged to Tokyo Bay and the Pacific Ocean through rivers. The area's climate is moderate and the population is enjoying four seasons. The rainy season and typhoon attack give the heavy rain, but in the other seasons droughts are experienced for which the water resource facilities had not been well prepared. In spite of 1500 mm of annual rain fall, the available water resource per capita accounts only 900 m^3/yr. This value is one third of the average value in Japan, because of the high population density in Tokyo. This situation leads to developing more water resource and promoting rain water harvesting and wastewater reuse (Sone, 2004).

Tokyo metropolitan government issued the plan, "Water recycle master plan" in 1999 to achieve effective use of limited water resource and to create adequate water cycle in this region. This master plan consist of four basic concepts including "developing effective water cycle in the urban area" and the seven goals. To realize these seven goals, seventeen measures are set. Among seventeen measures, the Bureau of Sewerage Works has responsibility to promote effective use of reclaimed wastewater and to develop the large area water reuse systems. The Bureau of Sewerage Works of Tokyo is operating twenty wastewater treatment plants and they treat 580 million m^3 of wastewater daily in 2009 from 1.3 million residents in approximately 1100 km^2 of planned area. Since treated wastewater is stable in terms of quality and quantity, the Bureau is promoting wastewater reclamation and reuse as a reliable source of water in Tokyo.

Brief history of water reuse development

The wastewater reuse in Tokyo has a long history and has started in 1955. A rapid sand filtration unit was constructed after activated sludge plant in Mikawashima wastewater treatment plant and reclaimed wastewater was supplied to industries to prevent over-drafting of groundwater in Tokyo Bay area. The capacity of this first reclamation plant was 15,000 m^3/d.

In the 1970s, Japan experienced severe droughts in wide areas of the country. For example, in 1978, the prolonged drought conditions in the City of Fukuoka forced citizens to accept serious water supply limitations for 283 days. These drought experiences led people to recognize reclaimed water as valuable alternative resource in urban areas. In the 1980s, the local governments undertook many planned water reuse projects and national government supported these projects by providing subsidies (Ogoshi *et al.*, 2001; Funamizu *et al.*, 2008).

The Ochiai wastewater treatment plant has started to supply 1400 m^3/d of reclaimed wastewater polished by rapid sand filtration to high-rise buildings in Shinjuku area for toilet flushing in 1984 (Figure 12.1). This project was the first milestone in Japan of water reuse in large urban areas. In 1990s, this type of area-wide water reuse systems were constructed in the New Water Front area (Figure 12.2), the East block of Shinagawa area, Ohsaki district and Shiodome area (Figure 12.3). In 2007 the system has been expanded to the Nagata-Cho area. Nowadays in Tokyo there are seven areas where the reclaimed water from wastewater reclamation plants are used for toilet flushing and their total area covers 1137 ha. In total, 9087 m^3 of reclaimed water are supplied as shown in Table 12.1.

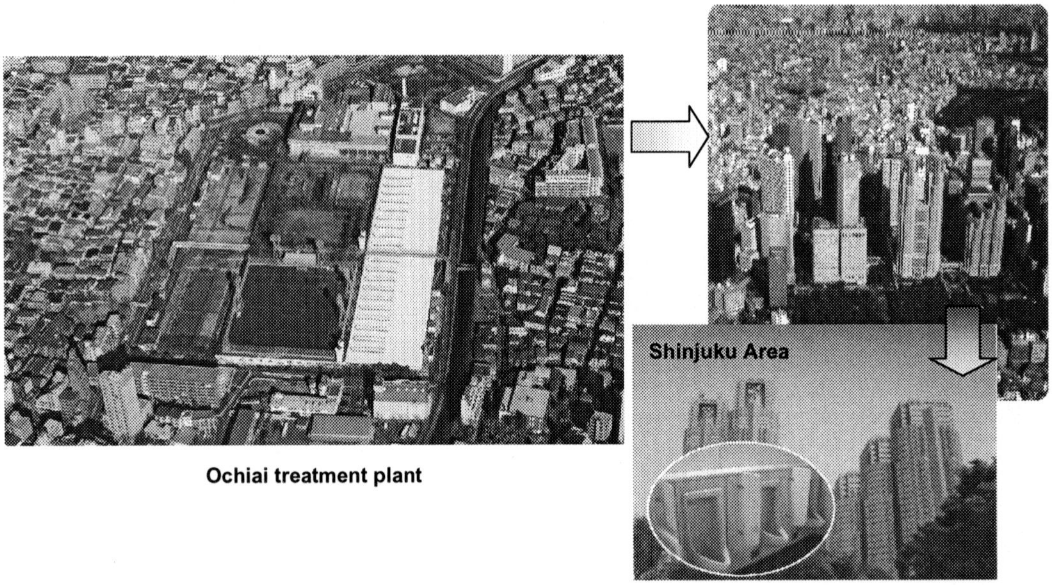

Figure 12.1 Views of the Ochiai Wastewater Reclamation Plant and the buildings supplied with recycled water in Shinjuku area (Photography credit: Tokyo Metropolitan Government).

Figure 12.2 Views of the Ariake Wastewater Reclamation Plant and the New Water Front area supplied with recycled water (Photography credit: Tokyo Metropolitan Government).

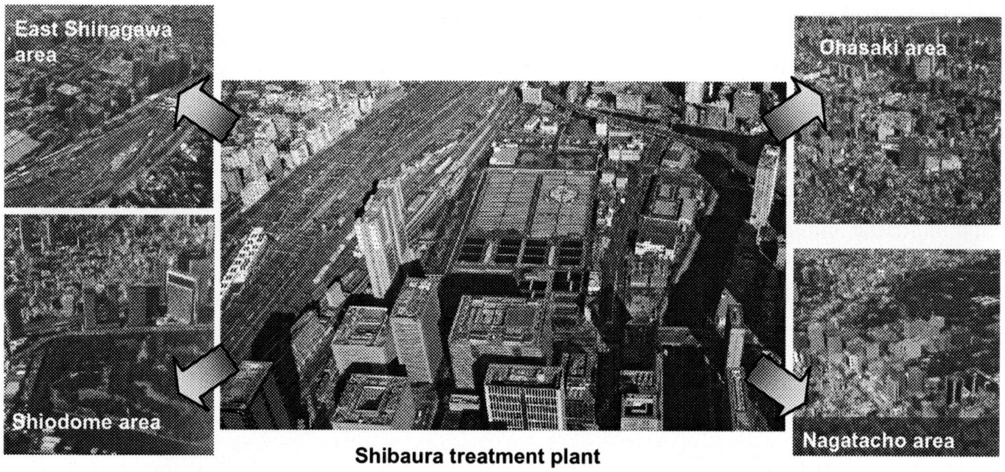

Figure 12.3 Views of the Shibaura Wastewater Reclamation Plant and four urban areas supplied with recycled water (Photography credit: Tokyo Metropolitan Government).

The reuse has been increasing, and the percentage reached about 9% of total volume of treated wastewater in 2009. The Tokyo Bureau of Sewerage Works is now promoting water recycling for toilet flushing by means of large area water reuse systems as shown in Figures 12.1 to 12.3.

The water quality criteria for water reclamation and reuse in Tokyo are summarized in Table 12.2. Since a dual pipe distribution system is an essential part for reuse of reclaimed wastewater for toilet flushing in a building, as a measure to promote this area-wide water reuse, the Tokyo Bureau of Sewerage Works asks the owners of building to install dual pipe systems at the initial design stage when they plan to construct large buildings having certain scale and floor space.

Table 12.1 Wastewater reclamation systems in Tokyo.

Area	Treatment plants	Water reclamation treatment trains	Supply (2009), m^3/d	Covered area, ha	Distribution network, km
Shinjuku	Ochiai	Rapid sand filter	3016	80	14.8
New water front	Ariake	Bio-filter + Ozone	2193	681	26.0
East Shinagawa	Shibaura	Rapid sand filter + Ozone + MF	2135	83	
Ohsaki				67	41.7
Shiodome			1120	31	
Nagata-Cho			222	138	
Yashio			401	57	
Total			9087	1137	82.5

Table 12.2 Reclaimed water quality criteria in Tokyo.

Parameter	Toilet Flushing	Cleaning, cooling fire fighting	Garden watering	Sustaining river flow landscape irrigation	Recreational uses
Total coliform counts	–	10/mL	50/100 mL	1000/100 mL	50/100 mL
E.coli	Not detected	–	–	–	–
Residual chlorine, mg/L	Trace amount	Trace amount	>0.4	–	–
Colour	Not unpleasant	Not unpleasant	Not unpleasant	<40	<10
Turbidity	Not unpleasant	Not unpleasant	Not unpleasant	<10	<5
BOD, mg/L	–	–	<20	<10	<3
Odour	Not unpleasant	Not unpleasant	Not unpleasant	Not unpleasant	Not unpleasant
pH	5.8–8.6	5.8–8.6	5.8–8.6	5.8–8.6	5.8–8.6

12.2 THE WATER REUSE PROJECT IN SHINJUKU AREA

General description and applications

The water reuse project in Shinjuku area started in 1984 and it supplied about 1400 m^3/d of reclaimed water to nine high-rise buildings in the beginning for toilet flushing. This project was the first milestone of the area-wide water reuse in Japan. Now its customers have increased to 30 high-rise buildings and the total amount of reclaimed water is 3000 m^3/d (Figure 12.4). In this project wastewater is treated and reclaimed at the Ochiai wastewater reclamation plant (see Figure 12.1).

The treatment train for the Shinjuku water reuse project is illustrated in Figure 12.5. The reclaimed water is transported to the water recycling centre located in the centre of the Shinjuku high-rise building district. The treated effluent of the plant is also used for restoration of river flow. This application of stream flow augmentation in Tokyo is described in Chapter 19.

Figure 12.4 Views of the high-rise buildings in the Shinjuku area in Tokyo supplied with recycled water for toilet flushing.

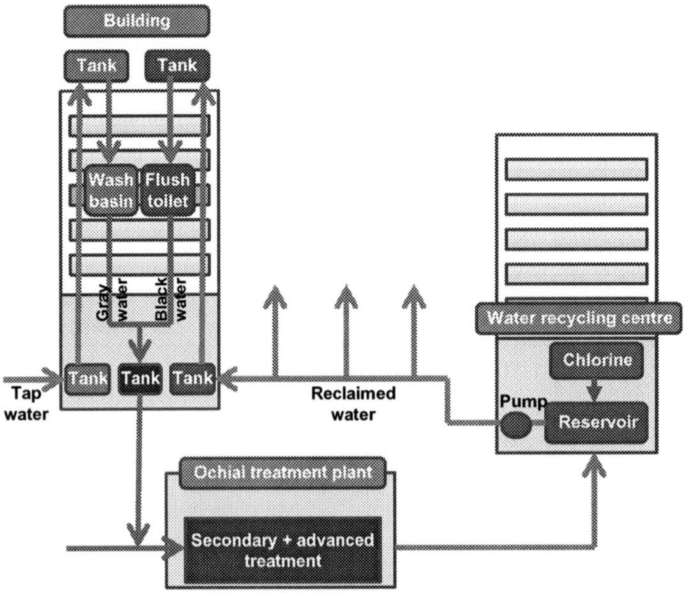

Figure 12.5 Schematics of the water recycling project of Shinjuku Area for toilet flushing in high-rise buildings.

The distance from Ochiai treatment plant to the water recycling centre is 9.1 km and these two facilities are connected by three pipe lines (Figure 12.6). The recycling centre is located at the basement of the Shinjuku International Building. At the water recycling centre, reclaimed water is disinfected by chlorine and stored in the distribution reservoirs with total capacity of 2909 m^3.

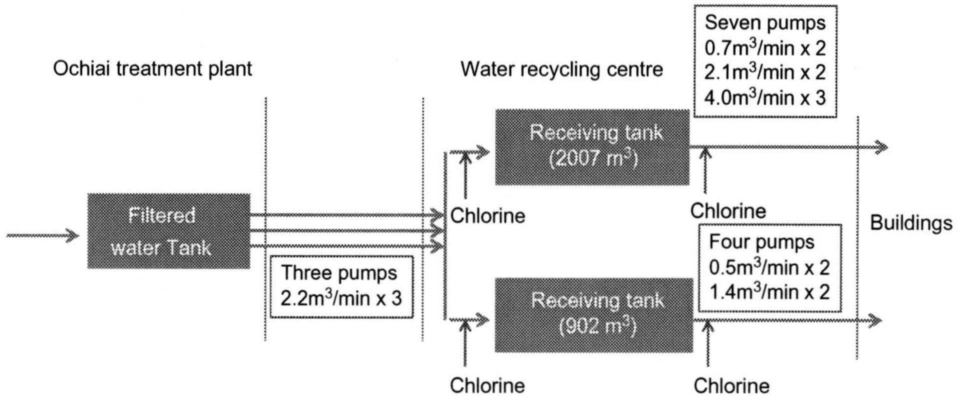

Figure 12.6 Diagram of reclaimed water transportation in Shinjuku Area.

The chlorinated water is distributed to 30 high-rise buildings by pumps from the recycling centre. The covered area is 80 ha. The buildings have a dual water distribution system consisting of receiving tank and distribution pipes. One distribution system is for tap water from Tokyo metropolitan water supply system and it covers wash basins, tea service area and restaurants in the building. The other system distributes reclaimed water for toilet flushing. The grey water from wash basin and restaurants and the black water from toilet are discharged to sewer pipe and transported to Ochiai Treatment plant.

Treatment train for water recycling at the Ochiai Treatment Plant

A conventional activated sludge system is used as a secondary treatment at the Ochiai wastewater reclamation plant and the secondary effluent is filtered by a sand filtration system as reclamation treatment. Ochiai treatment plant started their operation in 1964, its design capacity is 450,000 m^3/d and now it is treating an average daily flow of 380,000 m^3/d. The water quality of raw sewage, secondary effluent and recycled water after tertiary treatment is summarized in Table 12.3.

Table 12.3 Water qualities of raw water, secondary effluent and reclaimed water quality[1] (South series).

Parameter	Raw wastewater	Secondary effluent	Reclaimed water[2]
BOD, mg/L	180	2	1
Suspended solids, mg/L	150	3	<1
Total nitrogen, mg/L	33.1	12.3	11.4
Ammonia, NH_4-N, mg/L	19.9	0.3	0.1
Nitrate, NO_3-N, mg/L	0.2	10.4	10.8
Total phosphorus, mg/L	3.6	1.6	1.5
Orthophosphates, PO_4-P, mg/L	1.6	1.4	1.4
Total coliform count/mL	–	100	3

[1]average values in 2009.
[2]before chlorination.

Challenges in operation of reclaimed water distribution system

Three parallel pipelines with a length of 9.1 km are installed between the Ochiai wastewater reclamation plant and the water recycling centre (see Figure 12.6) in order to reduce the risk of shutting down the transporting system even during earthquake events. As mentioned previously, the water recycling centre is located at the basement of the Shinjuku International Building and is equipped with two water distribution reservoirs with a capacity of 2007 m^3 and 902 m^3, respectively, for compensating hourly variation of water demand for toilet flushing. The recycled water is distributed by nine distribution pumps and the capacities of these pumps are ranging from 30 m^3/h to 240 m^3/h (0.5 m^3/min to 4.0 m^3/min as shown in Figure 12.6). These different capacity pumps are operated and controlled for adapting the hourly variation of water demand coupling with storing water depths of the two receiving tanks. Since the most of the customers of the project are office buildings, the centre has to operate under big hourly variations of water demand. In addition, the recycling centre has two sets of chlorine dose systems and water distribution control unit for adapting to emergency situations.

In order to achieve a stable water supply, flow rate, pressure, water quality and residual chlorine concentration are monitored continuously and precisely controlled at the water recycling centre. Maintenance work of pipes is also well emphasized in this project. Since all accident either at the water transportation lines from the treatment plant to the recycling centre or water distribution pipes from the centre to the buildings can cause very big damage to their customers, water pressure and flow rate in water transportation and distribution pipes are monitored carefully to find immediately the failure of the system.

Challenges in operation of dual distribution system in high-rise buildings

As mentioned above, at the beginning of this project in 1984, reclaimed water was supplied to nine high-rise buildings. Since then the number of buildings supplied with reclaimed water gradually increased and thirty buildings are now receiving reclaimed water. Most of the buildings are high-rise office complexes, including the Tokyo Metropolitan Government building. Along with the increase in the number of connected buildings, the recycled water demand increased from 1400 m^3/d in 1984 to 3000 m^3/d in 2009 as shown in Figure 12.7.

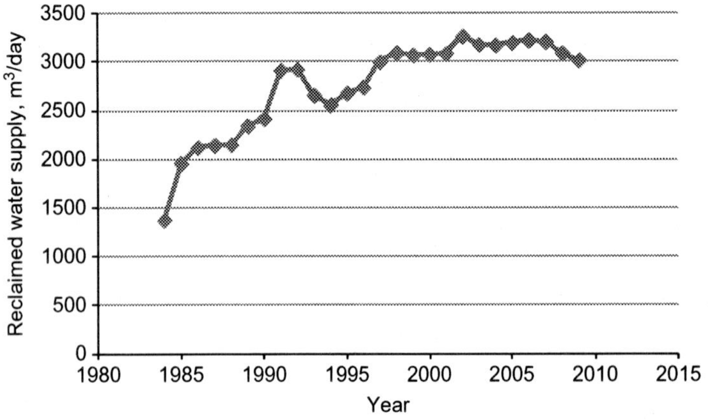

Figure 12.7 Increase in the total volume of reclaimed water supplied from 1984 to 2009.

The buildings receiving reclaimed water for toilet flushing have a dual water distribution system as shown in Figure 12.5. Installing such a dual distribution system is not costly if the dual distribution system is considered at the original design stage for the buildings. This is due to the floor plan of the office buildings that toilet and water service area are located at the same place in each floor and they are easily connected vertically. The zone of restaurants is planned separately from office working zones. The receiving water tanks for reclaimed water are equipped with an additional tap water line to supply tap water coping with an accident or emergency in reclaimed water system.

Pricing strategy of recycled water

The price of the reclaimed water in the Shinjuku Area is 273 Japanese yen (JPY)/m^3, equivalent to 3.41 US$/$m^3$. This tariff covers construction and operating & maintenance costs of advanced treatment unit in the wastewater reclamation plant, recycled water transport and distribution system. When the project started in 1984, the price system was 250 JPY/m^3 (3.12 US$/$m^3$) for less than 15,000 m^3/month of water use and 200 JPY/m^3 (2.5 US$/$m^3$) for more than 15,000 m^3/month. The income and outgo plan of the project was re-evaluated and the current price system was applied in 1995.

A comparison of water charges applied by the Tokyo metropolitan government for tap and reclaimed water uses are shown in Tables 12.4 and 12.5 for two cases depending on water consumption. The comparison of water charges shows that using reclaimed water for toilet flushing is cheaper than tap water use by about 23%.

Table 12.4 Comparison of total water charge for office building using 6,000 m^3/d of toilet flushing water.

	Tap water is used for all purposes		Reclaimed water is use for toilet flushing and tap water is use for other purposes	
Parameter	**in JPY***	**in US$**	**in JPY**	**in US$**
Monthly Total Charge	4,745,210	60,836	3,738,714	47,932
Items				
Charge for water supply	2,644,496	33,904	1,638,000	21,000
Charge for wastewater	2,100,714	26,932	2,100,714	26,932
Cost for 1 m^3 of water use	791 JPY/m^3	10 US$/$m^3$	623 JPY/m^3	8 US$/$m^3$

*1 US$ = 78JPY (in September, 2011).

Table 12.5 Comparison of total water charge for office building using 3000 m^3/day of toilet flushing water.

	Tap water is used for all purposes		Reclaimed water is use for toilet flushing and tap water is use for other purposes	
Parameter	**in JPY***	**in US$**	**in JPY**	**in US$**
Monthly Total Charge	2,385,860	30,588	1,832,964	23,500
Items				
Charge for water supply	1,371,896	17,588	819,000	10,500
Charge for wastewater	1,013,964	13,000	1,013,964	13,000
Cost for 1 m^3 of water use	795 JPY/m^3	10.2 US$/$m^3$	611 JPY/m^3	7.8 US$/$m^3$

*1 US$ = 78JPY (in September, 2011).

Since the distance of reclaimed water transportation is relatively short, e.g. 9.1 km from the Ochiai water reclamation plant to the reclaimed water use area, as well as to the high-rise buildings located in a smaller area of 80 ha, the initial capital cost was relatively low and maintenance and operation cost could be reduced. This is the main reason why the price of reclaimed water is cheaper than tap water (Funamizu *et al.* 1998).

Public acceptance and involvement

Public education and communication are considered as very important elements of this water reuse project. An exhibition space was opened at the water recycling centre in Shinjuku area in 1986 (Figure 12.8). This facility has several showpieces which include the explanation of the total system of the water reclamation and water transportation from Ochiai wastewater reclamation plant to high-rise buildings in Shinjuku area, as well as samples of raw wastewater, secondary effluent, and reclaimed water. Because of the advantages that the centre was located at the central part of Shinjuku area, this exhibition space received lots of visitors and contributed to promotion of water reuse in Tokyo. This exhibition space was closed in 2010 by ending the mission for more than 20 years as a bridge for information between people and water reclamation plant.

The owners of the buildings are satisfied with this project. There are three reasons for receiving user's satisfaction and acceptance. First one is well controlled water supply in terms of quality and quantity. Few claims are reported on water quality, such as, for example, colour or odour issues. The second reason is the price of water, because, as mentioned above, the cost for reclaimed water is cheaper than the tap water. The third reason is the initial cost for dual distribution system in the building. The buildings in this project area were planned to have a dual distribution system initially from the designing stage. This means that they have cost effective floor plan for the dual distribution system, which reduced the installation cost of dual pipe system significantly.

Figure 12.8 The exhibition space in the water recycling centre (ending the mission and closed in 2010).

The Tokyo metropolitan government is promoting water reuse by showing how much CO_2 reduction and cost for water will be reduced. Their estimate shows that 69,000 kg of CO_2 and 5.8 million JPY (74,360 US$) will be saved annually for 100 m^3/d toilet flushing use if the building uses the project reclaimed water instead of an individual water recycling system.

The metropolitan government set a guideline to request building owners to install several measures for efficient use of water and storm water management such as a dual distribution system, rain water harvesting for newly constructing buildings with total floor space greater than 10,000 m^2 or with footprint area greater that 3000 m^2. The Tokyo metropolitan government also is requesting building owners to install a dual pipe distribution system in the three areas listed in Table 12.1.

12.3 CONCLUSIONS: KEYS TO SUCCESS OF URBAN WATER REUSE IN TOKYO

One of the major keys to success of the water reuse project in Shinjuku area in Tokyo has been based on the win-win relationship between the building owners in the project area and the Tokyo metropolitan government. The buildings are able to receive stable water supply for toilet flushing continuously along with cheaper cost. On the other hand, the Tokyo metropolitan government can develop the Shinjuku area with very high density buildings without expanding the capacity of potable water supply system.

The building owners are very satisfied with the water reuse project results. The first reason is the well controlled water supply in terms of quality and quantity. The optimum control system, which includes two reservoir tanks and nine pumps, is ensuring a reliable reclaimed water supply even under conditions of very significant hourly variations of water demand of office buildings. In order to achieve a stable water supply, all the operating parameters are controlled on-line (flow rate, pressure, residual chlorine concentration) at the water recycling centre. Maintenance work on pipes is also emphasized in this project. The second reason of building owners satisfaction is the price of water. The Tokyo metropolitan government was set a lower price for reclaimed water expecting a reduction of 20% of the total water charge for the buildings. Such a lower price setting is possible because the capital and operation/maintenance costs of reclaimed water transportation and distribution were reduced due to the low transportation distances. The third reason is the relatively low initial cost for the dual pipe distribution system in the building due to the early planning since the building design stage.

The water reuse in Tokyo has a long history since 1955. To achieve the effective use of limited water resource and to create the adequate water cycle in this region, the Bureau of Sewerage of the Tokyo Metropolitan Government has responsibility to promote effective use of reclaimed water and to develop urban water reuse systems. With the greatest supports of public and decision makers, there are seven large urban areas in Tokyo where the reclaimed water is used for toilet flushing covering a total area of 1137 ha with the supply of 9087 m^3 of reclaimed water.

The following Table 12.6 summarizes the main drivers of the water reuse in Tokyo as well as the benefits and challenges.

Table 12.6 Summary of lessons learned.

Drivers/Opportunities	Benefits	Challenges
Willingness of the Tokyo Metropolitan Government to develop an effective water cycle in the urban area. Win-win relationship between decision makers and end-users.	Saving of potable water which avoided the expansion of the capacity of potable water supply system in congested urban areas. Cost saving for building owners.	Reliable recycled water supply despite the high variations of hourly water demand of high-rise buildings. Good aesthetic quality of recycled water to avoid complaints on colour or odour.

REFERENCES AND FURTHER READING

Funamizu N., Ohgaki S. and Asano T. (1998). Wastewater reuse and water environment. In: *Water Cycle and its Environment*, H. Takahashi and Y. Kawada (eds), Iwanami, Japan, pp. 211–239. (In Japanese).

Funamizu N., Onitsuka T. and Hatori S. (2008). Water reuse in Japan. In: *Water Reuse: An International Survey of Current Practice*, B. Jimenez and T. Asano (eds), IWA Publishing, London, UK, pp. 373–386.

Ogoshi M., Suzuki Y. and Asano T. (2001). Water reuse in Japan. *Water Science and Technology*, **43**(10), 17–23.

Sone K. (2004). Wastewater reuse in Tokyo, Japan. *Newsletter*, October 2004, 2–5, International Water Association, Specialist Group on Water Reuse.

WEB-LINKS FOR FURTHER READING

Tokyo metropolitan government. (2011). http://www.gesui.metro.tokyo.jp/odekake/sise_list.htm

13 Water reuse in the America's first green high-rise residential building – the Solaire

Yanjin Liu, Eugenio Giraldo and Mark W. LeChevallier

CHAPTER HIGHLIGHTS

The Solaire was a pioneer for green buildings as it was the first high-rise green building in the US. It was also the first LEED certified residential high-rise building in North America with an integrated design for the site, facility, landscape, and water management. The Solaire reuse system is located in the building basement and the recycled water is used for toilet flushing, cooling water, and landscape irrigation. The system has been able to consistently offset potable water use through recycling 100% of the treated water.

KEYS TO SUCCESS

- Regulatory mandate for sustainable development by the Battery Park City Authority
- Public support through grants and incentives
- Utilization of an emerging technology at that time – membrane bioreactor
- A team of people including the building manager, engineers and system operators who continue to optimize the system

KEY FIGURES

- Treatment capacity: 95 m^3/d
- Total annual volume of treated wastewater: 35,000 m^3/yr
- Total volume of recycled water: 35,000 m^3/yr
- Capital costs: US$560,000
- Recycled water tariffs: included in rental fee
- Water reuse standards: <1 fecal coliform/100 mL

13.1 INTRODUCTION

The 293-unit, Solaire Apartments in Battery Park City in lower Manhattan, New York City, was the first "green" residential high-rise building in the U.S. that incorporates advanced technologies to achieve water recycling and reuse within the building, and was one of the most significant projects that influenced the recent water and energy conservation campaign (Figure 13.1). The system is located in the building basement and the recycled water is used for toilet flushing, cooling water, and landscape irrigation. The system has adopted features that have set a model for the new developments in the future as populations continue to grow and water resources become more limited. It is also unique for such a system to be located in an urban setting as reuse systems are more commonly found in rural or suburban environments where access to public systems are often restricted.

Figure 13.1 The Solaire – America's first high-rise residential green building.

As municipal water supply and wastewater treatment costs continue to rise in the New York City, where the population is growing rapidly toward nine million people, and environmental sustainability becomes a more important focus, water reuse has proved to be a beneficial and economical alternative in green building design. The Solaire project was driven by an institutional commitment to sustainable design, as expressed in the Battery Park City Authority's environmental guidelines, an economic driver for water conservation provided by the New York City Department of Environmental Protection (NYC DEP), and as part of an integrated effort to reduce combined sewer overflows in the City (Carey, 2005; WERF, 2009).

In the New York City, approximately 70% of stormwater and domestic wastewater is collected through one common piping system, so-called combined sewer system, and then conveyed to the wastewater treatment plants for treatment (EPRI, 2010). During wet weather events, excessive stormwater and untreated raw wastewater may be discharged into New York Harbor, which is the main waterway of the City. The primary water management challenge facing the Battery Park City is the capacity of the combined sewer system and water quality problems associated with combined sewer overflows (CSOs). In the New York City, CSOs can be triggered by as little as 2.5 mm of rain and occur on average of once per week. Discharges of CSOs have adverse impact on the water quality of the New York Harbor. The discharge quality is routinely monitored by the New York City Department of Environmental Protection and the U.S. Environmental Protection Agency.

Treating collected stormwater and wastewater on-site creates the ability to reuse treated water for flushing toilets, irrigation and cooling towers; greatly reducing the amount of fresh water that is taken from a municipal water supply, eliminating the need to pump wastewater to a municipal plant, and reducing CSOs. So when the Solaire luxury high-rise building was planned for an already dense area like the Battery Park City in lower Manhattan, water recycling and reuse was considered as a strategic way to reduce CSO discharges and beneficially reuse the water resource to the most extent through collecting and treating stormwater and domestic wastewater within the building.

Main drivers for water reuse

The main drivers of in-building water recycling in New York are:

- Stringent water supply and increasing water demand,
- Combined sewer overflow reduction and control,
- Regulatory requirements by the Battery Park City Authority,
- Avoided costs in water/wastewater system expansion.

Brief history of the project development

The Battery Park City (BPC), located in the southern tip of Manhattan, is a mixed-use community of residential, commercial, and institutional properties, and covers 37.3 ha (92 acres) of land with a population of approximately 14,000. BPC did not exist fifty years ago. The land was built on swampland of the Hudson River using the dirt and rocks excavated during the construction of the World Trade Center and the sand dredged from Staten Island in the New York Harbor.

In order to mediate the CSO issues and build an environmentally sustainable community, BPC adopted a set of green development guidelines, the BPC Authority Residential Environmental Guidelines, in 1999. In the BPC Guidelines, it requires the developers to compete for new ideas to create environmentally sustainable building for new constructions. Water conservation is an important aspect of the BPC Guidelines and has been incorporated into the project. The Solaire became the first green building driven by the BPC Guidelines including the use of water conserving fixtures and appliances, rainwater capture and reuse, and wastewater reclamation and reuse.

The Albanese Development Corporation, the developer of the Solaire, won the bid to build the project out of nine top firms in 2000, and its construction was completed in August 2003. The Solaire water reuse system was designed, built and currently operated by Applied Water Management, a subsidiary of American Water. Recycling of "greywater" only, wastewater from kitchen sinks and showers, was originally considered for the Solaire. Recycling of "blackwater", wastewater from toilets, was incorporated into the design in a later phase and the wastewater discharge from the building was further reduced. Since the Solaire project, all buildings constructed in BPC have incorporated blackwater reuse in their design. Essentially, blackwater treatment has become an integrated part of the competitive process to win the right to be able to build in BPC.

Project objectives, institutional or financial incentives and water reuse applications

One of the objectives of the Solaire was to develop and implement a new practice of sustainable water management for urban development. This water management practice emphasizes all types of water including wastewater as valuable resources as opposed to burden to the environment. The collected wastewater from the building may be more than sufficient for the building use upon removal of pollutants and contaminants in the water. The recycled water can be used for toilet flushing, landscape irrigation, cooling water makeup and other cleaning purposes.

LEED, Leadership in Energy and Environmental Design, is the green building rating system developed and adopted by the U.S Green Building Council. At the time the Solaire project was originally planned and designed in 1999, LEED was not available for high-rise residential buildings. The BPC Authorities conducted extensive research and benchmarking to develop their own standards to attain a LEED certification with a level of Gold or above for the existing buildings and new buildings. The Solaire set an example and achieved a LEED Gold award in 2004 as the first green building in the U.S.

The NYC DEP has a comprehensive water reuse incentive program that offsets water reuse operational costs by providing a 25% reduction in rates for city water and sewer when at least 25% reuse is provided onsite. The Solaire building also received state tax incentives both for providing affordable housing and for offsetting the cost of some of the green building features in the design.

13.2 TECHNICAL CHALLENGES OF WATER QUALITY CONTROL

Treatment train for water recycling

The Solaire onsite wastewater treatment, storage and reuse system is located in the building's basement, and includes a series of common-walled, cast-in-place, concrete tanks (Figure 13.2). The plant is designed to provide high-level removal of organic material (measured as biochemical oxygen demand or BOD), total suspended solids (TSS) and nitrogen using a membrane bioreactor. Blackwater from the toilets and greywater from sinks and showers in the building are collected in an aerated 36 m^3 (9500 gal) feed tank and flows to a trash trap to remove larger non-biodegradable solids (American Water, 2008). A three-stage biological system consisting of an anoxic tank, an aerobic tank, and a membrane filter removes BOD, TSS and nitrogen. Phosphorus removal capability is provided to the treatment plant by addition of chemicals (aluminum salt).

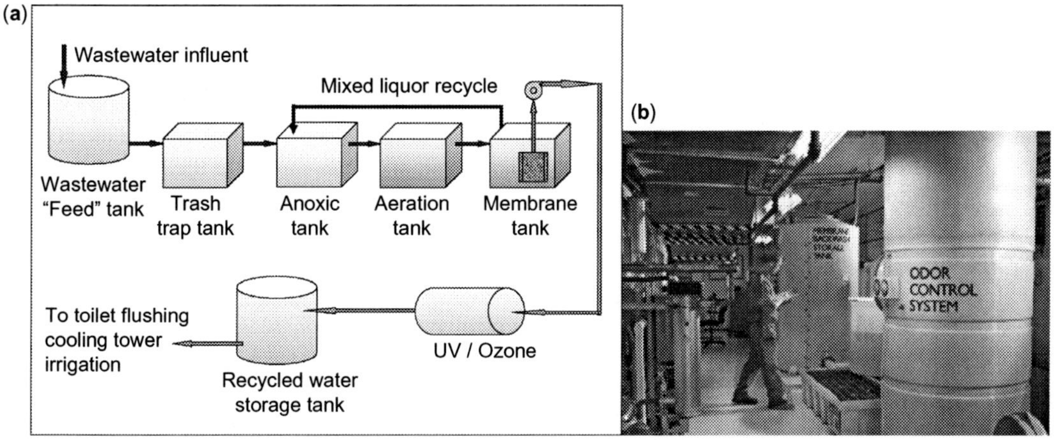

Figure 13.2 The Solaire wastewater treatment and recycling system schematic (a) and a view of plant interior (b).

The treated water is then further disinfected by ultraviolet lights. Any remaining colour is removed using ozone. The storage tanks serve as reservoirs for the treated water, which is used as flush water, make-up water for the cooling towers, and for irrigation. After water is extracted, biological sludge is sent to the city through the sanitary sewer system to aid in the municipal treatment process.

In addition, stormwater is collected and beneficially reused at the Solaire. The rooftop garden design incorporates a storm water retention system that retains rainwater using a mat below the soil surface. The rainwater can be saturated within the mat when it rains. Once the mat is saturated, excess rainwater spills over into drains on the roof's surface. The rainwater is then collected in a 38 m^3 (10,000 gal) tank in the building's basement and saved for reuse. The treated water is used for landscaping irrigation with the high-efficiency drip irrigation system just below the top of the soil throughout the roof garden. The gardens on the rooftop also provide added thermal insulation for the building and help reduce the heat island effect in the city.

Water quality control and monitoring

Plant influent wastewater sample is taken routinely to measure the wastewater characteristics for BOD, TSS, total phosphorus, orthophosphate, and TKN (total Kjeldahl nitrogen, sum of organic nitrogen and ammonia nitrogen). The typical concentrations of BOD and TSS measured in the influent are 146 mgO_2/L and 190 mg/L, respectively. The average water quality is presented in Table 13.1.

Table 13.1 Typical characteristics of raw wastewater and recycled water (treated effluent) of the Solaire.

Parameter	Units	Inlet raw wastewater	Treated recycled water
pH	–	–	6.5–8.0
Biological oxygen demand, BOD	mgO_2/L	146	<6
Total suspended solids, TSS	mg/L	190	<1
Total phosphorus	mgP/L	6.9	<1
Ortophosphates	mgP/L	3.3	
Total Kjeldahl nitrogen, TKN	mgN/L	37	
Turbidity	NTU	–	<0.2
Fecal coliforms	cfu/100 mL	–	<100
	cfu/100 mL	–	<1

The BOD and TSS concentrations in the treated wastewater effluent are typically less than 6 mgO_2/L and 1 mg/L, respectively. The pH of treated water is maintained between 6.8 and 7.2 by adding sodium hydroxide (NaOH) to the aeration tanks. Treated water is stored in a 34 m^3 (9,000 gal) storage tank. Although not for potable purpose, the typical turbidity of the treated water is between 0.1 and 0.2 NTU which is even less than the EPA's drinking water standard.

Main challenges for operation

Since the start-up of the system in 2003, Applied Water Management has worked closely with the building mangers to ensure the normal operation of this complex wastewater treatment and recycling system. They have continuously looked for opportunities to optimize and improve the efficiency and effectiveness of the system. However, there are still challenges for system operators to deal with such a compact and complex system while trying to minimize the impact on residents and the functioning of the building. The main challenges and lessons learned are briefly summarized below.

Typically, wastewater treatment plants, especially with biological treatment processes, show better performance if the influent flow is relatively consistent. But because all the influent flow to the treatment plant is generated from the building residents only, the flow at the Solaire can vary significantly depending upon the time of the day and the occupancy of the building. For example, summer and part of the winter are typically the vacation seasons when the residents leave the building, resulting in low flows. Sometime the flow is so low that the bacteria responsible for the removal of contaminants in wastewater are in jeopardy of starving due to lacking of food (typically carbon sources). When that happens, the experienced operators add food (sometime dog food) to the wastewater to keep the bacteria healthy and ready to work when needed.

A major concern with the operation of a wastewater treatment plant in a building basement is the potential for adverse impact on the residents. These issues include odour and noise which are common for a conventional wastewater treatment plant, but at Solaire precautions have been taken to reduce the odor and noise. The plant has an enclosed blower and each tank is sealed only with a hatch for operator to access the tank for inspection and maintenance. In addition, the air from the head space of the feed tank is vacuumed and piped to a carbon adsorption system located in the basement, which is then exhausted above the roof level to eliminate chances of odor from the plant escaping and entering the apartment building.

The original plant design did not include biological phosphorus removal, but chemical removal of phosphorus is now provided to minimize scaling in the building cooling system that uses recycled water. In some instances, recycled water is blended with city water to augment the flow. An on-going research project is taking place by a team of scientists and operators from American Water, with an aim to optimizing phosphorus removal while minimizing the use of the chemicals and energy.

13.3 WATER REUSE APPLICATIONS

The treated water is beneficially reused in the apartment building (Krogmann *et al.* 2007). Out of the 95 m^3/d (25,000 gal/d) that are recycled, 34 m^3/d (9000 gal/d) are used to flush toilets, 39 m^3/d (11,500 gal/d) go to the cooling tower, and 22 m^3/d

(6000 gal/d) are used for landscape irrigation. These applications of the treated wastewater required the approval from the City's Health Department. The Solaire only uses reclaimed water for non-potable purposes and does not supply the water for other purposes to prevent cross-contamination and drinking the reclaimed water. The treated stormwater is stored and used for irrigation of the green roof and rain garden.

Evolution of the volume of supplied recycled water

The influent flow to the Solaire wastewater system is continuously metered and the monthly data relative to 2009 are shown in Figure 13.3 (American Water 2010). The monthly wastewater volume varies significantly and ranges from 2300 to 4200 m^3. The low volume treated during the month of August 2009 was because the residents go on vacation during that period. Due to the beneficial reuse of the recycled water, the use of city potable water has been significantly reduced with only minor supplement when necessary. Figure 13.3 shows also the amount of city water needed to supplement flushing and cooling water use, which takes approximately only 15% of the total average water use.

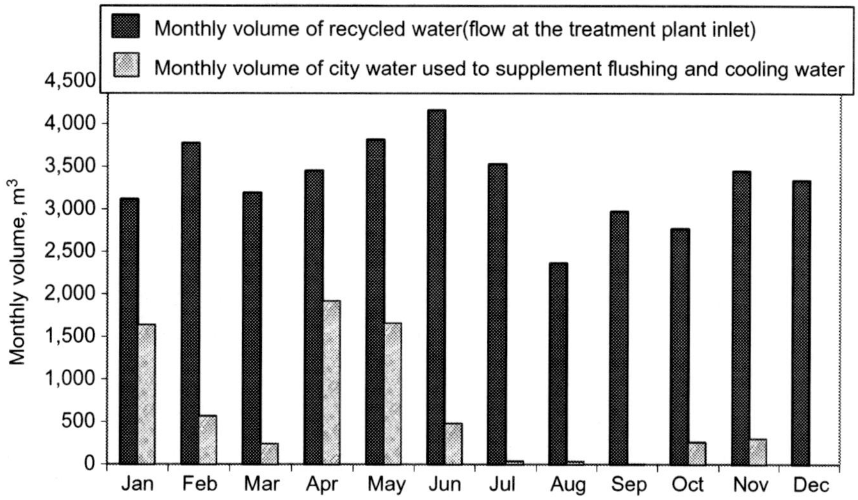

Figure 13.3 Monthly volumes of recycled water (measured at the plant inlet) and potable city water used to supplement flushing and cooling water in 2009.

13.4 ECONOMICS OF WATER REUSE

Project funding and costs

The Solaire project was initially financed through a construction loan and long-term financing. However, all the work was postponed after the September 11 World Trade Center attack in 2001. After the tragedy, all the financing had to be restructured. The new financing mechanism of the Solaire project used an innovative approach with public and private collaborations, including a private loan, and public sector grants from the New York State Energy Research and Development Authority (NYSERDA), the US Department of Energy, and the New York State Green Building Tax Credits.

The total capital cost of the Solaire project was US$114 million (Natural Resources Defense Council). The cost of the Solaire wastewater treatment and reclamation system was approximately $800,000. Along with energy efficient and other water efficient features in the building, the water recycling and reuse system help the project to meet stringent requirements to comply with the New York State Green Building Tax Credit, which was worth $2.8 million over five years. This tax incentive provided a means for the developer to offset some of the cost associated with the sustainable components through tax credits. Through its New Construction Program (NCP) NYSERDA provided a total of $628,079 to Solaire including a $100,000 grant for LEED design assistance and energy modeling, a $319,079 incentive for the reduced use of electricity and $119,000 for commissioning of the project. NYSERDA also provided a $90,000 grant for the purchase of the photovoltaic panels. In conjunction with NYSERDA, the U.S. Department of Energy contributed $100,000 to the commissioning (U.S. Green Building Council).

Applied Water Management (a subsidiary of American Water at the time this report was prepared) designed, built and currently operates the wastewater treatment and reuse system. The treatment system has a highly automatic operation which requires the plant operators to spend only an average of 4 to 6 hours a week on plant maintenance and operation.

Pricing strategy of recycled water

The fee for use of public water and recycled water is not listed as a separate charge to the Solaire residents since there are no dedicated meters for each residential unit. Instead, the water use and sewer fee is included in the overall rental fee. The savings resulting from the reduced water use and sewer discharge are reflected in the overall savings in operation for the building.

Benefits of water recycling

The water recycling at the Solaire has demonstrated economic, social, and environmental advantages over the conventional approach adopted by typical public water systems. The primary benefits of the Solaire's wastewater treatment and recycling system include:

- Reduction of water demand from the City – as a result of water reuse for toilet flushing, cooling water, irrigation water to an adjacent park, the Solaire is able to save more than 50% of water from the public water supply system compared to conventional apartment buildings.
- Reduction of the water fee from the City – although it was not the case initially, the City's Water Department has a lower rate for the Solaire due to the decreased use of water.
- Reception of grants and other incentives from public agencies to offset some of the capital investment.
- Reduction of discharges of stormwater and wastewater to the City's combined sewer system – stormwater runoff (approximately 650 m^3 of water per year) is collected in a storage tank in the building's basement to be used for irrigation of the rooftop gardens.
- Avoided cost to upgrade and expand the City's existing water and sewer systems due to the minimum impacts from the addition of the water demand and wastewater flow at Solaire.
- Increased market value as the first green building in the U.S. that uses both a stormwater and blackwater reuse system.
- As a model project to the other BPC buildings, the Solaire set up a series of standards and guidelines for the development of reuse systems for green buildings.

13.5 HUMAN DIMENSION OF WATER REUSE

Public education and communication strategy

Public involvement is one of the key components to the success of the Solaire project. The sustainable development concept was new to BPC in the early stage of the project. BPC Authorities organized a group of people to attend a USGBC meeting in the summer of 1999 in Chattanooga, Tennessee. Following the meeting, the BPC developed the initial Environmental Development Guidelines in one-year collaboration with a team of architects, engineers, and nongovernmental organizations (NGOs). External public organizations and city agencies were engaged to ensure that the guidelines would be accepted and embraced. It was challenging since the BPC Authorities had to educate and convince developers and construction contractors to build green buildings which were different from the well-developed conventional high-rise building guidelines. Through public involvement and outreach programs at a very early stage, BPC was able to identify key audiences and specific communities and offer information and opportunities for feedback.

Resident education is very important to the building management team. Staff members provide a tour for each person that moves into the Solaire, answering questions about green features and explaining the benefits and consequences of living in a green building. Hundreds of tours are given each year by the staff to educate the community and provide information on the project to the green building industry. Tours of the green roof and rain garden are also made available to residents and visitors.

The role of decision makers

The Solaire project received support from BPC Authorities, regulators and officials from the City and the State of the New York. During the opening ceremony of the building, George Pataki, the former Governor of the State of the New York spoke highly about Solaire: "This is sustainable living. This is environmentally sustainable living in a high-rise building in lower Manhattan. This is going to transform the way people live around the world".

Regulators helped establish clear permitting pathways for the Solaire project. BPC Authorities negotiated with developers and NYC departments to achieve approval of certain green project features. BPC Authorities also provided oversight during each phase of the design and construction process. In addition, the developers received approvals from the City Health Department so that there were two review processes: one at the BPCA level and one at the City level. The NYC Health

Department played an early role by visiting the plant during design and startup. The Solaire plant operators still send monthly sample results and quarterly and semi-annual reports to the NYC Health Department.

Public acceptance and involvement

The residents of the Solaire have fully embraced the green features provided in the building. Most of the people consider living in green buildings as a life style of the future. The use of recycled water and low flow fixtures have not caused any inconvenience to the residents, and to date, they have not complained about any operational issues of the wastewater treatment plant in their basement, although they have noticed that the temperature of the reuse water is different from the city water. Because the recycled water goes through all the treatment process within the building, the temperature of the reclaimed water is at about 20 to 25°C, which is about 5 to 10°C warmer than the city water.

13.6 CONCLUSIONS AND MAIN KEYS TO SUCCESS

The Solaire is one of six high-rise residential buildings in Battery Park City, an area adjacent to the Wall Street financial district, and is a leading example of urban environmentally sustainable development. The Solaire Apartments collect, treat, store and reuse the wastewater for toilet flushing, irrigation and cooling towers. This approach reduces the public water taken from the city's water supply by over 50% and significantly decreases energy costs as less drinking water is pumped from the city's treatment plant and wastewater is not transferred to the city's wastewater treatment system. The rainwater collection system irrigates 930 m^2 of rooftop gardens. Keeping with the "green" goal, the facility contains 50% recycled construction materials, consumes 35% less energy, and reduces peak demand for electricity by 65% using solar panel and other energy efficiency fixtures (American Water, 2008; Epstein, 2008).

The Solaire was a pioneer for green buildings as it was the first building to win the right to build on a site following the BPC Authorities Guidelines. It was also the first LEED certified residential high-rise building in North America with an integrated design for the site, facility, landscape, and water management. The Solaire set the goal for subsequent developers, becoming the green standard by voluntarily incorporating measures such as an onsite blackwater treatment plant with well-defined reclaimed water quality criteria.

In summary, some of the key factors that made the Solaire successful include the regulatory mandate for sustainable development by the BPC Authority, the public support through grants and incentives, the utilization of an emerging technology at that time – membrane bioreactor, and most importantly a team of people including the Solaire administration and Applied Water Management, who were open-minded to apply new technologies to meet innovation in green building development.

The following Table 13.2 summarizes the main drivers of this water reuse project as well as its benefits and challenges.

Table 13.2 Summary of lessons learned.

Drivers/Opportunities	Benefits	Challenges
Regulatory requirements	Reduction of water demand	Limited space in the building
Combined sewer overflow reduction and control	Reduction of the water fee from the City	basement for the reuse system
Stringent water supply	Reduction of discharges of stormwater and wastewater	Relatively high energy intensity for the membrane bioreactors
Population growth and increasing water demand	Avoided cost to upgrade and expand the City's existing water and sewer systems	
Population growth and increasing water demand	A show case of "green" building	

Acknowledgements

The authors thank the staff of the Applied Water Management, a subsidiary of American Water, especially Don Shields, Andy Higgins, Jim Huntington and John Tekula, for their outstanding work on the success of the Solaire and other Battery Park City's wastewater recycling systems. Special thanks to Michael Gubbins from the Solaire for his continued support of research and willingness to share information and expertise.

REFERENCES AND FURTHER READING

Carey H. (2005). The Solaire – Green by Design. Report of Battery Park City Authority. New York City.
Epstein K. (2008). NYC's living lesson. *High Performing Building*, Summer Issue, 57–65.

Krogmann U., Andrews C., Kim M., Kiss G. and Miflin C. (2007). Water mass balances for the solaire and the 2020 tower: implications for closing the water loop in high-rise buildings. *Journal of the American Water Resources Association*, **43**(6), 1414–1423.

Sustainable Water Resources Management, Volume 2: Green Building Case Studies. (2010). Report of Electric Power Research Institute. Electric Power Research Institute, Alexandria.

The Solaire Annual Report. (2010). Report of American Water. American Water, Voorhees.

WERF. (2009). When to Consider Distributed Systems in an Urban and Suburban Context? Water Environmental Research Foundation, DEC3R06.

WEB-LINKS FOR FURTHER READING

Case Study: The Solaire. Natural Resources Defense Council. http://www.nrdc.org/buildinggreen (accessed August 2011).

River Terrace – The Solaire. U.S. Green Building Council. http://leedcasestudies.usgbc.org (accessed August 2011).

The Solaire Wastewater Treatment System. (2008) www.amwater.com (accessed August 2011).

14 On-site water reclamation and reuse in individual buildings in Japan

Katsuki Kimura, Naoyuki Funamizu and Yusuke Oi

CHAPTER HIGHLIGHTS

In Japan, about 2500 individual buildings have on-site wastewater reclamation/rainwater harvesting systems. In these buildings, reclaimed water is used for a variety of purposes including toilet flushing, garden watering, cooling water, car cleaning, and fire protection. On-site wastewater reclamation/rainwater harvesting systems have been promoted mainly by local regulations and favourable taxes. Membrane bioreactor (MBR) is highly suitable for on-site wastewater reclamation, stimulating the widespread use of this technology.

KEYS TO SUCCESS

- Local regulations and favourable taxes.
- A design manual for on-site wastewater reclamation is available.
- Water quality requirements for reclaimed water are identified.
- Wastewater is divided into several fractions and is separately collected.
- Technological know-how of design and operation of wastewater reclamation system.

KEY FIGURES

- Number of installations: about 2500
- Typical total floor area of an example building: 563,800 m^2
- Day-time population in the building: 20,000
- Treatment capacity of the on-site reclamation system: 680 m^3/d
- Footprint of a treatment system: 600 m^2
- Number of operators: 2
- Frequency of monitoring: twice a week
- Water quality requirement for toilet flushing: Not detected *E.coli*/100 mL

14.1 INTRODUCTION

Main drivers for water reuse in individual buildings in Japan

In Japan, about 2500 individual buildings have on-site wastewater reclamation/rainwater harvesting systems. In these buildings, reclaimed water is used for a variety of purposes including toilet flushing, garden watering, cooling water, car cleaning and fire protection. Among these, toilet flushing is the dominant application.

Many benefits are expected from the implementation of on-site wastewater reclamation and reuse:

1. Reduction in tap water needs: on-site wastewater reclamation reduces the gap between demand and supply of water in drought conditions.
2. Reduction in loading of sewer system: by reducing the net amount of water and contaminants, on-site wastewater reclamation contributes to the conservation of natural water bodies. Old sewer systems generally have small carrying capacities, and therefore, are unfit for high-rise buildings that produce large amounts of wastewater. On-site wastewater reclamation can allow using old systems with new high-rise buildings.
3. Maintenance of public/economic activities even in disaster situations: Japan is prone to suffer from big earthquakes and preparation for them is essential. Installation of on-site wastewater reclamation in public buildings (e.g. city halls) can enable to use them as rescue centres when disasters occurs and the public water supplies do not work. On-site wastewater reclamation in business buildings can prevent shut-down of computer systems during disaster events being able to providing cooling water.
4. Recovery of heat from wastewater is also attempted.
5. Favourable publicity for companies: if companies install on-site wastewater reclamation systems in their own buildings, they can show to the public that they are environmentally-responsible.

Figure 14.1a shows the main applications of reclaimed water in on-site wastewater reclamation in Japan (Public Buildings Association, 2005). Toilet flushing is the most widespread application of on-site wastewater reclamation followed by garden watering, and cooling.

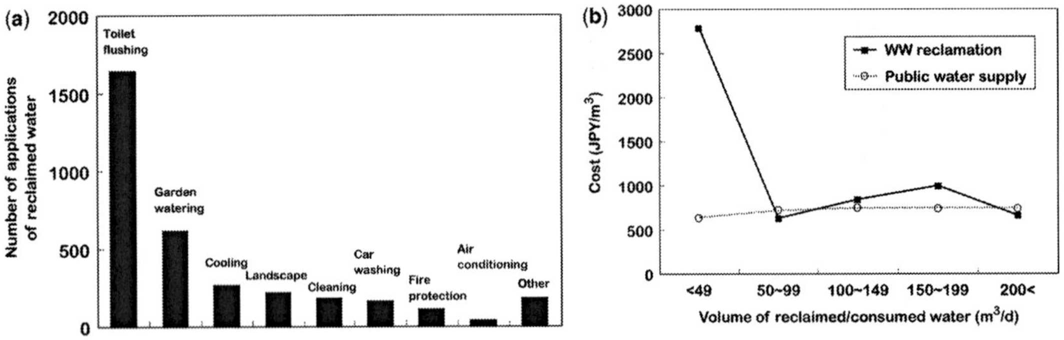

Figure 14.1 Applications of reclaimed water in on-site wastewater reclamation in Japan (a) and comparison of costs of wastewater reclamation and public water supply (b).

The cost comparison of wastewater reclamation and public water supply (Figure 14.1b) shows that costs of on-site wastewater reclamation are comparable to public water supply when water demand is higher than 50 m^3/d (Water Resources Department of Ministry of Land, Infrastructure and Transport Japan, 2000). This estimation was carried out on the basis of costs in Tokyo and included the fee for sewer discharge in public water supply.

From the 2500 buildings equipped with water reuse/rain harvest systems, 25.9% is public office buildings, 12.5% is business office buildings and 15.7% is schools. Hospitals, playgrounds, public parks and industrial factories also account for significant portions of water reuse implementations. The reason why public office buildings account for the largest portion is that they are expected to function as a base of rescue activities during disasters.

Historical background and institutional incentives of on-site water reuse

The history of on-site wastewater reclamation and reuse in individual buildings in Japan is long: Before 1965, several buildings already installed on-site wastewater reclamation systems. Through 1960′s and 1970's, increase in number of installations was steady but relatively slow. In the late 1970's or the early 1980's, the installation of on-site wastewater reclamation systems in individual buildings was accelerated. In 1978, a severe drought experienced in Fukuoka which is located in the southern part of Japan, was a big turning point for widespread implementation of water reuse in individual buildings.

Since 1980's, around 100 new installations per year have been reported in Japan, resulting about 2500 installations at present. These on-site individual reclamation systems are unevenly distributed within the country: there are two major areas including Tokyo and Fukuoka with high density on-site reclamation systems. The reason why many on-site systems are installed in these two cities is the local regulation that is promoting on-site reclamation. A new building must have a wastewater reclamation/rain water harvest system when its total floor area exceeds a determined value (30,000 m^2 in Tokyo and 5000 m^2 in Fukuoka). Water shortage was the major motivation for the cities to enact such regulations.

Other cities also provide favourable taxes to promote on-site wastewater reclamation in individual buildings. These efforts are still being continued and the number of installations of on-site wastewater reclamation system is increasing.

14.2 GENERAL DESCRIPTION OF ON-SITE WATER RECLAMATION IN INDIVIDUAL BUILDINGS

Category of wastewater produced in a building

Table 14.1 summarizes water quality requirements for reclaimed water used for toilet flushing (Public Buildings Association, 2005). To produce this quality of recycled water in an individual building, wastewater treatment for on-site reclamation should be efficient and the separation of wastewater is important. First, wastewater is divided into two fractions: black water (from toilet) and grey water. Grey water is further divided into three fractions depending on the degree of contamination: lightly contaminated grey water, moderately contaminated grey water, and heavily contaminated grey water. Lightly contaminated grey water includes blow down from cooling towers. Moderately contaminated grey water is collected from sinks, bathrooms, and office kitchens used for tea/coffee preparation. Heavily contaminated grey water is collected from restaurants and kitchens.

Table 14.1 Water quality requirements for reclaimed water used for toilet flushing*.

Parameter	Target value
pH	5.8~8.6
Odour	Not abnormal
Appearance	Almost colourless and transparent
E.coli	Must not be detected
Residual chlorine	0.1 mg/L (free) at tap 0.4 mg/L (combined) at tap
BOD	<20 mg/L
COD	<30 mg/L

*Wastewater collected in a building is assumed as raw water.

Lightly contaminated grey water is obviously suitable for on-site reclamation, while heavily contaminated grey water is difficult to be reclaimed because of high content of BOD, grease and surfactants. When a water balance in an individual building is considered, one big problem with on-site wastewater reclamation and reuse is generally found: the amount of lightly contaminated grey water is insufficient to match the amount of water used for toilet flushing. Typically, about a half of water introduced in an office building is used for toilet flushing when an on-site reclamation system is not installed. On the contrary, only less than 10% of the introduced water will become lightly contaminated grey water. Therefore, a part or all of the heavily contaminated grey water needs to be reclaimed to meet the demand of water for toilet flushing.

Selection of treatment processes for on-site water reclamation

The requirements for the treatment processes used for on-site wastewater reclamation in individual buildings can be summarized as follows: (1) small footprint, (2) easy maintenance and operation, (3) adaptable to fluctuation in inflow, (4) high quality of treated water, and (5) less production of sludge. The selection of treatment train is highly dependent on the raw wastewater.

In Japan, based on the past experience, typical wastewater qualities (Grey water A and Grey water B) are assumed for the design of on-site reclamation systems. Grey water A is assumed for grey water collected from washbowl; tea service, and so on. (i.e. lightly contaminated grey water). Assumed concentrations for this type of grey water are as follows: BOD 100 mg/L, COD 80 mg/L, and SS 100 mg/L. In contrast, Grey water B is assumed for heavily contaminated grey water: BOD 300 mg/L, COD 200 mg/L, and SS 250 mg/L. It should be noted that, in Japan, potassium permanganate is used for COD measurement instead of potassium dichromate. This is the reason why the value of COD in the model wastewater is lower than that of BOD.

According to the design manual that is available in Japan (Public Buildings Association, 2005), there are currently four standardized treatment trains (Figure 14.2).

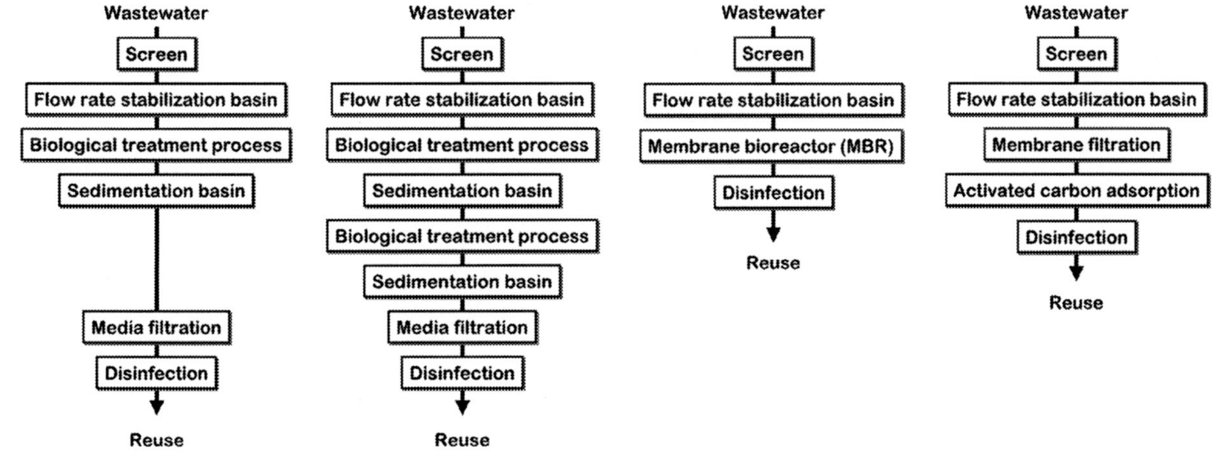

Figure 14.2 Flow sheets of the four standardized treatment process trains for on-site wastewater reclamation in Japan.

Biological treatment is usually the main part of a wastewater reclamation system in an individual building. Conventional activated sludge process, aerated biological filters, rotating biological contactors, and membrane bioreactors (MBR) are often used. Physical/chemical processes such as activated carbon adsorption or ozone oxidation are sometimes used to eliminate colour and odour from reclaimed water. Regarding harvested rainwater, a simple treatment (e.g. filtration) is generally sufficient for meeting the water quality guidelines.

The process train No. 1 in Figure 14.2 is suitable for Grey water A, while No. 2 is suitable for Grey water B. In the process train No. 2, biofilm processes such as aerated biofilters are often used as second biological treatment process. The membrane technology has been drawing a lot of attention recently in various fields of water/wastewater treatments. On-site wastewater reclamation systems are not an exception. MBRs have been already used in on-site reclamation systems as early as since 1980 in Japan. Actually, the use of MBR in on-site wastewater reclamation in individual buildings is a success, which stimulated the widespread use of this technology. The process train No. 3, which is based on MBR, is much simpler than No. 2, although higher water quality (e.g. free of suspended solids) is expected. This process train is becoming a main stream in on-site wastewater reclamation in individual buildings in Japan. The process train No.4, which also uses membrane, is suitable for Grey water A. In this case, the membrane is not used as a MBR, but only as a physical filter. This system flow does not include biological processes and is therefore very resistant to high fluctuations in hydraulic loading that is typical in high-rise business buildings (they are closed on weekends).

Operation and maintenance problems with on-site water reclamation

Fluctuations in quality and quantity of wastewater represent a severe problem from the start-up of on-site wastewater reclamation in individual buildings, and they are still a major challenge. The emergence of membrane technology would address this problem to some extent. It is demonstrated that MBR with very high concentration of biomass is more resistant to fluctuation in wastewater than conventional technologies. When membranes are just used as a physical barrier (e.g. Process train No. 4 in Figure 14.2), fluctuations in wastewater would not have significant impact on treatment performance.

In many existing on-site wastewater reclamation systems in individual buildings, the amount of reclaimed water is often insufficient to meet the demand. There are several reasons for this: fluctuations in the amount of wastewater produced in the building and/or overestimation during the design stage of the volume of the produced wastewater. In extreme cases, tap water is supplied to compensate the water shortage. To overcome this problem, the combination of wastewater reclamation and rainwater harvesting is becoming popular.

14.3 EFFICIENCY OF MEMBRANE BIOREACTORS USED FOR ON-SITE WATER RECLAMATION IN A BUSINESS BUILDING

Treatment capacity and water reuse applications

The efficiency of the MBR treatment for on-site water reclamation in high-rise buildings is demonstrated in the example of the business complex Tokyo Midtown (Figure 14.3a). The Tokyo Midtown, built in 2007, is a 54-story building and has a total floor area of 563,800 m^2. The day-time population in the building is about 20,000 people.

The treatment capacity of the on-site reclamation system is 680 m^3/d. In this case, reclaimed water is used only for toilet flushing. The inlet wastewater includes gray water from restaurants, grey water from offices, and blowdown from a cooling tower system. The black water from the toilets is not recycled, as the recycling of black water in an individual building is prohibited by the local regulation. Figure 14.3b shows the flow sheet of the wastewater reuse system.

Treatment train for water recycling

Both hollow fibre membranes and flat-sheet membranes can be used in MBRs. Due to the ease of maintenance, flat-sheet membranes are often preferred for application in small-scale systems such as on-site wastewater reclamation. In this case, a submerged-type MBR is used with 1800 flat-sheet membrane elements immerged in the aeration tank (Figure 14.4). The material of the membrane is chlorinated polyethylene with nominal pore size of 0.4 μm. The membrane flux is set at 0.4 $m^3/m^2 \cdot d$.

Compared to the MBRs used in municipal wastewater treatment (i.e. large-scale treatment), the mixed-liquor suspended solids (MLSS) concentration in the MBRs for on-site wastewater treatment tends to be higher (15–20 g/L).

The heavily contaminated grey water from restaurants, containing substantial amounts of oil/grease, is pre-treated by sequencing batch reactors (SBRs) before being mixed with the other wastewater.

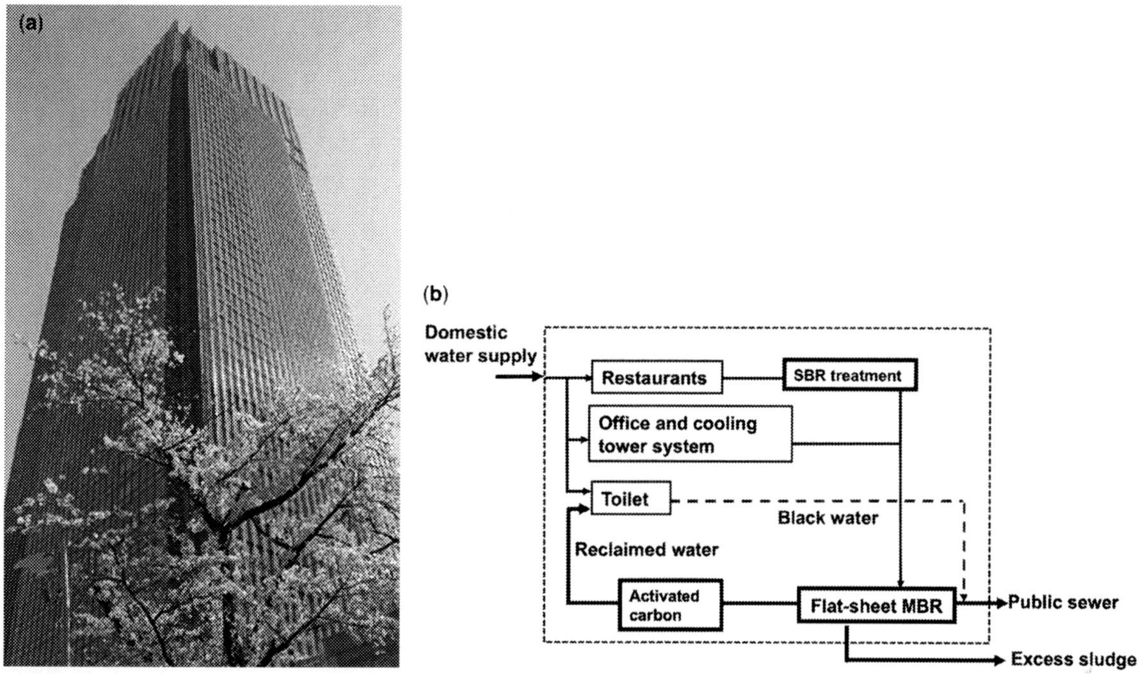

Figure 14.3 View of the Tokyo Midtown business complex building (a) and schematics of the on-site wastewater treatment system (b).

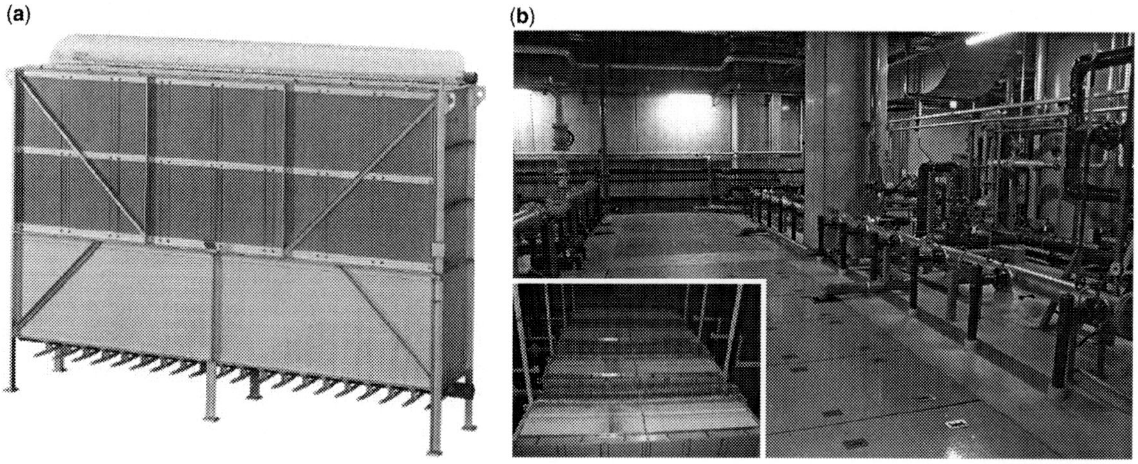

Figure 14.4 Views of a flat-sheet membrane module used in the MBR (a) and the plant in operation, installation of the membranes is shown at the lower left (b).

As a rule, effluents from MBRs are used directly for toilet flushing after chlorination, which is mandatory to guarantee the residual chlorine required by regulations (see Table 14.1). In this application, an additional treatment step is introduced to remove colour from the reclaimed water by activated carbon adsorption. The final chlorination before distribution to the toilet flushing system is carried out with sodium hypochlorite.

The operating parameters of the system (e.g. trans-membrane pressure in the MBR) are continuously monitored. As shown in Figure 14.4b, the footprint of this MBR system installed inside of the building is very small, 600 m^2, with a volume of the aeration tank of 214 m^3.

Water quality

The averaged values of water quality parameters obtained with the system are shown in Table 14.2. The design water quality of the effluent from the treatment system is shown in brackets. It should be noted that the raw wastewater represents a mixture of grey water from offices, blowdown from the cooling tower system, and effluents from the SBRs treating the wastewater from the restaurants.

Table 14.2 Water quality of the on-site MBR treatment system.

Parameter	Raw wastewater	Effluent
pH	6–8	7.7 (6–8)
Odour		Not abnormal
E.coli		Not detected
BOD (mg/L)	215	<1.0 (10)
SS (mg/L)	215	<1.0 (5)
n-Hexane (mg/L)	43	<1.0 (5)
Colour (colour unit)		4 (<10)
Turbidity (turbidity unit)		<1 (2)

Challenges for operation and lessons learned

The customer is satisfied with the net reduction of domestic water supply. The water quality performance of the MBR system has been satisfactory. The operation and maintenance of the MBR were found to be very easy. Two operators check the plant twice a week by monitoring the operating and water quality parameters including pH, water temperature, water levels in the reaction tanks, dissolved oxygen, redox potential, MLSS, flow rates, and aeration rate. Withdrawals of sludge and chemical cleaning of the membrane were carried out every week and every 4 months, respectively, and have been sufficient to maintain a stable operation of the system.

When possible, the use of grey water from restaurants as a source for water reclamation should be prevented because of the difficulty in their treatment. Unfortunately, the amount of "clean" grey water produced in the building, with an average volume of 250 m^3/d, is not sufficient to provide the amount needed for toilet flushing. To fill the gap, grey water produced in restaurants, with an average volume of 200 m^3/d, is also included as source of reclaimed water.

14.4 CONCLUSIONS: KEYS TO SUCCESS OF IN-BUILDING RECYCLING AND REUSE

In Japan, local regulations and favourable taxes have promoted on-site wastewater reclamation and reuse in individual high-rise buildings. Over the long history (>40 years) of it, water quality requirements for reclaimed water were identified and technological know-how has been accumulated. Membrane bioreactor is becoming a mainstream with on-site wastewater reclamation and reuse in individual buildings. Customers are satisfied with quality of reclaimed water and the net reduction of public water supply.

The following Table 14.3 summarizes the main drivers of in-building water recycling and reuse in Japan as well as the main challenges and benefits.

Table 14.3 Summary of lessons learned.

Drivers/Opportunities	Benefits	Challenges
Water shortages caused by droughts Local regulations encouraging water reuse and favourable taxes Good advertisement for companies: they can show that they are environmentally responsible Availability of technology including MBR	Reduction in volume of public water supply need Reduction in loading of sewer system Maintenance of public/economic activities in disaster situations	Combination with rainwater harvesting A water balance: the amount of lightly contaminated grey water cannot cover the amount needed for toilet flushing Treatment of heavily contaminated grey water containing grease and fats

REFERENCES AND FURTHER READING

Public Building Association (2005). *Design of Wastewater Reclamation and Rainwater Harvest Systems*, Tokyo. (In Japanese).

Water Resources Department of Ministry of Land, Infrastructure and Transport Japan (2000). *Investigation of the Actual Conditions in Wastewater Reclamation Facilities*, Tokyo. (In Japanese).

PART IV

Agricultural use of recycled water

Foreword

By The Editors

Background

Agriculture is usually the greatest water user in a region, but is often the "poor man" of water supply, least able to afford the development of new water resources. Many regions around the world are experiencing shortages of water for agricultural production due to growing demands, droughts and climate change. To maintain food production, there has been increased emphasis in recent years on irrigation efficiency (more crop per drop) but efficiency alone is not enough to overcome the problems.

Historical development of agricultural use

There is a long history of wastewater use in agriculture since ancient times. The largest scheme in the world is in the Mezquital valley in Mexico. Commencing around 1890, drainage canals were built to take wastewater from Mexico City to provide irrigation water for agricultural lands. The scheme now irrigates up to 90,000 ha of agricultural crops. An added benefit has been the recharge of groundwater aquifers in the region.

The first engineered systems of spreading urban wastewater for agricultural irrigation were implemented in the Paris (France) and Milan (Italy) urban areas. Following the extension and modernisation of the Paris sewers in 1850s, the first regulation on water reuse by spreading was published in 1899 and by the end of the 19th century, almost all sewage from Paris were used to irrigate about 5000 ha of land. The first irrigation consortium in Milan was constituted in 1881 for irrigation of 3500 ha with urban sewage. In parallel, the first handbooks on sanitary engineering were published in France and Italy.

In the 20th century, the development of effective sewage treatment systems opened the way to the safe use of reclaimed water for agricultural irrigation. Often these schemes involved small scale irrigation of pastures adjacent to treatment plants. In recent times water shortages and increasing competition for available fresh water supplies has prompted the development of numerous large agricultural reuse schemes around the world.

- At Monterey in California, development of a recycled water irrigation system overcame water shortages due to over-exploitation of the groundwater and saline intrusion into the aquifer. The Monterey Wastewater Reclamation Study for Agriculture published in 1987 demonstrated that filtered and disinfected recycled water could be used safely for irrigation of vegetable and salad crops in the area. The Monterey scheme now irrigates about 5000 ha of croplands.
- Since 1970 Israel has moved from small scale local reuse schemes for irrigation to large scale recycled water systems. The Dan Region project, which was constructed in the 1990s, uses soil aquifer treatment to produce recycled water for unrestricted irrigation. Israel now reuses about 75% of its treated wastewater, mainly for agricultural irrigation.

Table 1 Highlights and lessons learned from the selected case studies on agricultural irrigation with recycled water.

Project/ Location	Start-up/Capacity	Type of uses	Key Figures	Drivers and Opportunities	Benefits	Challenges	Keys to Success
Milan, Italy	2003–2004 Nosedo WWTP 430,000 to 1,300,000 m^3/d San Rocco WWTP 350,000 to 1,040,000 m^3/d	Irrigation of rice, corn, wheat, pastures and horticulture, Environmental use with restoration of river water quality, biodiversity and freshwater fishing	Total volume of treated wastewater: 241 ± 16 Mm^3/yr. Total volume used for irrigation: 86 ± 6 Mm^3/yr. Capital cost: 150 M€ (Nosedo), 132.6 M€ (San Rocco). BOT and DBO contracts. O&M costs: 0.115 €/m^3 and 0.139 €/m^3, Stringent standards for unrestricted irrigation <10 *E.coli*/100 mL	Strong negative impacts of untreated wastewater (Adriatic Sea). European, national and municipal policies for water resource management. National regulation on water reuse. Loss of agricultural production and activities.	Significant improvement of water quality level of surface water bodies with restoration of biodiversity and freshwater fishing. Valorisation of historical heritage. Clear institutional framework. Improvement of health safety, crop's yields and farmer's revenues. Amelioration of soil permeability. Fertilising value and reduction of the quantity of chemical fertilizers. Diversification of agricultural production with implementation of horticulture and organic farming.	Integration of wastewater treatment plant in the Agricultural park of Milan South with preservation of historic heritage. High level of treatment of high wastewater volumes. Very stringent water reuse standards. Funding for the construction and operation of wastewater treatment and recycling facilities. Supply of high-quality irrigation water to farmers free of charge. Energy consumption for pumping and its funding by farmers. Public education, communication and support of farmer's production.	Existence of a very old complex network of irrigation canals and agricultural activity in proximity of the city (fertile Po Valley). Delivery of high quality recycled water to farmers almost free of charge with an effective control of water allocation by two farmer associations. Financial equilibrium, high treatment efficiency and reliability of operation, ensured by high qualified staff and public-private partnership. Demonstrated environmental benefits for restoration of surface water, groundwater and biodiversity. Valorisation of historical heritage and peri-urban agriculture ("zero-kilometre products"). Public education programs and collaborations with non-profit organisations.
Noir-moutier, France	1980 8000 m^3/d		Total volume of treated wastewater: up to 1,400,000 m^3/yr Volume of recycled water: ~300,000 m^3/yr Capital costs: €5.6 million O&M costs: 300,000 €/yr Charge for recycled water: 0.3 €/m^3 + 190 €/ha · yr Water reuse standards: 1000 *E.coli*/100 mL	Maintain agricultural activity on the island. Reduce pollution emission and reserve the uses of seawater. Reduce over-exploitation of natural water resources.	Reliable, available and drought-proof water supply for irrigation. Economic benefits with over 50% reduction of water bill for farmers. Additional revenues from the expanded potatoes production in summer. Conservation and preservation of sensitive environment, biodiversity and economic and leisure activities.	Production of consistent recycled water quality despite the high variations in raw wastewater quality. High salinity due to seawater intrusion in sewers and salt leaching in ponds. Seasonal demand for irrigation and high storage needs. Requirements for high reliability of operation.	Proven economic efficiency. Unfailing commitment of farmers and local stakeholders. Implementation of a strong and reliable polishing treatment. Creation of large capacity storage ponds to guarantee sufficient reserves of water. Rigorous management of wastewater treatment and reuse infrastructures.

Australia	High growth since 1990 Over 270 recycling schemes for agriculture		Total annual volume of treated wastewater in Australia: ~2000 Mm^3/yr. Total volume of recycled water used in Australia: more than 420 Mm^3/yr. Total volume of recycled water used in agriculture: more than 280 Mm^3/yr.	Existing supplies fully allocated. Water shortages due to drought. Higher standards for discharges to the environment.	Water shortages overcome. Water and nutrients recycled. Crop production increased. River pollution reduced.	Sustainable irrigation methods to maintain soils and protect groundwater.	Development of national and state guidelines for use of recycled water. Development of guidance manuals to assist growers to use recycled water. A national program to assist growers to produce more with less water.

Agricultural reuse case studies

Part 4 of this book "Milestones in Water Reuse: The Best Success Stories" presents two selected case studies of little-known successful water reuse projects, as well as a review chapter on the Australian experience, showing how water authorities have used recycled water to meet the agricultural water needs.

Chapter 15. The water reuse project in Milan is an important milestone for agricultural water reuse in Europe, first because the long history of agricultural reuse in this area and second, because almost 30 years were necessary to approve the new scheme of wastewater treatment and reuse by local and government authorities. At present, this is the largest successful and beneficial water reuse project in Europe, producing high quality filtered and disinfected recycled water for agriculture and the restoration of polluted rivers in the Po valley. The recycled water is distributed through an existing network of canals and channels for irrigation of rice, corn, wheat and pastures. New projects for irrigation of horticultural crops are being developed. The benefits have included the revitalisation of peri-urban parklands, improved surface water quality and the restoration of biodiversity and freshwater fishing.

Chapter 16. The second case study illustrates how water recycling has helped resolve water shortages on the island of Nourmontier, a popular tourist area on the French Atlantic coast. The island has no freshwater reserves and freshwater is piped from the mainland. Local agricultural production which used any surplus capacity in the drinking water network was often short of water. A recycled water scheme has been developed and now supplies more than 90% for water needed for irrigation of local crops. Despite the long history of water reuse in France, the publication of the first water reuse regulations in 1889 and 1981, water reuse projects are facing many institutional and administrative challenges. This case study can be used as a good example on the implementation of low-tech and easy to operate water recycling schemes to secure the agricultural food production in rural areas and developing countries.

Chapter 17. The development of water recycling schemes for irrigation of crops has enabled Australian growers to increase production despite water shortages due to drought and increased competition for fully allocated river water supplies. A number of large recycled water irrigation schemes have been constructed in recent years including the major irrigation scheme of horticultural crops in Virginia, South Australia. The growth in recycled water use has been supported by development of appropriate guidelines for management of health and environmental risks together with extensive guidance materials to help growers to establish successful farm practices.

Keys to success

By providing a reliable source of water, recycled water has overcome water shortages due to drought, declining resources or competition between users. Agricultural reuse has enabled farmers to maintain and increase farm production, crop yields and improve their economic circumstances. The recycling of nutrients has reduced the need for chemical fertilisers and produced environmental benefits. Supplying recycled water for farm irrigation has made additional fresh water available to meet urban water needs. Keys to success have included government and regulatory support, stakeholder involvement, support for growers with education and guidance programs, easy to operate and effective treatment control to protect health, and low recycled water pricing. As with all forms of irrigation, care is required to ensure the recycled water irrigation is matched to crop needs, salinity is controlled, and both soil and groundwater conditions are kept sustainable.

The major highlights and lessons learned from the selected case studies are summarised in Table 1.

15 Production of high quality recycled water for agricultural irrigation in Milan

Roberto Mazzini, Luca Pedrazzi and Valentina Lazarova

CHAPTER HIGHLIGHTS

The water reuse scheme in Milan is the largest European project for agricultural irrigation using high quality filtered and disinfected water for indirect reuse in agriculture, restoration of the polluted rivers in the Po valley, as well as environmental enhancement, valorisation of historical heritage and restoration of biodiversity.

KEYS TO SUCCESS

- Existence of a very old complex network of irrigation canals and agricultural activity in proximity of the city (fertile Po Valley)
- Delivery of high quality recycled water to farmers almost free of charge with an effective control of water allocation by two farmer associations
- Financial equilibrium, high treatment efficiency and reliability of operation, ensured by high qualified staff and public-private partnership
- Demonstrated environmental benefits for restoration of surface water, groundwater and biodiversity
- Valorisation of historical heritage and peri-urban agriculture ("zero-kilometre products")
- Public education programs and collaborations with non-profit organisations

KEY FIGURES

- Treatment capacity of the two largest WWTRP: Nosedo 430,000 to 1,300,000 m^3/d; San Rocco 350,000 to 1,040,000 m^3/d
- Total annual volume of treated wastewater of 241 $\pm$ 16 Mm^3/yr, from which 86 $\pm$ 6 Mm^3/yr used for irrigation of rice, corn, grass and horticulture
- Maximum flow rate of 4 m^3/s supplied by each of the two facilities to four irrigation canals
- Capital cost: 150 M€ (Nosedo), 132.6 M€ (San Rocco), BOT and DBO contracts
- Operation and maintenance costs of 0.115 €/m^3 and 0.139 €/m^3, respectively for San Rocco and Nosedo WWTRPs (data for 2011)
- Stringent standards for unrestricted irrigation of $<$10 *E.coli*/100 mL
- Recognised environmental benefits and added value for agriculture

15.1 INTRODUCTION

The city of Milan is located in the Lombardy region of northern Italy, one of the wealthiest areas in the whole Italian territory in terms of waters: to the north lies the important crescent-shaped Alps system, with the Maggiore and Como lakes, and the city is situated among the three main Italian rivers, namely the Ticino, Adda, and Po rivers (Figure 15.1). Before the construction of the wastewater treatment and recycling plants, urban wastewater was collected by a grid of small streams and canals, particularly the Northern and Southern Lambro River and discharged by the Po River to the Adriatic Sea.

Agriculture and water reuse have a long history in the region of Milan. In the Ancient Roman era, an early land reclamation process was initiated by means of large marshes to protect the fertile territories in the so-called low plains from floods (Borasio, 1999). Later, in the Middle Ages, the various monasteries, as customary throughout Europe, particularly Cistercian monks, started to build an irrigation network for the lowlands. The Cistercian community from Chiaravalle (Figure 15.2) constructed the first elements of the current irrigation ditch network (Figure 15.3).

In the last millennium, the settlement strategies followed by the monastery communities stimulated an exceptional development in the Milan area, both qualitatively and quantitatively, with a high environmental quality. Until suppression of the order in the late 18th century, Cistercian monks were the officials in charge of the water system for this area. The arrangement of crops, and consequently the plotting of the irrigation system, originated from the need to regulate the abundant water flow in a constant interchange between the subsoil and surface waters. Agriculture was born as an economically advantageous instrument to steer the interactions between the "natural" water patrimony and human settlements.

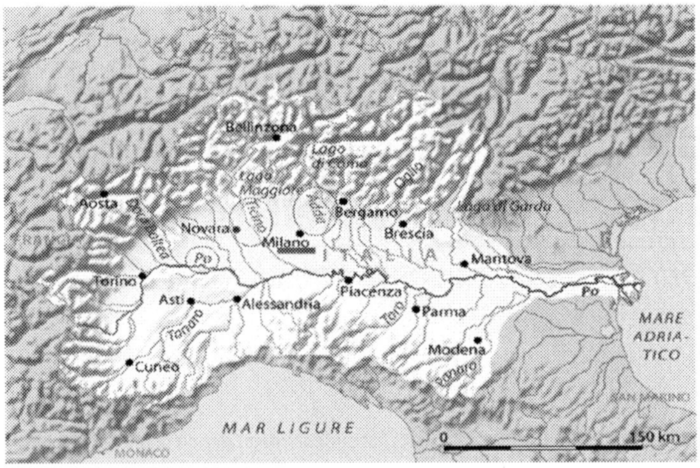

Figure 15.1 Location of the city of Milan in Northern Italy, with the Po valley and Po river flowing to the Adriatic Sea.

Figure 15.2 Aerial view of the Chiaravalle Abbey from the city of Milan.

Figure 15.3 Water meadow-flooded field "marcita" in front of the Chiaravalle Abbey in the late 1960s (source: Gentile *et al.* 1990).

The important relations between the city of Milan and the agriculture, including the role of the latter in purifying wastewaters, is described by Laccetti (1915) as one of the first sanitary engineering system in Europe. The landowners' union Vettabbia Consortium had a regular contract with the Milan town administration to use urban wastewater for irrigation of 3500 hectares (ha). This agricultural consortium is still functional and holds rights for the water use.

Another important project implemented in 19th century, was the construction of derivation canals from the lakes of Lugano and Maggiore with the objective to better control floods and provide irrigation water during the periods of drought to the agricultural lands south of Milano. The irrigation network was developed over an area of 3000 km enabling the doubling of agricultural production. The consortium of the consumers "Consortium Villoresi" was constituted in 1881 and at present is the second consortium in charge of water reuse under the name of "Consortium of East Reclamation Ticino Villoresi".

In the 20th century, Milan was the main driving force behind the Italian economy and experienced a high population growth, particularly in the 1960s–70s due to immigration of populations from poorer areas in southern Italy. The number of residents has doubled since the late 19th century, not counting an estimated one million daily workers. As a consequence, the quantity of wastewater increased with strong negative environmental impacts.

The agricultural lands south of the city, an extremely fertile area and primary source for the production of cereals, fodder, rice, dairy products and meat, were negatively affected by the huge amount of wastewaters flowing into the intermeshed canals and ditches from which irrigation water was taken. Land was subject to gradual loss of permeability and contamination. In addition, Milan was held responsible for 30% of the pollution in the northern portion of the Adriatic Sea south of Venice. In the late 1980s, that seashore suffered a devastating eutrophication phenomenon with an abnormal growth in algae, resulting in emphatic complaints against Milan, the main polluter, from the local authorities all along the Adriatic coast.

The Italian Act requiring the construction of wastewater treatment plants was issued in 1976. Consequently, since the end of the 1970s, the Town of Milan initiated a new project for the full treatment of the urban wastewater in three different locations: two large wastewater treatment plants, Milan Nosedo (1,2 million people equivalent, p.e.) and Milan San Rocco (1 million p.e.), as well as one treatment facility located close to the city, Peschiera Borromeo (0.25 million p.e.). The total length of the unitary sewers is 1430 km conveying wastewater from a surface of 12,000 ha (98% coverage).

The construction of the wastewater treatment plants faced strong opposition, especially for the purification plant in the Nosedo area, because of the proximity with the Chiaravalle Abbey and urban housing. Several committees were established with technical experts, architects and landscape engineers, with prolonged debates to find the best solution for the location and the choice of wastewater treatment processes.

Finally, the construction and operation of the two largest wastewater treatment and recycling plants of Nosedo and San Rocco (Figure 15.4), was a great success as the plants are located in a protected agricultural estate south of Milan named "Agricultural Park" and are fully compliant with the principle of sustainable development. The two plants are designed to treat exceptional storm water loads of 3 times the dry weather flows and practically all the treated water can be reused for agricultural irrigation via the existing complex networks of canals and ditches.

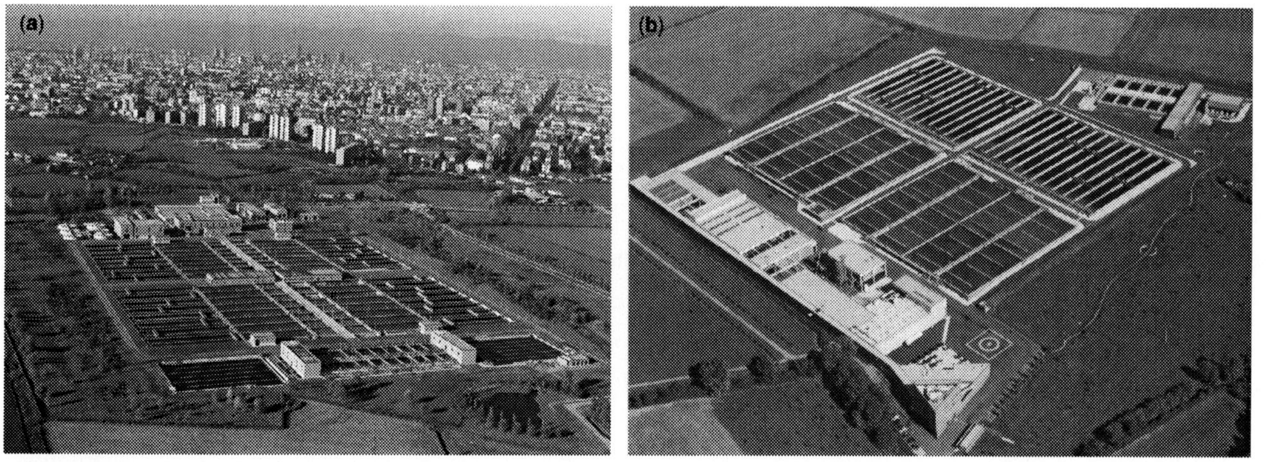

Figure 15.4 View of the two large recycling plants of Milan: a) Nosedo and b) San Rocco WWTRPs.

Main drivers for water reuse

The main drivers of the water reuse in Milan, Nosedo and San Rocco wastewater treatment and recycling plants, WWTRPs) are: (1) strong negative impacts of untreated wastewater to the environment, for example pollution of rivers of Po Valley and the Adriatic Sea (2) negative impacts on agriculture due to high water contamination and loss of soil permeability (3) European, national and municipal policies for water resource management, wastewater treatment and national regulations for water reuse.

Brief history of the project development

Several projects have been developed for the WWTRPs of Milan following a decision in the 1990s to consider water reuse for irrigation purposes as a fundamental and essential component. Nevertheless, these projects had a projected completion time of a further ten years, until the European Union threatened to enforce penalties upon the Italian Government. Therefore in 2000, the Extraordinary Commissioner and Mayor of Milan Gabriele Albertini started the planning and construction of the three wastewater purification facilities. In the spring of 2003, the first purified water from the city of Milan was produced by the Nosedo plant. In June 2004, after only two years of works, the San Rocco plant started to produce recycled water at full capacity.

Following relevant modifications to the initial projects, both on architectural and purified water quality profiles, as required by the Ministry of Historic Heritage and the Ministry of Environment, respectively, the investment costs increased substantially and the original design-and-build agreements for WWTRPs were converted to:

- BOT contract for Nosedo plant,
- Design, build and operation (DBO) contract for San Rocco plant.

The Vettabbia Irrigation Consortium is the reference Irrigation Authority which owns the recycled water from the Nosedo WWTRP and is in charge of its distribution to farmers. The management of recycled water from the San Rocco WWTRP is under the responsibility of the Consortium di Bonifica Est Ticino-Villoresi. The constant automatic monitoring of water level and flow in the canals allows the integrated and efficient management of the water resource with a substantial saving of time and improved efficiency of water use. During the irrigation period from May to September, the recycled water is distributed to farmers via the Roggia Vettabbia canals for Nosedo plant and for San Rocco facility via Roggia Pizzabrasa and Roggia Carlesca.

In summary, the project of the Municipality of Milan for wastewater treatment and reuse had two major objectives:

- Wastewater treatment for health and environmental protection
- Support and valorisation of agricultural production providing high quality water for irrigation

15.2 TECHNICAL CHALLENGES IN WATER QUALITY CONTROL

According to the quality level of purified water required by the new agreements, disinfection is required for the total capacity of the WWTRPs: 5 to 15 m^3/s (430,000 to 1,300,000 m^3/d) for Milan Nosedo and 4 to 12 m^3/s (350,000 to 1,040,000 m^3/d) for Milan San Rocco. Some chemical-physical parameters according to the contracts of operation are more stringent than those required for water reuse, for example, several heavy metals.

Treatment trains for water recycling

The two large wastewater treatment and recycling plants of Milan have similar treatment trains (Figure 15.5) including:

- Pre-treatment (large, medium and fine screening (3 mm), followed by sand and oil removal),
- Biological treatment using the step feed activated sludge process with nitrification-denitrification,
- Tertiary rapid sand filtration (Aquazur V) for removal of phosphorus and suspended solids,
- Disinfection (Figure 15.6) using peracetic acid for the Nosedo plant and UV disinfection for San Rocco with low UV dose before discharge in the river and high UV dose for agricultural reuse.

The treatment of excess sludge is also similar including thickening, filter press dewatering, biosolids valorisation in agriculture or to cementries after thermal drying.

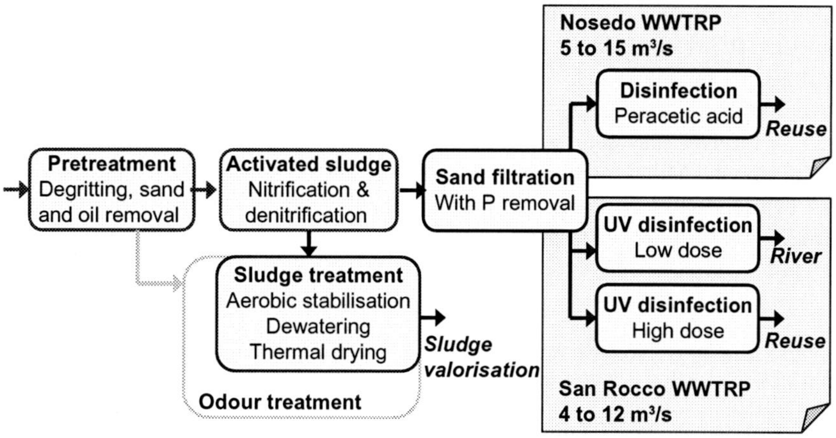

Figure 15.5 Wastewater treatment trains in Milan WWTRPs in Nosedo and San Rocco.

Figure 15.6 View of disinfection facilities in Milan before the use of recycled water for irrigation. (a) Peracetic acid contact tanks of Nosedo WWTRP, (b) UV disinfection of San Rocco WWTRP.

Both WWTRPs are equipped with online control of aeration (dissolved oxygen) and chemical dosage for phosphorus removal. An important additional advantage of on-line P control is to avoid overdosing of ferric chloride, which can affect UV disinfection efficiency (residual concentration of iron below 0.05–0.10 mg/L).

The tertiary sand filters are designed for an average flow rate of 7 and 15 m/h (max 26 m/h), respectively for Nosedo and San Rocco WWTRPs. The dose of peracetic acid is 2 to 5 mg/L with a minimal contact time of 30 min for the total flow of the Nosedo facility. Two different levels of disinfection are required in San Rocco plant, one for discharge in the Lambro Meridionale river in winter by means of a UV dose of 15 mJ/cm^2 (3 channels with one UV module each) and high disinfection for water recycling or river discharge in summer in 3 channels equipped each by 3 UV modules providing a minimal UV dose of 55 mJ/cm^2.

Water quality control and monitoring

The use of treated wastewater for irrigation in Italy was regulated, from 1977 until 2003, in the frame of the 1976 Water Protection Act (Annex 5, CITAI, 1977) with a very stringent disinfection requirement for unrestricted irrigation of <2 TC/100 mL (total coliforms). In June 2003, the Law decree n°152 (Ministry Decree, D.M. n°185/03) set new rules for water reuse with a target concentration of 10 *E.coli*/100 mL in 80% of the samples for unrestricted irrigation. In order to demonstrate the compliance with this stringent water reuse regulation, over 100 analyses per day are carried out at the two recycling plants of Milan in addition to the parameters that are monitored online. The analytical laboratories of the two recycled plants were equipped with state of the art advanced tools for routine monitoring of not only general water quality parameters, but also a number of priority substances such as heavy metals. Residual concentration of *E.coli* is controlled on a daily basis in the two recycling facilities.

The Table 15.1 shows the major parameters used for the control of recycled water quality, for example, the average values of recycled water in the two recycling plants since 2004 compared to target limit values. In addition to internal controls at the recycling plants, the Milan town administration, via its 100% controlled subsidiary Metropolitana Milanese, is carrying out its own controls and verifications. Furthermore, the public regional control authority ARPA (Agenzia Regionale Per l'Ambiente, Public Authority for the Environment) autonomously verifies and penalises any non conformity of water quality.

Table 15.1 Recycled water quality of Nosedo and San Rocco WWTRPs: comparison of average measured values since 2004 with Italian regulations for unrestricted water reuse and discharge in environment.

Parameter	Permit value (contracts)	Legal limits discharge (Leg. Dec. 152/06)	Legal limits for irrigation (Min. Dec. 185/2003)	Monthly average since 2004 Nosedo plant	Monthly average since 2004 San Rocco plant
pH	5.5–9.5	5.5–9.5	6–9.5	7.5	7.6
*Suspended solids, mg/L	10	35	10	<2 (1–4)	<2 (1–5)
*COD, mg/L	100	125	100	10 (6–20)	16 (5–38)
*BOD, mg/L	10	25	20	2 (1–6)	1.6 (1–9)
*Total N, mg/L	10	10	15	7 (4.1–9.1)	5 (3–8.7)
**Total P, mg/L	1	1	2	0.8 (0.3–1.1)	0.7 (0.4–1.3)
**E.coli*, CFU/100 mL (80%)	10	–	10	5 (80%ile 8, max 25)	4 (80%ile 7, max 9)
	5000	5000	–	–	2150 (100–4900)

*Daily sampling and monitoring; **Both on-line monitoring and daily sampling and control.

The agronomic parameters are very good with a low conductivity, consistently below 900 $\mu S/cm^2$, low sodicity (sodium adsorption index SAR of about 2.5) and low concentrations of toxic ions (boron, chloride, sodium) and toxic elements (heavy metals).

A study of the removal of pharmaceutical products was carried out at the Nosedo WWTRP demonstrating the high efficiency of removal (Zuccato *et al.* 2009). From the selected 29 drug substances, 7 were not detected, 7 were totally removed and for 16 from the 22 detected compounds, the removal efficiency was over 50%. Only one compound, atenolol (cardiovascular product) was detected with concentrations over 1 μg/L in the two inlet canals. Compared to other Italian facilities, the overall removal efficiency of this plant was the highest, 76%, probably due the complementary removal by tertiary sand filtration.

Main challenges for operation

The major challenge for the operation of the two water recycling facilities is to ensure high reliability of all treatment processes to consistently meet the stringent water quality parameters. During the irrigation period, for example, if any of the daily samples indicate more than 10 *E.coli*/100 mL, the supply of recycled water is interrupted. The few cases of presence of *E.coli* with maximum values below 25 CFU/100 mL recorded have been reported only during by-pass of tertiary filtration for maintenance. An additional challenge for operation was the management of the production of two quality recycled water in the San Rocco facility by UV disinfection. To better control disinfection efficiency and avoid UV lamp scaling due to the presence of iron from coagulation, online monitoring of residual phosphorus was installed ensuring iron concentrations consistently fall below 0.05–0.1 mg/L.

It is important to stress that the stringent requirements for water and sludge quality require the presence of qualified staff and the application of a rigorous operation and maintenance procedure. Preventive maintenance was very important with renewal and replacement of any aged or critical parts and equipment to avoid failure of treatment processes. The maintenance of the critical equipment, such as UV disinfection apparatus, was provided by respective manufacturers. Consequently, best operational practices, qualified staff and preventive maintenance are considered as the crucial conditions for the functional reliability of the two water recycling facilities without failures or malfunctions.

15.3 WATER REUSE APPLICATIONS

The high-quality filtered and disinfected water from the two recycling plants of Milan are used for (1) indirect agricultural irrigation via the existing complex networks of canals and ditches and (2) for environmental enhancement and restoration of biodiversity of water bodies (rivers and Adriatic Sea).

The major characteristics and management of agricultural reuse are summarised in Table 15.2. An indirect reuse scheme is adopted for the two recycling facilities as the treated water is distributed to farmers by means of the irrigation networks of open canals and ditches. During the irrigation period (April to September), the high-quality disinfected water of the Nosedo WWTRP is used for agricultural irrigation after injection in two points to the ditch Roggia Vettabbia Bassa (3.5 m^3/s) and can be partly sent to the channel Cavo Redefossi.

Table 15.2 Major characteristics of the irrigated fields and management of water reuse in the area of Milan.

Parameter	Nosedo WWTRP	San Rocco WWTRP
Farmers' association in charge of agricultural irrigation	Vettabbia Consortium (84 farmers, 90 farms)	Consortium di Bonifica Est Ticino Villoresi
Irrigated area	3700 ha (in the past 4400 ha)	24,630 ha (total irrigated area 114,000 ha)
Recycled water consumption	700 L/s max to irrigate 1700 ha (Vettabbia high); 3000 L/s for the remaining 2000 ha	30% of the water needed for the Cavi Litta area >100 km of canals and ditches
Type of irrigated crops	45% corn, 15% rice, 40% grasslands and wheat	45% corn, 15% rice and 40% meadows and grasslands (43% corn, 37% rice, and 20% grasslands for the total area)
Recycled water pricing	Farmers have a concession to use recycled water from the Lombardy Region upon a symbolic payment (1827 €/yr)	The two farmers associations in charge of the operation of the two canals have in charge the cost of pumping (about 27,000 €/yr)

The reuse capacity for agricultural irrigation of the San Rocco plant could theoretically cover the nominal dry weather flow of 4 m^3/s, which flows to an intermediate pumping station at a distance of 1.3 km to supply two canals, the Roggia Pizzabrasa (up to 3 m^3/s) and Roggia Carlesca (up to 1 m^3/s). The direct agricultural reuse of the total capacity is constrained by the need to maintain the minimum flow rate of the Southern Lambro river. The river discharge requires a lower level of disinfection (<5000 *E. coli*/100 mL).

Two main irrigation techniques are used in the area of Milan:

- high-pressure jets or low-pressure spray pivots, for wheat or corn (Figure 15.7),
- flooding, for rice and grass.

Figure 15.7 View of irrigation with recycled water by means of pressure jet and spray pivot near Milan.

In addition, several projects for sustainable agriculture for local population and the production of horticultural "zero kilometre" food crops are implemented or under development. The average water demand is 1000 m^3/ha·yr for the area of the consortium of Vettabbia and 5000 m^3/ha·yr for the consortium di Bonifica Est Ticino Villoresi, mainly because it is the largest area of rice production (1800 m^3/ha·yr for corn). Dry rice culture as also implemented by some farmers in addition to the traditional flooding rice cultivation.

Evolution of the volume of supplied recycled water

The evolution of total annual treated wastewater and recycled water supplied for agricultural irrigation is shown in Table 15.3.

Table 15.3 Evolution of total annual volumes of treated wastewater and supplied recycled water in Milan by the two recycling facilities of Nosedo and San Rocco.

WWTRP	Volume, Mm^3	2005	2006	2007	2008	2009	2010	2011
Nosedo	Treated water	143.8	143.4	137.9	136.3	148.6	157.4	149.0
	Supplied to reuse	82.9	76.5	73.2	70.0	75.4	74.7	79.0
San Rocco	Treated water	89.7	92.3	87.1	93.8	108.3	107.1	104.7
	During reuse periods	30.5	30.6	39.1	40.9	37.8	27.9	37.5
	Supplied to reuse	9.0	14.8	9.8	7.4	11.7	8.4	11.4

Since 2005, approximately 240 million m^3 par year (Mm^3/yr) of tertiary filtered and disinfected wastewater is produced in Milan, 60% of which by the Nosedo recycling facility (145 ± 8 Mm^3/yr). The average annual production of purified water in the San Rocco WWTRP for the last five years is 96 ± 9 Mm^3/yr.

Approximately 36% of the total treated volume, 86 ± 6 Mm^3/yr, is recycled for irrigation of rice, corn, wheat and grass. The major part of the recycled water for irrigation, about 88%, is distributed by the Vettabbia Consortium in the area of the Chiaravalle Abbey. During the peak irrigation period, farmers would like to have more irrigation water, but the increase of available volumes are limited by the lack of storage systems and the need to maintain a minimal flow in the Lambro river.

Relationships and contracts with end users

The Region of Lombardy and the Municipality of Milan (Milanese Metropolitana) remain the highest administrative decision-making entities. The build-and-operation contracts with the Municipality of Milan stipulate that the two Consortiums of Vettabbia and di Bonifica Est Ticino Villoresi are in charge of the delivering of recycled water, the contracts with farmers and the control of water allocation and use. No economical relationship binds the water treatment concessionaires or the Municipality of Milan with the farmers. However, excellent communication has been developed between farmers, local authorities and plant operators. Periodic meetings and other events are organised to discuss the challenges of water reuse, climate change and development of sustainable agriculture.

15.4 ECONOMICS OF WATER REUSE

Capital costs and funding

Following relevant modifications to the initial projects, both on architectural and purified water quality profiles, as required by the Ministry of Historic Heritage and the Ministry of Environment, respectively, the investment costs increased substantially and the original design-and-build agreements for WWTRPs were converted to:

- BOT contract for Nosedo plant, total capital cost of €150 million,
- Design, build and operation (DBO) contract for San Rocco plant, total capital cost of €136.2 million.

The total capital cost of €150 million for the construction of the Nosedo WWTRP was partially provided by the town administration (approximately 40%), while the remaining portion was covered by private funding. In order to build a proper structure for this operation, the contractors had to enlist established project companies: SPC (Milandepur SpA), EPC (Nosedo scarl) and O&M (Vettabbia scarl). The concessionaire group, via the dedicated Milandepur Company, signed a financing agreement with Banca Intesa and Royal Bank of Scotland. For the payment of the investment (e.g. interests, plant operation and maintenance, spare parts and renewals), the concessionaire group receives a semi-annual postponed fee comprising two components: an F component to repay the investment, locked to interest rates, and a G component to repay operation, maintenance, spare parts and renewal expenses, indexed on inflation and the volume of treated water.

The total capital cost of €136.2 million for the construction of the San Rocco WWTRP was totally provided by the Milan town administration. The design and build contract was in charge by a joint venture Degremont, CCC, Carlo Gavazzi Impianti and SO.GE.MA. After the plant construction, Degremont was in charge of the plant operation.

Operation costs

As illustrated in Figure 15.8, the distribution of operation and maintenance costs of these two large wastewater treatment and recycling facilities are quite different which is mainly due to the different disinfection processes.

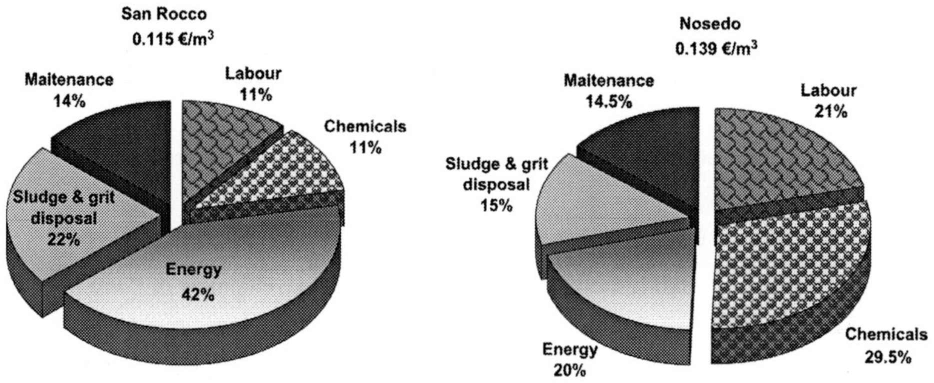

Figure 15.8 Major components of operation costs of the two recycling facilities.

The major component for the San Rocco plant is the energy consumption (42%), with a large contribution of UV disinfection, followed by sludge disposal (22%). Labour, chemicals and maintenance costs account for 11–14% of operation costs. The major component of operation costs of the Nosedo plant is chemicals, 29.5%, 50% of which being for the peracetic acid used for disinfection. Labour and energy consumption are the two other important components, with a contribution of 21% and 20%, respectively. Sludge disposal and maintenance are contributing to 15% and 14.5% of operation costs respectively. The operation and maintenance costs for 2011 are 0.115 €/m^3 for San Rocco and 0.139 €/m^3 for Nosedo.

Pricing strategy of recycled water

The Lombardy region is in charge of the water resource management. Water allocation to farmers has historically been free of charge, dating back to years when raw wastewater was diluted with the channel water. In a general context of declining agricultural activity, and especially those of the Province of Milan, there are an effort and institutional incentives in helping farmers to maintain and diversify their production activities. Consequently, farmers are using recycled water for free. Only the energy cost for pumping are covered by the farmers association in charge of the two canals distributing the recycling water from the San Rocco WWTRP (about 27,000 €/yr). The Vettabbia Consortium pays only a very low amount of 1827.42 €/yr to the Lombardy Region for the use of recycled water.

Benefits of water recycling

Starting from January 2006, a two year Environmental Monitoring Plan was implemented to evaluate the environmental impacts of the two recycling facilities. The results demonstrated that the operation of the two facilities had had a significant benefit on the improvement of the superficial receiving aquatic ecosystems without any negative impacts on the other investigated environmental matrices (e.g. groundwater, air and sediments). One of the most positive impacts of the high level of wastewater treatment was the health protection as the microbiological quality of the five receiving canals were of similar or better quality compared to the reference surface water stream. Water quality level of surface water bodies downstream was improved in terms not only of their chemical status, but also according to all biotic indexes of acute toxicity, extended biotic index, mutagenicity and algae growth test. Compared to artificial canals receiving mostly the high quality recycled water, the Lambro canal showed an intermediate ecological status due to dilution with upstream lower quality water.

Consequently the two major benefits are environmental enhancement and improvement of agricultural production.

The most important environmental benefits are as follows:

- Significant improvement of water quality of surface water bodies,
- Restoration of biodiversity and freshwater fishing,
- Valorisation of historical heritage with the construction of peri-urban parks and cycling areas, such as for example the Vettabbia park (100 ha).

The main benefits of agricultural water reuse, implemented since 2005 in the area of Milan are as follows:

- Supply of good quality irrigation water for agriculture free of charge, a reliable drought-proof resource which is available all year,
- Contribution to the development of sustainable agriculture, promoting a dynamic image of farming, diversification of crop production with implementation of horticulture and organic farming,
- Amelioration of soil permeability,
- Fertilising value of nutrient-enriched recycled water improving the crops' yield and reducing the quantity of chemical fertilizers.

It is also important to mention the positive social impacts of water recycling:

- Maintaining agricultural activity, increasing employment and enabling the development of "zero kilometre" farming,
- Supporting and collaborating with eco-education centres such as the Association Nocetum, founded by two Christian sisters with the rehabilitation of a peri-urban farm. This non-profit organization is collaborating with various institutions, including the Nosedo WWTRP, to create opportunities for awareness of the value of the environment, historical heritage, social integration, as well as remediation and redevelopment of farm land throughout the district.

15.5 HUMAN DIMENSION OF WATER REUSE

Ever since its operations began, the Nosedo and San Rocco purification plants have always been open for scheduled visits, particularly to schools or educational institutions and citizens from various local or non-local associations (Figure 15.9). Together with the universities of Milan (the Politecnico and Public University, Faculty of Biology and Environment Sciences) and the Mario Negri research centre, the issue of pharmaceutical products in wastewaters was addressed and analysed with monitoring of drug concentrations in the inlet and outlet of the purification plants.

Figure 15.9 View of sport events and visits organised in the Nosedo WWTRP.

Environmental awareness from citizens and schools is fostered through guided visits of the recycling plants. In particular, the Nocetum Association has developed, in cooperation with staff of the purification plant of Nosedo, an educational pathway related to the agricultural and food environment with visits to the plant, the Vettabbia Park and the Chiaravalle Abbey. Occasionally, farmers hold their meetings at the plant's conference room.

Several local politicians, representing the Milan town administration or province, or the Lombardy regional administration, hold meetings with enterprise unions, citizens, farmers or environmentalist associations, in order to discuss environmental requalification, agriculture development, food safety or energy reuse. Environmentalist associations also are organising their meetings in the plant's conference room to address issues regarding water and its reuse, as well as different environmental matters related to the research sector.

Total transparency and collaboration with large public, non profit organisations and stakeholders are realised through the following aspects:

- Constant presence of university students working on their graduation theses, in some cases assisted by the plant's staff,
- Meetings and debates held by cultural or environmentalist associations,
- Sports and entertainment events at the park and the plant's areas,
- Support of the activities of non-profit organisations and associations for environmental education, promotion of peri-urban and organic farming and social rehabilitation of people at risk of social exclusion;

The importance of communication and public education is well described by Callegari (2009) in an article presenting the wastewater treatment plant "as a friend". In addition, the Nosedo plant was awarded the Mediterranean Price of Landscape for the good environmental practices.

15.6 CONCLUSIONS AND LESSONS LEARNED

The water reuse project in Milan is a good example of successful cooperation between local authorities (Lombardy Region, Municipality of Milan), users (farmer's consortiums, non profit associations) and the management teams of the two recycling facilities. Even if plant's operators are not in direct contractual agreement with farmers, they are in charge of the control of recycled water quality and volumes delivered for irrigation. The project funding and economical viability are well ensured by means of public-private partnerships. Several environmental, economic and social benefits have been recognised, thanks to the high water quality and reliability of operation.

The following Table 15.4 summarizes the main drivers of the water reuse in Milan as well as the major benefits and challenges.

Table 15.4 Summary of lessons learned.

Drivers/Opportunities	Benefits	Challenges
Strong negative impacts of untreated wastewater to the environment(rivers and Adriatic Sea)	Significant improvement of water quality level of surface water bodies,	Integration of wastewater treatment plant in the Agricultural park of Milan South with preservation of historic heritage
European, national and municipal policies for water resource management and national regulation on water reuse	Restoration of biodiversity and freshwater fishing activities,	High level of treatment of high wastewater volumes
Negative impacts of pollution of water bodies on agriculture	Valorisation of historical heritage with the construction of peri-urban parks and cycling areas	Very stringent water reuse standards requiring almost total disinfection for agricultural reuse
Loss of agricultural production and activities	Clear institutional framework for the development of the project for the three wastewater treatment facilities, two from which were designed as water recycling facilities	Funding for the construction and operation of wastewater treatment and recycling facilities
	Improvement of health safety, crop's yields and farmer's revenues	Supply of high-quality irrigation water to farmers free of charge
	Amelioration of soil permeability	Energy consumption for pumping and its funding by farmers
	Fertilising value and reduction of the quantity of chemical fertilizers	Public education, communication and support of farmer's production with direct access of consumers to local products
	Diversification of agricultural production with implementation of horticulture and organic farming	
	Improvement of the image and social value of peri-urban farming and "zero kilometre products"	

REFERENCES AND FURTHER READING

Borasio M. (1999) Le valli dei monaci e Bernard de Clairvaux (The valleys of the monks and Bernard de Clairvaux). In: A.A.V.V., Studi per la progettazione di nuove unità ecosistemiche polivalenti in un'area del sud milanese: Chiaravalle, Regione Lombardia (Studies for the design of new units in a multi-purpose ecosystem of south Milan: Clairvaux, the Lombardy Region), Milan, Italy.

Callegari A. (2009) A Sewage Treatment Plant "as a friend". *Ingegneria ambientale (Environmental Engineering)*, **5**(May 2009), 205–209.

Gentile A., Braun M. and Spadoni G. (1990) *Viaggio nel sottosuolo di Milan tra acque e canali segreti (Journey into the Ground Water between Milan and Secret Canals).* Milan municipality, Milan, Italy.

Laccetti F. (1915) *Fognatura biologica – Depurazione delle acque luride (Biological Drainage – Purification of Sewage)*, Ulrico Hoepli (ed.), Editione Libraio Della Real Casa, Milan, Italy, p. 234.

Zuccato E., Castiglioni S. and Mazzini R. (2009) Environmental pollution by drug residues. *Ingegneria ambientale (Environmental Engineering)*, **7**/**8**(July–August 2009), 353–357.

WEB-LINKS FOR FURTHER READING

http://www.comune.milano.it/portale/wps/portal/CDMHome
http://www.depuratorenosedo.eu/it/

Views of the Vettabbia stream from the WWTP to the Chiaravalle Abbey and the croplands irrigated with recycled water.

16 Key to success of water reuse for agricultural irrigation in France

Antoine Fazio, Noël Faucher and Valentina Lazarova

CHAPTER HIGHLIGHTS

This case study illustrates why water reuse is essential for the preservation of tourist and economic activities of Islands with scarce water resources, as well as of a fragile and sensitive environment. Over 90% of irrigation demand for producing early crops, the most famous French potatoes, are covered by reclaimed water. Low cost and easy for operation tertiary treatment is implemented by means of polishing and storage ponds.

KEYS TO SUCCESS

- Proven economic efficiency
- Unfailing commitment of farmers and local stakeholders
- Implementation of a strong and reliable polishing treatment
- Creation of large capacity storage ponds to guarantee sufficient reserves of water
- Rigorous management of treatment infrastructures

KEY FIGURES

- Volume of wastewater treated annually: up to 1,400,000 m^3/yr
- Average annual volume of recycled water: 300,000 m^3/yr
- Capital costs: €5.6 million
- O&M costs: 300,000 €/yr
- Reclaimed water rates: 0.3 €/m^3 + 190 €/ha.yr
- Water reuse standards: 1000 *E.coli*/100 mL

16.1 INTRODUCTION

The island of Noirmoutier, located on the French Atlantic coast is a very popular tourist area (Figure 16.1). Other maritime activities such as oyster farming are also widespread around the island. Salt production is another significant local economic activity, with salt marshes covering a large part of the island as a part of a preserved environmental site Natura 2000.

Noirmoutier is also well known in France for its production of early crop potatoes which are greatly appreciated by consumers and sold at high price across the entire country.

The economy of the island is mainly supported by tourism which provides 70% of its income. The additional economic activities based on fishing, shellfish and salt production are also participating actively in the brand image of the island.

Several years ago, a sound water resource management policy has been implemented, which is enthusiastically supported by councillors through their shared interest in sustaining and preserving the island's activities and its fragile environment. At the heart of this policy is the concept of water reuse for irrigation of potato plantations.

Main drivers for water reuse

Like many other islands, Noirmoutier has no fresh water reserves. Almost all the island's drinking water needs are supplied by the Apremont plant on the mainland situated about 70 km away. Water is transported by two pipes inside the structure of the bridge which joins the island to the mainland (Figure 16.1). This water supply meets the island's seasonal water requirements, demand for which increases with summer population reaching nearly 80,000 people, compared to only 10,000 people over the rest of the year.

The cultivation of potatoes requires large volumes of fresh water, which can only be met with great difficulty by the drinking water network. In particular in summer and during spring holidays, the water demand is very high with strong constraints on water supply. Because of chronic water shortage and increasing restrictions on wastewater treatment and discharge, local stakeholders initiated this water reuse project. The main objective was not only to supply an alternative water resource for irrigation, but also to avoid wastewater discharge into the sensitive marine environment. In 1980, a water reuse agreement was signed between farmers and the local water board.

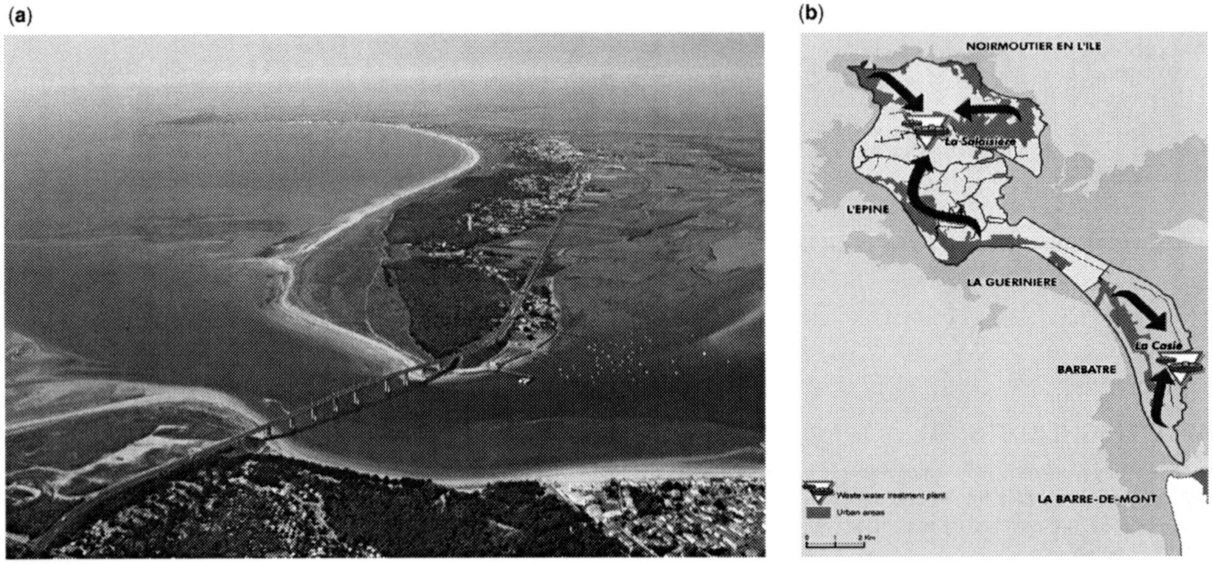

Figure 16.1 (a) Aerial view of the Noirmoutier Island and (b) location of the wastewater treatment plants.

Brief history of the project development

The water reclamation scheme was designed and set-up by the Noirmoutier' Community of Municipalities in such a way as to take account of the seasonal variations of population. The wastewater treatment system includes two water reclamation plants, one in the north of the island called "La Salaisière", which receives wastewater from three municipalities (Noirmoutier-en-l'Île, l'Epine and Guérinière) and the other in the south of the island, called "La Casie", which receives wastewater from the municipality of Barbâtre. The collection and transfer of wastewater takes place through a particularly extensive wastewater network including nearly 150 km of gravity-fed pipes, 69 km of pressurised pipes and 110 pumping stations. These installations are managed by the SAUR Company on behalf of the Community of Municipalities.

The project of water reuse for irrigation was established two years after the start-up of the activated sludge treatment on the "La Salaisière" wastewater treatment plant in 1980s (hydraulic capacity of 1600 m^3/d).

Three years later, with the increase of collected wastewater volumes, an "aerated lagoon", with a hydraulic capacity of 2200 m^3/d, was designed alongside the activated sludge treatment station.

Fourteen years later, in 1997, with another increase in wastewater volumes and new regulations, a new activated sludge plant was launched alongside the two other plants, with a hydraulic capacity of 4200 m^3/d. This is currently the only wastewater treatment and reclamation plant which is used throughout the year.

The smallest treatment plant "La Casie", which was initially designed as a natural wetland, is today refurbished to an activated sludge plant, and the natural wetland has been kept for polishing treatment before reuse or discharge. Because the low recycled water volume produced, this plant will be not presented and discussed in this chapter.

Project objectives, incentives and water reuse applications

The water reuse programme in Noirmoutier has become a crucial element in decision makers' overall strategy for integrated water resource management. This approach includes maintaining agricultural activity, limiting the importation of drinking water from the mainland for human consumption and reducing pollution discharge into the natural environment.

This water reuse project had been possible thanks to the strong involvement of local politicians and farmers who founded an association named ASDI (Trade Union Association of Drainage and Irrigation).

Currently, and since the project start-up, the only use of recycled water is the irrigation of potato plantations. The feasibility of water reuse for landscape irrigation and groundwater recharge have been also discussed and evaluated.

16.2 TECHNICAL CHALLENGES OF WATER QUALITY CONTROL

Taking into account the sensitivity of the receiving environment and its uses, even in absence of water recycling, it was necessary to establish an adequate recycled water quality with the removal of organic maters, nutrients and pathogens. The community and the public water board had also studied the suitability of the construction of a sea wastewater outfall for the

"La Salaisière" treatment plant. The study concluded that establishing a maturation polishing pond not only allowed consistent achievement of the water quality required for agricultural irrigation, but also enabled wastewater to be discharged into the marine waters used for shellfish production.

Treatment train for water recycling

The secondary treatment is a conventional activated sludge, designed to meet the discharge requirements of less than 30 mg/L of suspended solids, 25 mg/L of BOD_5 and 100 mg/L of COD. The tertiary treatment is performed in polishing maturation ponds, used also as storage reservoirs. The pond system (Figure 16.2) consists of a series of four lagoons, with depths ranging from 1.4 to 2.8 m and a total storage volume of 286,300 m^3. Secondary effluents are fed into the first basin (1.4 m deep) and flow by gravity from one pond to the next.

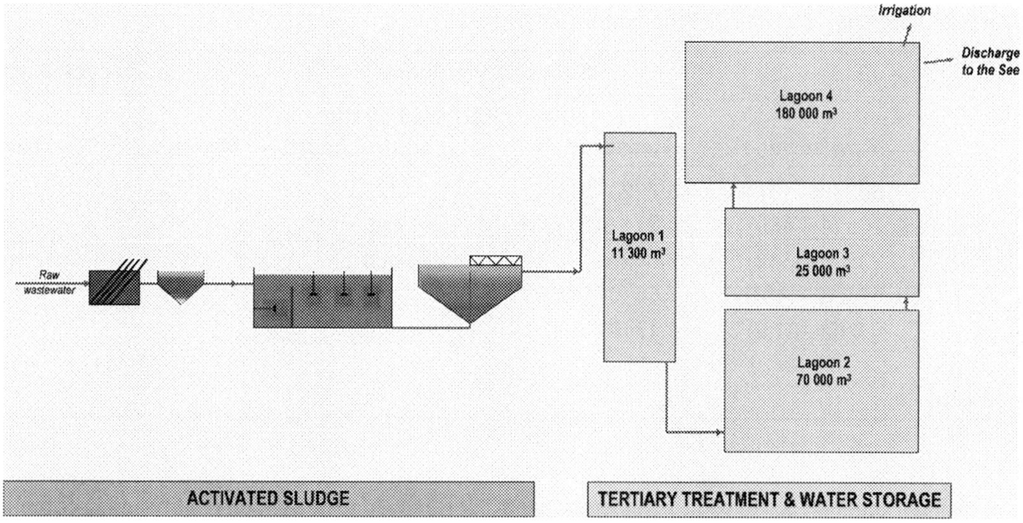

Figure 16.2 Schematics of the "La Salaisière" wastewater treatment plant.

This system was initially designed for storing water for irrigation with connections between the various ponds. In 1995, during a study to evaluate how the polishing ponds are operating, this system was transformed to enable water circulation compatible with a polishing lagoon system (Figure 16.3). After this tertiary treatment step, reclaimed water is channelled to the potatoes plantations through an underground distribution network. The plantations are served by connection terminals equipped with volumetric water meters.

Figure 16.3 (a) Views of the polishing ponds and (b) the intake of the recycled water pumping station.

Water quality control and monitoring

Water quality monitoring is carried out by the operator. Once or twice a month, at various points along the system, the following sampling points are monitored:

- Arrival of raw sewage,
- Entry into the first pond (secondary effluents),
- Exit from the polishing pond (water reused for irrigation / discharged into the sea).

The results of physical-chemical and bacteriological monitoring are summarised in Table 16.1 for the period from January 2000 to December 2009.

Table 16.1 Water quality characteristics of the water reclamation plant.

		Secondary effluent		Recycled water	
Parameter	**Raw sewage**	**Measured**	**Required by the water control board**	**Measured**	**Recommended**
COD, mg/L	503 (96–10)	45.7 (20–103)	<125	–	<125
BOD_5, mg/L	206 (44–489)	4.2 (1.5–37)	<25	–	<25
TSS, mg/L	238 (52–542)	7	<30	–	<150
Ntot, mg/L	57.9 (0–143.2)	7.54 (0–44.6)	–	–	–
Ptot, mg/L	9.6 (0–101.9)	1.13 (0–13.4)	–	–	–
E.coli/100 mL	–	–	–	81 (15–11,751)	≤1000
Œufs d'helminthes/1L	–	–	≤1	0	≤1

Despite the occasional slight breach of some parameters, the water is judged to be of good quality and in compliance with the regulations imposed by local health authorities.

The effluent which enters the plant presents a good biodegradability with a COD/BOD_5 ratio of about 2.4 (classic values between 2.0 and 2.5). However, the coefficients of variation for COD, BOD_5 and TSS show a high variability in terms of the quality of the raw wastewater, that is 41%, 47% and 43%, respectively. The pollution elimination during the secondary treatment is very good, with average values for the monitoring period of 96% for TSS, 97% for BOD_5 and 89% for COD. This consistency in terms of purification efficiency confirms the robustness of the wastewater treatment plant, which can cope with variations in load, as long as these remain within the admissible range. The plant's performance in terms of nitrogen and phosphorous removal is also high, consistently exceeding 80%.

In terms of disinfection, 86% of the results do not exceed the recommended value for *E.coli* of 1000 CFU/100 mL. This effectiveness depends principally on the hydraulic residence time, which can vary from 20 to 120 days, depending on the plant loads. Non conformity is observed, as a rule during summer periods, when the retention time required for correct decontamination is reduced, due to the increase of wastewater and irrigation volumes. Because the ponds are very attractive for birds, contamination with coliforms would also occur.

Main challenges for operation

Since the start of operation, no major challenges have been faced with the exception of a slight reduction in the microbiological quality in 2006, when the pond was out of service during work to extend the last polishing pond. A clear improvement in output was seen after this pond was put into service.

The only major problem identified concerns water salinity. Significant infiltration was observed in sewers with the intrusion of seawater. Moreover, the areas in which the polishing ponds are located have been used in the past for salt production. Increased salinity has an adverse impact on potato production, this was observed in some periods, notably in 1982, during the first years of operation of polishing ponds.

Currently, water conductivity is continuously monitored before distribution for irrigation, which is interrupted at high salt levels.

16.3 WATER REUSE APPLICATIONS

The potential irrigable surface area in Noirmoutier is around 710 ha. Currently, 380 ha are cultivated with potatoes (Figure 16.4), compared to 450 ha cultivated a few years ago. This reduction was not due to the availability of water resources, but to restrictions on the use of certain pesticides. Consequently, the number of varieties of potato has been reduced to four varieties, namely Sirtema, Lady Christl, Charlotte and la Bonnotte.

Figure 16.4 Views of the potato fields irrigated with reclaimed water.

An agronomical study conducted in 1997 by the Noirmoutier cooperative highlighted the fact that, without irrigation, the island's climatic conditions were only favourable to cultivating very early potato crops and early crops. This study also confirmed the value of establishing an irrigation system to ensure the production of these early crops and to cultivate potatoes during the summer months.

Currently, the annual production of potatoes is on average 10,000 tonnes. Based on information provided by the Noirmoutier agricultural cooperative, the lack of recycled water availability is resulting in a loss of 2000 tonnes of potatoes production in a normal year and 4000 tonnes in a dry year. Thus, maintaining the water reuse has become essential for maintaining the local agricultural economy.

Currently, the cultivation of potatoes represents an annual turnover of approximately €10 million per year.

Evolution of the volume of supplied recycled water

As shown in Figure 16.5 for a typical year, the peak in irrigation requirements does not coincide with the peak in reclaimed water production. In April, May and June, the water demand largely exceeds reclaimed water production. This deficit is generally compensated for by the occasional use of drinking water. The typical irrigation pattern is start-up in March-April, peak in May–June (April in a dry year) and end in September.

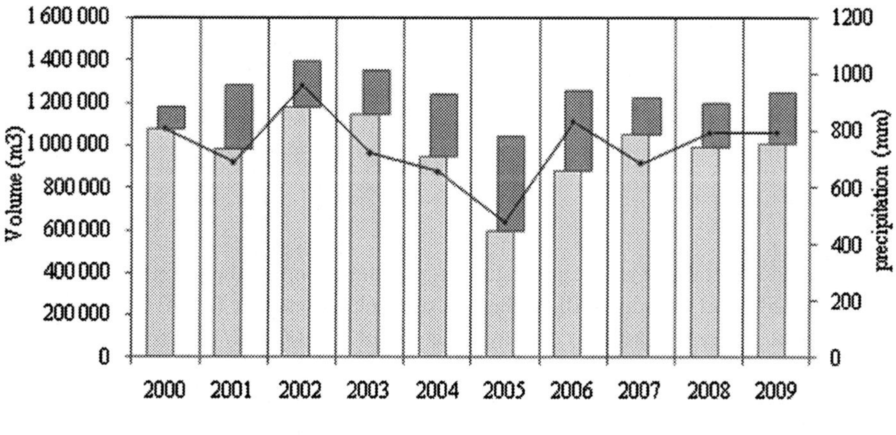

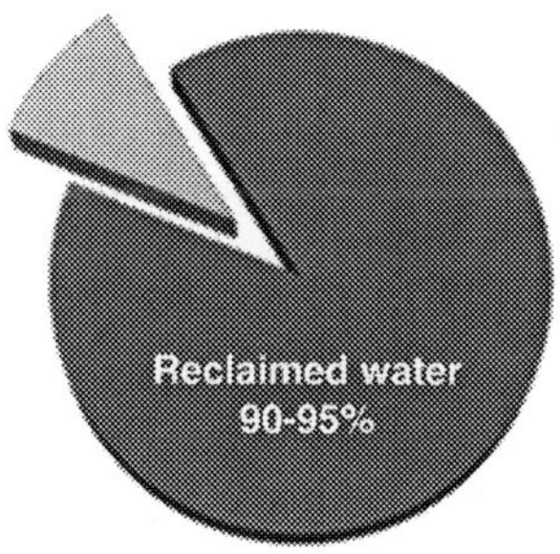

Figure 16.5 Evolution of the annual volumes of reclaimed water used for agricultural irrigation and percentage of reclaimed water in the annual irrigation water demand.

The percentage of water reuse on Noirmoutier varies from year to year. It is highly dependent upon climatic changes (Figure 16.5).

Between 2000 and 2009, an annual average of 22% of the produced reclaimed water was recycled for irrigation, corresponding to approx. 300,000 m^3/yr. This volume varies from 110,000 m^3, recorded in 2000, to a maximum of 448,000 m^3 in 2005 (a dry year).

In general, water reuse covers nearly 90–95% of annual water demand for potato production. This remaining 5–10% is covered by a supply from the drinking water network.

16.4 ECONOMICS OF WATER REUSE

Project funding, costs and pricing strategy

The water reclamation scheme of Noirmoutier includes:

- Lagoons and storage ponds managed by the Community of municipalities.
- A pumping station and irrigation network, which belongs to the irrigators' association.

The capital costs for polishing and storage ponds were covered by the Community with the financial assistance of the public water board and the Loire Bretagne Water Company.

The capital costs of the pumping station and irrigation network were covered by the farmers' association with the help of departmental grants.

As shown in Table 16.2, the total capital cost is €5.6 million or about 14,700 €/ha of irrigation. Operating and maintenance costs are around 170,000 €/yr, or about 450 €/ha of irrigated area.

Table 16.2 Capital and O&M costs.

Parameter	Capital costs, M€	Operation & Maintenance, €/yr
Lagoons and storage ponds	1.8	10,000
Pumping station and irrigation network	3.8	200,000
Total	5.6	300,000

At present, when the capital costs covered by ASDI have been paid off, the financial impact on the price of the reclaimed water accounts only for the operation, maintenance and renewal of pumping and distribution equipment.

Currently, the reclaimed water rates paid by the farmers are as follows:

- 0.3 €/m^3 of recycled water
- 190 €/ha·yr

Considering an average annual volume of reclaimed water of 300,000 m^3/yr and 380 irrigated hectares, the total specific cost is thus 0.54 €/m^3 (without any charge or employee).

Benefits of water recycling

The use of reclaimed water for irrigation instead of potable water is generating significant economic benefits for farmers using this irrigation system. The cost saving is about 50% on the water bill, as the cost of drinking water is 1.30 €/m^3 compared to 0.54 €/m^3 for reclaimed water. Thus, the total average annual cost saving is about 225,000 €/yr, or 22.5 €/tonne of produced potatoes.

Moreover, it should be noted that water reuse has enabled the production of potatoes to be extended over the summer period and thus the quantities produced to be increased by nearly 40%.

In addition, without recycling water, the community would have had to manage continuous discharge of wastewater throughout the year. Various impact analyses have shown the need not only for polishing treatment, but also for moving the discharge point far from the coast, which would have led to significant additional costs, estimated at approx. €2 million.

The main environmental benefits are the preservation of drinking water resources for domestic purposes and the reduction of the pollution being discharged into the sea, thus reducing the risks of affecting the quality of seawater for shellfish farming, swimming, shallow fishing and salt production.

16.5 HUMAN DIMENSION OF WATER REUSE

Communication strategy

Water reuse in Noirmoutier quickly grabbed the attention of scientists and researchers. Between 1993 and 1996, the European NORSPA LIFE programme, conducted by the SCE Group, conducted an initial assessment of the effectiveness of the system in terms of the recycled water quality for irrigation, and also in terms of polishing pond operation (control of hydraulic retention time). In the context of this programme, the design of the hydraulic flow between the ponds was revised. This programme and its results were the subject of several discussions notably with institutions, politicians and farmers.

The La Salaisière site was also one of the pilot sites for the European CatchWater programme (1999–2001), bringing together scientists, companies and regional groups from Italy, Spain, England, Israel, and France. The objectives of the programme were to evaluate the scope of application of water recycling in Europe and the conditions for developing a design methodology for water reuse projects, as well as to participate in the European effort to normalise the use of this alternative water resource.

In the context of this project, Noirmoutier was then chosen to host a workshop on the theme of "Integrated water management and water reuse" in September 2001, where participants discussed the results of their work alongside the experiences and expectations of users and technicians, public sector stakeholders and local groups in charge of managing water resources and water treatment (Figure 16.6a).

Figure 16.6 Views from an international workshop and a visit of decision-makers interested in water reuse.

It should be noted that the La Salaisière site is regularly visited by authorities and scientists from around France and also from abroad (Figures 16.6b and 16.7).

The role of decision makers

One of the keys to success of water reuse is the strong commitment and motivation of politicians and farmers. Farmers are contractually bound by a tripartite agreement with their operator. This agreement, updated in 2005, defines the responsibilities of the various stakeholders:

- The producer (the Community), owns the water reclamation system including the polishing and storage ponds, and makes the treated water available to irrigators. The producer is also obliged to guarantee the quality of the treatment.
- The user (the irrigants' association), owns the irrigation system (pumping structure and irrigation network), and must ensure it is operation, maintenance and renewal, as necessary.

- The operator of the infrastructures ensures the analysis of the physical-chemical and bacteriological quality of the reclaimed water.

Figure 16.7 Views of the building of irrigant's association and the Community of Municipalities and the potatoes packing hall.

16.6 CONCLUSIONS AND LESSONS LEARNED

The two major challenges in Noirmoitier that were resolved by the implementation of the water reuse are maintaining a profitable agricultural sector and reducing the pollution of sensitive environment to preserve economic and leisure activities. Noirmoutier has demonstrated this in an exemplary way for more than thirty years.

There are various keys to the success of this water reuse project. These include, in particular, the unfailing commitment of various local stakeholders; implementation of a strong and reliable polishing treatment; creation of large capacity storage ponds to guarantee sufficient reserves of water; rigorous management of treatment infrastructures, and, last but not least, proven economic efficiency.

The main drivers, benefits and challenges of water reuse in Noirmoutier are summarised in the following Table 16.3.

Table 16.3 Summary of lessons learned.

Main drivers	Benefits	Challenges
Maintain agricultural activity on the island Reduce pollution emission into the environment Preserve the uses of seawater Reduce overexploitation of natural water resources	Local production of an alternative water resource, reliable, available and drought-proof for irrigation Reduced dependence towards the supplier of drinking water Economic benefits with over 50% reduction of water bill for farmers Additional revenues from the expanded potatoes production in summer over 40% Conservation and preservation of sensitive environment and biodiversity Preservation of economic and leisure activities in the coastal area	Production of consistent recycled water quality despite the high variations in raw wastewater quality High salinity due to seawater intrusion in sewers and salt leaching in ponds Seasonal demand for irrigation and high storage needs High reliability of operation

REFERENCES AND FURTHER READING

Agence régionale de santé des pays de la Loire-Délégation territoriale de la Vendée. (2009). Qualité de l'eau distribuée en 2009—ile de Noirmoutier. http://pays-de-la-loire.sante.gouv.fr/envir/seep85_010_fichiers/000019.pdf (accessed 28 July 2011).

Lavison G. and Moulin L. (2007). Réutilisation des eaux usées: réglementation actuelle et paramètres d'intérêt. *L'Eau, l'Industrie, Les Nuisances*, **299**, 41–46.

Lazarova V. and Brissaud F. (2007). Intérêt, bénéfices et contraintes de la réutilisation des eaux usées en France. *L'Eau, l'Industrie, Les Nuisances*, **299**, 29–39.

Programme Environment and Climate (2000). Enhancement of Integrated Water Management Strategies with Water Reuse at Catchment Scale, Report ENV4-CT98-0790 of the Catchwater project.

Puil C. (1998). La réutilisation des eaux usées urbaines – Quelques exemples. http://www.u-picardie.fr/beauchamp/duee/puil.htm (accessed 16 Mar 2011).

Voujour C. (1997). Diagnostic agronomique de la culture de la pomme de terre de Noirmoutier à l'échelle de la filière, de l'exploitation agricole et de la parcelle. Conséquences de l'extension de la culture à Barbatre et de l'utilisation des eaux usées pour l'irrigation. Coopérative agricole de Noirmoutier.

Xu P., Brissaud F. and Fazio A. (2002) Non-steady-state modelling of faecal coliform removal in deep tertiary lagoons. *Water Research*, **3**, 3074–3082.

WEB-LINKS FOR FURTHER READING

http://www.cdc-iledenoirmoutier.com/
http://www.vendee-guide.co.uk/ile-de-noirmoutier.htm

Views of the famous potatoes from Noirmoutier

17 Irrigation of crops in Australia

Daryl Stevens and John Anderson

CHAPTER HIGHLIGHTS

The development of water recycling schemes for irrigation of crops has enabled Australian growers to increase production despite water shortages due to drought and increased competition for fully allocated river water supplies. The use of recycled water for irrigation of crops and pastures is the largest component of water reuse in Australia.

KEYS TO SUCCESS

- Development of national and state guidelines for use of recycled water
- Development of guidance manuals to assist growers to use recycled water
- A national program to assist growers to produce more with less water

KEY FIGURES

- Total annual volume of treated wastewater in Australia: About 2000 Mm^3/yr
- Total volume of recycled water used in Australia: More than 420 Mm^3/yr
- Total volume of recycled water used in agriculture: More than 280 Mm^3/yr

17.1 INTRODUCTION

Development of water recycling for irrigation

More than 100 years ago, before the development of ocean outfall discharges, some wastewater was used for irrigation of market gardens in Sydney and Melbourne. These schemes were abandoned early in the 20th century because of public health concerns and nuisance caused by the odours.

Over the years, recycled water from wastewater treatment plants was used for pasture irrigation on land adjacent to treatment plants in country areas and on a larger scale at Werribee in western Melbourne where lagooning, land filtration and grass filtration through overland flow formed part of the treatment process before discharge.

A number of drivers have encouraged the use of recycled water in Australia since 1980.

Economic drivers

The demand for water, particularly for agricultural activity has increased in response to commercial need of private companies and government economic policy. For example an increase in wine production was founded on expansion into international markets. The expansion of agriculture to supply national and international markets depends on the application of best practice production methods and the availability of adequate sources of water.

Water shortages

In Australia, water allocations for agriculture are subject to volumetric allocation schemes under which growers receive a percentage of their maximum allocations depending on season conditions and water available in storages. Growers may receive 100% of allocations in a wet year but reduced percentages in dry years.

Australia has moved to implement water markets where water allocations can be traded on a temporary or permanent basis to other users in the same river basin. Water prices are set by the market. An effect of water trading over time is to move water allocations to higher value irrigation enterprises. As a result of water shortages and water trading, the value of recycled water has increased.

Social drivers

Australians are becoming increasingly aware of the economic and environmental advantages of, and imperatives for, water recycling. Recycling of household wastes and green wastes has become common nationwide and focussed attention on the potential for water recycling.

Environmental drivers

There has been public interest and government policy action to reduce the impact of treated wastewater discharges on receiving waters. At the same time, water shortages due to drought, increasing competition for water and full allocation of existing supplies have generated intense public debate about the declining health of Australia's rivers, the adverse impacts of global warming and the sustainability of Australia's water resources. Measures to reduce agricultural allocations to provide baseline environmental flow allocations are intensely controversial because of the potential economic impact on rural towns and communities.

Growth in reuse for agriculture

There has been remarkable growth in recycled water use for agriculture since 1990. While approximately there are about 230 recycled water schemes supplying recycled water for urban uses, there are now more than 270 schemes supplying recycled water for agriculture in Australia. Agriculture uses the largest volume of recycled water, accounting for 66% of all recycled water used (280 Mm^3/yr). Most recycled water in agriculture (237 Mm^3/yr) is used for pastures and dairy farming but reuse in horticulture (fruits, grapes and vegetables) is growing. Golf courses, sporting grounds and parks are also a significant user (33 Mm^3/yr) (Stevens *et al.* 2006). Urban gardens are a relatively small user of recycled water (currently about 2 Mm^3/yr) but likely to grow to about 25 Mm^3/yr over the next 20 years.

Development of guidelines

Water reuse guidelines developed by Australian state authorities in the 1980s provided for irrigation of pasture and for food crops that are cooked before eating.

In 1993, New South Wales published Guidelines for Urban and Residential Use of Reclaimed Water. These guidelines provided for a filtered and disinfected recycled water grade similar to the California Title 22 regulation (NSW RWCC 1993). These guidelines established Australian acceptance of the use of high grade recycled water suitable for open public access and suitable for use on home gardens.

The 1996 National Health and Medical Research Council guidelines for use of recycled water defined recycled water quality in four grades A to D (Table 17.1). These grades are incorporated in most of the more recent state guidelines for irrigation with recycled water. A general indication of the grades and uses are indicated below:

Table 17.1 Recycled water grades commonly used in National and State guidelines.

Class	Range of uses (uses include all lower class uses)
A	Urban (non-potable): with uncontrolled public access. Agricultural: for example human food crops consumed raw. Industrial: open systems with worker exposure potential.
B	Agricultural: for example dairy cattle grazing. Industrial: for example wash down water.
C	Urban (non-potable) with controlled public access. Agricultural: for example human food crops cooked/processed, grazing/fodder for livestock. Industrial: systems with no potential worker exposure.
D	Agricultural: non-food crops including instant turf, woodlots, flowers.

Australia has now developed the Australian Guidelines for Water Recycling through a joint venture between the federal and state governments (EPHC/NRMMC/AHMC, 2005, 2006, 2008a, 2008b, 2009a, 2009b). The new Australian Guidelines use fit-for-purpose recycled water quality in place of the Class A to D system. The Guidelines have adopted a standard risk assessment and hazard analysis and critical control point (HACCP) system. HACCP is now the international standard for food safety. When the guidelines and best practice principles are followed, users and consumers can be confident that it is safe to work with recycled water, the food grown with recycled water is safe, and that the environment is not adversely affected by the use of recycled water.

17.2 CASE STUDIES: IRRIGATION OF CROPS IN AUSTRALIA

New South Wales

Shoalhaven REMS

The Northern Shoalhaven reclaimed water management scheme is located near Nowra, about 160 km south of Sydney. Recycled water from six wastewater treatment plants is transferred to a 0.6 Mm^3 storage before being supplied to irrigate 500 ha of dairy pasture on 20 properties. The scheme commenced in 2001.

The scheme was preceded by an extensive community consultation. The community expressed a strong preference for beneficial reuse over discharge to the ocean, and a willingness to pay higher charges to achieve this outcome. The Shoalhaven River flood plain, predominantly used for dairy farming was identified as the best reuse area with potentially 1500 ha suitable for irrigation. Barriers to reuse included relatively high rainfall (over 1000 mm/yr) and no prior experience of irrigation on the dairy farms. Keys factors leading the dairy farmers' decision to participate were an assured supply of water when required, modest financial incentives and first preference for additional supply (Moore, 2009).

Outcomes from the scheme have included:

- An 80% reduction in discharges to the ocean and elimination of discharges into the Jervis Bay Marine Park at an annual cost of around AU$50 (40 €) per household.
- Drought proofing the dairy farms has helped the dairy farms to improve their productivity and viability. During the 2002–2003 drought period the farmers made large saving in feed costs and milk production was maintained.

Victoria

Eastern irrigation scheme

Class A recycled water from Melbourne's Eastern treatment plant at Carrum are delivered into the Eastern Irrigation Scheme from a new ultrafiltration plant under a contract between Melbourne Water and the private sector contractor Earth Tech (Figure 17.1). Under the 25-year build-own-operate contract, Earth Tech delivers 15,000 to 18,000 m^3/d of Class A recycled water to growers and to new residential subdivisions in the Cranbourne area.

Figure 17.1 View of the ultrafiltration water recycling plant, eastern irrigation scheme (Photograph courtesy of Water Infrastructure Group, Melbourne).

Growers were finding it difficult to grow high quality produce with the decreasing quality and quantity of existing water sources in the area. Vegetable production was often curtailed in hot dry summer periods. Since the coming of the recycled water scheme, the local vegetable growers have had a reliable supply of high quality water for the first time ever. The constant supply of water has ensured full production over the whole growing season and the production of better quality vegetables. Local turf growers have been able to increase the amount of grass grown over summer.

Werribee irrigation district

Class A recycled water from Melbourne Water's western treatment plant at Werribee is delivered into the Werribee Irrigation District which is an important vegetable growing area. Over 400 growers produce lettuces, broccoli, cabbages and many other vegetables for local consumption and for export. Historically the vegetable growers have used water from the Werribee River and the underlying aquifer.

Following a treatment works upgrade and construction of a transfer pipeline, growers received the first deliveries of Class A recycled water in 2005. By 2009 more than 190 growers had signed supply agreements to use recycled water covering more than 90% of the irrigable land in the district. In the first 3 years of full operation the scheme used about 13 Mm^3 of recycled water which was previously discharged into shallow coastal waters in Port Phillip Bay.

Queensland

Mackay

The Mackay water recycling scheme was implemented by the Mackay Regional Council to protect the Great Barrier Reef from nutrients in treated discharges. The scheme recycles about 90% of Mackay's wastewater. It will also protect and rehabilitate overcommitted groundwater resources at risk from seawater intrusion by substituting 8.5 Mm^3/yr of recycled water for groundwater used by the local irrigation industry.

The scheme included the construction of a new wastewater treatment plant at Bakers Creek to produce Class A recycled water, a 2.2 Mm^3 balancing storage (Figure 17.2), a distribution pipeline network and 1.2 Mm^3 of on-farm storage on 27 sugarcane farms (Mackay Regional Council, 2011).

Figure 17.2 Recycled water storage, Mackay water recycling scheme (Photograph courtesy of Smart Water Australia, Canberra).

Wide Bay

Wide Bay Water, which serves the city of Hervey Bay in Queensland, faced public concerns about the environmental impacts of treated discharges into the shallow coastal waters offshore from Hervey Bay. In 1989 Wide Bay Water began a reuse scheme with the purchase of a 150 ha research farm to demonstrate agricultural reuse to local growers. The research demonstrated the use of recycled water for sugar cane irrigation and identified suitable hardwood eucalypt species for timber production. It was shown that sugarcane irrigation increased the cane yield by 62% and farm income by 80% due to higher sugar content in the cane, compared to non-irrigated cane.

The scheme has now been extended to more than 1000 ha of sugarcane plantations, eucalypt plantations (Figure 17.3), turf farms, sports fields and golf courses. A large recycled water storage has been constructed to hold winter and wet weather flows for later dry weather use. An advantage of the combination of sugarcane and eucalypt crops is that when sugarcane irrigation demands peak in hot dry periods, the eucalypt plantations are sufficiently hardy to tolerate reduced irrigation.

Figure 17.3 Wide Bay Water's eucalyptus plantation (Photograph courtesy of Wide Bay Water, Queensland).

Wide Bay Water has also increased the volume of recycled water available for irrigation by collecting stormwater in detention basins. This water is then pumped through the sewer system at night when sewage flows are low, treated, and delivered to the recycled water storages (Heron & Waldron, 2009).

South Australia

Virginia pipeline scheme

The 20,000 ha Virginia irrigation area on the North Adelaide Plains is a prime area for growing horticultural crops, providing vegetables, fruit and nuts for local and interstate markets. This industry relied on groundwater for irrigation. Annual water use was about 18 Mm^3/yr but the natural groundwater recharge rate was only about 6 Mm^3/yr. The groundwater resources in the region has become seriously depleted due to overuse resulting in falling groundwater levels, higher pumping costs and increasing risk of saline intrusion. At the same time, treated discharges from Adelaide's Bolivar treatment plant nearby were having serious environmental impacts on sea grass beds in the shallow waters offshore in the Gulf of St Vincent.

The Virginia Pipeline Scheme was commissioned in 1999. It includes a 110,000 m^3/d water recycling plant at Bolivar. The plant incorporates dissolved air flotation and filtration processes (Marks *et al.* 1998) and an extensive pipeline distribution system. The scheme has more than 240 supply agreements to supply up to 20 Mm^3/yr of recycled water (Figure 17.4). Recycled water use was 12.1 Mm^3/yr in 2004/05. Ultimately about 23 Mm^3/yr could be used (50% of flow through the Bolivar treatment plant. This could be increased significantly by aquifer storage and recovery which has been the subject of field trials.

The availability of additional water has enabled growers to increase production. A measure of the economic benefits can be assessed from the fact that good horticultural land in the region without water and improvements is valued at about AU $15,000/ha (12,000 €/ha) but with water allocation the land is worth AU$30,000/ha (24,000 €/ha). Census data shows an increase in agricultural employment in the Virginia area (Kelly, 2003).

New land has now been brought into horticultural production. A 20 km extension to Angle Vale was completed in 2009, increasing recycled water use from 15 Mm^3/yr to 18 Mm^3/yr.

Figure 17.4 Farm connection, Virginia pipeline scheme (Photograph courtesy of Arris Pty. Ltd., Melbourne).

Willunga scheme

In 1995, SA Water commissioned a study into the potential to use recycled water from the Christies Beach treatment plant south of Adelaide to supply water for irrigation of horticultural crops in the Willunga Valley. The region was already a renowned producer of quality wines. Although there was nearly 5000 ha of suitable land for viticulture, industry expansion was limited due to non-sustainable use of groundwater from the underlying aquifers. The study identified that 600 ha could be irrigated using the summer flow from the treatment plant and this could be expanded to 2400 ha by providing a balancing storage.

An agreement was negotiated between SA Water and the Willunga Basin Water Company, a consortium of 18 growers, landholders and winemakers for supply of recycled water. In 1999, the company constructed a pipeline transfer scheme with a capacity of 24,000 m^3/d. The use of recycled water was about 2.0 Mm^3 in 2001/02.

An extension to McLaren Vale is planned to supply vineyards, olive orchards and cut flower growers with recycled water in place of potable water.

Western Australia

Albany tree farm

At Albany on the south coast of Western Australia the community was willing to pay more for beneficial use of recycled water in place of ocean discharge to King George Sound. The WA Water Corporation established a major tree plantation in 1995 and delivers 4500 m^3/d of recycled water to irrigate 450,000 Tasmanian blue gums planted on a 300 ha site. Each year about 50 ha of trees are harvested and are processed at a local timber mill for export as woodchips. The return from the sale of woodchips and the land based wastewater reuse option makes tree plantations utilising recycled water both economically and environmentally sound.

The stored wastewater is applied to the trees mainly in the summer or when soil moisture content is low enough to allow irrigation. When it is too wet to irrigate the plantation, wastewater is stored in a large dam on site. The irrigation scheme is computer-controlled, with soil probes continuously monitoring the soil to prevent over-watering. Most nutrients are retained on-site, with about half of the nitrogen being removed by allowing the wastewater to flood irrigate 14 ha of pasture land, and the remainder being used by the trees. Phosphorus is captured by the clay soils of the site, which were especially selected for their capacity to absorb large quantities of phosphate. An additional 120 ha have been planted with the same fast-growing eucalypt. These trees are rain-fed and have been planted downstream of the irrigated plantation to intercept any runoff (WCWA, 2011).

Tasmania

Coal Valley

Clarence Valley East of Hobart in Tasmania provides recycled water to the South East Irrigation Scheme in the Coal Valley The scheme supports a variety of high value horticultural enterprises including vineyards, stone fruit for export markets, fresh salad products, opium poppies for the pharmaceutical industry, nurseries and commercial turf production. It also reduces treated discharges to the Derwent River estuary.

The total available recycled water supply is 2.85 Mm^3/yr which is distributed to 28 properties (19 farms, 3 vineyards, a plant nursery and 5 golf courses. The scheme commenced operation in October 2006. At present water is supplied to about 650 ha of land (Lane *et al.* 2009).

Since the scheme commenced operation, irrigation water requests have exceeded available supply in summer and there has been unused recycled water supply in winter. A 0.9 Mm^3 storage has now been constructed to carry winter flows over to the summer irrigation season and it is expected that 85% to 90% of the available recycled water will be used for irrigation.

17.3 GUIDANCE TO GROWERS

Over the last decade, the Australian irrigation industry has developed a wide range of guidance documents to assist growers (Figure 17.5). The guidance includes both general guidance for irrigation practices and specific guidance for irrigation with recycled water.

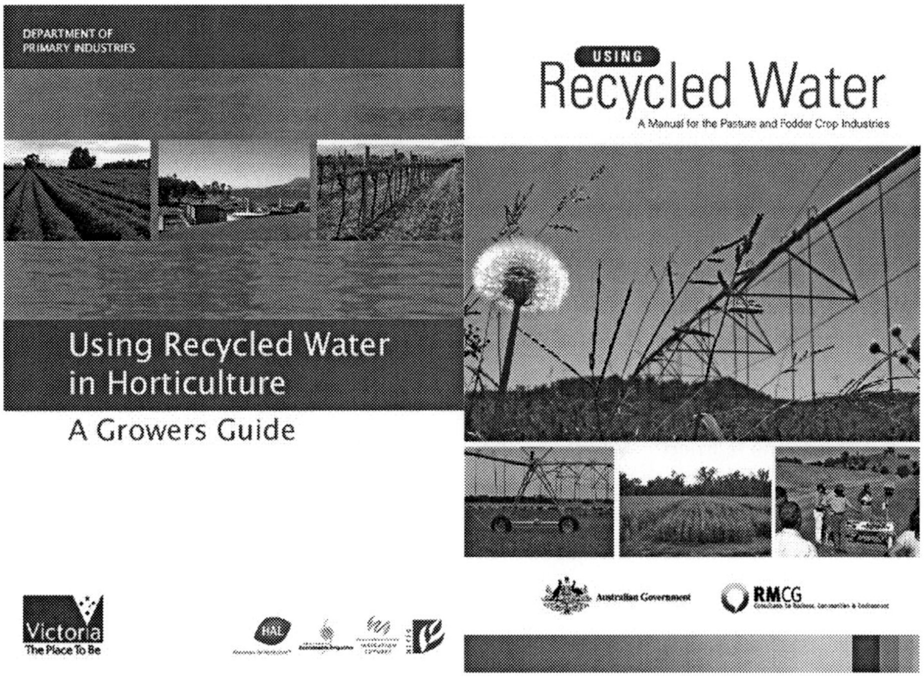

Figure 17.5 Grower guidance manuals.

National program for sustainable irrigation

The National Program for Sustainable Irrigation (NPSI) invests in research into and adoption of sustainable irrigation practices. The NPSI program collaborates with government and primary producer bodies and has been responsible for improved irrigation scheduling and application techniques. For many irrigators this has resulted in reduction of 40% or more of their water use over the last 20 years without loss of production.

The program is researching ways of coping with:

- Water scarcity
- Salinity and other soil factors
- Irrigation and nutritional requirements
- Biodiversity and production
- Plant physiology and water requirements
- The economics of irrigation

Beneficiaries have included a range of irrigation enterprises including rice, grain, cotton, vegetables, tree fruits, nuts, grapes, sugar, dairying and livestock. These irrigated industries account for more than a third of the nation's total gross value of agricultural production, so are significant to Australia's economy.

Guidance for use of recycled water

Horticulture Australia Ltd. And Federal and state agencies have funded the development of manuals, fact sheets and brochures to provide growers with specific guidance on the use of recycled water for irrigation.

Documents include

- *Water Recycling in Nurseries* (NPSI, 2005)
- *Growing Crops with Reclaimed Wastewater* (Stevens, 2006)
- *Using Recycled Water in Horticulture: A grower's guide* (Jarwal, Boland, Stevens & Faggion, 2006)
- *Using Recycled Water: A manual for the pasture and fodder crop industries* (Boland & Shanahan, 2008)

17.4 OVERVIEW

Since 1990, Australian has moved from small scale irrigation of pasture adjacent to treatment plants to major recycled water irrigation schemes. The development has been accompanied by development of appropriate guidelines for the management of health and environmental risks, together with extensive guidance materials to assist growers establish successful farm practices.

Without the use of recycled water many of Australia's urban and rural irrigation schemes would not be sustainable. There is a common theme running through many of the case studies in this report. Growers who were dependent on traditional water sources often experienced reduced production and economic losses in dry times. Since switching to use of a drought-proof recycled water supply, they have been able to maintain production, or even increase production in dry times.

The following Table 17.1 summarizes the main drivers for water reuse for agricultural irrigation in Australia as well as the benefits and challenges.

Table 17.1 Summary of lessons learned.

Drivers/Opportunities	Benefits	Challenges
Existing supplies fully allocated. Water shortages due to drought. Higher standards for discharges to the environment.	Water shortages overcome. Water and nutrients recycled. Crop production increased. River pollution reduced.	Sustainable irrigation methods to maintain soils and protect groundwater.

REFERENCES AND FURTHER READING

Boland A. and Shanahan M. (2008). *Using Recycled Water: A Manual for the Pasture and Fodder Crop Industries*. Worldwide Printing, June 2008. Available for download at www.recycledwater.com.au/index.php?id=96

EPHC, NRMMC & AHMC (2005). Australian Guidelines for Water Recycling: Managing Health and Environmental Risks – Impact Assessment. Environment Protection and Heritage Council, Natural Resource Management Ministerial Council & Australian Health Ministers Conference. Sep 2005.

EPHC, NRMMC & AHMC. (2006). Australian Guidelines for Water Recycling: Managing Health and Environmental Risks. Environment Protection and Heritage Council, Natural Resource Management Ministerial Council & Australian Health Ministers Conference. Nov 2006.

EPHC, NRMMC & AHMC. (2008a). Overview Document – Australian Guidelines for Water Recycling: Managing Health and Environmental Risks. Environment Protection and Heritage Council, Natural Resource Management Ministerial Council & Australian Health Ministers Conference. Mar 2008.

EPHC, NRMMC & AHMC. (2008b). Australian Guidelines for Water Recycling: Augmentation of Drinking Water Supplies. Environment Protection and Heritage Council, Natural Resource Management Ministerial Council & Australian Health Ministers Conference. May 2008.

EPHC, NRMMC & AHMC. (2009a). Australian Guidelines for Water Recycling: Managed Aquifer Recharge. Environment Protection and Heritage Council, Natural Resource Management Ministerial Council & Australian Health Ministers Conference. Jul 2009.

Heron D. and Waldron T. (2009). 20 Years of water reuse in Hervey Bay Queensland, Australia – turning waste into a clean green asset. Reuse09- IWA 7th Specialist Conf on Water Reclamation and Reuse, Brisbane, Sep 2009.

Jarwal S., Boland A., Stevens D. and Faggian R. (2006). *Using Recycled Water in Horticulture: A Grower's Guide*. Dept. of Primary Industries, Victoria, May 2006. Available for download at www.recycledwater.com.au/index.php?id=96

Kelly J., Stevens D. and White T. (2003). Achievements of the Virginia pipeline scheme: horticultural production with reclaimed water, *Proc Water Recycling Australia*, Aust Water Assn., Brisbane 2003.

Lane P., Peterson L. and Rae M. (2009). Managing the demand and supply of recycled water for agriculture in Coal Valley, Tasmania, Reuse09-IWA 7th Specialist Conf on Water Reclamation and Reuse, Brisbane Sep 2009.

Mackay Regional Council. (2011). Mackay Water Recycling Project, www.mackay.qld.gov.au/services/water_and_waste/mackay_water_recycling_project

Moore W. (2009). Shoalhaven Reclaimed Water Management Scheme: Innovative and effective partnerships lead to success, Reuse09- IWA 7th Specialist Conf on Water Reclamation and Reuse, Brisbane Sep 2009.

NPSI. (2005). Water Recycling in Nurseries, National Program for Sustainable Irrigation and Horticulture Australia, Fact Sheet NPSI 2005/3.

Stevens D. (ed.) (2006). *Growing Crops with Reclaimed Wastewater*. CSIRO Publishing, Melbourne, Australia, 2006.

Stevens D., Smolenaars S. and Kelly J. (2006). *Irrigation of Amenity Horticulture with Recycled Water: A Handbook for Parks, Gardens, Lawns, Landscapes, Playing fields, Golf Courses and Public Open Spaces*. Arris Pty Ltd., Melbourne, Australia. Available for download at www.recycledwater.com.au/index.php?id=96

Stevens D., Unkovich M. and Boland A. (2006). Water Recycling in Australia. Arris Pty Ltd., Melbourne, Australia. http://npsi.gov.au/files/products/national-program-sustainable-irrigation/px061130/px061130.pdf.

WCWA (2011). Albany waterwater treatment plant, Water Corporation of Western Australia www.watercorporation.com.au/W/wwtp_albany.cfm?uid=1187-6721-9241-1584

Irrigation of food crops in the Coal Valley, Tasmania (Photograph courtesy Brand Tasmania®).

Views of the Bolivar water recycling plant.

PART V

Industrial use of recycled water

Foreword

By The Editors

Historical development of industrial use

One of the earliest cases of industrial reuse in the USA was the use of chlorinated reclaimed water for steel processing at the Bethlehem Steel Company in Maryland for cooling of process equipment and steel products, commencing in 1942. In Japan, the Tokyo Metropolitan Sewerage Bureau supplied reclaimed water to industrial users from 1955 onwards.

The pioneer in closing the water cycle in industry is Germany, where rainwater and industrial water recycling have been practiced since the beginning of the 20th century, for example since 1938 by Volkswagen and in zero liquid discharge paper production by Smurfit Kappa Zülpich Papier since 1970.

Around the world there has been extensive use of recycled water for industrial process water and for heating and cooling in industrial plants, steelworks, oil refineries and power stations. Recycled water has also been used in pulp and paper mills, in textile processing and carpet manufacture.

The development of advanced membrane treatment systems has enabled both municipal and industrial recycled water to be used for high grade industrial applications. Examples include:

- In Australia, where a dual membrane system was built in 1994 to supply high quality recycled water for boiler feed in the 2400 MW Eraring power station near Sydney.
- In Singapore, where the NEWater water recycling plants use dual membrane treatment systems to produce high quality recycled water for semiconductor manufacture.
- In the USA, where the West Basin Municipal Water District water recycling plant built in 1995 produces six “designer” grades of recycled water to suit specific customer needs in various manufacturing processes and oil refineries (see Chapter 2). Water is also supplied for landscape irrigation and groundwater recharge.

In recent years in response to rising water prices, many industrial plant managers have implemented internal water recycling projects. By recycling process water within industrial plants, water and wastewater charges can be reduced substantially and there may be additional advantages through recovery of raw materials.

Industrial reuse case studies

Part 5 of this book “Milestones in Water Reuse: The Best Success Stories” presents a number of case studies showing how water managers have used recycled water to meet the industrial water needs.

Chapter 18. Mexico is well known for the reuse of untreated wastewater, but few people know that this country was the pioneer of developing “designer” recycled water production for various purposes, including industrial use of recycled municipal water to ensure the economic viability of water reuse. A good example is the success story of the Tenorio project in San Luis Potosi which produces several grades of recycled water to suit various uses, such as industrial uses, agricultural irrigation, groundwater restoration and environmental enhancement. The largest use is the supply of cooling tower make-up water to the 700 MW Villa de Reyes power station.

Table 1. Highlights and lessons learned from the selected case studies on industrial water reuse.

Project/ Location	Start-up/ Capacity	Type of uses	Key figures	Drivers and opportunities	Benefits	Challenges	Keys to success
San Luis Potosi, Mexico Industrial use of recycled urban water	2006 Treatment capacity: 90,720 m^3/d	Cooling make-up water for a power plant, Agricultural irrigation, Groundwater restoration, Environmental enhancement	Recycled volumes: up to 23.9 Mm^3/yr for irrigation purposes and 9.9 Mm^3/yr for industry Capital cost: US$67.4 million. BOT project Recycled water rate for industry: US$0.76 $/m^3$, free of charge for farmers Aquifer restitution in 6 years (48 Mm^3) Cost savings by the power plant: US$18 million in 6 years	Water scarcity and more frequent droughts. Need to increase wastewater treatment coverage. Conflicts for water between sectors. Local State Government Policy.	Alternative water resources for non-potable purposes. 70% of wastewater is treated (goal 85%) and 100% is reused. Support economic development in both industry and agricultural sectors. Long term sustainable water management to secure water supply and protect the environment. Economic feasibility of the project to ensure continuity and reliability.	To build infrastructure to secure distribution and continuous supply. Opposition of local farmer and industrial users has to be overcome. Enforcement of the implementation. Implement appropriate pricing strategies while industry covers the major part of the cost. Water exchange between the existing over-drafted aquifers.	Federal and State policies, water reuse regulation and political support. Creative project funding (BOT contract). Production of recycled water with a "Fit to Purpose" quality for agriculture and industry. Consistent water quality, adapted to customer's requirements and high reliability of operation. Public outreach and education.
Panipat, India Industrial water recycling	2006 Paniprat refinery recycling plant 650 m^3/h 2010 Naphtha Cracker plant 450 m^3/h	Boiler make-up water, Cooling tower make-up water	Total daily production: 26,500 m^3/d Total volume of treated wastewater: 10.9 Mm^3/yr Total volume of recycled water: 9.0 Mm^3/yr r Operation costs: 0.3 €$/m^3$ Water reuse standards: $< 10\ \mu g/L$ *silica;* $< 0.1\ mg/L$ *TDS*	Economic growth in India. Environmental protection targets. No proper recipient available. Industrial water supply security could be endangered in the future.	Environmental protection effects. Boost in industrial water supply security. Saving of large amounts of freshwater. Economic and social development.	Fluctuating raw water quality. Stringent reclaimed water quality targets. Zero liquid discharge target in the future.	Economic growth in India. Economic success of the investor. Environmental awareness in India and especially in the Panipat region. Employment of advanced technologies. Experienced plant operator.
RARE project, California, United States Industrial use of recycled urban water	2010 Treatment capacity: 13,249 m^3/d	Boiler make-up water, Cooling tower make-up water	Total volume of treated wastewater: 4.28 Mm^3/yr Water reuse standards: < 2.2 *E.coli*/100 mL	Commitment to sustainability and reliability. Willingness to diversify the water supply portfolio. Close proximity of the refinery to the WWTP.	Conservation of potable water and reduction the dependence on imported water supplies. Reduction of the threat of severe rationing during droughts. Conservation of energy avoiding pumping at long distance. Reduction of wastewater and pollutant discharges into San Francisco Bay.	High level of reliability in the treatment process. Disposal of the RO brines. Flow Equalization.	Successful public-private partnership. Proven technology. Redundancy and reliability.

Closing loops, Gemany Industrial water recycling	Since early 20th century and high growth since 1970s	All types of industrial uses, including cooling, boiler make-up, process water, cleaning, etc.	Examples of reduction of wastewater volumes: *Sugar industry*: 92% from 10 m^3/t sugar beets (0.8 m^3WW/t sugar beets) *Paper industry*: 50% from 29 m^3/t paper and 90% from 0–15 m^3/t paper *Textile industry*: 77,5% from 20 m^3/t of goods	Reduction of costs. "Green" image. In few cases water shortage.	Improved material utilization. Lower energy and water consumption. Lower wastewater production. Reduction of costs by the key benefits mentioned above.	Economic benefits from water recycling have to outbalance investment and operating costs. Stringent requirements for water quality. Needs for efficient and reliable wastewater treatment and reclamation processes.	Demonstrated economic benefits from closing loops. Technical and economic feasibility and high efficiency and reliability of water recycling schemes.

Chapter 19. High rate industrial water recycling has enabled expansion of the Panipat refinery in India. Advanced multi-barrier treatment systems have been employed to supply boiler feed and cooling water in the oil refining and petrochemical industries, as well as reducing fresh water use and boosting regional water security.

The project has increased regional economic activity and had positive environmental benefits by removing discharges to the Yamuna Canal.

Chapter 20 describes another successful water reuse project with advanced multi-barrier treatment system for the production of high-purity recycled water, but in this case from municipal effluents. The Richmond Advanced Recycling Expansion in the East Bay Municipal Utility District in California uses dual membranes (microfiltration and reverse osmosis) to supply high quality recycled water for boiler feed in the Chevron oil refinery.

Chapter 21 presents an overview of closing the water cycle loops in industrial water management in Germany. Internal measures for water reuse and recycling in industrial production processes are an effective tool for increasing water use efficiency, reducing wastewater volumes, and recovering raw materials or products. Examples are described from the food, drink, paper, textiles, metals and ceramics industries, enabling to decrease wastewater discharge by up to 92%.

Keys to success

The common themes running through these industrial reuse case studies are that industrial reuse often produces large savings in freshwater needs and delivers worthwhile economic and environmental benefits. Keys to success have included the development of reliable advanced treatment technologies, regulatory support, public-private partnerships, innovative project funding mechanisms, stakeholder involvement, water savings and identifiable economic and environmental benefits.

The major highlights and lessons learned from the selected case studies are summarised in Table 1.

18 The role of industrial reuse in the sustainability of water reuse schemes: The example of San Luis Potosi, Mexico

Alberto Rojas, Lucina Equihua and Valentina Lazarova

CHAPTER HIGHLIGHTS

The **Tenorio Project** in San Luis Potosi (start-up in 2006) is the first project in Mexico making possible the production of multi-quality recycled water for planned water reuse for different purposes, including industrial uses, agricultural irrigation, groundwater restoration and environmental enhancement.

KEYS TO SUCCESS

- Federal and State policies, water reuse regulation and political support
- Creative project funding (BOT contract) Production of recycled water with a "Fit to Purpose" quality for agriculture and industry
- Consistent water quality, adapted to customer's requirements and high reliability of operation
- Public outreach and education

KEY FIGURES

- Treatment capacity: 90,720 m^3/d
- Recycled volumes: up to 23.9 Mm^3/yr for irrigation purposes and 9.9 Mm^3/yr for industry
- Capital cost: US$67.4 million (BOT project)
- Recycled water rate for industries 0.76 US$/$m^3$
- Aquifer restitution in 6 years (48 Mm^3)
- Cost savings by the power plant: US$18 million in 6 years

18.1 INTRODUCTION

The San Luis Potosí Metropolitan Area is the 11th most populous metropolitan area in México with more than 1.3 million inhabitants. It is founded on a semi-arid region with less than 400 mm of annual rainfall. Its industrial and economic development has always been dependent upon water availability and water conservation efforts. Until 2005, the wastewater treatment coverage of the urban area was 32% and approximately 86,400 m^3/d of raw sewage was discharged to a 200 ha storm basin, called the "Tenorio Tank". This pond was functioning as degraded wetland and was used for agricultural crop irrigation without supplementary treatment. Due to water scarcity in the area, local farmers fought over access to this alternative water resource and its fertilising nutrients. Consequently, over 700 ha of land have deteriorated as a result of irrigation with low quality wastewater. This has had a marked negative impact on human health with a high rate of enteric diseases being recorded in the area.

During the 80's and 90's, the industrial and service sectors developed rapidly in this metropolitan area, increasing not only drinking water demand but the energy needs as well, which additionally aggravated the water availability. At that time, the local aquifer was the main source for meeting the water demand with approximately 127 Mm^3/year (million m^3 per year) of water withdrawals. Over drafting reached 43 Mm^3 in the year 2000 that is approximately 50% above the natural recharge rate. Groundwater provided 76 Mm^3/year of potable water, 35 Mm^3/year of irrigation water and about 16 Mm^3 of water for the industry with strong conflicts and competition between these sectors.

Therefore, due to water scarcity and aquifer overexploitation, the State Government of San Luis Potosi has decided to implement an Integral Plan for Sanitation and Water Reuse. The main objective was to substitute groundwater with recycled water for all non-potables uses, including agricultural irrigation, urban reuse (recreational parks, sports fields, golf courses, recharge of artificial lakes, fountains) and industrial uses, mainly for cooling purposes. As a consequence, seven wastewater treatment plants (WWTP) were built for this purpose, and another one is currently under construction. At present, 70% of San Luis Potosi's urban area wastewater is treated and 100% of treated effluents are recycled.

Within the city of San Luis Potosi's universal water reuse scheme, the largest wastewater treatment and recycling project is named Tenorio – Villa de Reyes. This WWTP, operated since 2006, treats 45% of the total wastewater of San Luis Potosi. This project is the first in Mexico to enable the production of multi-quality recycled water for multi-purpose planned water reuse, including industrial uses, agricultural irrigation, groundwater restoration and environmental enhancement.

Main drivers for water reuse

The main drivers of the "Tenorio" water reuse project are: (1) The willingness of the federal government to increase sewers and wastewater treatment coverage, which before the construction of this plant was only 30%, (2) The local government initiative to increase water availability through the production of recycled water as an alternative drought-proof resource for non-potable purposes, as well as to (3) Improve the quality of irrigation water as well as safeguarding the health of both farmers and the local population.

Brief history of the project development

The Integral Plan for Sanitation and Water Reuse of San Luis Potosi, conceived by the government and the State Water Commission (CEA), included not only the objective to achieve 70% coverage of wastewater treatment, but also the reuse of all effluents for irrigation, aquifer recharge and industrial use by the Thermoelectric Power Plant, located in Villa de Reyes. Since the beginning of the project, farmers were strongly against the allocation of recycled water to industry. The farmers' (approximately 250 users from 4 municipalities) main argument was that they were the rightful owners of the total amount of raw wastewater and that taking it away from them would affect both their economic interests and agricultural production.

The Project Tenorio water recycling scheme (Figure 18.1) includes the following elements:

- Construction and operation of the Tenorio municipal WWTP (Figure 18.2a) with a total capacity of 90,720 m^3/d;
- Five main sewage collectors (18.9 km);
- Bringing more than 500 ha of irrigation systems into water quality compliance;
- 38 km of purple distribution lines and one 4000 m^3 storage tank of recycled water for the cooling towers of the Villa de Reyes power plant;
- Restoration of the 200 ha storm basin called Tenorio to a controlled artificial wetland (Figure 18.2b) aiming to polish and disinfect most of the treated wastewater for agricultural reuse.

Figure 18.1 Aerial view of the Tenorio WWTP, reservoir and irrigated areas.

Figure 18.2 Views of the Tenorio wastewater treatment and recycling plant (a) and the wetland (b).

18.2 TECHNICAL CHALLENGES OF WATER QUALITY CONTROL

Treatment train for water recycling

The Tenorio WWTP is designed for a total capacity of 90,720 m^3/d and consists of two treatment lines as shown in Figure 18.3. Two qualities of recycled water are produced: polished and disinfected primary effluents for irrigation (57% of the total capacity) and high-quality tertiary treated recycled water for industrial uses in cooling towers (43% of the total capacity). The main treatment processes are as follows (Figure 18.3):

- Screening and advanced primary treatment enhanced with chemicals for the whole capacity (Figure 18.4a),
- Line 1 for agricultural reuse: naturally engineered treatment and polishing of at least 51,840 m^3/d by constructed wetlands (rehabilitation of the Tenorio reservoir),
- Line 2 for industrial reuse in cooling towers in the power plant of Villa de Reyes: secondary treatment of 38,880 m^3/d by activated sludge with nitrogen removal (Figure 18.4b), followed by tertiary treatment with lime softening, sand filtration, ion exchange softening and chlorine disinfection.

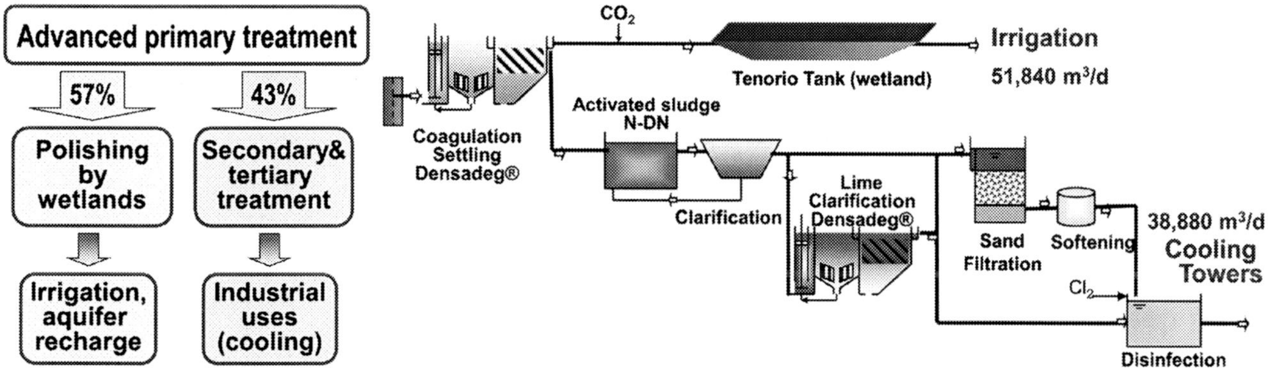

Figure 18.3 Flow diagram of the treatment process.

Distribution and storage of recycled water

The system for recycled water distribution includes a 39 km fiber glass conduction pipe and one intermediate storage tank that can be used to regulate the peak flow demand of the power plant (Figure 18.5). The recycled water is pumped to this storage tank and then flows by downward gradient to the power plant.

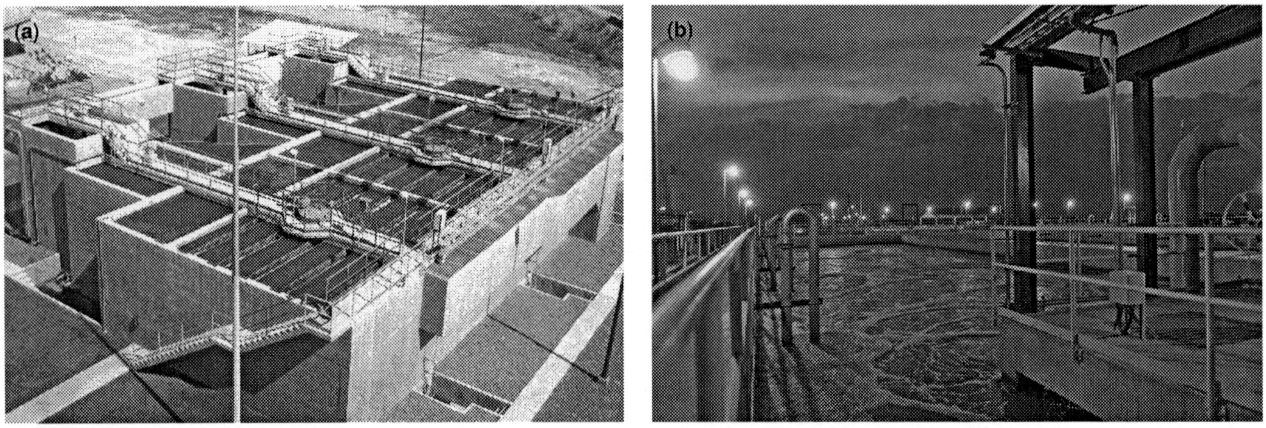

Figure 18.4 Views of the advanced primary treatment (a) and activated sludge aeration tanks (b).

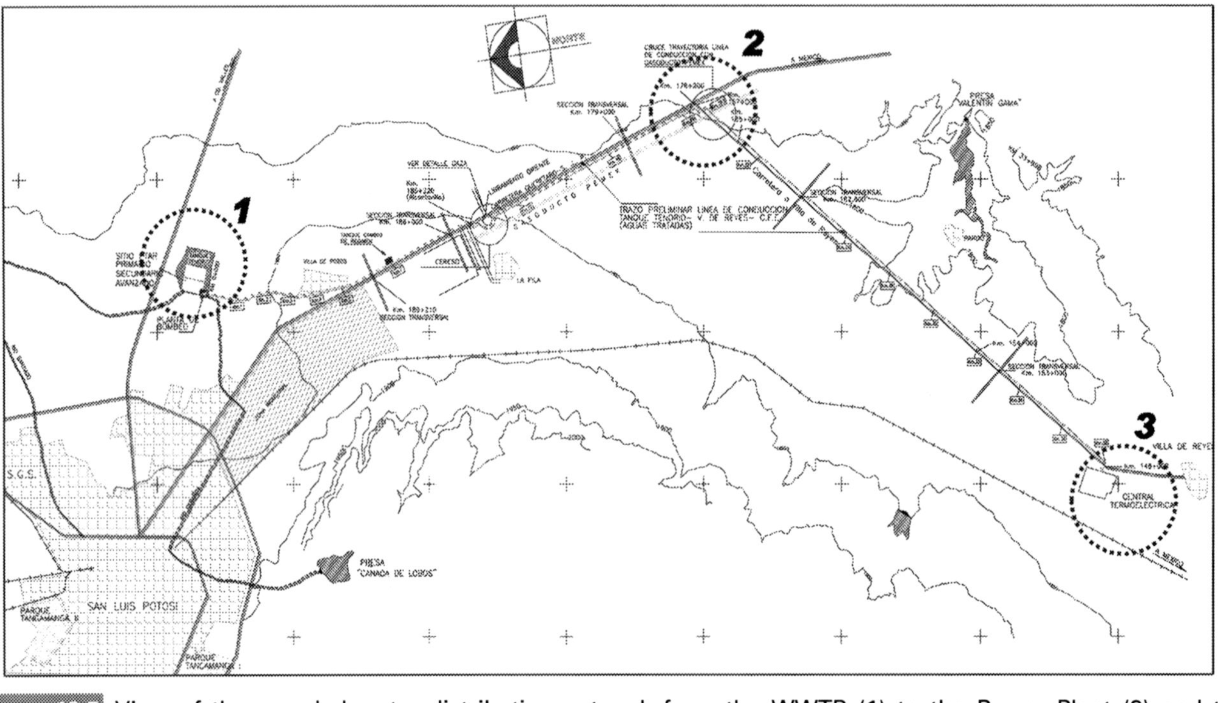

Figure 18.5 View of the recycled water distribution network form the WWTP (1) to the Power Plant (3) and the intermediate storage tank (2).

Water quality control and monitoring

The results of 4 years of water quality monitoring are summarised in Table 18.1. Despite the high variations of raw wastewater quality, advanced primary treatment ensured a good removal of suspended solids to an average value of 31 mg/L preserving the fertilising capacity of treated effluents in terms of carbon (average BOD_5 of 31 mg/L,) and nutrients (6.7 mgP_{tot}/L and 24.8 mgN/L). The additional polishing in the wetland (Tenorio tank) ensured a good level of disinfection with a resulting mean fecal coliform concentration of 160 FC/100 mL, which is largely below the WHO guidelines of 1000 FC/100 mL.

Salinity is one of the most important agronomic parameters as the high salinity can damage salt sensitive plants and decrease crop yield. The average salinity of the polished effluent remains relatively low, at about 870 μS/cm, with seasonal increases during the period of November to March up to 1320 μS/cm with values close to the permit level of 1200 μS/cm. Any negative impacts of recycled water o irrigated crops have not been reported.

Table 18.1 Characteristics of raw sewage and different effluents for reuse in agriculture and the power plant (2007–2011).

Parameter	Raw wastewater*	Tenorio Tank effluent to reuse in agriculture**	Criteria for agricultural reuse	Reclaimed water to power plant*	Criteria for industrial reuse in power plant
TSS mg/l	188 ($\pm$76)	28.8 ($\pm$10.6)	30	3.58 ($\pm$3.06)	10
BOD_5 mg/l	275 ($\pm$99.5)	31 ($\pm$7.3)	40	2.87 ($\pm$2.05)	20
COD mg/l	518 ($\pm$259)	84 ($\pm$19)	Not required	15.8 ($\pm$14.45)	60
P_{TOTAL} mg/l	8.7 ($\pm$3.9)	6.5 ($\pm$0.2)	15	1.3 ($\pm$0.9)	2
TKN mg/L	32.6 ($\pm$9.6)	22.3 ($\pm$5.1)	25	1.5 ($\pm$3.87)	15
Fecal coliforms/100 ml	$4.8 \cdot 10^9$ ($\pm 1.3 \cdot 10^2$)	161 ($\pm$402)	1000	18.4 ($\pm$16.6)	70
Total hardness mg/l	111.3 ($\pm$19.3)	Not measured	Not required	105.6 ($\pm$24.2)	120
Silica mg/l	104 ($\pm$20.3)	Not measured	Not required	64.9 ($\pm$9.3)	65

*Average (minimum-maximum) from 1525 daily composite samples.
**Average from 42 monthly composite samples.

Main challenges for operation

The major challenge for operation was the compliance and the reliability of the recycling plant for the supply of recycled water to the industrial user, the power plant. In this respect, the major problem was the average value of the conductivity, which was 5% to 10% above the 800 μS/cm guaranteed by the contract. After several months of investigation, a higher conductivity limit was accepted in exchange for a lower silica content of 65 ppm instead of the initial guarantee of 85 ppm. In order to meet this new guaranteed level, a 6 month test period was observed and the addition of ion exchange softening was implemented in order to account for an increase in hardness as a consequence of silica removal.

At present, after five years of operation, management at the power plant, aware that the reliability of operation with recycled water is very high, have expressed their interest in conducting a feasibility study on reverse osmosis in order to further reduce silica level to 10 ppm.

The other major challenges for operation are as follows:

- The power plant consumption was expected to be 38,880 m^3/d, but when the WWTP started the operation, the plant consumed only about 30%, that is 12,960 m^3/d. In 2010, the average consumption increased, but only up to 21,600 m^3/d. The operation was adjusted and monitoring of levels in the intermediate tank and basins of the cooling towers was implemented in order to be able to gradually increase the flow as required by the operation of the power plant.
- The control of residual chlorine and bacterial growth in the 39 km of distribution network and 4000 m^3 of storage reservoir were and remain a great challenge. Because of the high variations in recycled water demand, depending on energy demand, the detention time in the system can vary between 13 h and 60 h. Monitoring of residual chlorine and additional dosage of hypochlorite were implemented in the intermediate tank and in the inlet of the power plant to guarantee the target value of residual chlorine.
- It was also necessary to implement a biocide dosage to control biofilm growth in the cooling system of the power plant, despite that the microbial criterion guarantee being consistently met in the WWTP.
- It was necessary to implement on-line monitoring of the main parameters (silica, hardness, phosphorous, conductivity and residual chlorine) in order to control the evolution of the recycled water quality and the reliability of operation of the WWTP.

18.3 WATER REUSE APPLICATIONS

As mentioned previously, the two main applications of the recycled water remain agricultural and industrial with current water demand slightly different from the design flows (Figure 18.6):

- Industrial reuse of 21,600 m^3/d (up to 9.9 Mm^3/yr) in the 700 MW Power Plant Villa de Reyes for cooling purposes, currently saving 7.88 Mm^3 per year of potable water,
- Agricultural irrigation with an average daily flow of 60,480 m^3/d (up to 23.9 Mm^3/yr) for crops that are not consumed raw such as corn, barley and alfalfa,
- Environmental enhancement with the restoration of the wetland for wildlife habitat.

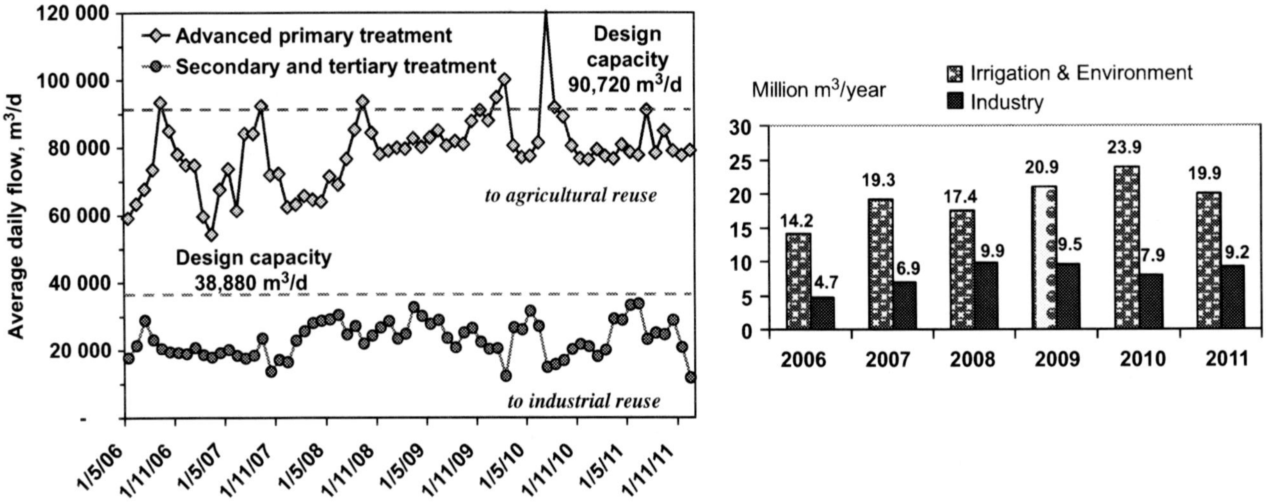

Figure 18.6 Evolution of the average daily production (a) and the annual consumption (b) of the two qualities of recycled water.

Once the artificial flow-control structures were completed at the Tenorio tank and it began to receive treated effluents from the Tenorio WWTP, naturally occurring plants started to develop, and a natural water treatment process was established, thus improving significantly water quality. At present, the tank is operated as a polishing wetland.

The productivity of the wetland reached approximately 20 Mm^3/year, providing over one third of the water supply for irrigation. This enables farmers to use advanced pressurized irrigation systems and to grow more valuable crops. In addition, the associated health risk to farmers and consumers was significantly reduced. The wetland, which in the past was only a polluted environment used for the storage of untreated wastewater, is now a safe habitat for plants and animals, improving thus the local biodiversity. An environmental study has been launched to document the different species of birds that currently nest in the Tenorio, which results and recommendations will be available in the last quarter of 2012.

Evolution of the volume of supplied recycled water

Over the years, the demand for recycled water increased for all applications, despite some initial opposition from all end-users (Figure 18.6b). As mentioned before, during the first months after beginning operations, the power plant managers were very reluctant to use recycled water. Their operators claimed that their cooling towers would be infested with algae and bacterial growth, which in turn would significantly reduce their global operating efficiency, thus increasing their water demands and the downtime periods for maintenance. As a consequence, a test period of several months was implemented without a real commitment to consume and pay for the recycled water.

At the end of the test period, the power plant confirmed their acceptance to use recycled water and signed an initial contract for one year, which was followed by the signature of 2-year contracts. It was demonstrated that the use of recycled water instead of groundwater facilitated the avoidance of severe problems due to silica and calcium carbonate scaling. At the same time, the industrial client was surprised by the high quality of the recycled water, the reliability of the supply and the absence of biofouling.

After a very difficult initial period, wherein most of the farmers were sceptical of the new quality of the irrigation water, it was also observed that the fertilising benefits of recycled water were the same as before. In fact, the selected treatment pathway enables the removal of pathogens without the complete elimination of nutrients. The majority of farmers are cultivating over 500 ha of fodder and the irrigation with recycled water allowed reduction in water related diseases and better acceptance of the produced crops.

Relations and contracts with end-users

The Tenorio WWTP was built under a BOT (Build, Operate and Transfer) contract incorporating 18 years of operation and maintenance. BOT contracts in Mexico have become a very important tool for financing infrastructure projects with long term contracts that guarantee a reliable operation and maintenance of the facilities during the contract period. To execute and finance the project the three development partners, Degremont, Sumitomo and Prodin, formed Aguas del Reuso del Tenorio, a Special Purpose Company in which they respectively hold 41%, 29% and 20% of the capital stock. All the operation and maintenance of the Tenorio project is executed by Degremont under a sucontract with ARTE which terms and conditions miror those of the BOT contract.

CEA signed two contracts, one with ARTE for the construction, operation and maintenance of the WWTP and one with Federal Commission of Electricity (CFE) for the supply of recycled water. The first contract between CEA and CFE was only one year in duration, in order to analyse the benefits and/or problems of operation with recycled water. Subsequent contracts were signed for two-year periods, after the reliability of the operation of the Tenorio WWTP had been demonstrated.

18.4 ECONOMICS OF WATER REUSE

Project funding and costs (capital and operation)

The capital cost of Tenorio-Villa de Reyes Project was US$ 67.4 million ($585.3 million Mexican pesos in May 2000). Mexico's Federal Government provided a non-refundable subsidy, equivalente to 40% of the total investment costs, through the Trust Fund for Infrastructure (FINFRA) which is managed by the Banco Nacional de Obras y Servicios Publicos. The remaining 60% of the capital investment was funded by ARTE thorugh debt and equity. The BOT contract establishes a remuneration mechanism through which ARTE is paid monthly during the 18 years operation period four rates: the flat Investment Amortization Rate (T1), the flat Fixed Costs Operation and Maintenance Rate (T2) and two rates that compensate for the Variable Operation and Maintenance Costs of the Primary and Tertiary treatment respectively (T3A and T3B). All the rates are indexed to inflation.

Pricing strategy of recycled water

After several negotiations with senior management personnel at CFE's Headquarters (Mexico City), the price of recycled water for industrial reuse was fixed at 67% on the charge for groundwater, as established by the Federal Law of Rights regarding water at availability zone II for industrial use. This pricing strategy provided significant economic benefits to the power plant, including 33% savings on the cost of water supply and a reduction of operation costs for the wells, used in the past to extract groundwater. It is important to stress that water cost for industrial uses in San Luis Potosi is amongst the most expensive in the country, 1.14 US$/$m^3$, while the recycled water is only 0.76 US$/$m^3$ (values calculated according to the change rate of the Mexican peso in February 2012).

Benefits of water recycling

Economic benefits

The major economic benefit of the Tenorio project is its compliance with the current legislation (Federal Decree of November 17, 2004). According to this new policy, the wastewater treatment in the municipality of State Capitals such as San Luis Potosi is becoming mandatory and the rights allocated for the discharge of raw sewage have been cancelled. Consequently, when the Tenorio WWTP started its operation, the debt value of the penalties for the discharge of raw wastewater was equivalent to the capital cost of the treatment facility.

Another economic benefit from the Tenorio project is the revenue from recycled water sale to the Thermal Electric Plant in Villa de Reyes, which is equivalent to the operation and maintenance costs.

Non-economic benefits

The major environmental benefit of this water recycling project is the rehabilitation of the overexploited aquifer of Soledad Graciano, a municipality within the San Luis Potosi Metro Area. In fact, initially, the water supply of the power plant was provided from this aquifer and the annual withdrawal of 9 Mm^3/yr resulted in social conflicts and complaints. At present, this "first" quality freshwater is mainly used for domestic potable water supply.

The use of recycled water with appropriate quality for agricultural irrigation enabled a significant reduction in levels of aquifer pollution due to infiltration of poor-quality wastewater. There are also recognisable health benefits: the mortality rate due to gastrointestinal diseases was reduced, as the polluted soil has been regenerated. The good-quality recycled water also provided new opportunities for farmers to upgrade irrigation techniques used in their farming zones with an access to financial support programs, and thus to improve their living standard.

Finally, one of the most beneficial environmental effects was the remediation of 200 ha at the location of the Tenorio tank, which nowadays has become a shelter for migratory birds, mainly from the north of the continent.

Environmental enhancement and preservation of biodiversity

An ecosystem monitoring program has been launched on the Tenorio tank that was deteriorated in the past due to the discharge of untreated wastewater. In 2006, the reservoir was modified to perform as a wetland. As a consequence of the improved water quality, a large diversity of migratory birds has been encountered over the last years. During the period from December 2011 to April 2012, three monitoring campaigns of the bird population, flora and fauna took place on the Tenorio tank. The major objectives of this study were:

- Having a baseline of information on the diversity of resident and migratory birds.
- Identify indicator species of environmental quality.
- Determine values of ecological importance with distribution patterns and temporal dynamics.
- Propose guidelines for management and conservation of the environment to regulate land use in the catchment area (control of the urban and industrial growth in the area).
- To determine the importance of wetlands as a basis for protection of biodiversity and the ecological status of the area, in accordance with state or national environmental policy.

As shown in Figure 18.7, the reservoir was divided at 10 observation points. During of each of the three monitoring campaigns, a total of 11 counts per sampling point were performed with a total of 110 counts. The results enabled the identification of 56 species of birds: 23 aquatic, 28 terrestrial and 5 predators. The 56 species belong to 23 families. The families with larger number of species were *Anatidae* (9), *Ardeide* (6) and *Tyrannidae* (5). Four species have been classified as protected birds by the Mexican Government. The polishing wetland is located in the route of several migratory species that have welcomed the site for regular nesting. Consequently, 48 from the observed species are migratory birds. The evaluated density of birds in the evaluated area (2703 m^2) was evaluated at approximately 1514 birds.

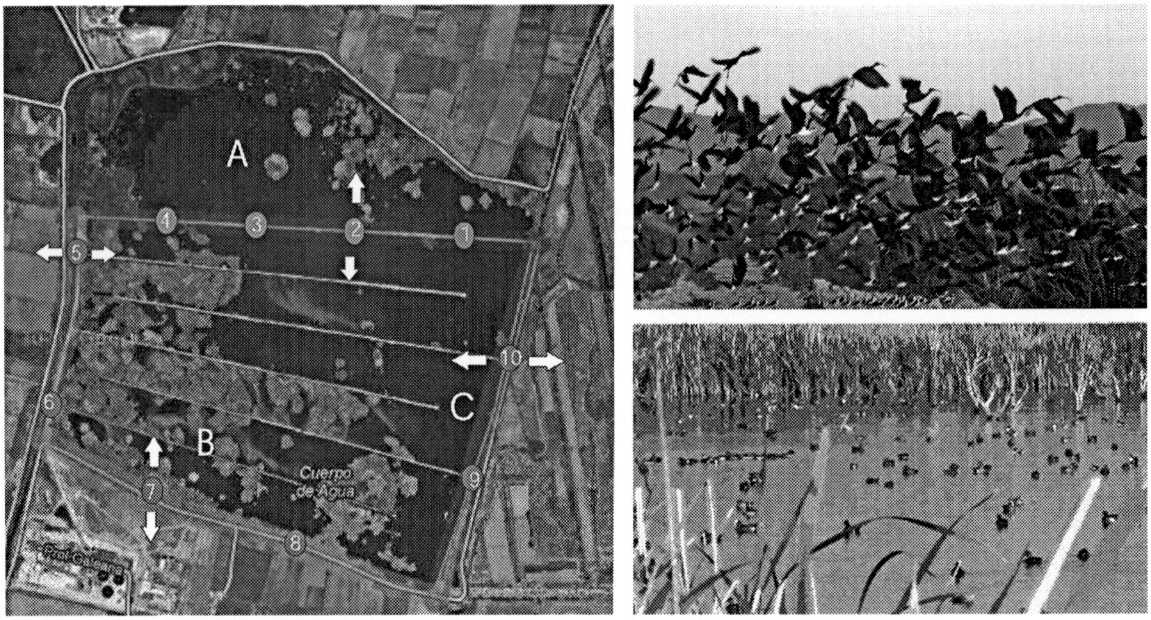

Figure 18.7 Location of the observation points of birds in the Tenorio tank and views of the bird populations of *Himantopus mexicanus* and *Anatidae* families.

This study demonstrated the role of the Tenorio tank for the enhancement of the biodiversity, in terms not only in increasing the bird population, but also for the improvement of flora and small mammal diversity. The birds are essential for maintaining the health of the surrounding ecosystem. Some species are also contributing to control pests and vectors of several diseases by eating insects and rodents. As a result of this study, the creation of a protected area with a mitigation zone in the surroundings is under consideration by local authorities.

18.5 HUMAN DIMENSION OF WATER REUSE

Public education and communication strategy

A public outreach program has been implemented since the beginning of the project, overcoming the initial opposition of farmers against industrial reuse of urban treated wastewater. In addition, the Tenorio project was presented and discussed at different regional and international forums.

Public education is the major tool to explain to the local population the value of water and the vital importance of water reuse. The Tenorio WWTP receives over 200 annual visitors, most of them coming from local schools and universities. Visitors from Federal and State Government Agencies also greatly appreciate the opportunity to learn about the keys implementing successful water reuse, and especially for a multi-quality water recycling project to achieve an economic viability.

The role of decision makers

The new water policy, adopted by the State Water Law of San Luis Potosi, greatly contributed to the feasibility and success of the Tenorio project. According to this policy, every industry (with the exception of the food industries) should implement treatment and reuse of their industrial wastewater, with internal water recycling, and the substitution of freshwater supply by urban recycled water, if available.

The strong engagement of politicians and decision makers plays an important role for the implementation of this relatively complex water recycling scheme. The contribution of the General Director of the National Water Committee (CONAGUA), CFE's General Director, the Federal Government support and the Head of the State Government, dating back to the signing of an intention letter in 1995, for the funding and construction of the Tenorio-Villa Reyes project were instrumental. Moreover, the technical and financial support of several national and regional entities was critical for good project execution and management.

Public acceptance and involvement

Despite the initial opposition of the farmers, the results from the first four years of irrigation with recycled water demonstrated that crop yield and quality are satisfactory and even better than expected. Currently, over 500 ha located in three communities (La Libertad, Los Gomez, Santa Rita) are irrigated with recycled water with a high fertilising capacity. The supply of an additional irrigation unit in La Libertad is in process, taking advantage of the Federal Government support programs, with the assistance of the College of Postgraduates at Chapingo University and the National Water Committee.

The reliability of supply and the high quality of tertiary treated recycled water provided to the power plant, convinced other potential industrial users on the benefits of water reuse. Currently, an industrial consortium is under negotiation to obtain the water surpluses that the power plant does not use for distribution to the industrial zone.

In addition to the satisfaction and recognition of the benefits of water reuse for cooling purposes at the Thermal Electric Plant, the Underground Water Technical Committee confirmed their support of the water reuse program, which contributed significantly to the rehabilitation of the over-drafted aquifer.

Consequently, the Tenorio water reuse scheme has become an example of a successful and viable water reuse project for the whole region.

18.6 CONCLUSIONS AND LESSONS LEARNED

Conceived as part of an Integrated Master Plan, the Tenorio Water Recycling Project has become a success story by meeting its original objectives of aquifer restoration and sustainable water management. It has also demonstrated that additional benefits can be obtained when the key priorities are well understood and established.

The long-term water quality monitoring and performance analysis demonstrated the feasibility and economic viability of integrated water management with various water applications by the implementation of appropriate treatment technologies, adapted to local conditions. The two treatment channels consistently produced recycled water with the required quality for agricultural irrigation, environmental enhancement and industrial uses.

The major challenge was the investigation and the improvement of recycled water quality for industrial reuse. Conductivity, silica and hardness were the mean parameters to be improved and a good balance was achieved by lowering silica levels to compensate the slightly higher salinity observed during certain periods.

The irrigation of 500 ha with good quality recycled water reduced the incidence of water related diseases and increased the public acceptance of the crops, preserving at the same time the fertilising capacity of wastewater. The aquifer balance has been recovered by the substitution of the groundwater with recycled water for industrial and agricultural uses.

The satisfactory operation of the power plant with recycled water encouraged the industry to explore new possibilities of improved treatment. Finally, other industries are interested in substituting their groundwater consumption with recycled water given the economical benefits and the demonstrated reliability of this water reuse system.

The following Table 18.2 summarizes the main drivers of the Tenorio Project as well as the benefits obtained and some of the challenges for the future.

Table 18.2 Summary of lessons learned.

Drivers/Opportunities	Benefits	Challenges
Water scarcity and more frequent droughts	Alternative water resources for non-potable purposes	Water exchange between the existing over-drafted aquifers in the area
Need to increase wastewater treatment coverage	70% of wastewater is treated (goal 85%) and 100% is reused	To build infrastructure to secure distribution and continuous supply
Conflicts between sectors, agriculture/urban/industrial	Support economic development in both industry and agricultural sectors	Opposition of local farmer and industrial users has to be overcome to gain public acceptance
Sand Luis Potosi State Government Policy	Long term sustainable water management to secure water supply and protect the environment	Enforcement of the implementation in compliance with the Integral Sanitation and Reuse Plan and Government Policies
Financial support for BOT projects	Economic feasibility of the project to ensure continuity and reliability	Implement appropriate pricing strategies while industry covers the major part of the cost

REFERENCES AND FURTHER READING

Courjaret C. and Díaz de León U. (2006). Saneamiento integral y reuso agrícola e industrial en San Luis Potosí, Foro del Agua, Mexico City, Mexico.

Diario Oficial de la Federación (2009). Actualización de la disponibilidad media anual de agua subterránea acuífero (2411) San Luis Potosí, Estado de San Luis Potosí, Mexico City, CONAGUA www.conagua.gob.mx (accessed September 2011).

Díaz de León U. (2008). Saneamiento integral y reutilización del agua en la ciudad de San Luis Potosi México, *Expo Zaragoza*, Spain.

Equihua L. O. (2006). Reuso de agua en la agricultura y en la industria en México: Caso del Proyecto San Luis Potosi, Foro del Agua, 2006, Mexico City, Mexico.

Medellín P. (2003). Tanque Tenorio, 10 años después, Pulso de San Luis Potosí, http://ambiental.uaslp.com/docs/PMM-AP030522.pdf (accessed April 2012).

Olivo S. and Martínez J. B. (2000). Saneamiento integral de las aguas residuales de la zona conurbada de San Luis Potosí-Soledad de Graciano Sánchez, S.L.P, Comisión Nacional del Agua, Gerencia Estatal en San Luis Potosí, San Luis Potosi, CONAGUA Congress 2000.

Rojas A. (2011). Experiencia de reuso del agua tratada en la Zona Metropolitana de San Luis Potosí., Jornadas técnicas sobre la recarga artificial de acuíferos y reuso del Agua, Instituo de Ingeniería UNAM, 9–10 June 2011, Mexico City, Mexico.

Views of the Tenorio WWTP landscape and the purified recycled water.

19 Recycling of secondary refinery and naphtha cracker effluents employing advanced multi-barrier systems

Josef Lahnsteiner, Srinivasan Goundavarapu, Patrick Andrade, Rajiv Mittal and Rajkumar Ghosh

CHAPTER HIGHLIGHTS

This water reuse case study represents a major contribution to Indian industrial development, as the high rate industrial wastewater recycling with zero liquid discharge in the mid-term was the critical factor for the implementation and expansion of the Panipat refinery (capacity of 15 million t/yr). Advanced multi-barrier systems are employed successfully for the recycling of boiler make-up in oil refining and petrochemical industries. The major benefits include positive environmental effects and a boost to the security of the industrial water supply and the industrial activity.

KEYS TO SUCCESS

- Economic growth in India
- Economic success of the investor
- Environmental awareness in India and especially in the Panipat region
- Employment of advanced technologies
- Experienced plant operator

KEY FIGURES

- Treatment capacity: 42,500 m^3/d
- Total annual volume of treated wastewater: 10.9 Mm^3/yr
- Total volume of recycled water: 9.0 Mm^3/yr of demineralized water
- Operation costs: 0.3 €/m^3
- Water reuse standards: <10 *μg/L silica; <0.1 mg/L TDS*

19.1 INTRODUCTION

Basically there are two resources available for industrial water reuse and recycling. Municipal secondary effluents comprise the first and own (in-house) effluents from industrial processes, the second. The evaluation of recycling and reuse of secondary effluents in advanced multi-barrier systems shows that these practices are technically and economically feasible (Lahnsteiner, 2010). Which of these two options is the most feasible depends upon several factors such as water management policy and legislation, price policy (e.g. price of fresh water from municipal networks), the hydrological situation (degree of water stress), the availability of proper recipients, pollutant concentrations in the reclamation plant inlet and reuse/recycling applications. Hespanhol (2008) discusses the two aforementioned industrial reuse options in a survey for the Brazilian State of Sao Paulo and comes to the conclusion that the reclaiming and recycling of in-house effluent is preferable due to the price policy of the public water suppliers. Furthermore, he concludes that the dissemination of water reuse in Brazil can only be achieved through a strong institutional decision, followed by the enactment of comprehensive federal legislation. In many cases, industry has to use both resources (in-house effluents and secondary municipal effluents) in its water management, especially when no other sources (ground and surface water) are available and reclaimed water from public sources is cheaper than fresh water from the public network.

In this chapter, two examples of reclamation and recycling by the Indian Oil Corporation Ltd (IOCL) are compared. IOCL is India's largest commercial enterprise and was ranked 125th on the Fortune Global 500 list in 2010. It owns and operates ten of India's 19 refineries with a combined refining capacity of 65.7 million tonnes per annum (t/yr). The two reclamation plants were realised in the Panipat refinery (12 million t/yr) and petrochemical complex. Panipat is located in Haryana State, 90 km northwest of Delhi.

Main drivers for water reuse

Annual precipitation totals approx. 500 mm, but over 70% of this rainfall occurs during the monsoon months of July to September. Nevertheless, there is practically no water shortage in the Panipat region due to the availability of sufficient surface water (river water). The Panipat Refinery (Figure 19.1a) is located in farmland, which is irrigated by the Yamuna canal (Figure 19.1b). The Indian Oil Corporation Ltd. Panipat had to build wastewater recycling plants as a response to the demand of the environmental authorities for zero liquid discharge. This request was made due to the fact that no proper recipient is available in the Panipat area.

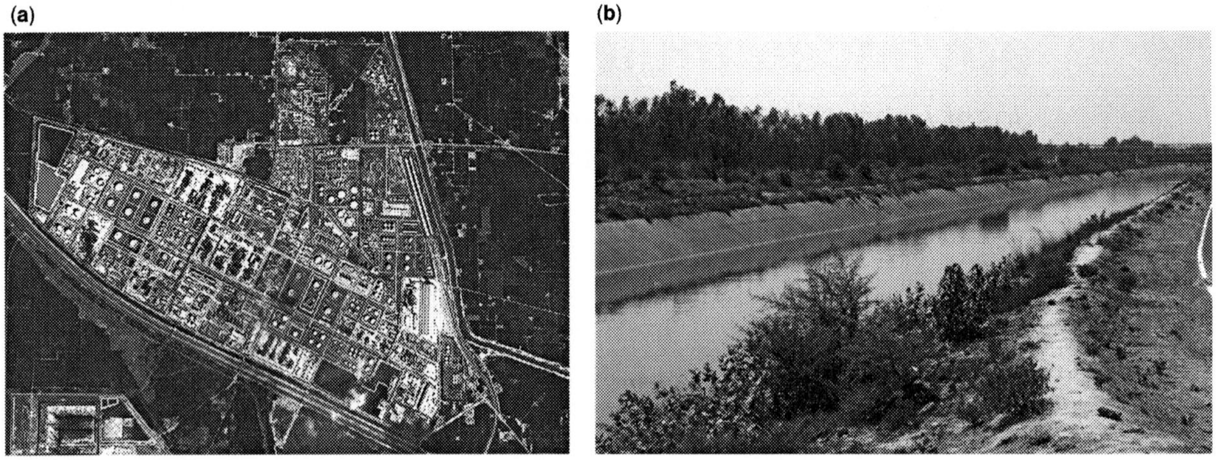

Figure 19.1 Views of (a) the Panipat Refinery (Google Earth, 2009) and (b) Yamuna canal.

The Yamuna canal is mainly used for potable water production and agricultural irrigation and thus cannot be employed as a recipient. Therefore, reclamation plants have been installed to meet stringent governmental regulations and prevent the pollution of nearby water bodies and not because of the unavailability of water for industrial use. Water losses during the refining processes are compensated for by the use of fresh water from the Yamuna canal. The reclamation of secondary municipal effluent was not considered due to the availability of sufficient volumes of fresh water, the distance between the municipal sewage treatment plant and the refinery, and the installation of zero liquid discharge facilities in the mid-term.

Brief history of project development

The original Panipat refinery (6 million t/y) was commissioned in 1998 and the refinery expansion (doubling of refining capacity) commissioned in 2006. This expansion also included the petrochemical production of paraxylene (PX) and purified terephtalic acid (PTA), which is the basis for producing polyester staple fibres, polyester filament yarns and other resins. The Panipat Refinery Expansion Water Reclamation Plant (PRE-WRP), which treats both secondary refinery effluents and various refinery/petrochemical process effluents, was commissioned at the end of 2006.

In 2009 the refinery was further expanded to 15 million t/yr. The naphta cracker and its downstream polymer units (Naphtha Cracker Complex) were commissioned in 2010. The Panipat Naphtha Cracker Water Reclamation Plant (PNC-WRP), which is virtually identical with the PRE-WRP (practically the same process and hydraulic capacity but with a different RO design) and reclaims process water from naphtha cracker secondary effluent, cooling water blow-downs and demineralisation regenerates, was commissioned in June 2010. In the mid-term, the RO brines have to be treated by evaporation and crystallisation at both reclamation plants (PRE-WRP and PNC-WRP) in order to meet the zero liquid discharge target requested by the authorities.

Project objectives and water reuse applications

The Panipat Refinery and petrochemical complex were built in the Haryana State, approximately 90 km north to Dehli for industrial policy reasons. The refinery not only meets the demands of Haryana, but also of the entire north-west region of India. The industrial complex is located in an agricultural area, and therefore, the environmental requirements are stringent especially with regard to water management, as no proper recipient is available. Therefore, the objective was high rate recycling and as previously mentioned, zero liquid discharge in the mid-term. This means that the existing reclamation plants have to be enlarged to include evaporation and crystallisation. The reclaimed water is reused mainly as boiler make-up and in the case of the PNC-WRP, also as cooling tower make-up.

19.2 TECHNICAL CHALLENGES OF WATER QUALITY CONTROL

The major technical challenges posed by water quality control involve the regular disruption of treatment performance in the biological wastewater treatment plants. These failures, which are caused by inhibiting or toxic substances from some refinery/petrochemical partial streams, reduce the quality of the treated effluent (inlet water of the reclamation plants) and subsequently cause severe fouling problems in the ultra-filtration steps of both reclamation plants. Another challenge is the stringent silica standard of less than 0.01 mg/L, which has to be met in the PRE-WRP. In future, a further challenge will derive from the fulfilment of the request from the environmental authorities for zero liquid discharge. The different wastewater streams treated in the reclamation plants (PRE-WRP and PNC-WRP) are shown in Table 19.1. The major design parameters of the blended wastewater flows (inlet to the reclamation plants) and target values of the reclaimed water are shown in Table 19.2.

Table 19.1 Wastewater streams of the Panipat Refinery and Naphtha Cracker Plant.

	Flow, m^3/h	
Wastewater streams	**PRE-WRP**	**PNC-WRP**
Secondary refinery effluent – ETP I	400	
Secondary refinery effluent – ETP II[1]	300	
PX[2]/PTA[3] effluent including cooling tower blow-down	272	
Demineralisation regenerate from Panipat Refinery I	60	
Demineralisation regenerate from Panipat Refinery II[1]	140	
Cooling tower blow-down from in-house power plant I	18	
Cooling tower blow-down from in-house power plant II[1]	50	
Secondary naphtha cracker effluent – ETP		150
Cooling Tower 1 blow down – NCU[4]		308
Cooling Tower 2 blow down – Polymer		253
Cooling Tower 3 blow down – CPP[5]		130
Blow down from CPP boiler		15
CPU[6] wastewater (intermittent)		45

[1]Expansion of refinery; [2]PX: Para-Xylene; [3]PTA: Purified Terephtalic Acid; [4] NCU: Naphtha Cracker Unit; [5]CPP: Cogeneration Power Plant; [6]CPU: Condensate Polishing Unit.

Table 19.2 Design parameters of PRE-WRP and PNC-WRP.

		Water reclamation plants			
		PRE-WRP		PNC-WRP	
Parameter	**Unit**	**Inlet**	**Reclaimed Water[1]**	**Inlet**	**Reclaimed Water[1]**
T	°C	15–35	15–35	25	25
TDS	mg/L	1786	<0.1	800	<0.1
Silica	mg/L	98	<0.01	100	<0.01
COD	mg/L	150	–	125	–
BOD_5	mg/L	10	–	10	–
Oil	mg/L	10	–	5	–
Q	m^3/h	900	764	871	620

[1]Demineralised Water after Mixed Bed Ion Exchanger.

Treatment trains for water recycling

Basically, the reclamation plants (design capacity of 900 m^3/h for PRE-WRP and 871 m^3/h for PNC-WRP) incorporate clarification (including silica adsorption on magnesium hydroxide), pressure sand filtration, ultra filtration (UF) and reverse osmosis (RO). The RO permeate is polished by mixed bed ion exchange filters and is then mostly recycled as boiler make-up water for the refinery and naphtha cracker power plants. Figure 19.2 shows the process flow diagram, which is more or less representative for both reclamation plants. The major difference is in the RO design (three stages in PRE-WRP, two stages in PNC-WRP).

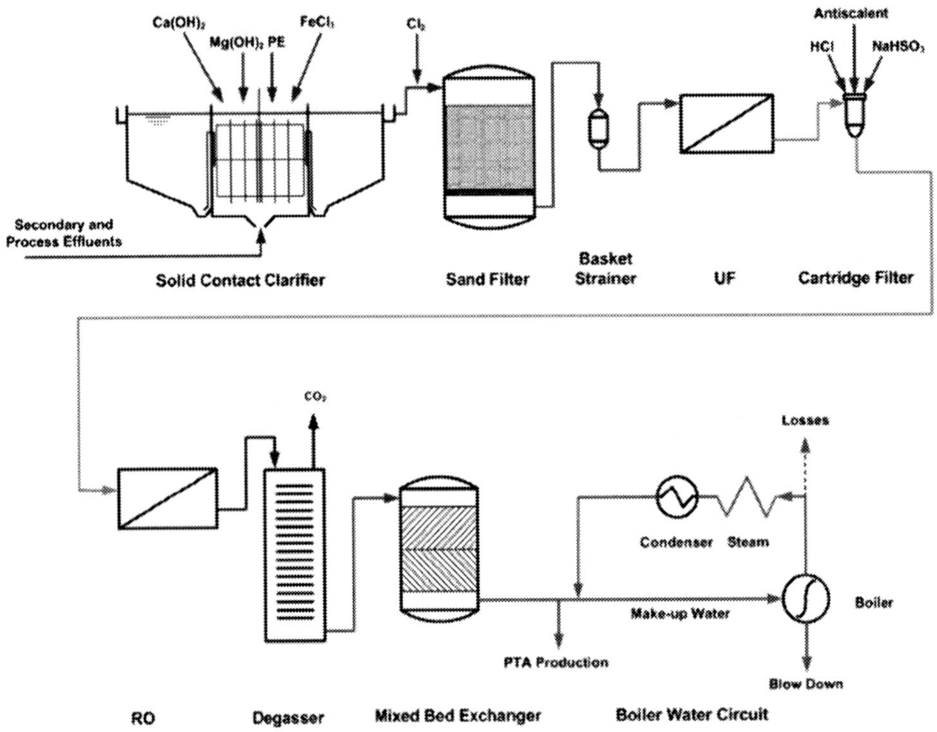

Figure 19.2 Water reclamation and recycling process diagram – PRE-WRP.

The same UF and RO membranes are employed in both plants. The ultra-filtration process steps consist of pressure-driven, inside-out, hollow fibre systems (X-Flow Xiga, Table 19.3). Both systems are operated in a dead-end mode. Figure 19.3a shows the PNC-WRP ultrafiltration process step, which consists of seven skids (6 + 1 stand-by) each with 18 pressure vessels.

Table 19.3 Ultrafiltration design parameters.

Membrane parameters	Unit	PRE-WRP	PNC-WRP
Membrane material		*Hollow fibre polyether-sulfone*	
Average feed flow	m^3/h	894	870
Design gross flux	$L/m^2 \cdot h$	54	50
Design permeate flow	m^3/h	760	766
Design net flux	$L/m^2 \cdot h$	46	44
Recovery	%	85	88
Skids	–	6+1 stand-by	6+1 stand-by
Pressure vessels/skid	–	18	18
Vessels total	–	108	108
Elements/vessel	–	4	4
Elements total	–	432	432
Membrane area/element	m^2	38	40
Membrane area	m^2	16,416	17,280

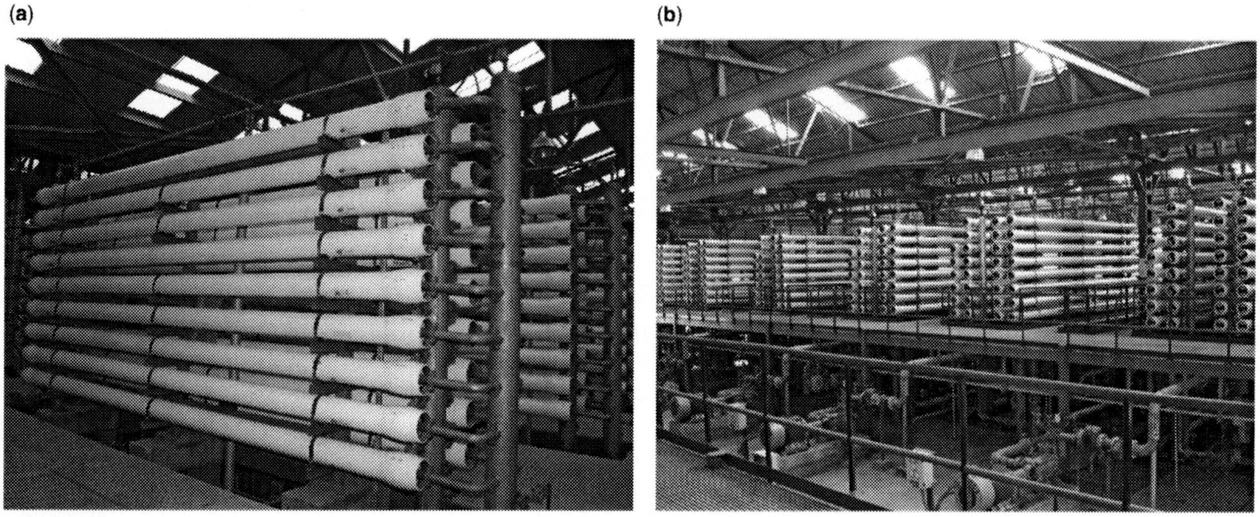

Figure 19.3 Views of the PNC-WRP ultrafiltration process (a) and the PRE-WRP reverse osmosis first pass (b).

The major task of the UF is the reduction of the silt density index (SDI) and the removal of turbidity, as well as suspended and colloidal matter, in order to minimise fouling of the downstream reverse osmosis process. The UF membrane fouling caused by the aforementioned impurities is removed by regular backwashing with permeate. The backwash is enhanced once a day in both UF plants using chemicals (chemical enhanced backwash – CEB with caustic NaOCl and HCl). As can be seen in Table 19.3, both UF systems were designed with a comparable flux ($46\ L/m^2 \cdot h$ and $44\ L/m^2 \cdot h$ net flux, respectively). The resulting membrane areas are $16{,}416\ m^2$ for the PRE-WRP UF and $17{,}280\ m^2$ for the PNC-WRP UF. The retentate of the PRE-WRP UF process step is recycled to the equalisation tanks upstream of the pre-treatment stages of the reclamation plant. The retentate of the PNC-WRP UF is reused in the fire fighting water system together with the dual media filter backwash water, the RO II brine and the mixed bed ion-exchanger regenerate.

In the refinery reclamation plant (PRE-WRP), a two-pass RO system is employed in combination with a brine concentrator. Figure 19.4 shows the process configuration. The UF permeate is fed to RO pass I (three internal stages, low fouling composite membranes, Hydranautics LFC 3, Figure 19.3b). The RO I permeate is further desalinated in RO pass II (three internal stages, low fouling composite membranes) and the RO I reject is fed to the brine concentrator (two internal stages, seawater membranes, Hydranautics SWC 3). The brine concentrator permeate is recycled to RO II. The recovery rate accomplished by this process configuration is 90%. The RO II permeate is degassed and in order to allow the further removal of dissolved solids, polished in mixed bed ion-exchangers containing strong acid cation and strong base anion resins mixed in a single vessel.

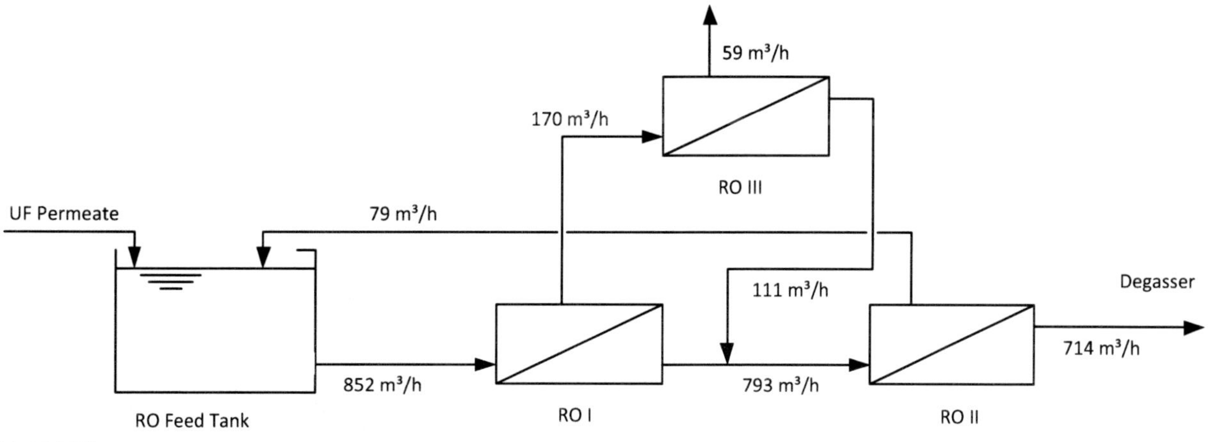

Figure 19.4 Reverse osmosis configuration – PRE-WRP.

As previously mentioned, the authorities have stipulated that the refinery has to achieve zero liquid discharge. However, it has temporary permission to dispose of the liquid waste produced in the reclamation process (RO brine). Currently 100% of the brine is used for refinery coke quenching. Other options are to blend the brine with fire-fighting water, or to use it for the irrigation (blended with low TDS water) of the free land along the refinery perimeter in order to provide green spaces (14 km long "green belt"). Nonetheless, in the mid-term this official request will be met through the installation and operation of evaporation and crystallisation.

Due to the lower design TDS concentration (800 mg/L) in the PNC-WRP, a single-pass RO system is utilised in combination with a brine concentrator. The UF permeate is fed to the first RO and the concentrate from this stage is further treated in the brine concentrator. The recovery rate accomplished by this RO process configuration is also 90%. Permeate from RO I (TDS <15 mg/L) is polished in the mixed bed ion-exchangers and then recycled as boiler make-up. RO II permeate (TDS < 292 mg/L) is blended with treated raw water (Yamuna Canal water, TDS 292 mg/L) and reused as process water (mainly cooling tower make-up). As previously mentioned, the brine from RO II is mixed with dual media filter backwash water, UF backwash water and the regenerate from the mixed bed ion exchangers and reused in the fire-fighting water system. The reclamation plant can be operated flexibly, for example, in case of poor effluent quality the plant can also be operated solely with raw water from the Yamuna Canal. Furthermore, the amounts of demineralised water (boiler make-up) and RO II permeate (process water) can be adjusted according to the demands on either source.

Water quality control and monitoring

Figure 19.5 shows a typical diagram of the major parameters (mean values of January 2009) TDS, silica and COD for the Panipat Refinery Expansion Water Reclamation Plant (PRE-WRP). TDS is mainly reduced in the reverse osmosis process step (from 1148 to 12 mg/L, 99.0%) and further cut in the mixed bed filter to less than 0.05 mg/L (total removal >99.996%). Silica is also mainly removed by RO (from 11.6 mg/L in the UF outlet to 1.4 mg/L in the RO I permeate and to 0.09 mg/L in the RO II permeate; 99.2% removal in both RO stages) and is then reduced further to 0.007 mg/L (7 µg/L, total removal 99.94%) in the mixed bed ion exchanger. This represents excellent removal efficiency, as 20 µg/L is the specified limit for boiler make-up water in various power plant guidelines such as the VGB (2011). Colloidal silica is zero, as it is completely removed in the reverse osmosis stages. In the pre-treatment steps (coagulation/sedimentation, sand filtration), COD is reduced from 69 mg/L to 48 mg/L (30.4%). It is then cut by a further 5 mg/L (from 48 to 43 mg/L) during UF (10.4%) and from 43 to 0 mg/L after reverse osmosis. These results are also representative for the periods from February to December 2009, 2010 and January to May 2011, except for TDS which was recently lower (600 – 900 mg/L) due to changes in the refinery processes.

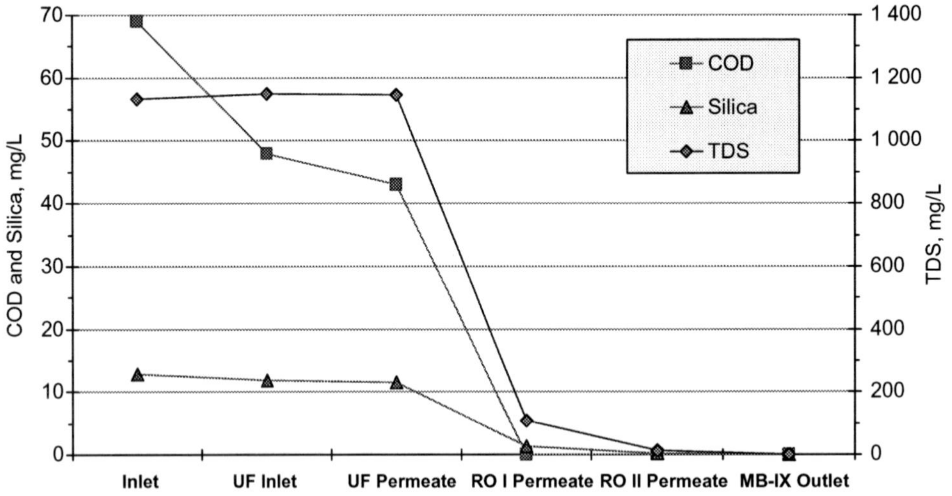

Figure 19.5 Removal of TDS, COD and silica – PRE-WRP.

The aforementioned COD reduction in the UF (5 mg/L or 10%) means that approximately the same amount can be attributed to both the high molecular fraction of the UF inlet (origin: refinery/petrochemical effluent; assumed macro-molecules: polyesters on terephtalic acid basis, etc.) and hydrophobic compounds, which are adsorbed on the polyether-sulfone membrane. The UF design gross flux of 54 $L/m^2 \cdot h$ and a COD reduction of 5 mg/L, result in a specific COD removal rate of 270 mg $COD/m^2 \cdot h$, which can be considered as moderate. Organic fouling can be removed relatively easily by CEB with caustic NaOCl (e.g. by saponification of esters), at least during regular operation. Since start-up in December 2006, the Panipat UF has been operated at design flux (and a little above). SDI outlet values have generally been within a range of 2.1 to 2.4 (inlet SDI is 6.7; the design outlet value is 3.0) and the turbidity values have been less than 0.1 NTU (limit of detection). An integrity tests conducted in December 2008 showed that there were only a few fibre breakages (e.g. in skid G: 36 in 72 membrane elements, i.e. 0.025‰ fibre breakages per year).

Figure 19.6 shows normalised (20°C) permeability values after six (May 2007) and eighteen months (May 2008) of operation. As can be seen in this figure, which is representative for the first 24 months of plant operation, there is practically no difference in the permeability values during the periods above. This means that stable filtration performance has been accomplished in the first two years. However, after 26 months of operation, a permeability loss of approximately 25% was observed. The main reason was that due to the request of the refinery, 50–100 m^3/h effluent from the Panipat Effluent Treatment Plant III (ETP III) had been additionally treated in the reclamation plant. This effluent contains increased COD concentrations in the 200–300 mg/L range, which fouled the UF membranes. The fouling was removed by more frequent and intensified CEBs but the original permeability values could not be recovered completely (as mentioned above: 25% permeability loss). In the last quarter of 2010 and second quarter of 2011, after four and four and a half years of operation respectively, permeability was still acceptable (70–180 $L/m^2 \cdot h$ bar). Despite these acceptable permeability values, at the client's request the rack A and B membranes were exchanged for spare membranes (rack A on November 25, 2010; rack B on April 14, 2011). The old rack A membranes were integrity tested with the result that after four years of operation there was still a relatively low specific fibre breakage rate (skid A: 119 breakages in 72 membrane elements, i.e. 0.041‰ fibre breakages per year; November 2010).

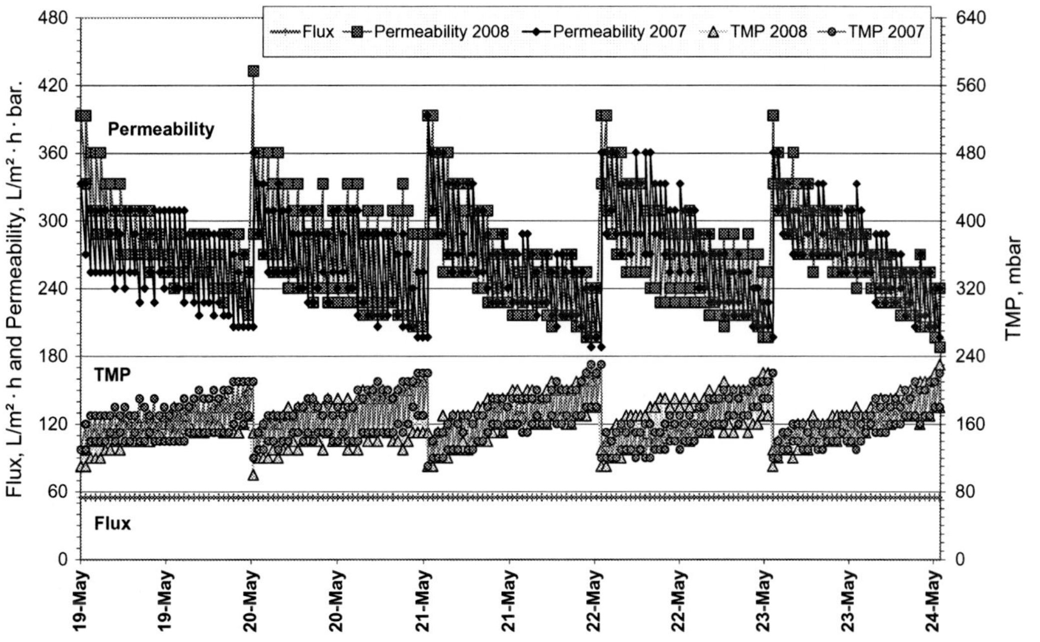

Figure 19.6 Panipat PRE-WRP UF filtration performance.

The removal of the major parameters TDS, COD, and silica in April 2011(average values) at the Panipat Naphtha Cracker Water Reclamation Plant are as follows: TDS in the UF permeate is reduced by reverse osmosis I from 393 mg/L (design value 800 mg/L) to 21 mg/L (94.7% removal), whereas silica is cut by RO I from 5.6 mg/L to 1.2 mg/L (78.6%). The COD is lowered from 26 to 16 mg/L (38.5%) by coagulation/sedimentation and filtration, and from 16 to 14 mg/L in ultra-filtration, that is a reduction of 2 mg/L. This means that approximately 12.5% of the UF inlet COD is of high molecular nature (origin: naphtha cracker and polymer unit effluents; macromolecules from polyethylene, ethylene glycol, polypropylene and polybutadien production). The design gross flux of 50 $L/m^2 \cdot h$ and a COD reduction of 2 mg/l result in a typical specific removal rate of 100 mg COD/m$^2 \cdot$ h, which is lower than in the PRE-WRP UF (normally >200 mg COD/m$^2 \cdot$ h). However, higher PNC-WRP UF COD removal rates (200–250 mg COD/m$^2 \cdot$ h) were also observed within the context of sub-optimum conditions in the PNC effluent treatment plant (PNC-ETP), which probably released higher concentrations of macromolecules.

Main operational challenges

As mentioned, from time to time poor treated effluent quality, caused by disruptive conditions in the biological treatment plants, leads to problems in the UF process steps of both reclamation plants. Figure 19.7 shows the permeability of the PNC-WRP UF from December 2010 to May 2011. The massive permeability loss in January and February 2011 was presumably caused by suboptimum conditions in the PNC-ETP. This poor performance can be traced to inhibiting or partly toxic substances from the petrochemical processes and/or low wastewater temperatures (15–19°C vs. up to 33°C in summer). It would seem that under these disruptive conditions, more macromolecules leave the ETPs, and subsequently cause fouling on the UF membranes. Fouling was exacerbated during this period, as the sodium-hypo-chlorite dosing pumps had to be serviced, that is the dosing pumps were not in operation during all the chemically enhanced backwashes. This resulted in non-optimum cleaning performance with regard to organic foulants.

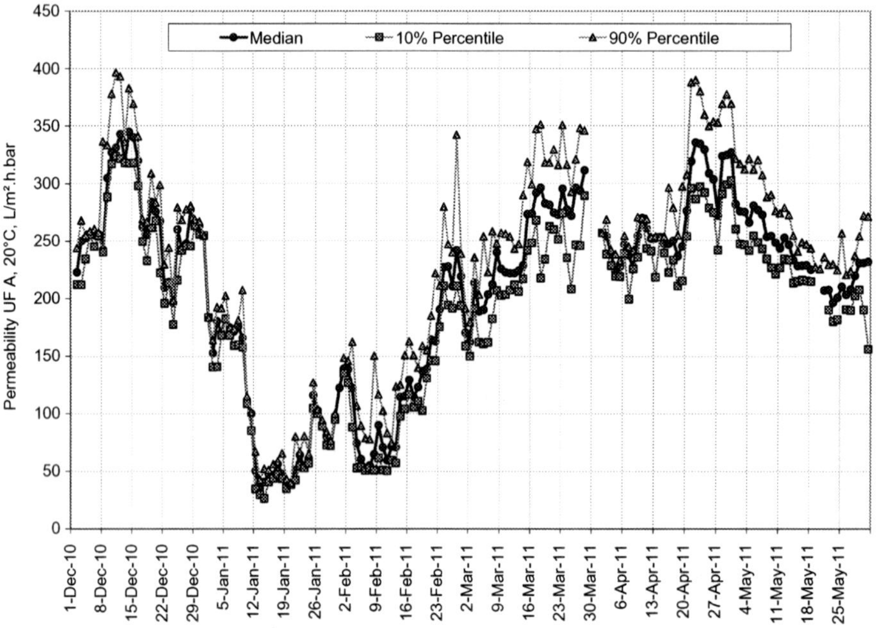

Figure 19.7 Evolution of the permeability of the PNC-WRP UF for a six month period.

19.3 WATER REUSE APPLICATIONS

The reclaimed water of the PRE-WRP is used mainly as boiler make-up water and additionally as process water for the production of purified terephtalic acid (PTA), which is employed in the textile industry as a substitute for dimethyl terephthalate (DMT). The manufacture of PTA demands high-quality water, for example, zero colloidal silica and low TOC for the preservation of the catalyst, which is needed for the chemical reaction, and practically absolute softened water for the end product quality (textile elasticity). The reclaimed water from the PNC-WRP is reused as both boiler make-up water (demineralized RO I permeate) and cooling tower make-up (RO II permeate).

Evolution of the volume of supplied recycled water

Figure 19.8 shows the volumes produced in the PNC-WRP and PRE-WRP for the period from June 2010 to May 2011. Only demineralised water is produced in the PRE-WRP (400,000–500,000 m^3/month), while in the PNC-WRP, apart from demineralised water (200,000–350,000 m^3/month) RO II permeate is also recycled (approx. 50,000 m^3/month).

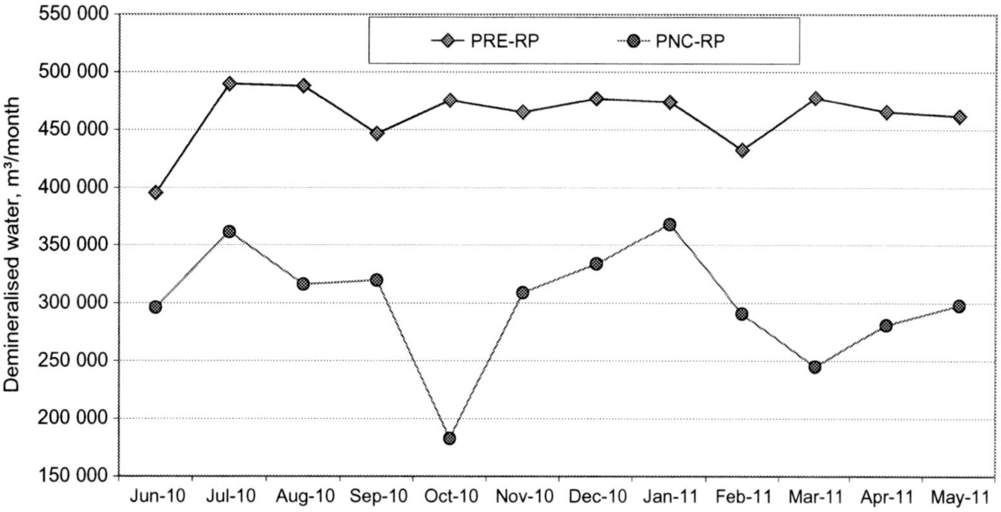

Figure 19.8 Volumes of demineralised water produced in PRE-WRP and PNC-WRP (June 2010–May 2011).

19.4 ECONOMICS OF WATER REUSE

Project funding and costs

The Indian Oil Corporation Limited covered the capital costs of the water reclamation plants within the framework of its refinery and naphtha cracker complex investment programme. The operating costs of both reclamation plants are presented in Table 19.4.

Table 19.4 Operating costs of PRE-WRP and PNC-WRP.

		Reclamation plants	
Cost type	Unit	PRE-WRP	PNC-WRP
Power	€/m^3	0.160[1]	0.154[2]
Chemicals		0.11	0.141
Man Power		0.022	0.031
Maintenance, Admin., etc.		0.003	0.004
Total Operating Cost	€/m^3	0.295	0.330

[1]1.41 kWh/m^3 (0.114 €/kWh); [2] 1.07 kWh/m^3 (0.144 €/kWh).

As can be seen in Table 19.4, the reclamation costs (OPEX) for the production of demineralised water (boiler make-up water) amount to 0.30 €/m^3 for the PRE-WRP and 0.33 €/m^3 for the PNC-WRP, that is the OPEX of the PRE-WRP is lower in spite of the higher TDS. This can be explained mainly by the fact that the power price is higher for the PNC-WRP (0.147 €/kWh vs. 0.116 €/kWh for the PRE-WRP), by the lower specific chemicals demand and the higher degree of capacity utilisation of the PRE-WRP. However, according to the higher TDS, the specific power demand of the PRE-WRP (1.4 kWh/m^3) is higher than that of the PNC-WRP (1.1 kWh/m^3).

Benefits of water recycling

There is no direct economic benefit from water recycling, as the cost of boiler make-up water production from treated Yamuna Canal water is lower than the cost of water reclamation. However, there are indirect monetary and non-monetary benefits due to environmental protection effects. Furthermore, it can be concluded that the entire investment programme and the subsequent economic and social development of the Panipat region would have been impossible without water recycling (and other environmental protection measures).

19.5 HUMAN DIMENSION OF WATER RECYCLING

Water recycling is one of the decisive factors in the industrial and subsequent economic and social development of the Panipat region. Apart from the creation of thousands of jobs, IOCL has initiated a community development programme in order to improve the life of the people in the vicinity of the refinery. Major initiatives of the programme include health and hygiene, clean drinking water and education.

19.6 CONCLUSIONS AND LESSONS LEARNED

The realisation of the Panipat Refinery, Refinery Expansion and Naphtha Cracker projects is representative of India's rapid industrial growth and economic development ("rise of the sleeping tiger"). Water management (process water production, effluent treatment, water reclamation and recycling) is a key factor in this context.

At Panipat, no proper recipient is available. The only surface water body (Yamuna Canal) has to be used for both potable water production and agricultural irrigation. Therefore, water reclamation and recycling are essential and in order to further increase water recovery rates (in the refinery and naphta cracker plant), the environmental authorities have requested that zero liquid discharge (by evaporation and crystallisation) be implemented in the near future. Advanced technologies such as membrane filtration are of major significance, especially where the functioning of industrial (refinery and petrochemical) processes has to be guaranteed. With this in view, it is also important that the advanced multi-barrier systems employed are operated and monitored by an experienced water technology specialist. Apart from environmental protection and the reliable provision of process water (mainly boiler make-up), large quantities of freshwater can be saved. This boosts the security of the industrial water supply, which could be endangered in the future due to increased demands for freshwater for agricultural and potable purposes. Last, but not least, it can be concluded that the water recycling practice described constitutes a major factor in sustainable development (environmental, social and economic) in the Panipat region.

The following Table 19.5 summarizes the main drivers, benefits and challenges of the water reuse in Paniprat.

Table 19.5 Summary of lessons learned.

Drivers/opportunities	Benefits	Challenges
Economic growth in India	Environmental protection effects	Fluctuating raw water quality
Environmental protection targets	Boost in industrial water supply security	Reclaimed water quality targets
No proper recipient available	Saving of large amounts of freshwater	Zero liquid discharge target in the future
Industrial water supply security could be endangered in the future	Economic and social development	

Acknowledgement

The authors wish to thank the Indian Oil Corporation Ltd. Panipat, and the VATECH WABAG colleagues, L. Umayal, B. K. Singh, T. Selvakumar, C. Walder, A. Proesl and N. Hoffmann for their support in the preparation of this chapter.

REFERENCES AND FURTHER READING

Hespanhol I. (2008). A new paradigm for urban water management and how industry is coping with it. IWA – Scientific and Technical Report No. 20. In: Water Reuse – An International Survey of Current Practise, Issues and Needs, B. Jimenez and T. Asano (eds), IWA Publishing, London, UK, pp. 467–482.

Lahnsteiner J. and Mittal R. (2010). Reuse and recycling of secondary effluents in refineries employing advanced multi barrier systems. *Water Science and Technology*, **62**(8), 1813–1820.

VGB POWERTECH (2011). VGB-S-010-T-00. Feed Water, Boiler Water and Steam Quality for Power Plants/Industrial Plants. VGB PowerTech Service GmbH Essen, Germany.

20 High purity recycled water for refinery boiler feedwater: the RARE project

Alice Towey, Jan Lee, Sanjay Reddy and James Clark

CHAPTER HIGHLIGHTS

The RARE Water Project provides numerous environmental benefits, most notably by reducing demand for limited potable water supplies. In addition, the project features an innovative design that uses advanced water treatment technology to produce the high level of purity required by refinery boilers. The Project is also a unique example of collaboration between a public agency and a private company in California. The financial arrangement – wherein Chevron pays for all capital and operations and maintenance costs – is a model for the water recycling industry.

KEYS TO SUCCESS

- Successful public-private partnership
- Proven technology
- Redundancy and reliability

KEY FIGURES

- Treatment capacity : 13,249 m^3/d
- Total annual volume of treated wastewater: 4.28 Mm^3/yr
- Water reuse standards: <2.2 *E.coli*/100 mL

20.1 INTRODUCTION

In an effort to conserve potable water by expanding the use of recycled water in its service area, the East Bay Municipal Utility District (EBMUD) recently brought the 13,250 m^3/d (3.5 million gallon per day, MGD) Richmond Advanced Recycled Expansion (RARE) Water Project online. The Project is the result of several years of collaboration between EBMUD, Chevron, and the West County Wastewater District (WCWD). This case study will present the challenges associated with a public private partnership and the ultimate win-win for all parties.

Background

The East Bay Municipal Utility District is a publicly-owned utility located in the eastern San Francisco Bay area in California. The District provides water to approximately 1.3 million customers in Alameda and Contra Costa counties, in addition to providing wastewater treatment for approximately 640,000 people. The total service area is 842-km^2 (325 square-mile).

As part of its commitment to sustainability and reliability, EBMUD has an extensive water recycling program. Prior to the implementation of the RARE project, the volume of recycled water supplied for irrigation and industrial uses was about 20,800 m^3/d (5.5 MGD). The water reuse program was implemented in this service area in 1971. The EBMUD Board of Directors recently reaffirmed its commitment to pursuing the water recycling policy during the development of the Water Supply Management Program 2040 (WSMP, 2040), which establishes the plan for meeting its water needs through the year 2040. The Board elected to pursue the highest level of water recycling which was proposed, setting a goal of delivering 75,700 m^3/d (20 MGD) of recycled water by 2040.

The Chevron refinery in Richmond is one of EBMUD's largest water customers. The refinery processes over 240,000 barrels of crude oil each day in the manufacture of petroleum products and chemicals. Before the RARE project, the District had already been supplying approximately 15,900 m^3/d (4.2 MGD) of recycled water to the refinery for use in its cooling towers since 1996. Secondary effluent from West County Wastewater District is treated at the North Richmond Water Reclamation Plant (NRWRP) via coagulation and sand filtration and pumped to the refinery. Currently, the recycled water is used in three cooling towers.

Project planning

Looking to expand water reuse, in 2005 the District undertook a Feasibility Study to identify additional water recycling projects in the Richmond area. This study consisted of several main components. A customer assessment was completed to identify potential recycled water users. A study of the available supply of secondary effluent was conducted, and potential treatment technologies were reviewed. A siting analysis was conducted to compare different potential locations for a new treatment facility.

The Feasibility Study recommended a project to supply high purity recycled water to Chevron for use as boiler feed water makeup, due to the close proximity of the refinery to West County Wastewater District facilities, and the constant high demand for boiler make-up water. The study also evaluated potential treatment technologies for supplying the quantity and quality of recycled water needed for Chevron's boilers. The Chevron refinery includes both low pressure boilers (<450 psi) and high pressure boilers (≥450 psi). The water quality requirements are stringent for both, especially the high-pressure boilers. Constituents of particular concern include total suspended solids (TSS), total dissolved solids (TDS), hardness, silica, and certain metals. The presence of these contaminants in the make-up water can lead to corrosion or scaling, resulting in premature failure of the boilers or downstream equipment.

Possible treatment technologies were broken down into pretreatment (primarily for removal of TSS and organic foulants) and TDS removal. Pretreatment options included granular media filtration, microfiltration (MF), ultrafiltration (UF), precipitate softening, and granular activated carbon. Of these, MF was chosen for its relatively low costs and highly efficient particle removal process. Reverse osmosis (RO), electrodialysis reversal, ion exchange, and thermal distillation were all considered for TDS removal. RO was selected with the caveat that a second pass of RO or additional treatment would be needed to meet the stringent requirements of the high pressure boilers.

A pilot test program was developed and conducted from July 2005 through October 2005 to demonstrate the ability of the MF and RO technologies to meet the boiler water quality requirements. Both secondary effluent from West County Wastewater District and tertiary effluent from the North Richmond Water Reclamation Plant were evaluated. The pilot study showed that all of the MF and RO membranes evaluated could meet the water quality requirements. The MF system recovery rate was 92.5% for the secondary effluent and 94% for the tertiary effluent. The RO system was able to operate at a recovery rate of 85% and meet the water quality requirements for both secondary effluent and tertiary effluent sources.

The project was executed in compliance with the California Environmental Quality Act. EBMUD prepared an Environmental Impact Report for the RARE Project which was certified in May 2007.

In 2008, a Product Supply Agreement and Ground Lease Agreement were committed between EBMUD and Chevron to define the responsibilities for design, construction, and operation of project facilities and spell out water quality requirements. In addition, Chevron agreed to pay all capital and operations and maintenance costs for the 25 year life of the project. EBMUD also signed an agreement with West County Wastewater District for the supply of secondary effluent for use at the North Richmond Water Reclamation Plant and RARE. Modelling indicated that during the dry season there would be times when all of West County Wastewater District's effluent flow would be captured for reuse.

20.2 TECHNICAL CHALLENGES AND WATER QUALITY CONTROL

Figure 20.1 shows the facilities of the Richmond Advanced Recycled Expansion (RARE). EBMUD was responsible for the treatment plant, influent pump station, and backup potable water meter. Chevron was responsible for pipelines through the refinery, provision of utilities to the treatment plant and IPS sites, and retrofits to Chevron's existing RO system.

Treatment trains for water recycling

Figure 20.2 shows the basic treatment process that was selected. Secondary effluent from West County Wastewater District is diverted from the plant outfall and pumped via a new pump station located on the north edge of the refinery to the treatment plant site. A 3800 m^3 influent tank provides flow equalization. As suggested by the pilot study recommendations, 0.5 mm autostrainers are the first step in the treatment process. These strainers remove particles and debris that could clog or damage the MF membranes.

The MF pretreatment provides a stable, high-quality feed stream for RO systems, while minimizing colloidal and particulate fouling of RO membranes. Individual MF membrane modules are arranged on discrete skids with separate feed, filtrate, backwash, cleaning and air supply connections, forming a unit. The project uses a total of five Pall Microza MF membrane units with a total surface areas of 15,000 m^2 (161,400 ft^2). Four units are needed to meet the flow requirements under the initial phase of the project, and the fifth unit serves an on-line standby. Additional space was left for future expansion of the plant capacity to 15,000 m^3/d (4.0 MGD). Figure 20.3 shows the microfiltration and reverse osmosis systems.

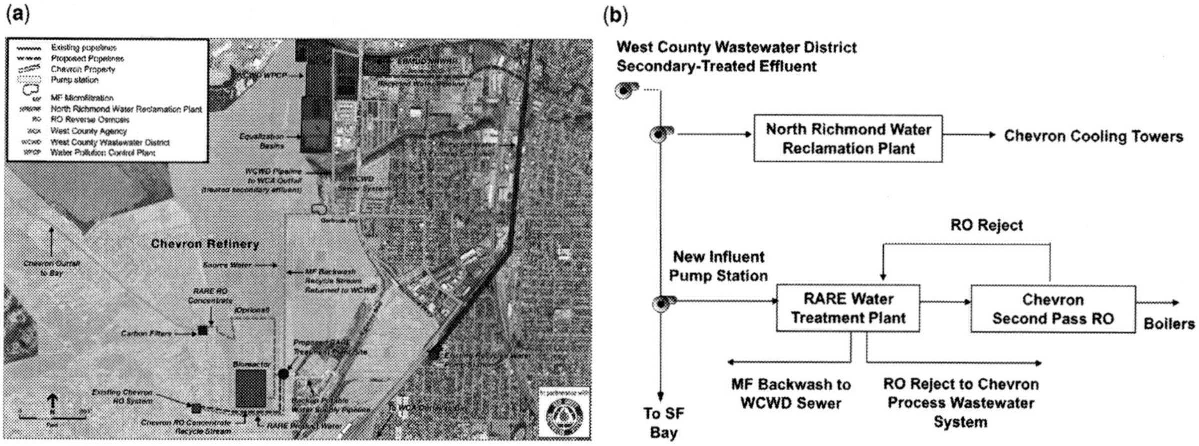

Figure 20.1 Map (a) and scheme (b) of the facilities of the Richmond Advanced Recycled Expansion (RARE) project.

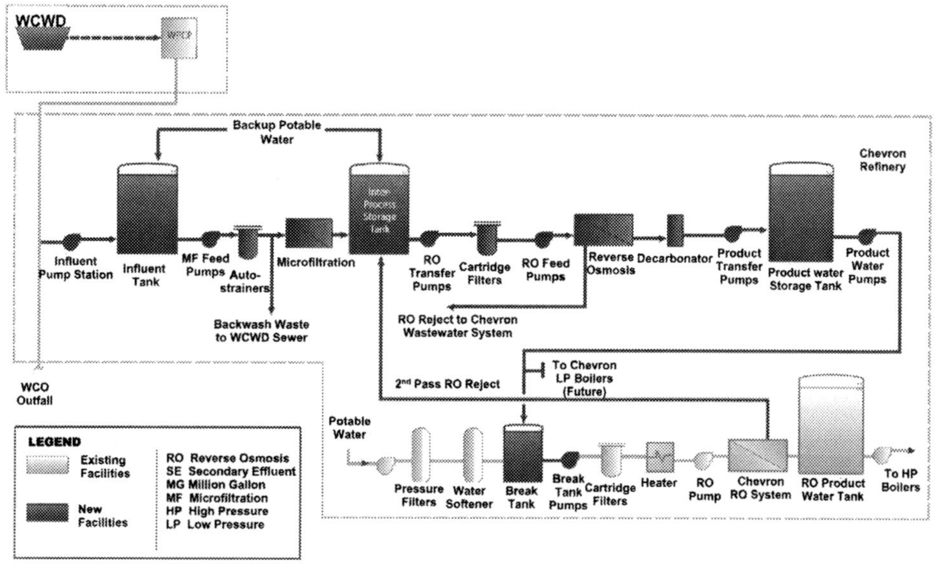

Figure 20.2 The RARE process schematic.

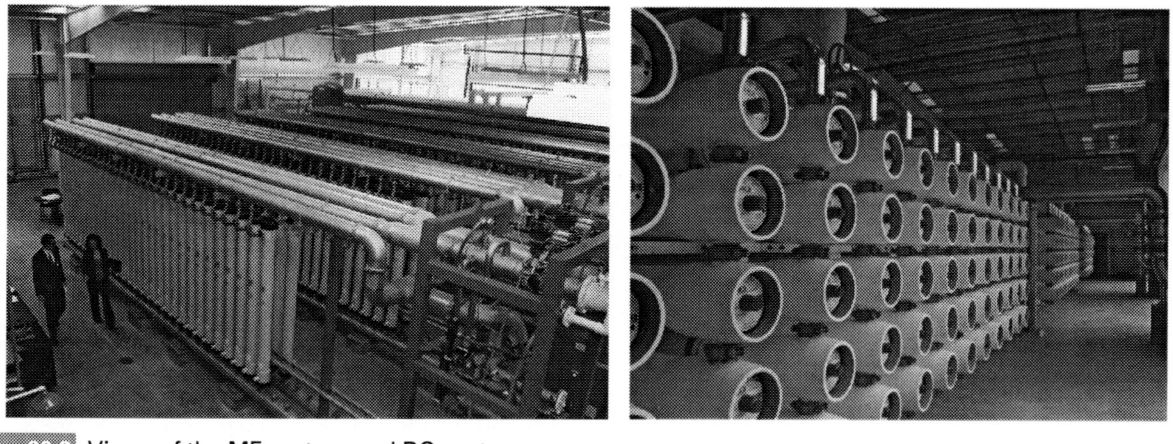

Figure 20.3 Views of the MF system and RO system.

The MF membranes cleaning is very important for reliable operation and maintenance of a high membrane flux. During normal operation, the MF membranes undergo two cyclical, regenerative processes to recover membrane permeability: reverse filtration and enhanced flux maintenance. The reverse filtration process is initiated on a time or volume basis (typically about 15 min or 45 m^3 (12,000 gallons) of production, whichever occurs first) and includes reverse flow of filtrate through the membrane modules coupled with an air scour. The enhanced flux maintenance process is a longer procedure usually conducted about once a day on each unit, consisting of the recirculation of a heated high strength chlorine solution. In addition, a Clean-in-Place (CIP) system is used periodically for cleaning of the MF membranes in-situ when membrane permeability decreases until a target value. The 6-hour CIP process uses caustic and acid solutions to remove foulants from the membrane surface and is triggered by a rise in Trans-Membrane Pressure (TMP). When clean, the MF membranes will have a TMP of roughly 34.5 kPa (5 psi). A membrane CIP sequence is initiated when TMP rises and stabilizes near 207 kPa (30 psi), or at a minimum once per quarter.

A 592-m^3 (156,500 gal) inter-process storage tank between the MF and RO systems serves to equalize the MF filtrate prior to the RO system. The storage tank has a backup potable water fill connection similar to the influent tank. Normally, potable water should be treated by the MF system prior to RO treatment. However, if the MF system is out of service, potable water can be fed directly to the RO system on a short-term basis. In addition, RO brine from Chevron's second pass RO system (used to further product water from RARE for the high pressure boilers) is fed into the interprocess storage tank.

Three (2 duty, 1 standby) RO transfer pumps pump the MF filtrate stored in the inter-process storage tank to the cartridge filters and inline RO feed pumps. The two cartridge filters, located upstream of the RO system, protect the RO membranes from debris, suspended solids, dirt, and other particulate matter.

After the RO feed stream passes through the cartridge filters, high pressure RO feed pumps boost the pressure to the RO membranes. Each RO train is served by a dedicated feed pump, thereby allowing for different operation conditions on the individual membrane trains depending on the degree of membrane fouling.

The purpose of the RO system is to remove dissolved solids from the plant feed water, conditioning it for use as high purity industrial process water. The RARE RO membranes are made of sheets of a polyamide composite material, capable of rejection rates in excess of 99.5% total salinity, arranged in spiral wound cartridges. Feedwater enters through the exposed edge of the bundle at the feed end of the element and proceeds cross flow through the tail end. En-route, permeate is extracted across the active membrane surface and collected in a central tube as purified product (permeate). Individual membrane elements are arranged in tubes (pressure vessels) housing a column of 7-elements in series. The recovery (volume of permeate extracted per volume of feedwater) of a single 7-element pressure vessel is limited to slightly over 50%. To increase recovery and maximize production, a second bank of pressure vessels is employed, and residual feed water (concentrate) from the first bank of vessels is collected and delivered as feed to the second bank of vessels.

Permeate produced by all of the vessels is combined and forms the train permeate stream. Concentrate or reject from the second bank of pressure vessels is collected and forms the final concentrate stream from the train. The rate of final concentrate flow is controlled by a high performance modulating valve which, in concert with the train feed pump controlling product flow, controls the overall recovery of the train. Design recovery of the RO system is approximately 85%, to minimize concentrate production while maintaining acceptable saturation levels of dissolved constituents.

Three RO trains are used in the Project. Two trains are required to meet the product capacity requirement of the facility and the third provides an installed standby. Each train has 40 pressure vessels in the first stage and 20 pressure vessels in the second stage. The vessels are installed on welded steel frames for support.

The RO membrane cleaning processes include a membrane flush and CIP. When an RO train is shut down, residual feed water is retained in the membrane elements. If the train remains off-line for a period in excess of 30 min, the membranes are flushed with permeate (product water) to prevent fouling of the membranes. Similar to the MF system, periodic cleaning of the RO membrane elements via a chemical CIP is required to restore permeability.

The RO permeate is stored in a stainless steel product water tank with a capacity of 7,600 m^3 (2 million gallons), equivalent to 12 hours of storage. Chevron pumps product water from this tank directly to its break tank. The product water pump station consists of 2 pumps; the main pump is steam driven, and the backup pump is electric. The product water pump station is connected to Chevron's Plant Control System (PCS) and operated by Chevron.

Flow equalization

A review of historical data from West County Wastewater District showed that during the summer months, the effluent may not always be sufficient to meet the demands of both RARE and the North Richmond Water Reclamation Plant. RARE was therefore designed with provisions to access a backup potable water supply when necessary to produce the required product water flow.

Based on extensive modelling, the size of the RARE influent tank was increased – initially set at 2300 m^3 (0.6 million gallons) during the conceptual design – to 3800 m^3 (1.0 million gallons). Combined with the 6500 m^3 (1.7 million gallon) influent tank at the North Richmond Water Reclamation Plant, this provides capacity to capture water during periods of higher flows for use during periods of lower flow.

The North Richmond Water Reclamation Plant is effectively "upstream" of the influent pump station for RARE. Operators allow a set flow to bypass the North Richmond Water Reclamation Plant for use at RARE. The water level in the North Richmond Water Reclamation Plant influent tank is allowed to vary in relation to the West County Wastewater District effluent flow. The set point for the amount of flow allowed to bypass to RARE varies on a seasonal basis.

Reliability of operation

The Chevron refinery boilers operate 24 hours a day, seven days a week and require a constant supply of high quality water. The project agreement requires that the production of recycled water meeting the quality specifications more than 99.5% of the time during a given contract year.

The high dependability requirement, combined with remote operations of the facilities, made it important to have a high level of reliability in the treatment process. The MF and RO systems were designed to include more trains than were needed to meet the production requirement so that the facility could remain fully operational with one train down for maintenance or cleaning. Similarly, redundant pumps were also provided in many areas of the plant.

During final design, a criticality assessment was completed analyzing the entire treatment process to identify critical components and their potential failures. The criticality assessment then recommended design changes or preventative maintenance activities to mitigate potential critical component failures. As a result of the criticality assessment, some changes were made to the design to incorporate increased redundancy. On the recommendation of the criticality assessment, the construction specifications were modified to include numerous spare parts and boxed spares.

To provide additional reliability, the product water tank was sized at 7600 m^3 (2.0 million gallons), or the equivalent of over 12 hours of storage (Figure 20.4). Additionally, the influent tank and interprocess storage tank were plumbed so that they can be fed potable water if secondary effluent is not available. The potable water connection in the interprocess storage tank also allows the plant to continue producing high purity water even if the MF system is not functioning.

Figure 20.4 Aerial view of completed RARE treatment plant with the storage reservoir.

Since the project came online in June 2010, RARE has successfully met the demand of the Chevron boilers without going offline. Backup potable water has been used to supplement the secondary effluent supply from West County as needed, either due to insufficient effluent supply in the summer months or maintenance of the MF system.

RO brine disposal

One of the major issues that had to be addressed in this water reuse project was the disposal of the RO-reject. After considering several alternatives, it was decided that the RO-reject would be sent to Chevron's process wastewater system for disposal. However, by using recycled water in place of potable water in several processes, the waste streams produced have higher concentrations of some constituents that have numeric limits in Chevron's permit. Chevron therefore negotiated with the

Regional Water Quality Control Board to allow to take "credits" for its use of recycled water, since any constituents in the secondary effluent would have been discharged if not reused.

Chevron submitted a technical report to the Regional Water Quality Control Board in 2006 providing justification for the use of recycled water credits. The report analyzed the results of the pilot test program to show that the use of recycled water credits would not impair the vicinity of the discharge, or cause a zone of toxicity to aquatic organisms. The Regional Water Quality Control Board granted Chevron the right to calculate mass and concentration credits for its use of recycled water. EBMUD conducts regular sampling in conjunction with Chevron to calculate mass and concentration credits.

Plant performance and recycled water quality

The RARE Water Project began delivering high purity recycled water to Chevron on July 21, 2010. Table 20.1 summarizes the recycled water production at RARE to date. In its first year of operation, RARE delivered approximately 4.2 million m^3, Mm^3 (1.1 billion gallons) of recycled water to Chevron.

Table 20.1 RARE production during first year of operations.

Month	Monthly total Mm^3 (million gallons)	Daily average Mm^3 (MGD)
July 2010 (beginning July 21)	0.12 (30.55)	0.011 (2.78)
August 2010	0.29 (77.74)	0.009 (2.51)
September 2010	0.35 (93.12)	0.018 (3.10)
October 2010	0.34 (89.48)	0.011 (2.89)
November 2010	0.25 (64.83)	0.008 (2.16)
December 2010	0.38 (100.44)	0.012 (3.24)
January 2011	0.42 (109.90)	0.013 (3.55)
February 2011	0.37 (98.55)	0.013 (3.52)
March 2011	0.36 (94.10)	0.011 (3.04)
April 2011	0.37 (97.67)	0.012 (3.26)
May 2011	0.39 (102.75)	0.013 (3.31)
June 2011	0.26 (103.59)	0.013 (3.45)
July 2011 (through July 20)	0.25 (66.65)	0.013 (3.33)
Total Production	**4.28 (1129.3)**	
Annual Average (MGD)		**0.017 (3.09)**

The average daily production during this period was 11,740 m^3 (3.1 MGD). Although the facility has the capacity to produce up to 13,250 m^3/d (3.5 MGD) of recycled water, actual production varies depending on Chevron's demand and on the quantity of second pass RO reject the refinery sends to RARE.

Water quality control and monitoring

There are three main sampling programs at RARE. First, EBMUD conducts daily coliform monitoring on the product water as required by California Title 22. Second, EBMUD also tests product water at RARE for compliance with the water quality requirements in the agreements with Chevron. Lastly, EBMUD conducts extensive sampling on the RO reject sent to Chevron's process wastewater treatment system in support of Chevron's NPDES permit. To date the product water quality from RARE has been very high, and has met the requirements of the agreement with Chevron and the requirements of Title 22.

Main challenges for operation

There were several operational challenges during the start up and early operations of the RARE Water Project. The most significant operational issues were attributable either to equipment problems or variable water quality in the secondary effluent supply from West County.

During the first year of operations, staff noticed that the aqua ammonia storage tank and feed system was not functioning properly. Staff observed high discharge pressure buildup and frequent clogging of the flowmeter strainer. Inspection revealed that the internal tank had corroded somewhat during construction. To prevent further clogging and discharge pressure buildup, staff installed a recirculation pump and filter to remove remaining particulate matter.

Similarly, during startup the cartridge filters upstream of the RO system became clogged. Staff determined that the material clogging the cartridge filters was primarily construction and startup debris and not a long-term problem.

Perhaps the most serious issue to arise so far in operations has been periodic rapid fouling of the MF membranes. Although staff continues to investigate the cause of this issue, it is currently believed to be related to variable water quality from West County.

Operations staff first noticed the MF fouling issue in January 2011. The transmembrane pressure across the MF membranes would rise rapidly. Conducting a clean in place sequence would drop the transmembrane pressure back to a normal level, but the pressure would quickly rise again. Staff attempted to resolve the issue by experimenting with changes in the clean in place sequence and "recipe," but the problem persisted. This led to decreased production from the MF system due to racks frequently being taken offline for cleaning. Supplemental potable water was needed to make up production to meet Chevron demands. A variety of analyses were conducted to attempt to determine the exact source of the problem. However, the water quality improved significantly in April 2011, before the investigation could be completed. The fouling problem recurs periodically and seems to be somewhat correlated to higher suspended solids in the secondary effluent from West County Wastewater District. EBMUD continues to investigate this issue.

20.3 ECONOMICS OF WATER REUSE

The finances behind the RARE Water Project are unusual. As part of EBMUD's agreement with Chevron, Chevron is responsible for all capital and operations and maintenance costs for the project. As such, the project has almost no impact on EBMUD's rates for its other customers.

Total capital costs for the project, including planning, design, construction, and construction management, amounted to approximately $55 million. Operations and maintenance cost approximately $2.3 million, including a $400,000 annual contribution to a Major Equipment Repair and Replacement fund.

20.4 BENEFITS OF WATER REUSE

Recycled water is an important component of EBMUD's diverse water supply portfolio. In 2009, EBMUD's Board of Directors set an ambitious goal of delivering 20 MGD of recycled water by the year 2040. The RARE Water Project is an important component of that goal; by delivering 3.5 MGD of recycled water to Chevron, EBMUD is able to conserve an equivalent amount of potable water.

Water recycling can help conserve energy as well as water. A significant portion of California's energy use is for pumping water at great distances – recycled water utilizes a local resource (wastewater) to significantly reduce these impacts.

It is also anticipated that RARE, in combination with EBMUD's North Richmond Water Reclamation Project, will use all dry weather flows from WCWD. For part of each year, WCWD will be a "zero discharge" wastewater treatment plant, thereby reducing wastewater and pollutant discharges into San Francisco Bay.

Future phases

The initial phase of RARE was designed to supply up to 3.5 MGD of recycled water. EBMUD's recent WSMP 2040 identified a variety of projects that EBMUD could implement to meet its goal of producing 75,700 m^3/d (20 MGD) of recycled water by 2040. Two projects were identified to expand the RARE Treatment Plant to increase recycled water deliveries to Chevron. The RARE Phase 2 Project will add additional MF modules to increase recycled water production to 15,140 m^3/d (4.0 MGD). The RARE Future Expansion Project would increase the size of the treatment plant to produce a total of 18,930 m^3/d (5.0 MGD) of recycled water. All RARE facilities were designed to accommodate potential future expansion. Implementation of these phases would require an increase in available effluent supply, either due to increased flows from West County Wastewater District or the use of an additional source, such as effluent from Chevron's process wastewater system.

20.5 CONCLUSION: A SUSTAINABLE SOLUTION

The implementation of high quality recycled water for industrial purposes in petroleum industry has conferred numerous benefits to EBMUD, Chevron, and the environment. By delivering recycled water to Chevron, EBMUD is able to conserve an equivalent amount of potable water. Water recycling projects reduce the District dependence on imported water supplies and reduce the threat of severe rationing during a drought.

All dry weather flows from the municipal wastewater treatment plant is reused in the two recycling facilities. For part of each year, West County Wastewater District will be a "zero discharge" wastewater treatment plant, thereby reducing wastewater discharges into San Francisco Bay.

For Chevron, the RARE water project represents a drought-resistant water supply for its boilers. In addition, Chevron has been able to reduce the amount of water treatment needed, as the high purity recycled water from RARE exceeds potable water quality.

The successful collaboration between a public agency and a private company has enabled EBMUD to conserve potable water at virtually no cost to its ratepayers, making this project an excellent model for the water industry. As potable water supplies in arid parts of the world become scarcer, projects like RARE will become more important. Water agencies will need to continue to encourage water conservation, diversify their water supply portfolios and find efficient ways to reuse water.

The following Table 20.2 summarizes the main drivers, benefits and challenges of the RARE project.

Table 20.2 Summary of lessons learned.

Drivers/Opportunities	Benefits	Challenges
Commitment to sustainability and reliability Willingness to diversify the water supply portfolio Close proximity of the refinery to the wastewater treatment facility	Conservation of potable water Reduction the dependence on imported water supplies Reduction of the threat of severe rationing during droughts Conservation of energy avoiding pumping at long distance Reduction of wastewater and pollutant discharges into San Francisco Bay	High level of reliability in the treatment process Disposal of the RO brines Flow Equalization

WEB-LINKS FOR FURTHER READING

EBMUD's website: www.ebmud.com

21 Closing loops – industrial water management in Germany

Karl-Heinz Rosenwinkel, Axel Borchmann, Markus Engelhart, Rüdiger Eppers, Holger Jung, Joachim Marzinkowski and Sabrina Kipp

CHAPTER HIGHLIGHTS

Plant-internal measures for water reuse and recycling in industrial production processes are an efficient tool in terms of reducing wastewater volume and load and recovering input material or product. Thus material utilization may be improved, aquatic pollution can be prevented, energy consumption, and therefore, energy and wastewater treatment costs can be reduced. In this chapter, industrial water management measures and concepts in Germany are presented. Examples from food and drink, paper, textile, metal and ceramics industries are described where measures for water reuse and recycling have been successfully implemented.

Key factors for the success of such measures are their technical and economical feasibility and efficiency. On the one hand the benefit a company expects to gain by implementing a water reuse and recycling system is essential. Cost reduction through improved material utilization, lower energy and water consumption and wastewater production has to outbalance investment and operating costs. On the other hand, technical limitations and requirements have to be considered, for example, concerning the quality of the recycled water depending on its designated use in respect to chemical, biological and physical aspects.

In order to evaluate the performance of plant-internal measures for water reuse and recycling several different key figures can be taken into account: reduction of water consumption/ wastewater volume, reduction of wastewater load (e.g. COD, resources, salt content), reduction of energy consumption or costs induced by the implementation of plant-internal measures.

The following table illustrates the reduction of water consumption and wastewater volume as key figures. Some examples of industrial water management measures in Germany are shown that are described in detail in this chapter.

KEYS TO SUCCESS	KEY FIGURES	
By type of industry	**Average wastewater (WW) volume without reuse**	**Reduction of wastewater volume with water reuse**
Food and drink industry		
Water recycling in sugar industry	10 m^3/t sugar beets (DWA, 2007)	92% (0.8 m^3 WW/t sugar beets)
Paper industry		
Circuit water and effluent treatment by combined fine filtration and ozonisation	29 m^3/t paper (Öller & Offermanns, 2002)	50%
Integrated water treatment using membrane bioreactor and reverse osmosis	0–15 m^3/t paper (DWA, 2011)	90%
	average freshwater consumption without reuse	**reduction of freshwater consumption with reuse**
Textile industry		
Water (and Heat) Recovery from Reactive Rewashing Using Hot-Nanofiltration	20 m^3/t of goods	77.5%

21.1 INTRODUCTION

Industrial water resource management and recycling are serious alternatives to end-of-pipe-technology for saving water resources. In industries water is needed for production, cleaning and several other purposes, so that water management systems are necessary to determine the part streams to be recycled as well as the applicable technology required for each stream treatment. For that purpose a decision support system for water resource management has to take into account economic and ecological components of treatment and recycling technology. In every case, possibilities for internal measures have to be identified and verified such as recycling, product recovery, reduction of loads, nutrient content, temperature, cleaning and disinfecting agents. In general, reduction of wastewater load by in-plant internal measures at its origin is superior to end-of-pipe treatment of organic or inorganic loads in municipal or industrial wastewater treatment plants.

21.2 DEVELOPMENT OF INDUSTRIAL WATER DEMAND IN GERMANY

In Germany, there is a long-standing experience involving substantial available data for different production technologies and specific water volumes and wastewater loads. These data compare the different water volumes and wastewater loads for different industries and different technologies, and may be utilised to identify the best available technologies for a production process. The reasons for saving water in industry in Germany are mainly cost-orientated, but in some cases there are image issues to be considered. In general, water availability in Germany is much higher than water consumption. However, in several instances and regions, there is, from time to time, not enough water for industrial production.

Water balance

The overall balance of water availability in Germany indicates an unutilised surplus of about 156 billion m^3 per year (Gm^3/yr). This is true as long as the use of virtual water by imported goods is not taken into account. Hoekstra *et al.* (2004) calculate a demand of virtual water for Germany of 127 Gm^3/yr, not taking into account water demand for exported or re-exported goods – another 70 Gm^3/yr. Of this 127 Gm^3/yr, about 37 Gm^3/yr is water from domestic sources for municipal and industrial use. The largest part by far of 67 Gm^3/yr is water from imported agricultural and industrial sources, 50 and 17 Gm^3/yr.

Altogether, considering water demand for exported goods of 32 Gm^3/yr, the balance shows a net water import of 35 Gm^3/yr – a number of the same dimension as the use of domestic resources, most of it for imported agricultural goods. Fundamentals and figures for the different shares vary between the sources (Statistisches Bundesamt, 2010; Hoekstra *et al.* 2004) so they cannot be harmonized entirely, but these numbers illustrate the dimension and relations.

German mining and processing industry itself, as the largest consumer apart from thermal power stations (Figure 21.1), exhibited a consumption of approximately 7.2 Gm^3/yr during 2007, most of this water being surface water and 28% groundwater. Chemical industry with its diverse structure is predominant, followed by metal, paper and food industry with a more uniform composition of the applied industrial processes within each branch. Examples from these industries and some additional branches like textile and car industry will be presented in this chapter.

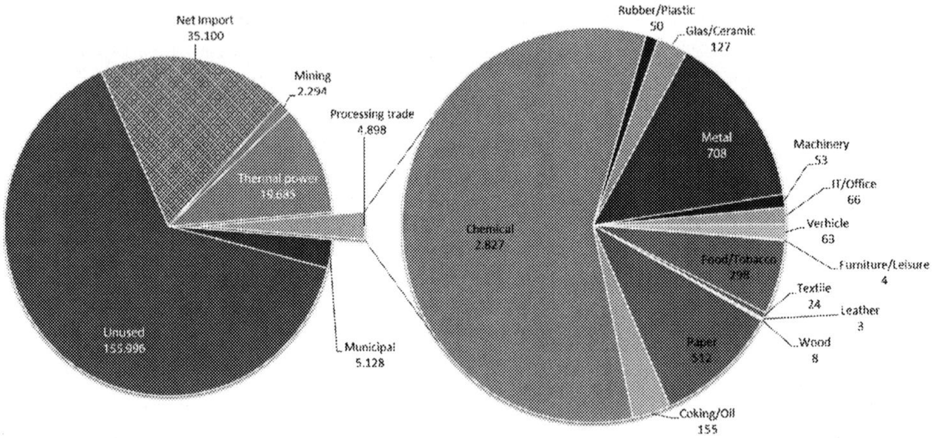

Figure 21.1 Consumption of water resources in Germany (left) and consumption of the different branches of processing trade (right) (*Source*: Statistisches Bundesamt, 2010; Hoekstra *et al.* 2004).

Water demand and reuse

Within the last 30 years, the water demand in mining and processing trade has dropped by more than 30%, only increasing due to German reunion in 1991 (Figure 21.2). This trend did not develop due to economic decline and less production in consequence of this, but due to increase of water reutilisation in industries. The utilisation factor for water is defined as the quotient of water demand divided by the water supplied. This factor more than doubled in this period, reaching a level of 5.8 in 2007.

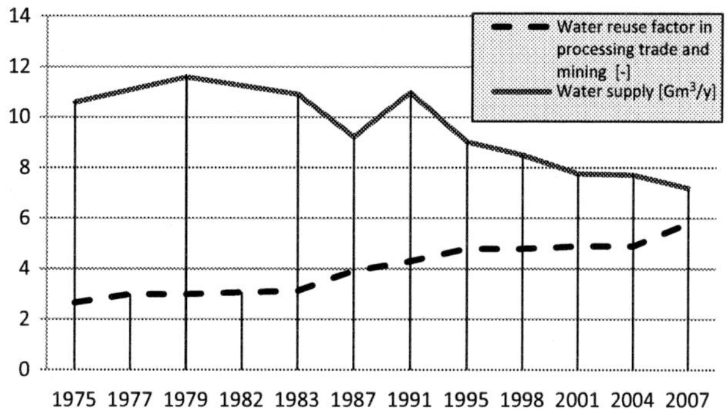

Figure 21.2 Development of the utilisation factors and water demand in mining and processing trade.

21.3 DRIVERS FOR IMPLEMENTATION OF WATER SAVING TECHNIQUES

The rational usage of process water, which is defined as "water with different quality, including drinking water, for commercial, industrial, agricultural or similar purposes" is gaining more and more importance for industry. Reasons for this are, on the one hand, the existing limitations of water production or supply with simultaneously growing output, and on the other hand, the effort to minimize the costs of water. The costs of supply arise from drinking water and intake fees as well as charges for extraction conditioning and transport. In the disposal of wastewater, the main costs incur by treatment, pumping or fees and surcharges. These expenses can often be reduced by multiple or looped use of water, however resulting in additional costs for internal treatment and distribution. Balancing between charges and discharges, it is necessary to find the optimum considering legal and regional restraints as well as availability and evolution of production.

Criteria and water quality

The first step for implementation of water saving technologies is a detailed determination of production demand and load, in terms of production scheme and water demand in different streams, entailing the calculation of specific demands and loads. These calculations and measures have to be divided into different load components, like carbon, nitrogen, phosphorus, suspended solids, heavy metals, and so on. After the calculation of the specific demands and the loads for comparison with other companies in this branch, a proposal for in-plant measures has to be prepared. The third step aims at the calculation of efficiency of the extensive measures for recycling that has to be carried out in order to determine the efficiency of the water conservation measures.

In general, there are a lot of proposals for reduction measures of water consumption, like training personnel, as well as daily measurements and cost calculations. However, in addition there are many possibilities that could be integrated into the process to reduce water demand, like changing transport systems or changing the cleaning system and integrating water cycles, or changing the cooling system. Water recycling continues to gain in significance and is also taken into consideration in more sensitive areas of industrial production for economic reasons. The treated water is fully recirculated and is maintained at the same level of quality by means of permanent treatment.

In this context, a significant problem is the monitoring of the quality of process water. Biological contamination is only evident after a period of at least 48 hours through conventional processes. Preventative measures like chlorination or the addition of biocides, when permitted by law, are often undesirable, since these substances adhere to or affect the products, and in addition could substantially contaminate the wastewater streams. Additional components for water quality control could require substantial additional costs, which may be not justified by the small enhancement of the security level. In particular, UV-lamps in the medium and high pressure range, as well as oxidation (advanced oxidation process) are utilised. Unfettered application is however not possible, since poor water quality such as for example high turbidity, may require an extremely high UV dose. During oxidation, operating costs rise considerably according to increased organic content, measured by the chemical oxygen demand (COD).

Fields of application

The fields of application of water recycling in industry differ by section in the particular process and their target performance. While adapted and well designed production engineering is able to reduce water, wastewater and waste loads, internal water circulation can only cut the amount of water and wastewater for the specific production step, while increasing the concentrations of contaminants. On the other hand, product recovery and (pre)treatment are mainly focused on minimizing concentrations or loads. If, however, pre-treated wastewater is reused, water demand and wastewater discharge can be diminished as well. Considering this, reduction of water and wastewater management costs start by implementing intra-plant measures. Some of the most common methods are:

(a) Use of production technologies with non or little wastewater discharge
(b) Product recovery
(c) Production cycles with multiple usage of water, for example:
- Cascading or washing water cycles
- Counter-current washing
- Brine and acid cycle (CIP – cleaning in place)
- Cooling water cycles

(d) Wastewater (pre)treatment

21.4 EXAMPLES OF CLOSING LOOP CYCLES

Food and drink industry

In terms of food industry, in Germany there had been a stable utilisation factor of approx. 3.5 since 1980. The reason for this is to be found in the requirements for the water quality in the food industry, combined with the costs of extensive measures for treatment to achieve a high hygienic standard for recycling. The increase in the last 10 years of water utilisation factor to the actual level of 4.9 can be attributed to more affordable technological solutions, for example, membrane technology and improved water management systems.

Recycling of rinsing water from PET-bottle washing

Nowadays, PET (Polyethylene terephtalate) bottles for beverages can be found everywhere. Consumers predominantly value reduced weight and improved robustness when compared to glass packaging. Basically PET bottles are produced from preforms through a blow moulding process at elevated temperature directly before filling. Once the PET bottle has been formed it needs to be sterilized using chemical disinfectants before any beverage is filled in. There are different ways to sterilize the bottles in the filling line. During hot sterilization, gaseous H_2O_2 is used. The H_2O_2 vapour then condenses in the bottle and is removed using sterile air (Kuhn, 2004). During wet sterilization, a mixture of peracetic acid, H_2O_2 and steam is injected in the bottles. After a certain reaction time, the disinfectant residue is removed from the bottles in the aseptic rinser where sterile water is used (Krones, 2010). Because of low impact on bottle integrity and form stability, wet sterilization is a proven concept. Major drawback is the considerable use of sterile water for flushing, which may reach up to 20 m^3/h subject to filler capacity. One measure to reduce freshwater intake for bottling lines is the recycling of rinsing water as a separate stream. This concept has proven to be successful, when limited water resources prevent further increase of production capacity or sterile water has to be produced from tap water including a costly desalination process, for example, ion exchange.

Water recycling has to take into account any remaining disinfectant residues in the rinsing water and hygienic needs for reuse at the bottling line. For this purpose, reverse osmosis is used followed by neutralization of the rinsing water (Kuhn, 2004). During RO treatment, residual organic matter, germs and soluble salts are rejected by the membrane. Around 20% of reject concentrate is discharged and has to be replaced using fresh water. Around 80% of permeate is re-directed to the rinsing station after having been subjected to a further standard UHT (ultra-high temperature) sterilization procedure. Because direct reuse of rinsing water is quite sensitive and might impact product quality, all technical equipment has to be constructed from stainless steel and according to food production standards. Membranes also need to be food safe. CIP cleaning of the installation and regular hot sanitization of membranes have to be incorporated into the process.

Several factors influence the economics of rinsing water recycling:

- Source of fresh-water
- Costs for rinsing water preparation
- RO pre-treatment costs
- Impact of chemical products on membrane durability

- Cleaning and sanitization intervals
- Operation stress at the bottling line

Therefore recycling of rinsing water still is not a standard application at bottling lines. Main drivers for the future application will be reduced water and carbon footprint compared to conventional single-use treatment options.

Water demand and recycling in sugar industry

The water demand in sugar industry decreased nearly to zero in Germany within the last 20 years. The only wastewater volume emanates from the beets, currently approximately 0.8 m^3/t. During production up to 1.7 m^3 of water per tonne of beets is required. This total can be divided into the following types as indicated in Table 21.1.

Table 21.1 Water and wastewater types in the sugar industry (Rüffer *et al.* 1991).

Water type	Consumption (L water/t beets)
Flume water	500–800
Sweet water	150–200
Hot well water	400–600
Fresh water for juice extraction	30–40
Cooling and sealing water	20–100
Total consumption	600–1740

This water consumption is not covered by fresh water but through recirculation of process water and supplementation of losses with process condensates and water originating from dewatering the beet pulp.

Recirculation of soaking water in malting industry

At malt factories, a variety of soaking processes are used. Depending on the specific process, the volumes of wastewater and its contamination vary considerably (Table 21.2).

Table 21.2 Contamination of soaking water (Ahrens, 1998).

	Immersion soaking process		
Parameters	Box	Germinating line	Spray soaking process
Specific volume (m^3/Mg barley)	1.8 to 2.5	1.73	0.2
BOD (mg/L)	600 to 880	410	1800
COD (mg/L)	925 to 1732	823	2720
Sediments (mL/L)	1.8 to 2.9	2.9	13

In the malting industry an integrated process water treatment system could be used, as reported by Kraft (Kraft, 1997), utilizing a microfiltration unit as treatment technology. The treated water, virtually disinfected through the membranes, can be returned into the production process, so that only a small input of fresh water is necessary for evaporation losses. The BOD from the effluent is approximately 5 mg/L and the ammonia concentration lower than 1 mg/L. Further advantages are the high reduction rate of microbial germs, a low footprint of the technology and low sludge-production.

Paper industry

Water is one of the key components in papermaking. Using more than 1 billion cubic meters of water per year (1 Gm^3/yr), paper industry in Europe had been challenged to reduce their impact on water resources as one of the most important industrial water consumers. Legislation, stringent discharge standards as well as process and product demands force industry to ensure higher water quality corresponding to increasing costs. For the water consuming industry, water is no longer regarded as a consumable or utility but as a highly valuable asset.

Attention to water scarcity and pollution results in new legislative directives, forcing industries to reduce water-use and pollution, and motivating them to implement innovations and observe carefully the impact of measures. The Water Framework Directive (WFD) is one of the main drivers for sustainable water use in Europe, which forced the member states to pay more attention to sustainable and efficient water use.

Competent decision making at the top management and well trained and motivated staff delivered substantial progress in reducing the water consumption in the paper industry. Due to its high competence in closing water circuits paper industry nowadays counts among the most advanced industrial sectors in water efficiency. Multiple recirculation of process waters substantially supported by internal process water treatment as well as process modelling and automation has lead to a decrease of the average specific effluent volume in the German pulp and paper industry from 46 m^3/t to approximately 10 m^3/t of product produced over the last decades (Jung & Pauly, 2011).

Although, the increasing closure of water loops involves many problems, it is bound to be part of the paper industry's future development. In the following sections three milestones of this development will be illustrated.

Closed water circuit with integrated biological water treatment

Several paper mills producing packaging grades from recycled are operating their process with a closed water circuit. Reasons for closed water circuits are among others a reduction of the water related costs (e.g. effluent fees), enabling higher process temperatures and becoming less dependent on water supply. On the other hand a closure of the water circuits results in an increased loading of the process waters in terms of dissolved and colloidal material that cause severe quality deterioration and a drop in productivity. If the effluent volume is to be reduced successfully, the impact of such measures on the papermaking process must also be taken into consideration (Jung & Pauly, 2011).

Most important example for the installation of a biological water treatment in a closed water circuit is Smurfit Kappa Zülpich Papier. Smurfit Kappa is producing corrugated base paper from 100% recycled fibres and has been operating its paper production completely effluent free since 1970. Forced by an increasing amount of odorous compounds both in the process water and the produced paper, in combination with the upcoming plans of an expansion of the production capacity, it was decided to install an integrated biological water treatment plant consisting of an anaerobic and aerobic stage in 1995 (Diedrich *et al.* 1997). Approximately 4 m^3/t of paper is treated under mesophilic conditions and recirculated to the papermaking process. This treatment leads to a significant improvement of the process water quality (Habets & Knelissen, 1997). A significant elimination of the odour forming volatile fatty acids in the process water is achieved leading to a significant improvement of the paper quality. In addition, a reduction of dissolved organic compounds, water hardness and conductivity is achieved (DWA, 2011).

The anaerobic treatment of the water generates biogas. The utilisation of this biogas in the power plant together with the avoidance of effluent fees and the minimised fresh water consumption compensates the operating costs of the biological treatment plant. However, a reliable and optimised operation of the plant requires an increased technological, personnel and organisational effort including the need for a team of experts, patrol tours and trouble shooting for all shifts and integration of the water treatment plant in the process control system of the paper machine (DWA, 2011).

Smurfit Kappa has been a forerunner for the biological treatment of process water in a closed water circuit and several other installations followed. Nevertheless, especially scaling affecting the efficiency of the bioreactors and disturbing the paper making process is still a major problem for the paper mills (Möbius, 2010).

Reduction of water demand using ozonisation

The Bavarian paper mill Büttenpapierfabrik Gmund, which produces high-quality wood-free and mostly coloured paper, is discharging its effluent to the Mangfall River via a municipal treatment plant that later flows through one of the catchment areas that supplies Munich with its drinking water. For this reason, the mill has always been aware of its responsibility to continually reduce its environmental impact.

Therefore Büttenpapierfabrik Gmund has integrated the world's first circuit water and effluent treatment plant based on a combined fine filtration and ozone treatment stage. Effluents from the flushing and cleaning process necessary during colour changes are treated for decolourisation and fibre recovery so that they can be recirculated into the process to be used instead of fresh water for further flushing and cleaning during subsequent production cycles, regardless of the paper colour produced without any adverse effects on the process or product quality (Öller & Offermanns, 2002). Therewith the mill succeeded in reducing its specific effluent volume by about 50%.

Owing to its innovative and environmentally friendly character, the project has been financially supported by the German Environmental Foundation (DBU) with €350,000. In addition to the ecological relief achieved, cutting the effluent volume by half also amounted to a financial savings due to a decreased effluent fee and a lower additive consumption that must be used for effluent treatment prior to the start-up of the new plant. The Büttenpapierfabrik Gmund has broken new grounds by the implementation of this new and innovative integrated water treatment plant. In the past years, further ozone treatment plants have been installed for upgrading and extending the wastewater treatment capabilities in paper mills (Kaindl & Liechti, 2008).

Integrated water treatment using membrane bioreactor and reverse osmosis

Forced by a drastic increase in the costs for fresh water and effluent discharge and increasing energy costs the German cardboard producer Köhler Pappen installed the world's first membrane bioreactor with a subsequent reverse osmosis in a paper mill. The membrane bioreactor, a compact aerobic biological treatment unit combined with an ultrafiltration membrane, produces solid free biologically treated water which can be used in the production process for the replacement of fresh water. A part of this water is furthermore treated with the reverse osmosis to reduce the amount of inorganic dissolved materials like salts which can be used for more sensitive water consumers (Junk *et al.* 2008). To avoid scaling problems at the ultrafiltration membrane and the subsequent reverse osmosis, a softening unit is installed primary to the biological treatment. Both the reversed osmosis and the softening unit enable the mill to keep the salt concentrations on an acceptable level for the paper production.

The installation of the membrane bioreactor in combination with the reverse osmosis enabled the paper mill to reduce its effluent volume by 90% leading to a significant reduction of the water related costs. Due to the reuse of treated warm effluent in the production an increase of the process temperature could be achieved leading to a reduction of the steam demand for water heating and board drying and consequently to a reduction of the CO_2 emissions.

Membrane treatment is evolving into a key technology for continued water savings in the paper industry. Beside the above described example there are several further applications of the membrane technology in paper industry with a huge field of applications like fresh water and process water treatment, treatment of coating colour effluents, wastewater treatment by means of a membrane bioreactor and tertiary wastewater treatment downstream of a biological effluent treatment system (Simstich & Öller, 2010).

Outlook

The pressure on the paper industry to further narrow its water circuits has increased significantly during the past decade, although approximately 90% of the mill's water intake is returned to its source. Since the technologies of the past can no longer meet the requirements of the future, saving fresh water or lowering the organic loads of their effluents requires the increasing use of innovative internal and/or external treatment technologies and concepts. Paper industry is leading the issue water and numerous companies are active: for example, Holmen paper in Madrid is switching to 100% municipal waste water (Presas, 2011). Several European research projects like AquaFit4Use or ALBAQUA aim on the development of new treatment technologies to support the European sustainability policy, like reducing the use of scarce fresh water, improving the water quality and sharing corresponding experiences with other sectors.

Textile industry

Textile finishing processes are connected with a high demand of fresh water and energy and the resulting amount of effluents. The foremost reason to treat effluents is often dyestuff – otherwise the burden is limited. Membrane processes seem to be most appropriate to treat this kind of effluents; but are still rarely used because of the effort. A different approach to introduce membrane technology could be used to overcome this drawback. Placing one membrane plant close to one specific production process or production plant, thus closing a small loop, has several advantages:

- A minor effort to clean the effluent because the input is less varying and the output is supposed to be used in the original process,
- Auxiliaries can be recycled with the permeate,
- Permeate can be recycled at a temperature close to the original process, thus saving energy to heat fresh water.

The integration also bears the opportunity to re-design the production plant and recycle the permeate in a manner that also enhances the product quality, or if the quality is sufficient, allows for the same quality with less effort.

In textile industry, difficult general requirements lead to high demands on the cost-effectiveness of membrane techniques. Thus, a promising approach is to integrate them into existing production processes and thereby accomplish multiple-use when recycling the filtrate.

The approach to identify the recycling potential includes three steps:

1. Investigation of the processes/process steps (material input, temperature, heat energy, gas/electricity consumption, amount of water-/wastewater, key figures and their temporal change).
2. Optimising the process/process steps (environmental/economic).
3. Establishing measures to conserve resources close to the process.

Water and heat recovery from reactive rewashing using hot-nanofiltration

Reactive dyeings of cellulose fabrics need to be washed at 95°C to remove hydrolysed dyes. The optimization of the washing process of an open-width washing machine was realized by establishing a vacuum suction unit for dewatering and extraction of the fabric. Instead of using 20 L/kg (L washing water per kg of goods) and two washing passages, the process could be reorganized to one passage with a water consumption of only 4.5 L/kg of goods. Energy savings amounted to 80% and time savings to 50%. The hot nanofiltration (NF) was only applied for half of the washing water and only in the second half of the washing machine. The permeate was recycled at the end of the washing machine in counter current to the product flow while keeping the process temperature (80–90°C). The concentrate was used as washing water in the first half of the washing machine. A NF ceramic module (HITK) was used as membrane. The permeate flux was approximately 120 $L/m^2 \cdot h$ at 10 bar with a recovery rate of 97% and a retention rate of 90%, expressed as difference in colouring (PIWATEX 2).

Water recycling from polyester yarn dyeing

Disperse dyes that are used to dye polyester are not completely retained by nano- and ultrafiltration membranes, respectively. The particle size of the unfixed disperse dyes in the process water at the end of dyeing is lower than 0.01 μm. In addition, the process water contains the complete amount of the highly water-soluble dispersing agent. Due to the low residual dye concentration in the filtrate, it cannot be used in a subsequent dyeing process, if this process water is treated by hot nanofiltration. Thus, ultrafiltration was applied as an open loop treatment during the reductive post-treatment and subsequent rinsing. Thereby, three production steps could be substituted by just one. The concentrate was discharged as wastewater and replaced by fresh water. The permeate performance added up to 20 L/m^2·h.bar and a ceramic membrane with a cut-off MWCO = 20 kD was used (PIWATEX 2).

Recycling and optimization of biological degradability of dyeing water using ozonisation

For oxidative decolourisation of dye baths – originating from cold pad batch dyeing of cotton woven fabrics with reactive dyes – an ozone treatment was investigated in pilot-scale. For this purpose, the concentrated bath from the dyeing padder and the first washing baths are joined and decolourised in batches in the ozone reactor. The products of the oxidative degradation are colourless and show good biodegradability. Thus, the pre-treated process water is available for internal reuse (water cycle).

The conventional treatment of process water containing water-soluble dyes was carried out by precipitation and flocculation procedures, but these techniques lead to a shifting of the environmental burden. By omitting these procedures, the amount of precipitation sludge is decreased by 60 m^3/yr (Table 21.3). Subsequently, the amount of dye which is absorbed on sludge in the biological waste water treatment decreases by 205 kg/d.

Table 21.3 Changes in flows due to the replacement of the coagulation/flocculation by ozonation.

Parameter	Before conventional: Coagulation/flocculation	Afterwards innovative: Ozonation	Delta
Flocculation agent	17,500 L/yr	–	−17,500 L/yr
Sludge	65 m^3/yr	5 m^3/yr	−60 m^3/yr
Wastewater	5130 m^3/yr	–	−5130 m^3/yr
Ozonation: energy- and oxygen consumption	25,000 kWh	85,000 kWh 46,500 Nm^3 O_2	+60,000 kWh +46,500 Nm^3 O_2

The wastewater of 80% of the applied colour batches (medium to dark/black colours) of the first washing section and the wastewater of all applied colours of the dyeing padder are decolourised in the ozone reactor. Thus, the total amount of ozonated water (residual liquors + washing water of the first section of dark colouration) adds up to 14,800 L/d, whereas the amount of dye, which is decolourised by oxidation, is 205 kg (91% of the daily load). This results in an average dye concentration of 13.85 g/L. Due to the fact that neither additional salt originating from precipitation, nor excessive flocculation agents are contained in the decolourised process water, recycling of the treated water is possible.

The ozonisation is executed according to a procedure of the company ITT. Ozone is created from oxygen in the ozone generator. The product gas is passed into the reactor for batch treatment of the highly coloured wastewater. Residual ozone is destructed catalytically before the exhaust fumes are discharged. For the oxidation of 205 kg/d of dye, an ozone amount of 1.5 g O_3/m^3 in relation to Δ FZ (a weighted average of three SAC values for different wave lengths) is needed. When achieving a decolourizing of maximum Δ FZ = 2,500 m^{-1} and approximately 15 m^3/d, the ozone consumption adds up to around 40 kg O_3/d. This is equivalent to a gas volume of 280 m^3 O_2/d and an energy consumption of approximately 200 kWh/d (Constapel, 2009).

Water and heat recovery in industrial laundries

Water and energy consumption is also a major issue for industrial laundries. Fresh water (tap water or well water) has to be pre-treated (e.g. by filtration, softening) and heated up to process temperature (60–90°C) prior to use. Generated wastewater has to be treated according to local and national standards before usually being discharged to municipal sewers. As cost for freshwater intake and wastewater discharge differ from location to location, but may reach 3–5 €/m^3 (combined). Consequently, the use of recycled water is one obvious choice for large scale laundries. Because of elevated washing temperatures, as also known from textile finishing processes, energy recovery from spent process water is a viable option, too.

Used process water after washing is loaded with pollutants of various nature, like spent detergents, suspended solids, hydrocarbons and some heavy metals. Criteria for reuse of recycling water in the washing process are manifold since water quality directly influences dosage of detergents and cleaning results. Therefore, recycling water has to meet organic and inorganic parameters as well as basic hygiene. Water shall be free of odours and colour, low in microbiological counts (<100 CFU), free of pathogens, low in organic (COD < 200 mg/L) and inorganic pollution (Fe < 0.1 mg/L, Cu < 0.05 mg/L, Mn < 0.03 mg/L) at constant pH, hardness and conductivity, just to name a few. Some additional limits for indirect discharge to municipal sewers are given in the Table 21.4.

Table 21.4 List of discharge limits for indirect discharge of laundry effluents (Engelhart, 2011).

Parameter	Unit	Discharge limit
Max. Temperature	°C	35
pH	–	6–10
Sludge Volume	mL/L	10
Hydrocarbons	mg/L	20
AOX	mg/L	2
Sulfate	mg/L	600
P total	mg/L	50
Ag	mg/L	0.1
Al	mg/L	10
As	mg/L	0.1
Cd	mg/L	0.1
Cr total	mg/L	0.5
Cu	mg/L	0.5
Fe	mg/L	10
Hg	mg/L	0.05
Ni	mg/L	0.5
Pb	mg/L	0.5
Zn	mg/L	2

Cost considerations led to the development of a new recycling concept for a commercial laundry in Germany where washing of 4.1 million towel rolls and 1 million floor mats generates around 53,000 m^3 spent process water per year (Müller, 2010). A combination of biological treatment in a membrane bioreactor (MBR), followed by reverse osmosis (RO) and disinfection was chosen in order to meet the water quality demands (Figure 21.3). This process combination has become a standard configuration for modern water recycling plants, but had to be adapted according to the laundries specific needs. A two-stage vibrating sieve removes suspended solids and filaments before the MBR. Influent to the MBR is cooled down to around 30°C in a heat exchanger simultaneously heating up RO permeate before re-use.

Figure 21.3 Water recycling facility in a large scale laundry (courtesy of EnviroChemie GmbH).

Filtered biological effluent is acidified, conditioned with biocide for reduction of biofouling and afterwards collected and led to the RO plant. RO permeate is stored in an insulated storage tank and re-directed to washing facilities via the heat exchanger. Chlorine dioxide is used for disinfection. Disposal of suspended solids, surplus sludge and RO concentrates also had to be addressed. A disposal contractor takes care of sievings and surplus sludge. Concentrates of RO are further treated by selective ion exchange and then discharged to the municipal sewer according to discharge limits as given in Table 21.4.

For this kind of recycling technology in laundries, an overall water recovery of 70–80% was demonstrated. RO recovery rate strongly depends on feed water salinity, therefore optimisation of washing processes and use of cleaning agents was conducted in parallel. Heat recovery leads to savings of maximum 1.9 GWh/yr of energy. Energy recovery makes a major contribution on pay-back time. The recycling plant was placed directly near the washing facilities in a storage hall. In its true sense it is a production integrated facility.

Metal and ceramics industry

Water management at Volkswagen – a long tradition of water recycling

Sustainable water management and the responsible use of natural resources are an important part of environmental policy at Volkswagen. This is reflected by the foresight shown when the Wolfsburg factory was founded in 1938. Indeed, provision was already made at this early date for the use of rainwater and efficient recycling of purified industrial water. At the heart of the water recycling system is a storage reservoir, which collects industrial wastewater and rainwater. It is an important part of the flood control measures of Wolfsburg city. The storage system is comprised of four interconnected basins which provide a habitat for many species of fishes and birds. It holds around 1.8 million m^3 of water and covers an area of 307,405 m^2, equivalent to 60 football pitches (Figure 21.4).

Figure 21.4 Aerial view of the Volkswagen factory in Wolfsburg and its storage reservoir.

The separate sewer systems installed at the site, the independent drinking and industrial water networks, and Volkswagen's own biological treatment plant complete the recycling system. Drinking water is supplied by the company wells and water from the Harz region.

Cooling water, rainwater from the factory site and urban areas and purified wastewater from the company's biological treatment plant are channelled into the company's industrial water storage reservoir. After a five to seven day retention period the industrial water is filtered to the required quality in a treatment plant and fed into the industrial water supply system for reuse. It is mainly used to cooling processes and for all toilet flushing systems on the factory site. Around 23 million m^3 of industrial water are recycled each year. Statistically, the water circulates up to seven times before being diverted into the receiving watercourse, the Aller River.

Centralised and decentralised facilities for preliminary and secondary wastewater treatment are important components in the industrial water cycle (Figure 21.5). Depending on its constituents, industrial water from the production processes is chemically, physically or thermally pre-clarified at source in suitable pre-treatment facilities to selectively remove contaminants.

All wastewater from the body preparation and paint shops is treated in the central physico-chemical treatment plant at "Abwasserzentrum Mitte" (AZM). This includes the removal of heavy metals from the wastewater. Wastewater containing specific organic contaminants, such as landfill leachate, is treated in a membrane bioreactor. Oily wastewater is treated thermally at "Abwasserzentrum Ost" (AZO) in a vacuum evaporator using energy-efficient vapour compression. The high-purity distillate is recirculated into the industrial water cycle via "Abwasserzentrum Mitte" and "West", and any resulting oil concentrates are used in Volkswagen's own power station to generate electricity and heat. The pretreated wastewater is subjected to further purification together with the sanitary sewage in the company's own biological treatment facility at "Abwasserzentrum West" (AZW), after which it can once again be used in industrial processes.

The advantages of this water management concept are the security of the water supply in the long term, a significant reduction in environmental pollution, and the conservation of water resources. The industrial water cycle has made investment in the construction of two cooling towers, which would have to be fed with higher quality drinking water, unnecessary to date. The efficient use of resources has always been both an obligation and an objective. It represents an important economic and sustainable contribution to the protection of our environment.

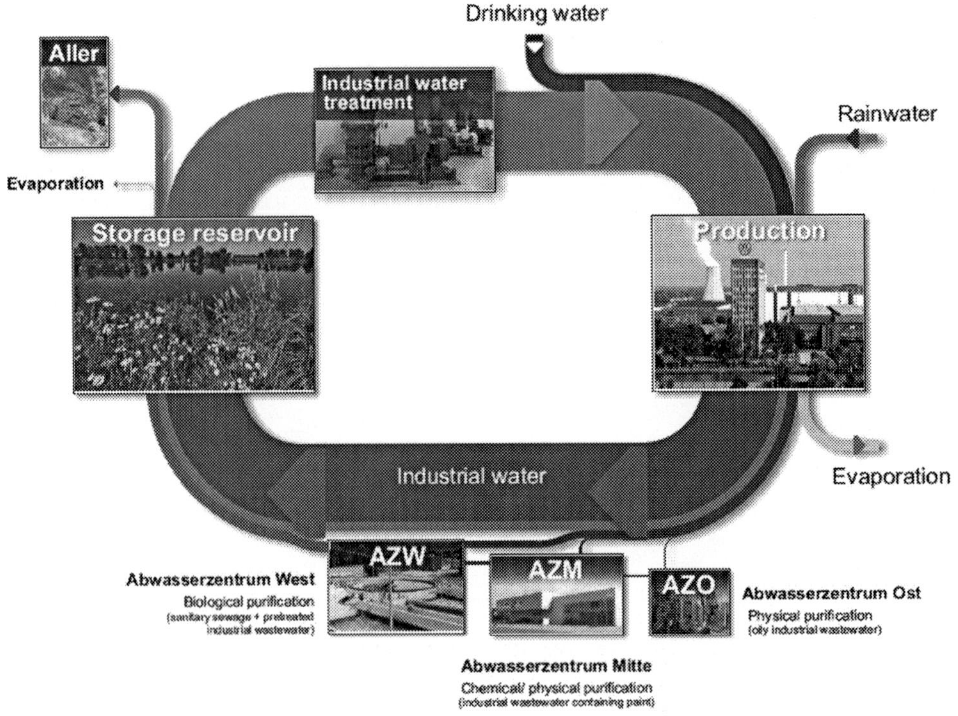

Figure 21.5 The industrial water cycle at the Volkswagen factory in Wolfsburg.

Recycling of rinsing water from chromium-plating of plastic surfaces

Chromium plating of plastic surfaces is carried out in a sequence of process steps interspersed with rinsing processes. A possible order of process steps for ABS-parts can be described as follows. First, the surface of the plastic part is roughened in an etching bath with chromic acid and sulphuric acid. After nucleatation with palladium, an electroless nickel layer is applied, followed by an electro-plated nickel layer. Next, a copper- and a further nickel layer are applied. As a finishing step, the ultimate chromium layer is applied by electro-plating with a thickness of about 1 µm only. There are different methods to achieve desired effects and for different surface materials. Thus, the facilities may contain up to 100 bath positions. When transferring the workpieces from the process baths to the following rinsing baths, carryover of the electrolyte solution occurs. Despite ideal dwell time for dripping over, this cannot be prevented.

In most of the larger electroplating shops, closed loops for processing of rinsing water are established. The so-called circulation water is treated via a bypass. The treatment processes are arranged to remove metal cations and various anions, for example chromic acid, and therefore consist of cation and anion exchange systems. Often, the pH value of rinsing water is a criterion to select the suitable treatment steps. Consequently, those cations and anions which were not adsorbed by the ion exchanger are distributed over all joined rinsing baths. Thus, perfluorinated surfactants, which are only applied in the etching bath and chromium electrolyte, are found in all electroplating baths.

Chromium (VI) salts (CrO_3, chromium trioxide, chromic acid anhydride), which are used in the etching bath and chromium electrolyte, are very toxic and additionally classified as carcinogenic. They show high water solubility and form yellow-red chromic acid. Chromium electrolytes contain up to 350 g/L chromium trioxide and 1% sulphuric acid (98%) in relation to chromium trioxide amount.

Due to reduction of surface tension and by forming a layer of foam during the electrolytic process, surfactants prevent the release of aerosols containing chromic acid. However, the hydrogen gas generated during electrolysis must be able to exhaust. Too much foam is an evidence for too much surfactant. Because of the aggressive effects of chromic acid, only chemically very stable surfactants can be applied. An oxidation resistant surfactant is added to both the etching bath and the electrolyte with concentrations of at least 80 mg/L perfluorooctanesulfonic acid (PFOS) and 500 to 800 mg/L fluorotelomer sulfonate, respectively. Until recently PFOS was used, but now a partly fluorinated surfactant (fluorotelomer sulfonate 6:2 FTS) is used as an alternative in plastic plating industries. Both types additionally contain perfluorobutanesulfonic acid (PFBS) as a solubilizing agent. Even several years after the change to alternative products, PFOS is found in internal wastewater systems. It is assumed that the reason may be the ability of PFOS to adsorb strongly on surfaces and then desorb gradually and thereby contaminate the water for years. It is assumed that especially when a thick layer of foam was formed, the carryover of surfactants increases as the foam only runs off insufficiently from the workpieces due to its high viscosity.

PFOS in particular matches criteria defining PBT substances: Persistence (P) and bioaccumulation (B). Further, a high toxicity (T) is suspected. Due to the combination of these properties, those substances are assumed to pose a high risk for humans and environment. The European regulation for chemicals REACH provides the basis for EU-wide restrictions for these products. The application is only allowed if no alternative products exist and the socio-economic advantages are proved without doubt. This means that occupational health and safety needs to be guaranteed, which is not necessarily given for alternative products. Due to the fact that PFOS is classified as a PBT substance, the EU restricts the use and placing on the market since 27th of June, 2008 (EU directive 2006/122EG). Among the acceptable purposes and specific exemptions from these restrictions are electro-plating applications, if for example, best available techniques for reduction of carryover are applied. Additionally, PFOS is listed in the Annex B of the UN Stockholm convention on persistent organic pollutant, which means the use is restricted on a global scale also.

To eliminate these surfactants during the treatment of circulation water, the adsorption of the surfactants of interest on ion exchange materials under practical conditions was studied in a research project. The adsorption rate amounted to >95% for a macroporous, weakly basic anion exchange resin. After regular regeneration, PFOS still remained on the ion exchange resin and can thereby effectively be prevented from entering the environment. The alternative surfactant 6:2 FTS and particularly PFBS which is a component in both PFOS- and 6:2 FTS-based products showed a lower adsorption rate under the same experimental conditions.

So far, ion exchange materials with adsorbed perfluorinated surfactants are treated with high temperature combustion to ensure a complete elimination, and no recycling approach is established. The elution of fluorinated surfactants from ion exchanger materials cannot be achieved by conventional aqueous solution. The weekly regeneration of the anion exchange resins – for example with sodium hydroxide – results in the recovery of chromic acid. In the process water, which should be treated by ion exchange techniques, the molar ratio of PFOS to chromium is approximately 1:500,000. Thus, many loading and regeneration cycles can be accomplished before the ion exchange material must be disposed and combusted. Polyfluorinated surfactants and short-chain perfluorosulfonic acids are not adsorbed completely by anion exchangers during the cycling of process water and moreover are desorbed when regenerating the ion exchange resins (EU project GALVAREC).

When an adsorption of per- and polyfluorinated surfactants on anion exchange materials is included in the process, this is so far positioned at the end of the internal wastewater treatment. However, investigations of the University of Wuppertal indicate that anion exchangers are effective only for a short period of time for a mixture of process water containing chromic acid and process water from other process steps. Particularly sulphate and anionic additives, as well as anionic flocculation polymers, result in a faster clogging of the exchange resins. These substances also promote caking of reverse osmosis membranes, which were studied with respect to retention of perfluorinated surfactants of internally treated process water. One of the rinsing baths after the chromium electrolyte, respectively etching bath, was found to be the best position for elimination of perfluorinated surfactants. The water cycle of rinsing water for those baths should be separated from other process water flows to allow optimum recycling of chromic acid and surfactants. As an additional effect, saving of water and the omission of subsequent rinsing steps can be achieved. Therefore, this procedure should be further developed for practical use. The results gathered in the project GALVAREC point out that a two-step process consisting of ion exchange or activated carbon and nanofiltration is a suitable option for treating the rinsing water after chromium plating. It is possible then to recycle chromic acid, nickel and copper cations as well as perfluorooctane sulfonic acid.

Recycling of dilution water in ceramics production

Production of ceramics makes use of water for preparation and dilution of stock and powdered raw material (clay, chalkstone, oxides etc.). The essential components of a glaze are silica, fluxes and alumina. Organic ingredients like gum may be added to prevent settling of suspended matter. White glaze is used as robust outer coating, for example, for tableware or sanitary ceramics. In modern production of ceramics liquid glazes – suspensions of various powdered minerals, and metal oxides – may be applied on the soft-fired piece by spraying it onto the surface with an airbrush or similar tool in a spray booth. Surplus glaze will drip from the surface of the piece and will be collected in the bottom section of the booth and flushed with additional water as waste glaze. Waste glaze usually is discharged together with various other production effluents to a treatment plant. Since effluents mostly contain inorganic suspended solids, coagulation and precipitation with final sedimentation is the choice for treatment prior to discharge. Sludge is dewatered and disposed of on landfills or used for production of lower quality products. Expensive raw material and water are lost this way.

In order to prevent the loss of raw material and water a filtration process may enable direct reuse of dilution water for preparation of white glaze but also recreates recycling glaze from waste glaze, ready for use in the production process

(Bohner, 2008). Filtration therefore allows for a completely closed circuit of water and glaze in side stream process, when waste glaze is collected separately in close vicinity to the spray booths.

Main challenges for filtration of glaze are the high apparent density which needs to be met for reuse together with the high abrasiveness of the inorganic suspension. Waste glaze usually shows an apparent density of 1100–1200 g/L, whereas ready-to-use recycling suspension needs to be concentrated up to around 1700 g/L. Increase in density is attended by exponential increase in shear viscosity as can be seen in Figure 21.6.

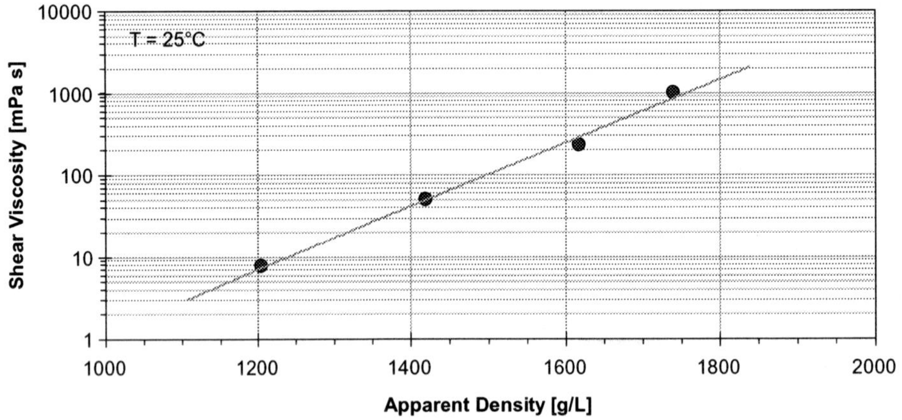

Figure 21.6 Exponential rise of shear viscosity with increasing apparent density for white glaze (for comparison only: shear viscosity of glycerine 1490 mPa · s at 20°C) (Engelhart, 2011).

Following in-depth pilot studies, a cross-flow microfiltration unit was introduced in the production of sanitary ceramics as demonstration plant already in 2007. Specially tempered membranes with a pore size of 0.1 µm are used for this process. Waste glaze is fed to a working tank and filtered in semi-batch mode at temperatures <60°C in order to lower viscosity. Microfiltration permeate is stored for direct reuse in glaze preparation. Apparent density of the glaze suspension is monitored continuously during each filtration run in a by-pass line using on-line measurement. When final density criterion is met, the working tank is emptied and recycling glaze is also re-directed to the preparation units.

Despite initial concerns regarding robustness, the first set of membranes still operate at full capacity in the filtration plant. Recycling of glaze is the main financial benefit of the process, leading to savings of 180,000–240,000 €/year because of lower input of raw material. Lower chemicals consumption in the final effluent treatment plant of the production facilities leads to further reduction in costs. Because of positive experience, new developments may deal with recycling of diluted cream of clay which is used as basis for the ceramic piece. Since different pieces require various characteristics of raw material and particle size distribution, filtration of cream of clay still needs to be investigated in more detail.

21.5 PERSPECTIVES

Since water recycling and reuse gain increasing significance in industrial processes, strengthening of wastewater treatment becomes obligatory to keep up production itself. This results in industrial wastewater treatment becoming integrative part of the production process and being no longer just additive technology.

In most industries, water is used for different purposes as resource, for transport, additive or for the staff. The types of use are differentiated between single, multiple and closed cycle, Figure 21.7 illustrates the most common measures for reduction of water demand and their potential within the next 10 years, according to a survey by Hillenbrand (2008) for different branches of industry. In almost every branch – according to the own valuation of the respondents – membrane technology will play an important role for wastewater treatment in the future.

Future development of water demand and its reduction for different industrial branches in Germany are also predicted by Hillenbrand (2008), as shown in Figure 21.8.

With the potential for a reduction of water demand of up to 50%, paper industry is heading towards a production without any external water demand. Nevertheless, in all considered branches of industry there will be a potential of at least another 20% reduction and more research and development in the future to provide adapted technologies.

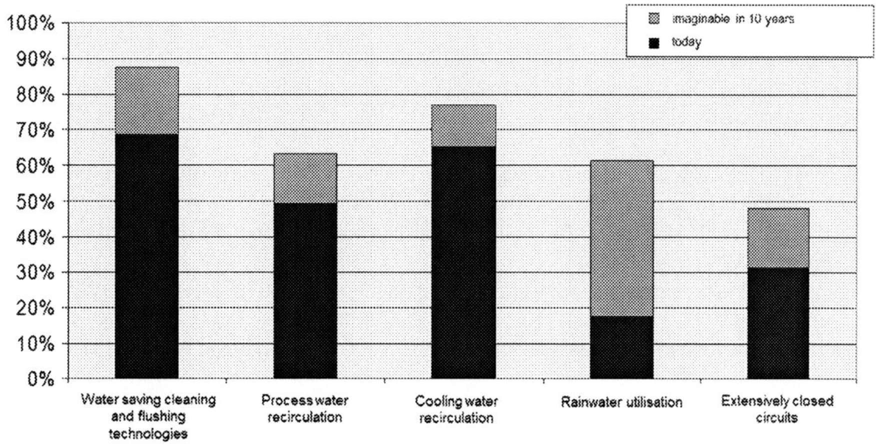

Figure 21.7 Results of a survey for measures and future trend in water saving techniques in Germany (Hillenbrand, 2008).

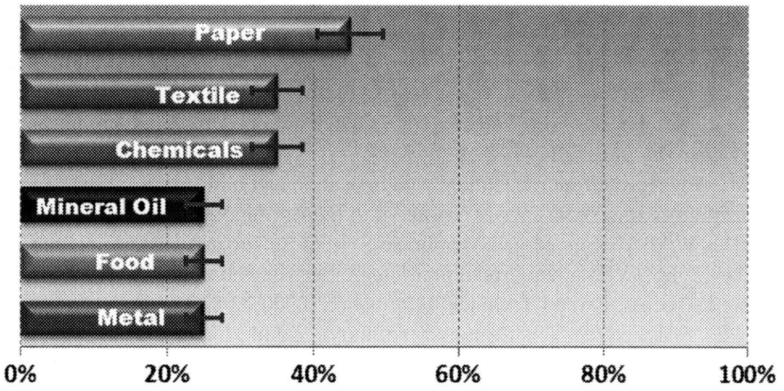

Figure 21.8 Prognosis of the reduction in water demand in German industry until 2020 (Hillenbrand, 2008).

21.6 CONLCLUSIONS AND LESSONS LEARNED

In the last decades, the trend in German industries shifted from end-of-pipe technologies towards in-plant measures and integrated solutions. Using adapted treatment and recycling technologies for a part of the water stream prior to dilution can minimize water demand and may result in the recovery of valuable product traces and reduction of demand for cleaning agents as well as additives. In this context, economic benefits and ecological improvements often coincide with each other.

It can be securely stated that there are technical possibilities for water recycling for almost any application. The decision in favour of water recycling depends on a number of factors and aspects. The most important boundaries to be considered are hygienic aspects, regional water supply, provision of energy and, in some cases, the availability of waste heat. All these factors have to be considered in an economic and ecologic evaluation, especially for food and drink industry. For this purpose, the application of a decision support system is necessary to determine the best solution for every case.

Considering water demand and water reuse, major advances have been made in Germany, but the production process itself often remains untouched. Even process integrated measures only resemble end-of-pipe technology at a smaller scale of the relevant production step. It is due to rethink or actually reinvent the product itself, its ingredients and the additives used, while simultaneously cutting emissions, by restructuring the production process. At best bearing in mind the whole life cycle of the product, since the responsibility for recycling is increasingly transferred to its manufacturer.

In addition to that, and due to growing social awareness, these measures improve the company's image and are increasingly acknowledged and rewarded by the consumer. In an even less altruistic approach, the obtained technological advance to competitors results in a strengthened market position, when it comes to foreseeable tightening of legal requirements. More and more companies even encourage the legislator to raise the minimum standards, promoting a trend towards sustainability among their competitors.

REFERENCES

Ahrens A. (1998). Abwasserreinigung und Vermeidung in der Brau-, Malz- und Getränkeindustrie. *Brauwelt*, Heft 15/16, S. 679.

Bohner C. (2008). Membrane technology for recycling and recovery of resources in industrial water and wastewater applications – from lab testings to production experiences. Proceedings of 10th World Filtration Congress, 2008/04/14–18, Leipzig, Volume **II**, pp. II-111–II-114.

Constapel M., Schellenträger M., Marzinkowski J. M. and Gäb S. (2009). Degradation of reactive dyes in wastewater from the textile industry by ozone: analysis of the products by accurate masses. *Water Research*, **43**, 733–743.

Diedrich K., Hamm U. and Knelissen J. H. (1997). Biologische Kreislaufwasserbehandlung in einer Papierfabrik mit geschlossenem Wasserkreislauf. *Das Papier*, **6A**, V153–V159.

DWA Deutsche Vereinigung für Wasserwirtschaft, Abwasser und Abfall e. V (2007). Merkblatt DWA-M 713 Abwasser aus der Zuckerindustrie, Hennef.

DWA Deutsche Vereinigung für Wasserwirtschaft, Abwasser und Abfall e. V (2011). Merkblatt DWA-M 731 Abwasser und Abfälle aus der Papierherstellung, Hennef.

Engelhart M. (2011). *Anwendungen von Membrantechnologie zum Recycling im Non-Food-Bereich.* DWA WasserWirtschafts-Kurse N/5 Behandlung von Industrie- und Gewerbeabwasser, 2–4 March, 2011, Kassel, Germany, pp. 386–416.

Habets L. H. A. and Knelissen J. H. (1997). In-line biological water regeneration in a zero discharge recycle paper mill. *Water Science and Technology*, **35**(2–3), 41–48.

Hillenbrand T. and Böhm E. (2008). Entwicklungstrends des industriellen Wassereinsatzes in Deutschland. *KA: Abwasser, Abfall*, **55**(8), 872–882.

Hoekstra A. Y. and Chapagain A. K. (2004). Water footprints of nations. Value of Water Research Report Series No. 16, UNESCO-IHE Delft.

Jung H. and Pauly D. (2011). Water in the Pulp and Paper Industry. In: *Treatise on Water Science*, P. Wilderer (ed.), Academic Press, Oxford, UK, vol. **4**, pp. 667–684.

Junk H.-H., Dörfer T., Strätz K. and Lausch B. (2008). Weltweit erste Membranbioreaktoranlage mit nachgeschalteter Umkehrosmose bei Köhler Pappen, *Allgemeine Papierrundschau*, **4**, 44–46.

Kaindl N. and Liechti P. A. (2008). Advanced effluent treatment with ozonation and biofiltration at the paper mill SCA-graphic laakirchen AG – Austria: design and operation experience. *Ozone: Science & Engineering*, **30**(4), 310–317.

Kraft A. and Mende U. (1997). Das WABAG Submerged Membrane System für Prozesswasserreinigung und -recyling am Beispiel der Mälzereiprozesswasser-Aufbereitung. Colloquium produktionsintegrierter Umweltschutz – Abwasserreinigung, Bremen, 1997.

Krones (2010). Krones PET-Asept L Process, Krones Technical Information 0 900 22 857 7, 06/10.

Kuhn M. (2004). Wirtschaftlich PET Flaschen reinigen, Getränke! Technologie und Marketing, 03/2004, pp. 14–16.

Marzinkowski J. M., Hildenbrand J., Medrow M., Rabhi A. and Papazachou S. (2008). Prozessintegrierter Umweltschutz durch Aufarbeitung von Abwasserströmen und Rückführung in den Prozess. Abschlussbericht zum 2. Teilvorhaben PIWATEX (2008), F 06 B 401 Tech-nische Informationsbibliothek (TIB) Hannover, Germany.

Marzinkowski J. M., Schmitz O. J. and Gäb S. (2011). Teilstrombehandlung chromathaltiger Abwässer; Schlussbericht zum Verbundvorhaben GALVAREC. Bergische Universität Wuppertal (2011); BMBF-Vorhaben KMU-innovativ, FKZ: 0330860 A, Projektträger Jülich GmbH.

Möbius C. H. (2010). *Waste Water of Pulp and Paper Industry*, 4th edn, Revision December 2010, Augsburg: http://www.cm-consult.de.

Müller C. (2010). Energie aus Industrieabwasser, wwt wasserwirtschaft wassertechnik, 09/2010, pp. 32–34.

Öller H.-J. and Offermanns U. (2002). Successful start-up of the world's 1st ozone-based effluent re-circulation system in a paper mill. In: Proc. Int. Conf. Advances in Ozone Science and Engineering: Environmental Processes and Technological Applications, N. J. D. Graham (ed.), 15–16 April, The Hong Kong Polytechnic University and The International Ozone Association, Hong Kong, pp. 365–372.

OXITEX. (2004). Innovatives Abwasserrecycling durch gezielte oxidative Entfärbung von Färbereiabwasser, gestützt durch den Nachweis der chemischen Abbauprodukte und deren Wirkung auf die Veredlungsprozesse. Abschlussbericht (Kurzfassung) zum BMBF-Forschungsvorhaben, Sächsisches Textilforschungsinstitut, Chemnitz (November 2004) FKZ: 0339937.

Presas T. (2011). Water – let's lead this issue, *ipw*, **6**, 3.

Simstich B. and Öller H.-J. (2010). Membrane technology for the future treatment of paper mill effluents: chances and challenges of further system closure. *Water Science & Technology*, **62**(9), 2190–2197.

PART VI

Environmental and recreational use of recycled water

Foreword

By The Editors

Historical development

In water-short regions, recycled water provides an extra water resource to meet environmental and recreational needs. Around the world there are numerous projects where recycled water has been used to restore river flows, create artificial lakes or wetlands and to restore natural wetlands. The most common types of environmental and recreational water reuse are restoration of wetlands, stream flow augmentation, replenishment of lakes and ponds and snowmaking. The use of recycled water in water bodies used for swimming, boating and fishing, for example, allowing a direct contact with public, requires stringent water quality with total disinfection.

As well as producing environmental benefits, many of the projects have also created recreational opportunities. Examples include:

- The Santee Lakes near San Diego in California were created in disused sand and gravel pits. Reclaimed water from the local treatment plant percolates through infiltration basins into the lakes which were opened for public recreation in 1965. Following a treatment plant upgrade in 1995 the water in the lakes is suitable for swimming.
- The City of Arcata in California has used recycled water to create a constructed wetland and wildlife habitat which was completed in 1986. The wetland supports research, educational and recreational activities.
- At San Luis Obispo in California, recycled water has been used to restore environmental flows and aquatic life in the San Luis Obispo Creek. The system was commissioned in 1994.
- In Australia, recycled water is used for water features and cleaning of the animal enclosures at the Taronga Park Zoo in Sydney.
- In Australia recycled water has been used for artificial snowmaking on the ski slopes at the Mount Buller and Mount Hotham ski resorts.

Although this type of reuse is practiced since years in the United States with specific regulations and well recognised benefits in the United States, Australia and some other countries, its development is only emerging in Europe. This could be considered paradoxical because the European Water Directive is probably the most adequate regulation aiming to preserve natural water resources and restore biodiversity. Only the new Spanish water reuse regulation includes a definition and water quality requirements for environmental and recreational uses of reclaimed water. Nevertheless, some recent successful projects would be able to unlock and promote these very important water reuse applications not only in Europe, but also in all countries, as shown by the successful case stories here below (see also Chapter 15 and Chapter 18).

Table 1 Highlights and lessons learned from the selected case studies of environmental and recreational use of recycled water.

Project /Location	Start-up/Capacity	Type of uses	Key Figures	Drivers and Opportunities	Benefits	Challenges	Keys to Success
Tokyo, Japan	1984 Nobidome Yosui/Tamagawa Jousui Project 248,200 m^3/d 1995 Jonan Three Rivers Project 450,000 m^3/d	Restoration of urban streams	*Nobidome Yosui/Tamagawa Jousui* Total volume of treated wastewater: 57.8 Mm^3/yr Total volume of recycled water: 9.1 Mm^3/yr: Capital costs: €43.2 million *Jonan Three Rivers Project* Total volume of treated wastewater: 138.6 Mm^3/yr Total volume of recycled water: 29.2 Mm^3/yr	Dried up steams due to urbanization and sewerage. People expectations for a better environment, particularly the presence of clean water.	Restoration of flowing water is improving living environment. Preservation of important historic treasure of Tokyo. Transported reclaimed water can be used in the emergence situation such as earthquake.	Good aesthetically pleasing quality of reclaimed water. Controlling the growth of attached algae by nutrients removal. Controlling the flow rate during storm event by continuous river flow monitoring as well as by weather forecast.	The restoring live stream project has received strong public support. The historic Nobidome Yosui/Tamagawa Jousui were constructed in 1654 and had served for supplying water for 300 years, and preservation of this historic stream generated a string public support. The technical improvement of the water reclamation process switching from chlorination to ozonation to reduce odour and colour had positive public acceptance.
Olympic Park, Beijing, China	2007 BeiXiao He MBR plant 60,000 m^3/d QingHe Tertiary Reclamation Plant 80,000 m^3/d	Creation of new water bodies and recreational areas	Total volume of treated wastewater: 18 Mm^3/yr (BeiXiaoHe); 25 Mm^3/yr (QingHe) Total volume recycled to Olympic Park's dragon shaped water system: 1.8 Mm^3/yr each (BeiXiaoHe and QingHe) Recycled water charge: 0.12 €/m^3	Increasing population and decreasing water availability. Water scarcity and frequent droughts. Political willingness to improve urban infrastructure.	Improvement of water supply security. Positive environmental protection effects. Improvement of leisure quality.	Water quality management in artificial water bodies. Maintain the water quality of artificial water bodies in a long-term through water ecology regulation. Establish stable ecosystems in the artificial water bodies with good self-purification.	Absolute will of all political institutions to organise successful green Olympics. Alleviating the contradiction between ecological environment of urban landscape and water shortage. Positive economic development and national pride. International co-operation.
Mexico City, Mexico	Since 1980s Treatment capacity: 103,680 m^3/d	Restoration of the Texcoco Lakes	Total volume of treated wastewater: 37.8 Mm^3/yr Total volume of recycled water: 37.8 Mm^3/yr Capital costs: US$1130 million Benefits: ~US$1.4 million per year in 1998 (avoidance of damage from urban floods, respiratory diseases and the closing of the airport) Water reuse standards: $<$100 fecal coliforms/100 mL	Urban Floods. Urban dust storms. Restoration of local biodiversity. Better urban planning.	Avoidance of damage to urban infrastructure. Reduction of respiratory illness. Creation of a recreational area. Halting the growth of slums in risky urban areas.	Providing suitable storage capacity in a soil with poor building capacity. Controlling soil erosion through reforestation and not afforestation. Restoring the original environmental conditions as much as possible. Encouraging communities to move to safer places to live.	Detailed planning of all the elements of the project. A sound technical background. Recognition that by reusing wastewater to restore the lake, dust storms, urban floods and soil erosion in Mexico City could be controlled. Utilization of a reuse project as a tool to educate citizens on Mexico City's history.

Environmental and recreational reuse case studies

Part 6 of this book "Milestones in Water Reuse: The Best Success Stories" presents three selected, little-known case studies, showing how water managers have used recycled water to meet the environmental and recreational water needs in both developed and emerging countries.

Chapter 22 describes the restoration of urban streams in the Tokyo Metropolitan area to restore natural environmental conditions, preserve valued historic features and provide public recreation activities. Use of advanced treatment to produce high quality recycled water has allowed a wide range of recreational activities.

Chapter 23 presents the creation of new water bodies and recreational areas in the Beijing Olympic Park. Advanced treatment has been used to produce high quality recycled water. In addition, there is a water circulation system incorporating wetlands to maintain water quality in the lakes. As well as providing recreational opportunities, the lakes are the centrepiece of public education programs about water management and recycling.

Chapter 24. In Mexico City wastewater has been used to restore the Texcoco Lakes which date from the Aztec period. The restoration has eliminated dust storms, controlled urban flooding and created an important site for the conservation of native bird species. A reforestation program around the shores of the lakes has helped to control erosion. The lakes are used as a valuable education resource for school students.

Keys to success

The common themes running through these case studies are that using recycled water for environmental purposes has restored river, lake and wetland environments, preserved culturally important features, enhanced aquatic life and wildlife habitat, and created new recreational opportunities for communities. Keys to success have included government and regulatory support, stakeholder involvement, production of good quality recycled water, detailed project planning, development of community support through education, and recognition of the economic and environmental benefits delivered by these projects.

The major highlights and lessons learned from the selected case studies are summarised in Table 1.

22 Restoration of environmental stream flows in megacities: the examples in the Tokyo metropolitan area

Kiyoaki Kitamura, Naoyuki Funamizu, Shinichiro Ohgaki and Kingo Saeki

CHAPTER HIGHLIGHTS

Several water reuse projects for environmental purposes within urbanized areas have been developed in Japan using reclaimed water for restoring live streams and lakes. This chapter presents two projects in Tokyo to revive the neglected and dried up urban streams, preserving thus the environment and an important historical treasure by introducing highly treated reclaimed water. Wastewater treatment processes included secondary treatment by activated sludge process, tertiary treatment by sand filtration, and advanced treatment by A_2O nutrients removal process followed by chlorine disinfection and UV or ozonation for colour and odour removal. The public accepted this water enthusiastically as a live stream and enjoying the water environment created.

KEYS TO SUCCESS

- The restoring live stream project has received strong public support and acceptance.
- The historic Nobidome Yosui/Tamagawa Jousui were constructed in 1654 and had served for supplying water for 300 years, and preservation of this historic stream generated a string public support.
- The technical improvement of the water reclamation process switching from chlorination to ozonation to reduce odour and colour of the reclaimed water had positive public acceptance.

KEY FIGURES

Nobidome Yosui/Tamagawa Jousui Project (start up 1984)

- Treatment capacity: 248,200 m^3/d
- Total annual volume of treated wastewater: 57.8 Mm^3/yr
- Total volume of recycled water: 9.1 Mm^3/yr
- Capital costs: JPY4,182 million (€43.2 million)

Jonan Three Rivers Project (start up 1995)

- Treatment capacity : 450,000 m^3/d
- Total annual volume of treated wastewater: 138.6 Mm^3/yr
- Total volume of recycled water: 29.2 Mm^3/yr

22.1 INTRODUCTION

Many of the urban streams in large cities in Japan lose their flow in dry weather due to an increase in the impermeable area within their basins as the result of urbanization. On the other hand, people in cities have come to put more and more emphasis on a better environment, particularly an environment with clean water. Reclaimed water is a stable water resource within cities that can be used to restore dry and neglected urban streams, as well as augment surface water bodies for various beneficial purposes.

Several water reuse projects for environmental purposes within urbanized areas have been developed in Japan using reclaimed water for restoring live streams and lakes. Consequently, in 2008, 61.3% of reclaimed water from publicly owned wastewater treatment and reclamation plants were used for environmental purpose, including restoration of stream flows. In this chapter, two important projects for restoring stream flow by reclaimed water in the Tokyo Metropolitan Areas are introduced and the reasons why these projects have succeeded are discussed.

In Tokyo, the coverage percentage of sewage service area reached 100% in 1994. Since spreading of sewer pipe installations in Tokyo and the urbanization caused the decrease in the flow rate of streams in urban area, the use of reclaimed water to sustain stream flows has become attractive. The Tamagawa-Joryu wastewater treatment plant (this plant is now designated as Tamagawa-Joryu Water Reclamation Center) started to supply reclaimed water to Tamagawa Jousui and Nobidome Yosui channels in 1984, and Senkawa Josui channel in 1989. In 1995, the Ochiai wastewater treatment plant (now Ochiai Water Reclamation Center) started to supply recycled water after tertiary treatment via sand filtration to Johnan three rivers as shown in Figure 22.1.

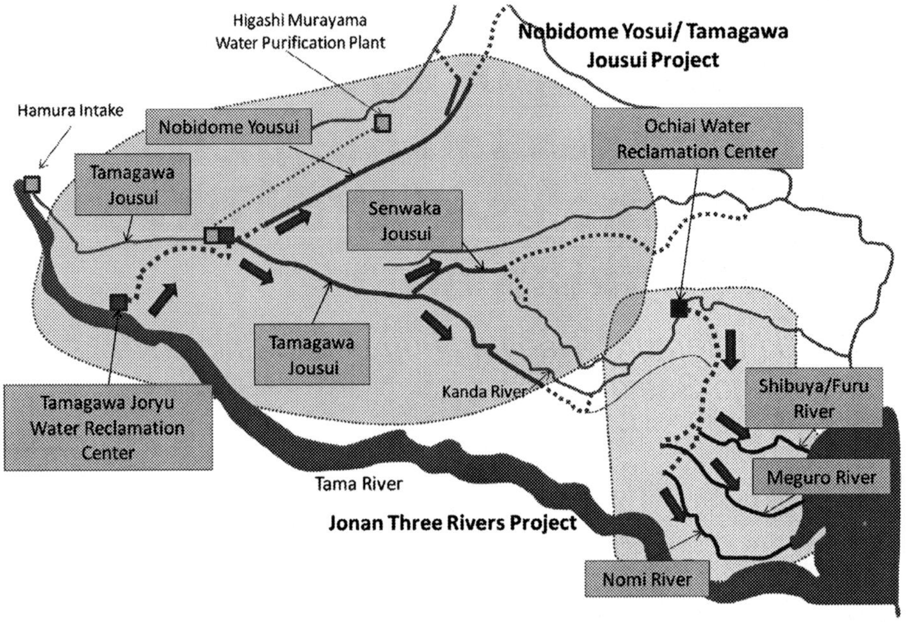

Figure 22.1 Schematics of the two projects for restoration of river flow in Tokyo.

22.2 NOBIDOME YOUSUI/TAMAGAWA JOUSUI PROJECT IN TOKYO METROPOLITAN AREA

Brief history of the project development

Tamagawa Jousui channels were constructed in 1654 and used to be the aqueducts for supplying potable water and irrigation water for Edo city, the old name for Tokyo. Figure 22.2, an old map, shows that Tamagawa Jousui was a gravity flow aqueduct starting from Hamura Intake, passing through the ridgeline of Musashino hill and terminating at Yotsuya Ohkido. The channel is 43 km long with 92 m difference in altitude. In 1898, the Yodobashi water purification plant started to purify the water transported by Tamagawa Jousui. The Tamagawa Jousui had served for supplying water from Tama River for about 300 years. Water flow in Tamagawa Jousui had diverted to 33 channels which included Nobidome Yousui and Senkawa Jousui.

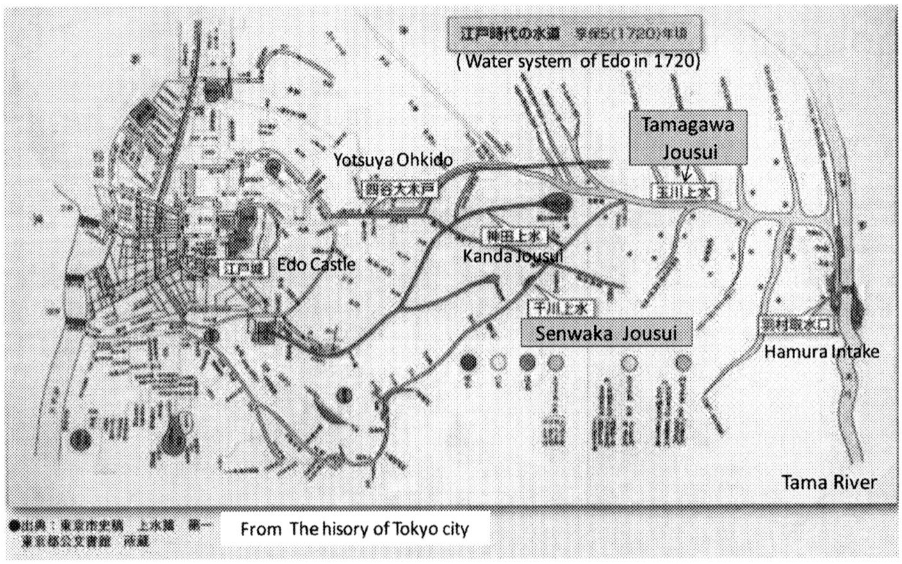

Figure 22.2 The old map showing an aqueduct system of Edo (Tokyo) metropolitan area in 1720.

However, the water supply system in Tokyo has had to be changed to deal with the rapid growth of Tokyo Metropolitan area and an old water purification plant in Tokyo central area, which was supplied by Tamagawa Jousui, was replaced by a larger water purification plant, Higashimurayama Water Purification Plant in 1965 (see Figure 22.1). Since then, Tamagawa Jousui had lost its flow (Ohgaki & Sato, 1991). Because Nobidome had started to receive the wastewater from the catchment area instead of clean water from Tama River, the channel changed to be an open sewer. The construction of sewage works around Nobidome catchment area had made Nobidome dried up in 1977 (Eto,1986).

Nobidome Yosui/Tamagawa Jousui Water Reuse Project

The Tokyo Metropolitan Government started in 1974 to study the feasibility to restore these stream channels for meeting the demand of citizens to revive the ornamental environment with water courses. After various potential water resources were examined for restoring these channels, the reclaimed water from the Tamagawa-Jouryu wastewater treatment plant (Tamagawa Water Reclamation Center) has been introduced into Nobidome in 1984. In 1986, Tamagawa Jousui channel came to life again by reclaimed water from the treatment plant.

The reclaimed water is pumped 8.7 km (24 m height) from the sewage treatment plant to a pressure release tank by a pipeline (diameter 700 mm). The reclaimed water through the pressure release tank is introduced to Nobidome and Tamagawa Jousui channels with design flow rates of 20,000 m^3/d (0.23 m^3/s) and 23,200 m^3/d (0.29 m^3/s), respectively. The capital construction costs up to 1984 for restoring Nobidome were 4182 million yen (Approx. U.S.$53.6 million), including construction of pipes 3339 million yen ($42.8 million), sand filters 590 million yen ($7.56 million), repairs of channel 147 million yen ($1.88 million) and other costs of 106 million yen ($1.36 million). The additional construction costs were 1795 million yen, equivalent to $23 million (Kanai & Yokota, 1987).

Nobidome channel has about 9.6 km length with 1.5–2 m width and 10–30 cm water depth in Tokyo Metropolitan administrative district. The flow enters Saitama Prefecture and joins a stream. Tamagawa-jousui channel has about 18 km length with 5 m (average) width and 10–30 cm water depth. After 18 km run, the flow of Tamagawa-jousui channel is introduced into Kanda River which flows through the Tokyo central district (see Figure 22.1). Figure 22.3 shows the situation of Tamagawa-Jousui channel before and after the project.

Figure 22.3 Views of the Tamagawa Jousui Stream (a) before the water reuse project and (b) after the project.

Treatment train for water reclamation

Current treatment train in 2011

The treatment train for water reclamation is shown in Figure 22.4. The secondary effluent from the Tamagawa-Jouryu Water Reclamation Center is filtered by upflow sand filters after the in-line coagulation with PAC, and then the filtered water is treated by ozonation. This tertiary treatment train, with a design capacity of 43,200 m^3/d, ensures further treatment of odour, colour, suspended particles and pathogens. The Tamagawa-Jouryu treatment plant has two series of treatment trains for secondary treatment, a conventional activated sludge process with a design capacity of 150,000 m^3/d and an A_2O process with 98,200 m^3/d of capacity.

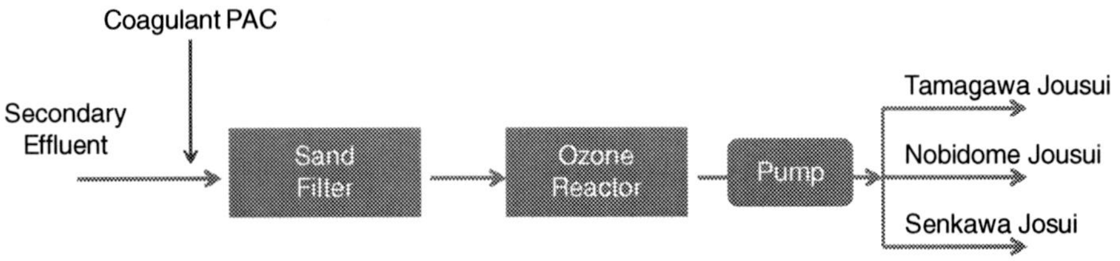

Figure 22.4 Treatment train for restoration of river flow in the project.

The water quality of raw sewage, secondary effluent and reclaimed water by tertiary treatment is shown in Table 22.1. The water quality criteria in Tokyo are given in Table 22.2.

Table 22.1 Water quality[1] of raw water, secondary effluent and reclaimed water.

Parameter	Raw wastewater	Secondary effluent	Reclaimed water[2]
BOD, mg/L	220	2	2
Suspended solids, mg/L	160	1	1
Total nitrogen, mg/L	28	9.7	9.4
Ammonia, NH_4-N, mg/L	16	<0.1	<0.1
Nitrate, NO_3-N, mg/L	–	8.2	8.8
Total phosphorus, mg/L	3.8	1.2	0.4
Total coliform count/mL	–	120	1

[1]average value in 2009.
[2]before chlorination.

Table 22.2 Reclaimed water quality criteria in Tokyo.

Parameter	Sustaining river flow, landscape irrigation	Recreational uses	Cleaning, cooling, fire fighting	Toilet flushing	Garden watering
Total coliform counts	1000/100 mL	50/100 mL	10/mL	–	50/100 mL
E.coli	–	–		Not detected	–
Residual chlorine, mg/L	–	–	Trace amount	Trace amount	>0.4
Colour	<40	<10	Not unpleasant	Not unpleasant	Not unpleasant
Turbidity	<10	<5	Not unpleasant	Not unpleasant	Not unpleasant
BOD, mg/L	<10	<3	–	–	<20
Odour	Not unpleasant	Not unpleasant	Not unpleasant	Not unpleasant	Not unpleasant
pH	5.8–8.6	5.8–8.6	5.8–8.6	5.8–8.6	5.8–8.6

Progress in the treatment train

At the beginning of the project, sodium hypochlorite was dosed before and after filtration for bacterial growth control in the filters and for disinfection. The dose rate was controlled to maintain the residual chlorine of reclaimed water to be less than 0.1 mg/L for preservation of fishes and aquatic organisms in the water channels.

According to the results of questionnaire survey on the impression of Tamagawa Jousui to the people in the neighbourhood within 200 m distance from the channel (Takiguchi *et al.* 1988), half of the people in the residential area from the discharge point to about 3 km down stream felt the odour from the channel, while only 10% of people in the area of lower reaches was aware of odour emissions. Although the odour was not strongly unpleasant, the introduction of ozone treatment into the reclamation processes was installed.

Since the secondary treatment at the Tamagawa Jouryu treatment plant was a conventional activated sludge process, at the beginning of the project no advanced treatment of nitrogen and phosphorous was done. The heavy growth of attached algae due to high concentration of phosphorus and nitrogen has not been observed because the channels are shadowed by woods and a soil deposit covers the greater part of the channel bed (Kawahara, 1987; Tsukui, 1989). At present, the A_2O nutrients removal process is installed in the Tamagawa Jouryu Water Reclamation Center to improve secondary effluent quality in terms of lowering nitrogen and phosphorous concentrations.

22.3 RESTORTION OF JONAN THREE RIVERS IN TOKYO METROPOLITAN AREA

Brief description of the project

This environmental water reuse project has started in 1995 to restore Jonan three rivers: Shibuya/Furu River, Meguro River and Nomi River. The project supplies 17,460 m^3/d of reclaimed water to Shibuya/Furu River, 27,410 m^3/d to Meguro River and 35,230 m^3/d to Nomi River. The overview of the project area is shown in Figure 22.5.

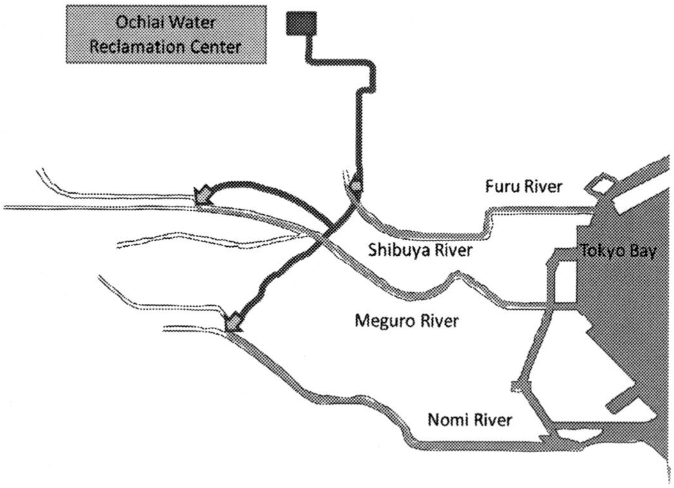

Figure 22.5 The Jonan Three River restoration project.

The basins of three rivers are highly populated area as shown in Figure 22.6. Before starting the project, these three rivers did not have enough flow as shown in Figure 22.7 (left). Peoples living in this area requested to Tokyo Metropolitan Government to have their living environment improved with the river water front. The current situation of these rivers is shown in Figure 22.7 (right). And, the transported reclaimed water is planned to use in the emergency situation such as earthquake.

Figure 22.6 Bird view of three rivers' basin in Tokyo Metropolitan area.

Figure 22.7 Views of the Meguro River before the water reuse project and after the project.

Treatment trains for water recycling at Ochiai Water Reclamation Center

A conventional activated sludge system is used as a secondary treatment in the Ochiai Water Reclamation Center and the secondary effluent is filtered by a sand filtration system as tertiary treatment. Ochiai Water Reclamation Center started its operation in 1964 with a design capacity of 90,000 m^3/d. The current average daily wastewater flow is 380,000 m^3/d. The water quality is summarized in Table 22.3.

Table 22.3 Water quality[1] of raw wastewater, secondary effluent and reclaimed water (South series).

Parameter	Raw wastewater	Secondary effluent	Reclaimed water[2]
BOD, mg/L	180	2	1
Suspended solids, mg/L	150	3	<1
Total nitrogen, mg/L	33.1	12.3	11.4
Ammonia, NH_4-N, mg/L	19.9	0.3	0.1
Nitrate, NO_3-N, mg/L	0.2	10.4	10.8
Total phosphorus, mg/L	3.6	1.6	1.5
Orthophosphates, PO_4-P, mg/L	1.6	1.4	1.4
Total coliform count/mL	–	100	3

[1]average values in 2009.
[2]before chlorination.

Challenges of operation and water transportation

The reclaimed water is transported to the three rivers by water transporting pipe as shown in Figure 22.8.

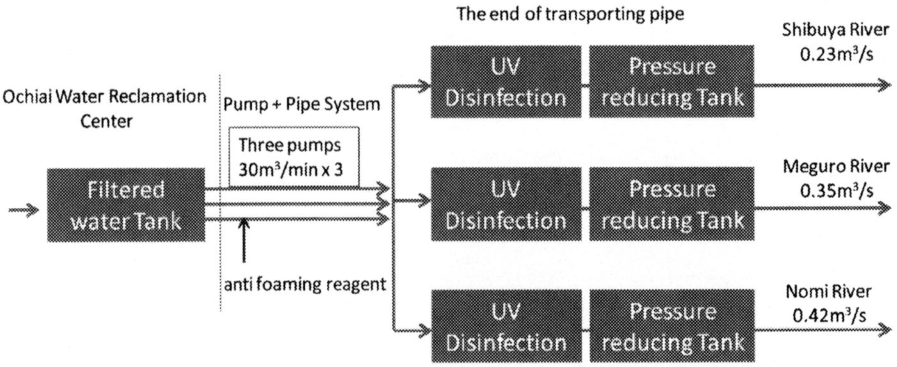

Figure 22.8 Schematics of the reclaimed water transportation system from the Ochiai water reclamation center.

The total length of the pipeline is approximately 17.5 km, made from earthquake-proof pipes and joints. At the end of the transporting pipe, reclaimed water is disinfected again by UV irradiation before reaching the pressure reducing tank. These tanks are equipped for releasing reclaimed water to river in natural (unpressured) way. The pumps are set and operated from the Ochiai Water Reclamation Center. During storm events, the Centre control the flow rate based on the information from weather forecast and continuous river flow monitoring system.

The pipe is installed deep in the underground to pass the subways and other underground infrastructures as shown in Figure 22.9. It took five years to install these pipes.

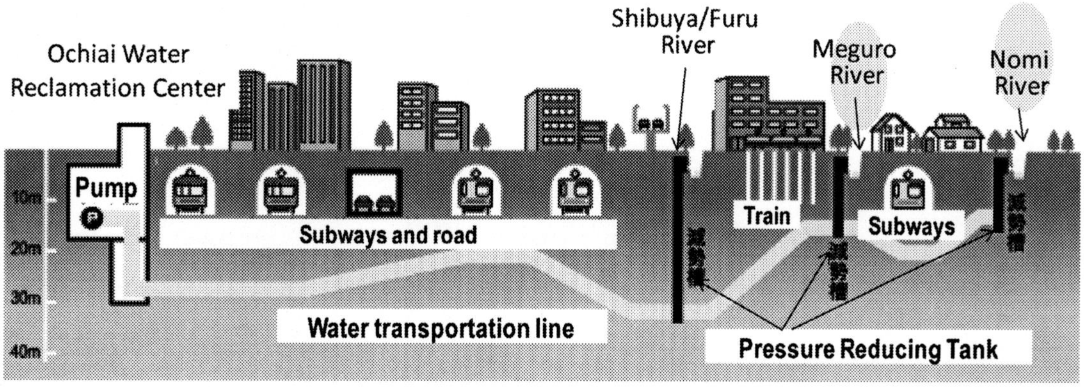

Figure 22.9 Location of transportation pipes set in underground to pass the subways.

22.4 PUBLIC ACCEPTANCE

Nobidome Yousui/Tamagawa Jousui Project

The project location lies on a low ridge and hilly area in Tokyo, which has no big stream. This geographical condition was the reason why the route for an aqueduct from west to east in Tokyo was selected on the location of Tamagawa Jousui and why the development of Nobidome had been required for irrigation in the 17th Century. The loss of flow in these streamsmeant the loss of scenery with clean water flow from the area because of no alternative streams for the people in the neighbourhood. Since the public had enjoyed the scenery with clean water flow of Nobidome and Tamagawa Jousui over 300 years until 1965, the restoration of flowing water would have improved the living environment of the surrounding area even if reclaimed water is used for the river flow (Ohgaki & Sato, 1991).

According to the questionnaire survey on Nobidome channel by Tokyo Metropolitan Government (1986), 95% of people in the neighbourhood 1 km distance from the channel knew the restoration project. According to the survey results, the public seems to accept the reclaimed water reuse for Nobidome. Newspapers and other media have provided favourable reports on the project. As mentioned above, the geographical and historical background helps the public acceptance and support.

In 2003, Tamagawa Jousui is nominated as a national historic site. Since the Tamagawa Jousui contributed to the development of Edo and Tokyo metropolitan area for more than 300 years, it has been recognized as one of the important historic infrastructures to be preserved. Corresponding to this nomination, Tokyo metropolitan government has started the new water reuse project for maintaining and making best use of Tamagawa Jousui since 2009. The new project is aiming at preserving the Tamagawa Jousui as an important treasure of Tokyo metropolitan and to inherit "water and green area" to next generation. Figure 22.10 shows current three major activities of this project: (a) bank protection and maintenance works, (b) maintenance of "Koganei cherry blossom street" and (c) facilities for easy access to historic site.

Jonan three rivers project

Historically in Tokyo, rivers in Edo and Tokyo were used for transportation and the river water front was a place where people enjoy fishing, dabbling in water, and viewing cherry blossom. Now lots of people are enjoying river water front again. Figure 22.11 shows the map of Meguro river basin for walkers prepared by Tokyo Metropolitan government. This map gives visitors the information not only on Meguro River but also sightseeing spots and historical sites. Similar maps are also available for Sibuya/Furu River and Nobu River. These maps are widely distributed and these rivers have become a kind of visitor's spots not only for local people but also for tourists.

Figure 22.10 Views of the Tamagawa Jousui current project.(a) Bank protection and maintenance, (b) Koganei cherry blossom street and (c) Facilities for easy access to historic spot.

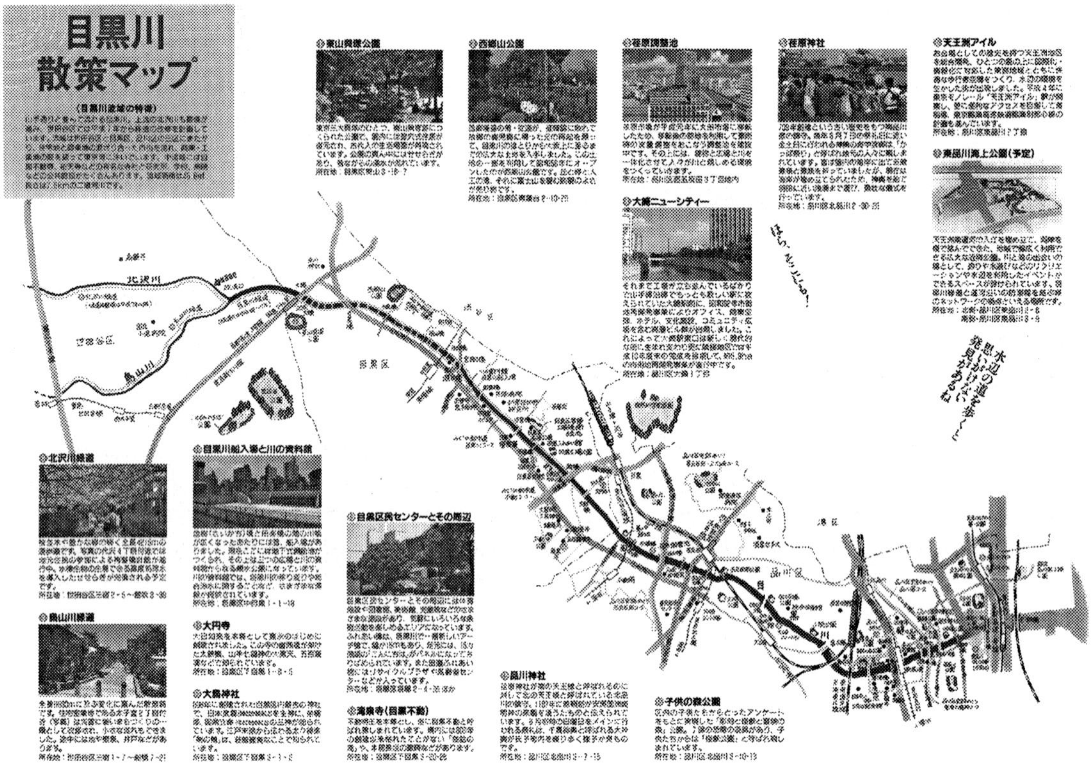

Figure 22.11 Walk map of the Meguro River published by Tokyo Metropolitan Government.

22.5 MAIN KEYS TO SUCCESS

The restoring the Nobidome and Tamagawa Jousui channels and Jonan three rivers in Tokyo Metropolitan area have already received strong public support and acceptance. The expanding of urbanization area in Tokyo has resulted in the decrease of flow rate in the live streams in Tokyo because of sewage works construction and increase of pavement area. Okubo et al. (1987) showed the correlation between increase of sewage works service area and decrease of stream flow rate of 23 streams in Tokyo during the period from 1977 to 1983. Therefore, the restoration project of these channels by means of water reuse has given significant benefits to all the population in Tokyo.

As to the Nobidome and Tamagawa Jousui, geographical and historical backgrounds also help the public support. The Tamagawa Jousui had served for supplying water to Edo or Tokyo from Tama River for about 300 years. The water flow of Tamagawa Jousui had diverted to 33 channels which included Nobidome Yousui and Senkawa Jousui and the public had enjoyed the scenery with clean water flow of Nobidome and Tamagawa Jousui over 300 years until 1965. Nomination as a national historic site is also promoting this project. People are thinking now that the Tamagawa Jousui has more than 350 years history and national treasure even if the reclaimed water is flowing in it now.

Technical improvement of the reclamation process from chlorination to ozonation is another key point in this project enabling to reduce the odour and colour of the reclaimed water in response to people's complaints.

Jonan three rivers project were supported by people who would like to improve their living environment, especially life with river water front. Now, peoples are able to enjoy walking along with rivers, viewing cherry blossoms in the banks of three rivers and fishing in three rivers. People with this life style are main supporters of the projects. There are three technically important points in this project to obtain the public acceptance: UV disinfection, pressure reducing tanks, and river flow rate control system. The UV disinfection unit, located at the end of water transportation pipe along with chlorine disinfection unit at the treatment plant enable to consistently control health risk from pathogens without damaging river aquatic ecosystem. Pressure reducing tanks are used for smooth and natural release of reclaimed water to rivers. During storm events, the water recycling centre controls the flow rate based on the information from weather forecast and continuous river flow monitoring system in order to prevent flooding from these rivers.

The following Table 22.4 summarizes the main drivers, benefits and challenges of these two water reuse projects for environmental restoration in Tokyo.

Table 22.4 Summary of lessons learned.

Drivers/Opportunities	Benefits	Challenges
Urbanization and sewer pipe system had made many of urban streams dried up. People in urban area have come to put more and more emphasis on a better environment, particularly an environment with clean water.	Restoration of flowing water will improve living environment for the people. The project will preserve the important historic treasure of Tokyo. Transported reclaimed water can be used in the emergence situation such as earthquake.	Good aesthetically pleasing quality of reclaimed water with respect to colour and odour. Controlling the growth of attached algae by nutrients removal via A_2O process. Controlling the flow rate during storm event by continuous river flow monitoring as well as by weather forecast.

REFERENCES AND FURTHER READING

Eto J. (1986). Restoration of Nobidome channel using reclaimed wastewater. *Kankyo Gijutsu*, **15**(6), 456–460. (In Japanese).

Kawahara H. (1987). Rehabilitation of the amenity of a small urban channel by the introduction of secondary treated wastewater. *Journal of Japan Sewage Works Association*, **24**(279), 45–53. (In Japanese).

Ohgaki S. and Sato K. (1991) Use of reclaimed wastewater for ornamental and recreational purposes. *Water Science and Technology*, **23**, 2109–2117.

Okubo T., Nakasugi O. and Ohgaki S. (1987). Impact of urbanization on flowrate and water quality of streams in Tokyo, *21st Annual Conference of Japan Society on Water Pollution Research*, 167–168. (In Japanese).

Takiguchi H., Matsuo T. and Hanaki K. (1988). A Questionnaire survey on the reuse of treated sewage in Tamagawa channel. *Environmental System Research*, **16**, 74–79. (In Japanese).

Tokyo Metropolitan Government (1986). Questionnaire survey on Nobidome channel.

Tsukui T. (1989). A revival of the rivers – reviving clean streams in the Tamagawajousui Waterway and other rivers by using treated sewage water. *Japan Journal of Water Pollution Research*, **12**(7), 417–420. (In Japanese).

View of a cherry blossom in the landscape area of the Tamagawa Jousui project.

23 Creation of a new recreational water environment: the Beijing Olympic Park

Ying-Xue Sun, Hong-Ying Hu, Josef Lahnsteiner, Yu Bai, Yi-Ping Gan and Ferdinand Klegraf

CHAPTER HIGHLIGHTS

The Beijing Olympic Park water reuse case is outstanding as it combines advanced reclamation technology and eco-biological recycling systems in order to create a "leisure paradise", that is a great recreational water environment, which substantially improves the population's quality of life.

KEYS TO SUCCESS

- Absolute will of all political institutions to organise successful green Olympics
- To alleviate the contradiction between ecological environment of urban landscape and water shortage
- Positive economic development and national pride
- International co-operations

KEY FIGURES

- Treatment capacity: 60,000 m^3/d (BeiXiaoHe MBR) and 80,000 m^3/d QingHe Tertiary Reclamation Plant
- Total annual volume of treated wastewater: 18 Mm^3/yr (BeiXiaoHe); 25 Mm^3/yr (QingHe)
- Total volume recycled to Olympic Park's dragon shaped water system: 1.8 Mm^3/yr each (BeiXiaoHe and QingHe)
- Recycled water tariff: 0.12 €/m^3
- Water reuse standards: national standard

23.1 INTRODUCTION

China's Beijing Olympic Park was the site of major activity during the 29th Olympic and Paralympic Games in 2008. While medals were being awarded to top competitors, a landscape garden was also constructed as a leisure area that is emotively Chinese and embodies both the long cultural history of Beijing and the characteristics of a harmonious coexistence between humankind and nature. The finishing touch to the park was provided by a newly constructed dragon-shaped water system, replenished by reclaimed water which forms the heart of the entire project.

Due to its geographical location and climate, Beijing, the capital of China, is now in a state of emergency with an availability of water resources per capita below 300 m^3/cap, which is far less than the water scarcity index of 1000 m^3/cap. The drought, which started in 1999, was the longest and most severe since the new Chinese state was founded.

Because of frequent droughts and water shortages, urban water management is facing numerous major obstacles. With economic development, urban expansion and improved living standards, the contradiction between water shortage and urban liveable environment quality is becoming increasingly acute.

Main drivers for water reuse

The main drivers of water reuse in Beijing are: (1) increasing population and decreasing water availability (2) water scarcity and frequent droughts (3) the willingness of the municipality to improve urban infrastructure for the Olympic Games in the summer of 2008.

Brief project development history

Water is an important part of an urban environment, as it not only enhances environmental quality, enriches open spaces and raises living standards, but also increases the atmospheric humidity in residential areas, reduces airborne dust and improves the microclimate. Moreover, it can also create an atmosphere that gives people a sense of returning to nature, providing them with spiritual pleasure and thus playing an important role in raising the quality of life of urban populations.

However, in the years up to 2005, only 60–80 million cubic meters of water per year (Mm^3/yr) were supplied to the urban rivers and lakes in the urban area of Beijing. This amount was only just sufficient to maintain the basic environmental function of these important waters, while replenished water for the rivers and lakes constituted 2% of the city's total water consumption.

In the meantime, in order to effectively alleviate Beijing's water shortages, reclaimed water has become the city's second-largest water source. In 2010, reclaimed water volumes reached 680 Mm^3, thus accounting for 19% of the city's water supply, a figure that was 11% in 2005. Reclaimed water is now used extensively in industrial manufacturing, agricultural irrigation, urban greening, rivers and lakes, the environment and other fields.

The dragon-shaped water system in the Beijing Olympic Park (Figure 23.1) was built in the period after 2005 as a present for the 29th Olympic Games and generally stores a water volume of 55–60 Mm^3. Furthermore, this water must be replenished constantly in order to compensate for the losses caused by evaporation, seepage and the transpiration of aquatic plants, which cannot be met by natural precipitation. Consequently, it was decided to supply reclaimed water from two Wastewater Treatment Plants in Beijing to bring the water system up to its designed water level in 2007. This strategy fulfilled the concept of sustainable development and provided a solution to the water shortage problem.

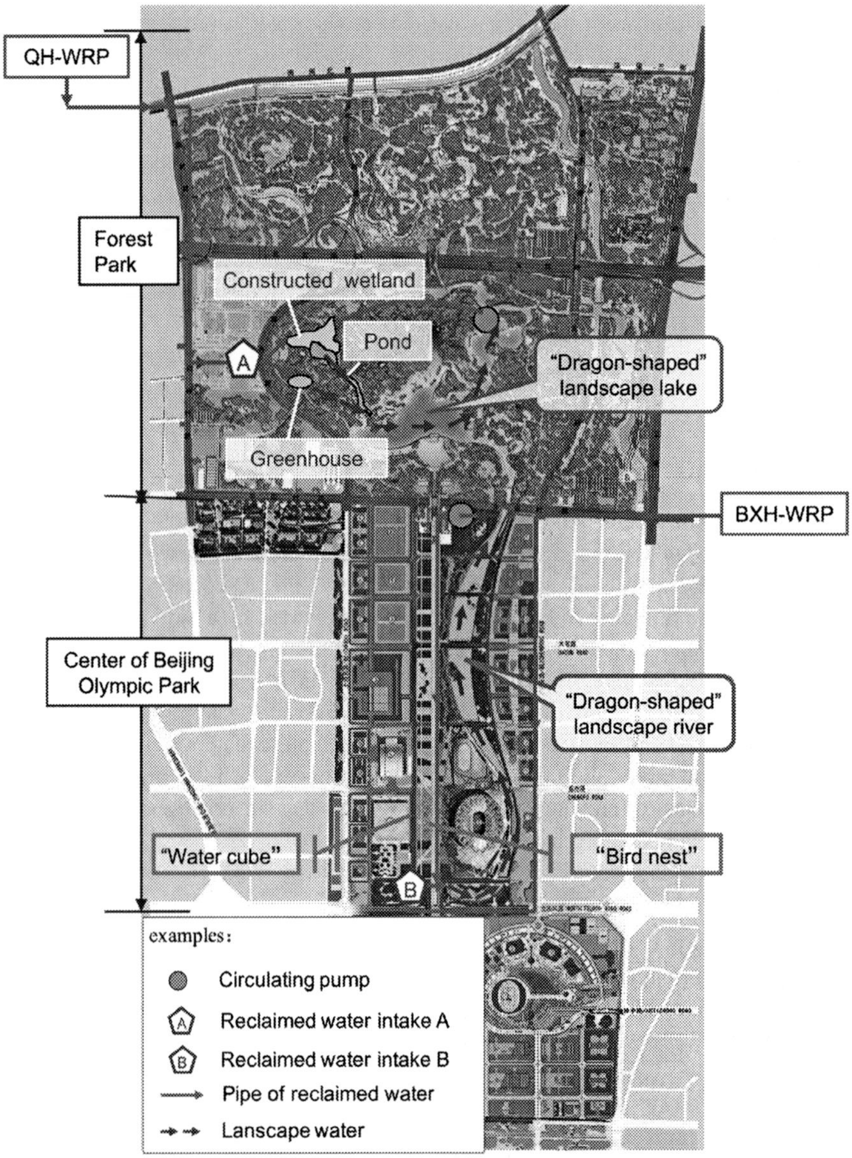

Figure 23.1 View of the dragon-shaped water system in the Beijing Olympic Park. (*Source*: Beijing Tsinghua Urban Planning & Design Institute).

Project objectives and financial incentives

The aim of the realisation of the dragon-shaped water system, which is replenished with reclaimed water, was to provide an ecologically acceptable urban environment in condition of water scarcity. Accordingly, in line with the principle of cost compensation and fair return, the Beijing municipal government fixed the price of the reclaimed water at a level that was considerably lower than that for potable water. This reflected the quality of the reclaimed water and was also intended to encourage its use for various purposes.

In addition, the dragon-shaped water system was to offer a leisure venue for Beijing's population and serve to improve the city's ecological balance and perfect its urban infrastructure. The project was also envisaged as a demonstration of how a new recreational water environment can be created using reclaimed water.

23.2 TECHNICAL CHALLENGES OF WATER QUALITY CONTROL

Description of the recreational dragon-shaped water system

The constructed recreational water system in the shape of a dragon winds through the full length of the Beijing Olympic Park, extending from the "head" in the Beijing Olympic Forest Park (in the north of the Park) to the "tail" near the Bird's Nest stadium (Figure 23.1). Overall, the *dragon-shaped water system* consists of a landscaped lake and river, which are separated into two blocks with differing flow patterns and ecological functions.

The landscaped lake of the dragon-shaped water system is a newly constructed artificial area of water that forms the "head" of the dragon and is of standard shallow design. It has a surface area of approximately 20 ha, a depth of 1–3 m and a water volume of 400,000 m^3. Like most very shallow lakes (Ruley & Rusch, 2004), the landscaped lake exhibits a polymictic nature with weak, daily stratification and mixing patterns induced mainly by surface winds. There are recreational features such as a musical fountain and boating and fishing facillities (Figure 23.2). In addition, a constructed wetland and an aquatic plant pond have been added to the landscaped lake, which thus constitute an integrated, scenic water system. The only supplementary water source is provided by the Beijing QingHe Water Reclamation Plant (QH-WRP): the reclaimed water intake can be seen in Figure 23.1 (reclaimed water intake A). Moreover, the water in the scenic lake is recycled around the constructed wetland and the aquatic plant pond. In addition to beautifying the environment, these features provide an animal habitat, purify the water from the scenic lake, and also gradually furnish the water of the artificial landscape with natural ecological functions.

Figure 23.2 Views of the landscaped lake of the dragon-shaped water system.

The landscaped river of the dragon-shaped water system meanders through the center of the Beijing Olympic Park twisting and turning in the shape of the dragon's "body". The shallow, fish-filled scenic river has a total length of 2.7 km, a depth of 0.4–1.5 m, and a width of 20–125 m. Every day more than 3000 m^3 of water is needed to compensate for losses due to evaporation and ground seepage. The reclaimed water provided by the Beijing BeiXiaoHe Water Reclamation Plant (BXH-WRP) is put into the system at the "dragon's tail" near the Bird's Nest stadium (reclaimed water intake B, Figure 23.1).

The landscaped river has been designed as an area of non-contact recreational water and is divided into nine zones. A large musical fountain is found in zones 6 and 7 with a total of 2008 nozzles, which correspond with the date of the 2008 Olympic Games. Furthermore, different kinds of aquatic and ornamental plants have been planted in the other zones, changing in a curve along the line of the river banks. The image of the landscaped river blends in with the National Stadium "Bird's Nest" and thus presents a harmonious and beautiful composition (Figure 23.3).

Figure 23.3 Views of the landscaped river of the dragon-shaped water system.

Advanced membrane treatment for water recycling

The supplementary water for the recreational water system in the Beijing Olympic Park consists of reclaimed water produced by two water recycling facilities: the QingHe Water Reclamation Plant (QH-WRP; Figure 23.4a) and the BeiXiaoHe Water Reclamation Plant (BXH-WRP; Figure 23.4b) operated by Beijing Drainage Group Co., Ltd.

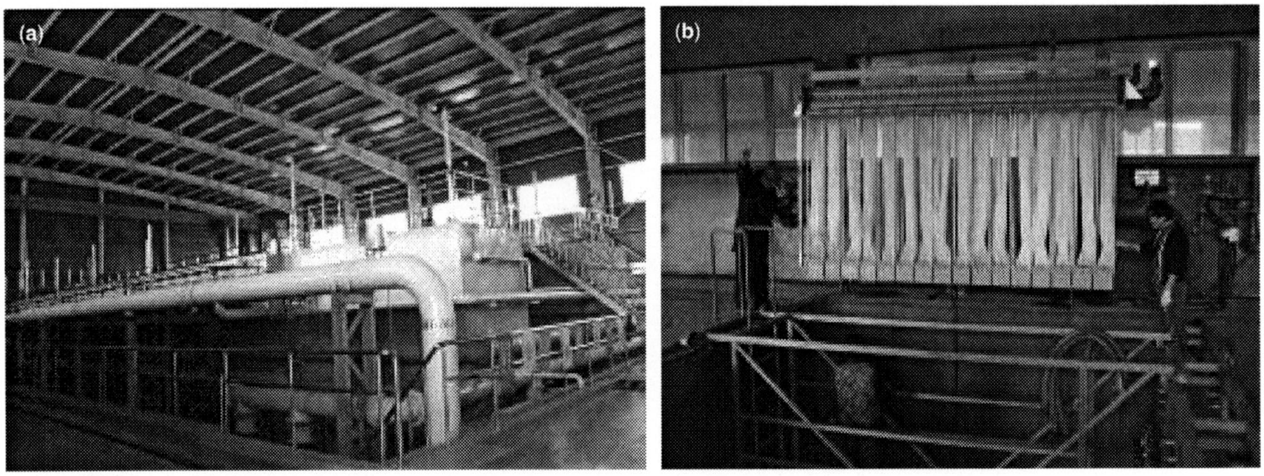

Figure 23.4 Views of the QingHe water reclamation plant (a) and the BeiXiaoHe water reclamation plant (b). (a) Tertiary filtration – ZeeWeed 1000 membrane tanks (b) MBR – installation of a Memcor Memjet module.

QH-WRP, a water reclamation plant with a membrane treatment (capacity 80,000 m^3/d) has been built close to the QingHe Wastewater Treatment Plant (QH-WWTP). QH-WWTP occupies an area of 30.1 ha. It has a capacity of 400,000 m^3/d and currently serves a total area of 15,942 ha and a population of about 814,000 people. The secondary effluents are produced by an anaerobic/anoxic/oxic (AAO) process in QH-WWTP, and then pumped to QH-WRP.

Recycled water is produced in the QH-WRP by an advanced process that includes screening of the secondary effluent by a 300 μm strainer, sequential ultrafiltration (Figure 23.4a, ZeeWeed 1000; six membrane trains with a net production capacity of 13,333 m^3/d each), ozone oxidation, and chlorine disinfection. Of the total capacity of 80,000 m^3/d, 60,000 m^3/d is used as a water supply for landscaping the Beijing Olympic Park, and about 3200 m^3/d of this volume is further treated in a constructed wetland and an indoor eco-system (greenhouse). Both effluents, from constructed wetland and indoor ecosystem, are subsequently polished in an aquatic plant pond before being fed to the dragon shaped water system (into the landscape lake "head" of the dragon). This is necessary in order to maintain good water quality through further nutrient removal (N, P) to minimise the risk of algae bloom in the lake. The remaining 20,000 m^3/d are supplied to the municipality for road washing, toilet flushing and other purposes. The water quality design parameters of the QH-WRP can be seen together with those of the BXH-WRP in Table 23.1.

Table 23.1 Water quality target concentrations and typical outflow values of the QH-WRP and BXH-WRP.

		QH-WRP			BXH-WRP				
		Design		Actual	Design			Actual	
			Outflow	Outflow		Outflow		Outflow	
Parameter	Unit	Inflow	Tert. UF	Tert. UF	Inflow	MBR	RO	MBR	RO
BOD_5	mg/L	20	6	<2	280	6	4	<2	<2
COD	mg/L	60	30	40–50	550	30	20	15–20	<10
TSS	mg/L	30	1.5	–	340	2	–	–	–
NH_3–N	mg/L	15	1.5[1]	–	45	1.5[1]	–	<1	–
N_{total}	mg/L	30	15	<20	65	15	2	<15	<0.5
P_{Total}	mg/L	1	0.3	<0.5	10	0.3	0.2	<0.3	<0.01
Turbidity	NTU	–	0.5	–	–	0.5	0.5	<0.1	–
Coliforms	CFU/L	10^4	3	–	–	3	3	–	–

[1]At a design temperature of 13°C.

BXH-WRP is the expansion and modernisation of the BeiXiaoHe treatment plant with the installation of an MBR system alongside an AAO process (Figure 23.5). The wastewater is pumped into a joint pre-treatment stage consisting of coarse and fine screening, and aerated sand removal process. Following this joint pre-treatment, 60,000 m^3/d are conducted to the MBR process and 40,000 m^3/d to the AAO process. At present, the AAO process is also being upgraded to an MBR and will be commissioned in 2012.

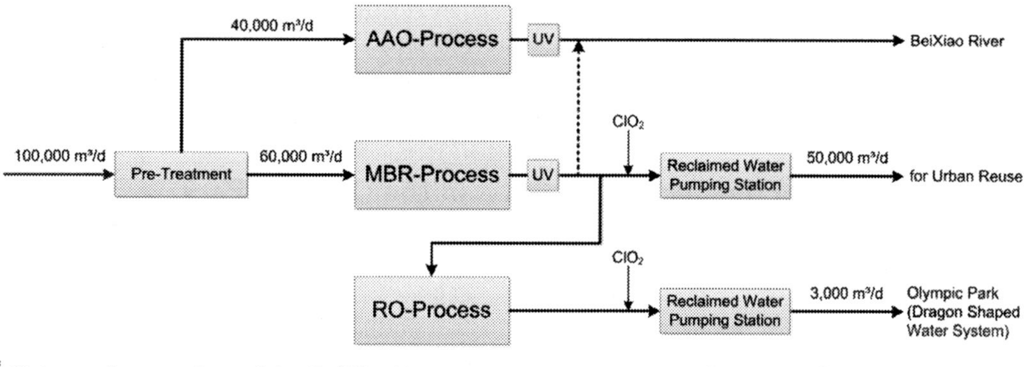

Figure 23.5 Schematic overview of the BeiXiaoHe wastewater treatment and water reclamation plant.

The wastewater treated in both the AAO and MBR processes is disinfected by means of UV-light radiation and subsequently the cleaned water from the MBR plant is subjected to additional chlorine dioxide disinfection. The feed water for the reverse osmosis (RO) process is taken from the MBR permeate flow, after UV disinfection. The RO permeate is disinfected with chlorine dioxide.

The MBR system has a hydraulic output of 60,000 m^3/d of which 50,000 m^3/d is mainly reused for irrigation in the Olympic Games Village and in the Datun Development Zone which has a total area of 23.8 km^2. A partial flow of 4000 m^3/d (10,000 m^3/d after extension of the RO) is further processed by means of reverse osmosis technology. The permeate (3000 m^3/d) is then separately drawn off and reused for toilet flushing and landscape irrigation in some of the

Olympic venues, respectively, and as replenishment water for the dragon-shaped water system (for the landscape river, i.e. "tail" and "body" of the dragon) in the Beijing Olympic Park. As mentioned previously, the water quality design parameters of the BeiXiaoHe water reclamation plant is summarized in Table 23.1.

Wastewater pre-treatment of the BXH-WRP consists of screening, grit and grease removal, and fine sieving with a 1 mm mesh. The pre-treated wastewater is fed to the four-line activated sludge stage of the MBR process. Internal re-circulations are installed between both nitrification and de-nitrification and between de-nitrification and the anaerobic stage, which is controlled on-line with the inflow. The design parameters for the activated sludge tanks are a maximum load of 60,000 m^3/d, sludge age of 17 d and an average MLSS concentration of 9 g/L. In spite of biological phosphorous removal, simultaneous precipitation utilising poly-aluminium chloride (PACl) is necessary in order to achieve the required phosphorous outlet concentration. Instead of the homogenised tanks using submersible motor agitators employed in biological phosphorous elimination, denitrification and the vario-stage (denitrification and nitrification volumes can be varied according to specific process requirements), aerated nitrification is carried out in circulation tanks. These are fed by the backflow from the membrane operating system (MOS), which results in a uniform sludge concentration. Fine bubble aeration of the activated sludge tanks and, if required, the vario-zone, takes place using disc aerators. Process air input in the nitrification phase is controlled automatically via dissolved oxygen measurement. Hollow fiber ultra-filtration membranes from the Memcor Company are installed in eight external, downstream filtration cells (Figure 23.4).

A total membrane area of 182,886 m^2 has been installed, which is divided into eight cells. Each of these is fitted with 38 racks, which in turn contain 16 membrane modules. A membrane module has a membrane area of 37.6 m^2. Under maximum hydraulic load, the net design flux amounts to 19.5 L/m^2·h at the reference wastewater temperature of 18°C.

The filtration cells are fed with 500% of the average through-flow by pumps with a result of 400% overflows. This overflow is collected in a joint outflow channel and is then fed back to nitrification. Permeate extraction takes place using pumps in a suction mode. The clean water volumes required for membrane cleaning are stored in a reservoir.

The MBR system was commissioned in winter 2007/2008 and since the spring of 2008 has been operating in line with the design requirements (Lahnsteiner *et al.* 2011). In addition to water quality characteristics, starting from fine sieving and ending with the outflow prior to UV light disinfection, process energy consumption was consistently approximately 0.73 kWh/m^3. The plant has now been in continuous operation for more than four years and is currently running at 75–92% of the hydraulic capacity.

Due to enhanced treatment in both the MBR (high performance biological system) and the RO (desalination, removal of residual nutrients, organics and micro-pollutants), the quality of the reclaimed water from the BXH process is superior to the quality of the reclaimed water produced by tertiary treatment in the QingHe Water Reclamation Plant.

As can be seen in Table 23.1, the total phosphorous and total nitrogen content in the RO permeate of the BXH-WRP amount to less than 0.01 mgP/L and less than 0.5 mgN/L, respectively. The concentrations in the QingHe tertiary effluent are far higher (UF permeate: $P_{tot} < 0.5$ mg/L, $N_{Total} < 20$ mg/L). Therefore, the BXH reclaimed water can be reused directly in the landscape river of the dragon shaped water system (the "tail" and "body" of the dragon), while as previously mentioned, the QingHe reclaimed water has to be further treated (in the artificial wetland, indoor eco-system and aquatic plant pond) before it can be supplied to the landscape lake.

Eco-biological purifying and recycling systems in the Olympic Park

The supplementary volume of reclaimed water required for the landscaped lake of around 3200 m^3/d from QH-WRP are further polished by a constructed wetland, aquatic plant pond and UV disinfection unit to control the risk of algae bloom and maintain the water quality of the landscaped lake. In addition, various ecological communities and complex ecosystems were constructed inside the lake to improve its self-purification, stabilizing and buffering capabilities. A schematic diagram of the water recycling process is shown in Figure 23.6.

The indoor eco-system for the feeding water purification (Figure 23.7) was installed in the Greenhouse located to the west of the park on 2200 m^2 of land (see Figure 23.1). In addition, the functions of the Greenhouse incorporate the high-efficiency collection and utilization of solar energy, intelligent management and real-time monitoring in order to ensure water quality under all types of weather conditions.

Natural eco-system elements have also been included in the Greenhouse project including landscaped rockeries and islands, a variety of aquatic plants and various kinds of aquatic animals. All these elements combine to form an integrated aquatic ecosystem. Moreover, the Greenhouse provides an opportunity for visitors to learn how the aquatic eco-system works and understand the significance of environmental protection. This project combines scientific knowledge, scenic interest and educational functions. The project is considered as an engineering model and provides an example for the establishment of modern water purification and maintenance systems in urban landscape water bodies.

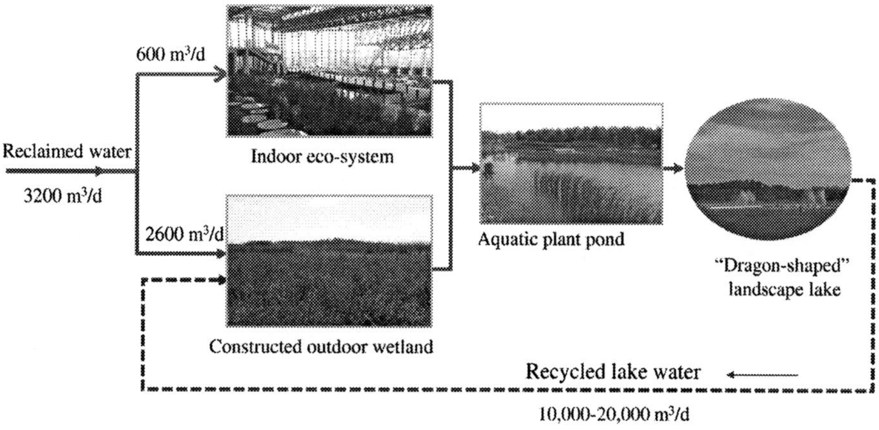

Figure 23.6 Water purification and maintenance for the landscaped lake.

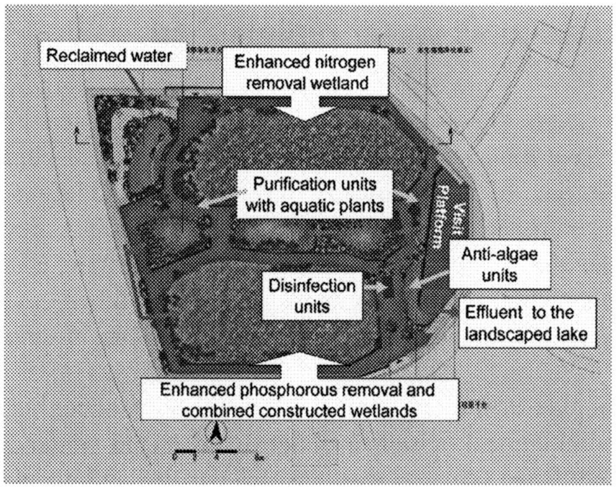

Figure 23.7 The indoor water purification eco-system and a view inside the Greenhouse.

The constructed wetland is located in the northwest of the Beijing Olympic Forest Park (see Figure 23.1) with six units and total area of 4.15 ha. The constructed wetland has two functions. The first function is the further purification of around 2600 m^3/d of reclaimed water performed by the first wetland unit, and the second function (performed by the other five units) is to purify the recycled water (10,000–20,000 m^3/d) from the landscaped lake to remove the pollutants produced by decaying organics, dust and rainfall. The effluent from the constructed wetland flows through the aquatic plant pond before finally entering the landscaped lake as feed water (see Figure 23.6). In addition, the constant circulation of lake water between the wetland and lake accelerates the water fluidity of the landscape water and to a certain degree inhibits algae overgrowth. The design of the constructed wetland discards fixed rectangular basins, channels and regular geometry and instead adopts bank-line processing according to the geomorphic conditions of the Beijing Olympic Park in order to avoid artificiality and be as natural as possible. It is thus integrated into the scenery of the park as a whole (Figure 23.8).

The landscaped river (the "tail" of the dragon) is fed by around 2000–3000 m^3/d from BXH-WRP. The water flow pattern imitates that of natural rivers and is so designed that the water level elevation at the south end is higher than that at the north end with an altitude difference of 2.6 m. The water is then recycled to the south end through pipes by circulating pumps. Water purification and maintenance for the landscaped river is achieved mainly by means of aquatic plant purification coupled with the hydraulic circulation. The aquatic plants can absorb, transform, accumulate inorganic nutrients, and can also inhibit the growth of algae through allele-chemicals (Li & Hu, 2005). In practice, the hydraulic circulation of the landscaped river can inhibit algae overgrowth to a certain degree. Moreover, in order to attain sustainable water purification and build various attractive natural waterscapes, a benign aquatic ecosystem in line with the law of natural succession has been constructed in the landscaped river by means of breeding aquatic animals and inoculating and immobilizing microorganisms together with aquatic plants.

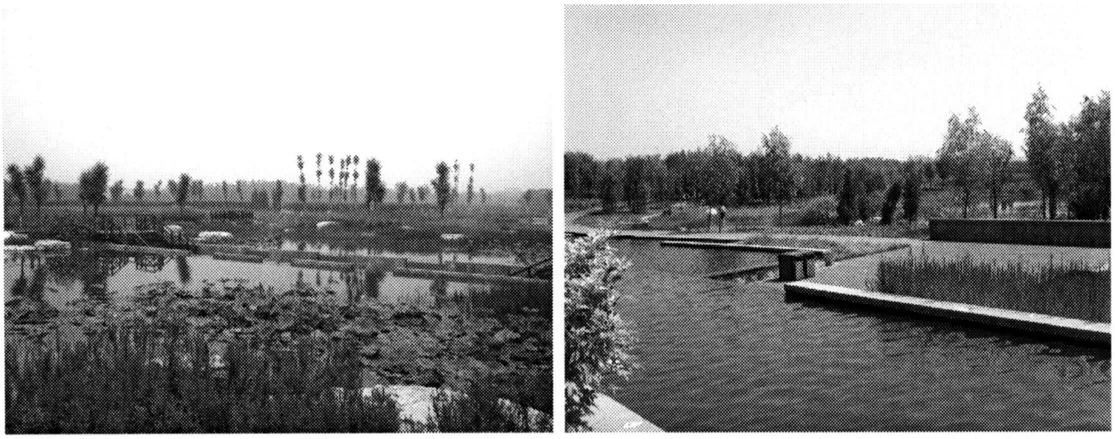

Figure 23.8 View of the constructed outdoor wetland.

Main operational challenges relating to artificial scenic water replenished by reclaimed water

Compared with large natural lakes or rivers, artificial scenic waters are usually shallow, closed or semi-enclosed and stagnant (Han & Park, 1999; Suh *et al.* 2004), and difficult to have robust ecosystems with functions of self-purification. Therefore, artificial scenic waters are always subject to the risk of water quality deterioration and algae bloom, especially when supplemented with reclaimed water containing nitrogen and phosphorus (Sun *et al.* 2009). Because of the rich nutrients, the micro-algae will grow excessively under the higher light intensity and temperature in summer. Thus, the control of the concentrations of nitrogen, phosphorus and other nutrients in the reclaimed water is a major requirement. Water purification and the maintenance of scenic water in the case of water bloom in summer are also important.

The dragon-shaped water system in the Beijing Olympic Park replenished by reclaimed water has been functioning for over three years. The reclaimed water is further purified by the eco-biological system including constructed wetland, indoor eco-system and aquatic plant ponds. The water quality of the landscaped lake and river are also purified and maintained by the eco-biological purifying and recycling systems. After the treatment, the water quality was at a relatively stable level, especially in terms of total nitrogen and phosphorus (Figure 23.9). In general, good water quality water recycling and urban landscape benefits have been attained.

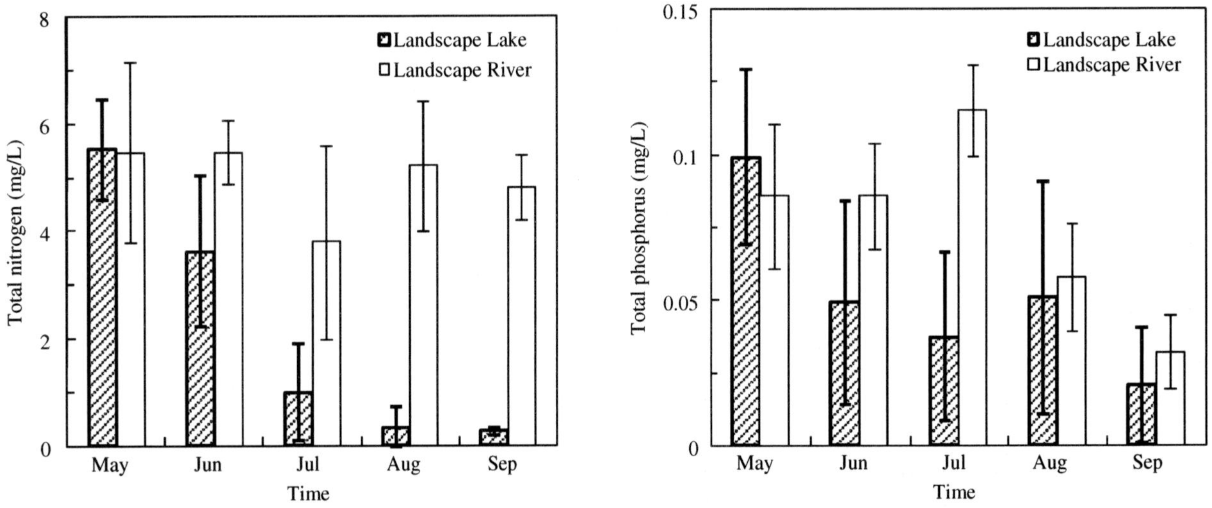

Figure 23.9 Variations of total nitrogen and total phosphorus in the scenic water after eco-biological purifying and recycling systems in 2008–2010.

However, there are still few challenges, such for example how to maintain the water quality in the long-term through water ecology regulation, and how to switch from conventional manual maintenance to a natural self-maintained ecosystem. These issues represent important problems that directly influence the prospects for the utilization of reclaimed water in urban landscapes.

23.3 WATER REUSE ECONOMICS AND BENEFITS

The Beijing municipal government has taken positive steps to encourage the use of reclaimed water that including the fixing of the price of reclaimed water at 0.12 €/m^3 (1.0 Yuan/m^3) and the exemption of the recycled water users from water and sewage treatment fees. Moreover, the price of potable tap water has been raised to promote reclaimed water utilization. At present, the price of tap water in Beijing used for industrial and commercial enterprises and urban services is 0.75 €/m^3 (6.21 Yuan/m^3). The total reclaimed water volume recycled to the Olympic Park's dragon shaped water system is 1.8 Mm^3/yr each, so about 113,800 €/yr (9.38 million Yuan/yr) can be saved comparing to tap water use.

Moreover, the dragon-shaped water system in the Beijing Olympic Park is of major significance for the improvement in the ecological environment in the northern part of Beijing, better city infrastructure and enhanced urban water environment quality. The park has become a leisure area for local residents and the inhabitants of Beijing, the graceful and clear landscaped water body providing a large waterside area that has brought considerable environmental benefits.

23.4 HUMAN DIMENSION OF WATER REUSE

At present, the Chinese government is developing a policy of environmental protection, sustainable development and education to improve the whole nation's consciousness of the importance of environmental protection as a strategic task. Education regarding the closing loops of water resources has been introduced on a wide scale in kindergartens, primary and middle schools and universities. The population is also educated about the utilization of reclaimed water by means of public service advertising and television commercials. The reclaimed water ecological treatment system in the Beijing Olympic Forest Park Greenhouse is a model project that integrates science, enjoyment and entertainment. The Greenhouse uses advanced reclaimed water purification technology and education displays on environmental protection. In fact, it has become a place for visitors to learn about and understand ecosystem functions and take part in the environmental protection (Figure 23.10).

Figure 23.10 Views of Greenhouse visits.

A survey regarding "the support rate of Beijing residents for the use reclaimed water" (investigation by the students of Tsinghua University) has shown that landscaped water features with reclaimed water as a supplementary water source are gradually gaining public acceptance and that the support rate among Beijing residents has reached over 95%. However, the utilization of reclaimed water in other fields is not seen in such a positive light and the support rate is much lower in the areas of car washes, agricultural irrigation and residential toilet flushing. The major issue that still makes residents uncertain is the securing of the reclaimed water quality.

23.5 CONCLUSIONS

The dragon-shaped water system in the Beijing Olympic Park has become a successful example of recreational landscape waters replenished by reclaimed water, which brings clear economic and environmental benefits. In this context, the employment of advanced water reclamation technology such as ultrafiltration in combination with ozone oxidation and membrane bioreactors in combination with reverse osmosis can significantly reduce the nutrients and other trace contaminants in treated wastewater. The further purification of reclaimed water and the circulating treatment of the lake water system based on aquatic plant purification and constructed wetland play important roles in maintaining the water quality of the scenic water. Good water quality and urban landscape benefits of the dragon-shaped water system replenished by reclaimed water have been created. The Beijing Olympic Park has become a leisure area for the public, where the graceful and clear landscaped water body provides a great recreational water environment. Table 23.2 summarizes the main drivers, benefits and challenges of this project.

Table 23.2 Summary of lessons learned.

Drivers/Opportunities	Benefits	Challenges
Increasing population and decreasing water availability Water scarcity and frequent droughts Willingness of the municipality to improve urban infrastructure for the 2008 Olympic Games	Improvement of water supply security Positive environmental protection effects Improvement of leisure quality	Water quality management in artificial water bodies Maintain the water quality of artificial water bodies in a long-term through water ecology regulation Establish stable ecosystems in the artificial water bodies with good self-purification

REFERENCES

Han M. W. and Park Y. C. (1999). The development of anoxia in the artificial lake Shihwa, Korea, as a consequence of intertidal reclamation. *Marine Pollution Bulletin*, **38**(12), 1194–1199.

Lahnsteiner J., Klegraf F., Prasad Y. and Mittal R. (2011). Sustainable water resource management with urban and potable reuse. Proceedings of the IWA Conference on Sustainable Water Resource Management, Nagpur, India, January 2011.

Li F. M. and Hu H. Y. (2005). Isolation and characterization of a novel antialgal allelochemical from phragmites communis. *Applied and Environmental Microbiology*, **71**(11), 6545–6553.

Ruley J. E. and Rusch K. A. (2004). Development of a simplified phosphorus management model for a shallow, subtropical, urban hypereutrophic lake. *Ecological Engineering*, **22**(2), 77–98.

Suh S. W., Kim J. H., Hwang I. T. and Lee H. K. (2004). Water quality simulation on an artificial estuarine lake Shiwhaho, Korea. *Journal of Marine Systems*, **45**(3–4), 143–158.

Sun Y. X., Hu H. Y. and Li X. (2009). Seasonal water quality changes in a newly constructed artificial lake fed by reclaimed water: a case study of Main Lake in the Beijing Olympic Forest Park, In: Proceedings of 7th IWA World Congress on Water Reclamation and Reuse, September 2009, Brisbane, Australia.

24 Improving the air quality in Mexico City through reusing wastewater for environmental restoration

Blanca Jiménez-Cisneros

CHAPTER HIGHLIGHTS

Part of Mexico City's wastewater has been successfully reused to restore a lake dating from the Aztec period. In this way dust storms and urban floods within the city were controlled and the area environmentally restored. The Texcoco Lake is now considered an important site for bird conservation and it hosts native species that were close to extinction.

KEYS TO SUCCESS

- Detailed planning of all the elements of the project
- A sound technical background
- Recognition that by reusing wastewater to restore the lake, dust storms, urban floods and soil erosion in Mexico City could be controlled
- Utilization of a reuse project as a tool to educate citizens on Mexico City's history

KEY FIGURES

- Treatment capacity : 103,680 m^3/d
- Total annual volume of treated wastewater: 37.8 Mm^3/yr
- Total volume of recycled water: 37.8 Mm^3/yr
- Capital costs: US$1,130 million
- Benefits: Around US$1.4 million per year in 1998 due to avoidance of damage from urban floods, respiratory diseases and the closing of the airport
- Water reuse standards: <100 fecal coliforms/100 mL

24.1 INTRODUCTION

Since the time of the Aztecs, Mexico City has been a megacity. In 1521, when the Spanish arrived, the City had 200,000 inhabitants, the same population as Paris; at that time London had only 50,000 inhabitants and Los Angeles – which at that time was a possession of Mexico – had less than 5000 (Wikipedia, 2011). Mexico City was located on a small island, surrounded by 5 lakes. Today the city has expanded and the lakes have been almost completely dried. The city suffers from a lack of water to supply the 21 million inhabitants and from urban floods. Originally the Mexico Valley was a closed basin and was artificially opened up in the XVII century to dispose of the water from lakes, rainwater and wastewater to avoid floods. At the present time, the total volume of the 7.7 m^3/s of wastewater that is treated using public facilities is reused: 11% to fill lakes, 54% for agricultural irrigation, 31% for industrial cooling and 5% for the urban solids wastes and car washing.

Water reuse is a response to the high water demand (86 m^3/s), the overexploitation of the local aquifer and the need to import water from sources that are 130 km away and 1100 m below the level of the city (Jiménez, 2008).

In addition, Mexico City suffers from severe air pollution. During the 1970s, one of the major problems of the City was air pollution by suspended particles. To overcome this problem, one of the biggest water reuse projects in the city's history was implemented. It consisted of restoring one of the lakes from the Aztec era, the Texcoco Lake, to control the dust storms that were polluting the urban air. This chapter describes that project, which besides controlling the particle content in Mexico City's air conferred other environmental and safety benefits on the city. The project also illustrated some limitations to the reuse water project, linked to the local environmental conditions.

Historical background

In order to understand the development of the project it is necessary first to review why the Mexico Valley Lakes were dried. This was a planned activity that took around 360 years to achieve.

The Aztec era

It was the Aztlan tribe that created the Aztec culture. The Aztlan people lived hundreds of kilometres north of Mexico City, in an arid area. They believed their gods told them to travel south until they reached a place where an eagle would be found devouring a serpent on a cactus in the centre of an island. This occurred in the Mexico Valley, where they founded the city of Tenochtitlan. In the valley, the Aztecs had water for municipal and agricultural supply, they could fish in the lakes and hunt in the forests and use the timber. Several lakes were located in the valley; there were five important ones: Xaltocan and Zumpango located to the north, the Xochimilco and the Chalco Lakes to the south and the Texcoco Lake to the east. The Texcoco Lake was the largest, had highly saline water and was artificially separated from the others by means of an earth dike known as the "Albarradón de Neztahualcoyotl" in honour of the Texcoco King who designed and built it between 1440 and 1503. Several species of birds lived in the Texcoco Lake, some of them migratory. Birds made up part of the Aztecs' daily diet and feathers were used for adornment and as money (Diaz del Castillo, 1998). Since they were surrounded by lakes (Figure 24.1), the Aztecs mastered the management of water. Besides hydraulic infrastructure to control floods, they also built infrastructure for agricultural irrigation, fluvial transport, domestic and municipal supply (Rojas Rabiela, 2004; Cruickshank, 1998; Leon Portilla, 1963).

Figure 24.1 Tenochtitlan city (*Source*: Tomas Filsinger, 2006) and illustration of agriculture in the Mexico Valley during the Aztec era (*Source*: http://aztecgroup.blogspot.com/p/aztec-environment.html).

Colonization

When the Spanish arrived at Tenochtitlan in 1521, they were amazed by the cleanliness of the city and the ample use of water for personal hygiene, municipal fountains, urban use and domestic and public fountains and gardens (Jiménez and Birrichaga, in press). However, they did not understand the importance of the Aztec hydraulic infrastructure and they destroyed it along with many other important architectural constructions. This resulted in severe flooding of the city in 1555 after it rained continuously for 24 hours, forcing the Spanish to restore part of the hydraulic infrastructure. This was insufficient, and another flood occurred in 1586. The streets were then re-levelled and the rivers and canals dredged. In 1604, another large flood led to the decision to drain the lakes. This task was undertaken over the next 360 years, beginning in 1607 just after a new flood. The Viceroy, Luis de Velasco, asked Enrico Martínez to build a drain to take water away from the city. The project consisted of a channel and a tunnel known as the "Tajo de Nochistongo". Due to a lack of money the tunnel was constructed slowly and poorly and therefore it malfunctioned. This led to a new flood in 1629. This time the city remained flooded for six years and it was even considered that the capital should be moved to Puebla. Instead, it was decided that an open channel should be built. The purpose of this channel was to effectively drain the lakes through the Tequisquiac tunnel along with the excess pluvial water and wastewater. It ended in the Tula Valley (Mezquital Valley) where the non-treated wastewater began to be used for irrigation.

The 20th century

In 1805, the city once again was flooded. During this period the city authorities spent their time constantly improving the sewer system, but also committed to three wars. At this time in the city still existed fluvial canals that went from Xochimilco to Chalco which were used for transport and tourism. In 1840, the traditional boats known as "trajineras" were replaced with steamboats.

In 1856, a further severe flood revealed that the sewer system once again had reached its limit. In 1857, the engineer Francisco de Garay proposed the building of the Gran Canal. Construction began in 1866 and was not finalised until 1900.

The first half of the 20th century

In 1901, it was once again decided to drain the valley, this time completely. For this it was decided to drain the Texcoco Lake, which by this time had already been reduced from an original size of 207,600 ha to only 27,000 ha. The idea was to transform 15,482 ha into an agricultural area and the rest was to be used for the building of houses. For the first purpose, a tank to store the irrigation water was to be built plus two drainage systems, one to manage the irrigation water and the other to be used as a sewer. In total, 9 m^3/s of water would be used to reduce the salinity of the land. This project was never completed. The city was once again flooded in 1920. Between that time and 1938, all of the canals of the city and most of the rivers were closed and/or covered to be used as streets. This explains why many important streets in the city are named after rivers. In 1940, there was further flooding (Figure 24.2a, Cruickshank, 1998). The reason was the reduction of the sewer capacity caused by the sinking of the city due to groundwater overexploitation (Figure 24.2b). Two large floods occurred in 1942 and in 1945. The population of Mexico City began to migrate to the southern and western parts of the city, with only the poors remaining in the east. The capacity of the Tequisquiac Tunnel was once again increased, but in the early 1950s a storm of around 120 mm covered 2/3 of the city with 2 m of water, wastewater and sediments causing severe economic damage. Construction of a new Tequisquiac Tunnel of larger dimensions was begun, and was completed in 1954. Between 1954 and 1957 the most important rivers, the Churubusco, Remedios, Consulado and de la Piedad were channelled and/or covered and became part of the most important avenues of the city.

Figure 24.2 The problem of floods in Mexico City: a) view of floods and b) effect sinking on the drainage.

The second half of the 20th century

In 1967 the construction of the Deep Drainage System began. It basically consisted of two enormous pipelines and an open channel. The Emisor Central Sewer was a 60 km tunnel, had a capacity of 200 m^3/s and was up to 200 m below the surface. In 1980 the Interceptor Poniente Sewer, 16.5 km in length, was built and the Gran Canal sewer was reinforced.

The 21st century

The sewerage system continued to grow. At the beginning of the XXI century it consisted of 12,764 km of pipelines, between 0.3 and 7.6 m in diameter, 96 pumping stations with a total capacity of 670 m^3/s, 91 bridges carrying 14.3 m^3/s, 106 marginal collectors, 12 storm tanks with a capacity of 130,000 m^3, several inverted siphons to allow the metropolitan railway to pass by, 3 (ex) rivers, 29 dams, the Gran Canal Sewer, 47 km in length, and the Deep Drainage System, 155 km in length and with pipe diameters ranging from 3 to 6.5 m and depths between 20 and 217 m. In 2009, as Mexico City continued to suffer from floods, construction of a new sewer was begun, to be completed in 2012. This is the TEO (Tunel Emisor Oriente) Sewer, a tunnel 62 km in length and 7 m in diameter, with depths of between 150 and 200 m. Its capacity is 150 m^3/s of wastewater and the estimated cost is around 1,130 million US$.

Main drivers for water reuse

By 1920, due to the construction of the sewerage system, the Xaltocan, Zumpango, San Cristóbal and Chalco lakes had disappeared, the Xochimilco Lake had decreased to a small area used nowadays as tourist site and the Texcoco Lake, to less than 12,000 ha. In addition, the Texcoco Lake was used as a discharge site for part of Mexico City's wastewater and in the dried areas, settlements for the lower classes were created and several slums sprang up. The population growth rate for this part of the city amounted to 7.8% (Cruickshank, 1998). Where no settlements were located, a desert of fine saline clay appeared. Drying the Texcoco Lake resulted in different environmental effects, such as:

(a) *Ecosystem effects*. The terrestrial flora and fauna were noticeably reduced and some species, such as migratory and native birds, almost disappeared. The aquatic fauna was lost and the willow (*Salix*) trees that surrounded the lake were replaced with eucalyptus, "pirul" (*Schinus molle*) and pine.

(b) *Soil erosion*. The sediment from the bottom of the lake was exposed. It was 1.5 m in depth, and classified as *solonchak, gleic* and *mólic*, which are scientific terms used to describe soils that are highly saline, unstructured and highly water absorbent. Basically these consist of very fine clays mixed with organic sedimented matter and a saline content twice that of the sea. As this soil was exposed for a long period of time it dried (see Figure 24.3), rendering the surface highly erodible. The non-urbanized area soon became a saline desert with no vegetative growth.

(c) *Dust storms*. In the dry season, from January to March, the winds picked up the fine soil particles, forming dust storms. With wind speeds of 7–9 m/s, the dust storms passed into the city. In addition, if convective winds were formed during the hottest hours of the day (12 a 16 h), the dust was carried to higher altitudes in a phenomenon similar to dust storms observed in the Sahara Desert near the Algerian coast. By the end of the 1960s there were around 68 dust storms per year which were the source of several respiratory and gastrointestinal diseases, and the cause of economic losses of 20 million pesos (52,000 US$ in 1975). In addition, dust storms caused the airport to shut down 24 times per year. The cost of closing the airport for 1 hour was 1 million de pesos (2,631 US$ in 1975, Cruickshank, 1998).

(d) *Unhealthy conditions*. The ex Texcoco Lake area became a place where not only wastewater but solid wastes were disposed of. This became a source of foul odours and vector-borne diseases.

Project objectives

In 1965 two engineers, Nabor Carrillo and Gerardo Cruickshank, began work on the idea of restoring the Texcoco Lake. Mr. Carrillo died in 1967, and therefore when the President created the Texcoco Lake Project Mr. Cruickshank led it. The objectives were:

(a) To treat part of Mexico City's wastewater, to reuse it to restore the lake, control dust storms and recover the ecosystem.

(b) To use the lakes as storm control infrastructure to avoid urban floods.

(c) To reuse the treated wastewater for agricultural irrigation, afforestation and eventually for industries and other public and commercial services.

(d) To stop the growth of slums in high risk urban areas.

The Project was planned in two phases. In the first, one the hydraulic infrastructure needed was to be built and the region was to be "reforested". During the second phase, different complementary economic and social programs would be set up. The first phase was achieved fairly successfully, but several adjustments to the objectives had to be made for the second phase.

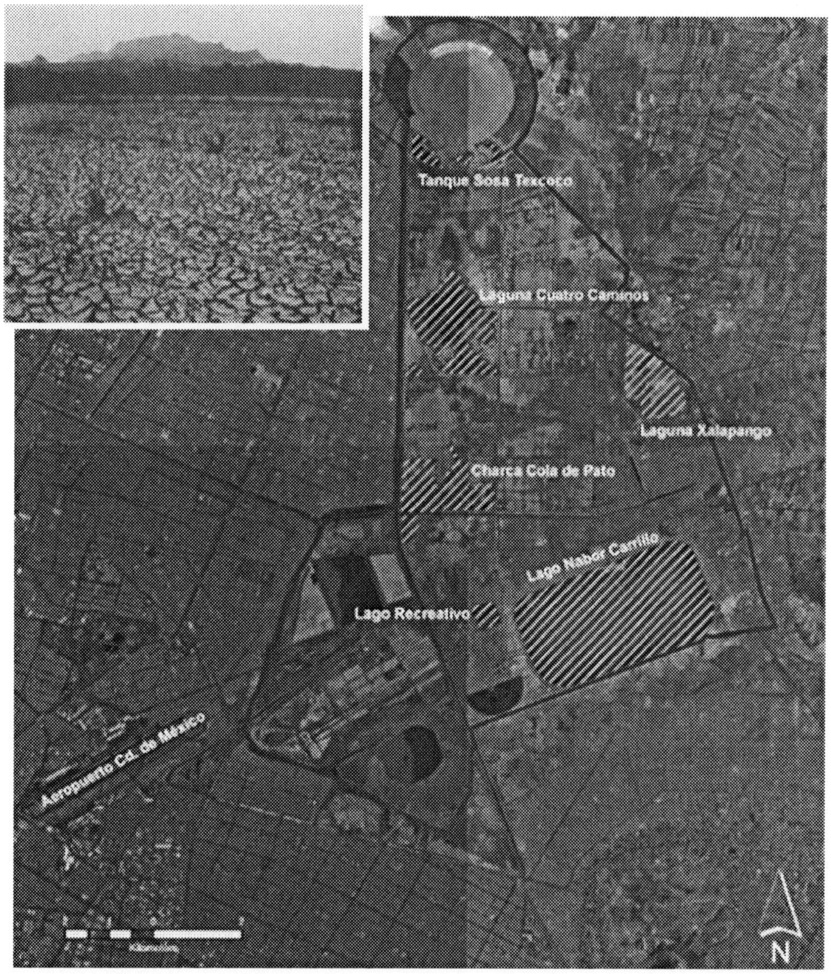

Figure 24.3 Lakes built in the ex Texcoco Lake area with a view of exposed soil from the Texcoco Lake.

24.2 TECHNICAL CHALLENGES OF WATER REUSE AND ENVIRONMENTAL RESTORATION

The area of the project is located to the west of Mexico City. The weather is semiarid temperate with hot summers. The mean annual temperature is 14–15°C, with minimum and maximum values between −5°C and 2°C during January and to up to 38°C in July. Rains last from May to October, with precipitation values ranging from 460–600 mm. In contrast, evapotranspiration is very high: 1750 mm in mean conditions but with maximum values reaching 2519 mm per year as result of the high temperatures and wind speeds.

Three sewers (previously rivers) arrive to the project site from the city: the Churubusco River (with a mean flow in 1995 of 10.4 m^3/s, but with maximum flows of 70 m^3/s during rain storms); the de la Compañía River (4.22 m^3/s) and the Los Remedios River. In addition, 10 rivers discharge to the eastern part. Most of these rivers only convey water during the rainy season, while those that flow year-round only do so because they are used as sewers. All of the rivers provoked erosion in the mountains surrounding the ex Texcoco Lake area during the rainy season. One of the rivers, the Papalotla, with a mean flow of 0.4 m^3/s, conveys up to 124 m^3/s during the rainy season.

Technical challenges for the project implementation

Due to the local soil conditions and the sinking of Mexico City, the construction of the lakes posed a considerable challenge, in particular to create a depression to store the water since an elevated lake was not feasible. In order to build the lakes (as construction of a single one was not feasible), three construction procedures were tested: (a) extracting the water from the subsoil using shallow wells to compress the upper soil layers; (b) mechanical excavation using floating dredges and (c) liquefying the clays with explosives. In total 5 lakes were built (Figure 24.3):

(a) The Nabor Carrillo Lake (Figure 24.4). It has a 36 million m^3 (Mm^3) capacity and a rectangular surface of 1000 ha. It can be clearly observed during take-off or landing at Mexico City airport. It is 3.2 m in depth and has walls 4 m in height. It began to operate in 1982, and was partly built to extract water from 180 shallow wells over 6 years. It is fed with treated wastewater and plays a major role in controlling dust storms in the city.
(b) The Hourly Regulation Lake. It has a capacity of 4.5 Mm^3 and a surface area of 150 ha. It was built by dredging up to a depth of 3.5 m. Its purpose is to regulate the storm water from Churubusco River that flows from the southeast part of the City, where it rainfall is greatest.
(c) The Churubusco Lake. It has a capacity of 5.10 Mm^3 and a surface area of 270 ha. To build it, the water from the subsoil was extracted over 4 years. It is used to store pluvial water from the Churubusco and de La Compañía rivers. It began to operate in 1983.
(d) The Xalapango Lake. It has a capacity of 3.6 Mm^3 and a surface area of 240 ha. This was the only natural depression that could be used as a lake. It is fed with water from the Papalotla, Coxcacoaco and Xalapango rivers. It has operated since 1982 and discharges to the Nabor Carrillo Lake.
(e) The Recreational Lake. It has a capacity of 0.375 Mm^3 and a surface area of 25 ha. It was the lake in which the method of pumping groundwater water was tested. It began to operate in 1982 and is filled with well water.

To allow the lakes to function, the Churubusco, de los Remedios and de la Compañía Rivers were modified. On the remaining rivers, small dikes to store water, control erosion and increase water infiltration to the soil were built.

Figure 24.4 General view and a floating dredge at the Nabor Carrillo Lake (*source*: Loa, 2011; Cruickshank, 1998).

Treatment trains for water recycling

Of the 16 m^3/s of wastewater that in total reached the Texoco area in 1971, 1.2 m^3/s were to be treated and reused. Reuse was to fill the lakes, for agricultural and green area irrigation and reforestation. Specifically, the agricultural reuse programme aimed to exchange first use water (fresh water) with reclaimed water. A wastewater treatment plant (WWTP) of 1 m^3/s capacity, was built but due to its poor design, related to the inability of the soil to support any type of construction, many problems were experienced during its operation. The tanks simply "floate" on the soil and they did not properly function hydraulically. After several procedures to re-level and balance the tanks, the plant finally functioned at 0.6 m^3/s.

The treatment process is activated sludge treatment with an 8 h retention time. The effluent is sent to two stabilization ponds, the first with 9 days of retention time, and the second with 14 days. The total surface area of both ponds is 20.6 ha. Part of the effluent (50 L/s) is treated to a tertiary level, by removing detergents by coagulation-flocculation coupled with a high rate lamella settler, sand and anthracite filtration, and absorption on activated carbon. The effluent is injected to the groundwater. A small part of this effluent is sent to a nanofiltration process to produce drinking water as a demonstration project. The secondary treated wastewater is send to the stabilization ponds where, more than 50% of the treated water is lost due to evaporation. The sludge produced in the facilities is treated by aerobic stabilisation (Cruickshank, 1998).

24.3 WATER REUSE APPLICATIONS

As mentioned previously, the project was very ambitious. The advantages and limitations soon became evident.

Dust storm control

The treated effluent began to be used to wash down soil salinity. Unfortunately, the soil has very poor drainage capacity and soil washing is a very difficult task. Other methods were tested to overcome soil salinity, such as the addition of enzymes, acids and gypsum combined with sulphur, but without success. To turn the soil from the bottom of the lake into a productive agricultural area entailed a tremendous effort, and therefore, the programme was abandoned. The soil was covered using local native saline grass and romeritos (*Suaeda torreyana*) that were successful in controlling dust storms.

Flood control

This is the other important ecosystem service provided by the water reuse project. Thanks to construction of the lakes, 3 million people and 550,000 households in Ciudad Netzahualcóyotl, Ecatepec and area of the airport were protected from floods. Considering only 20% of the area vulnerable to floods, 11,320 ha of urban area, 20,100 ha of agricultural area and the 750 ha of the airport were protected. This represented savings of 500 million pesos per year (1.3 million US$ per year in 1998).

Flora and fauna restoration

Along with construction of the lakes, flora and fauna have been restored, partly artificially and partly naturally. Six thousand hectares of land were covered once again with halophyte grassland (*Egagrostis obtusiflora* and *Distichlis spicata*). The aquatic flora recreated in the lakes is largely composed of bird populations (Table 24.1, Figure 24.5). Some species close to extinction, such as the Mexican duck (*Anas diasi*) and "chichihuilotes" (*Charadrius wilsonia*) were recovered. Populations of yellow fish and even the *Spirulina* bacteria – highly proteinaceous bacteria used as source for food – also recovered (Cruickshank, 1998). In 1985 there was a total population of 35,000 birds. This increased to 300,000, encompassing some 134 species by 1998. Of these, 80% are migratory birds that arrive during autumn and winter from Canada and the USA in order to reproduce. The Texcoco Lake area is now listed as important site of bird conservation.

Table 24.1 Examples of bird species that inhabit the lakes: local names are shown in parenthesis.

Actitis macularía (Alzacolita)	*Fulca Americana* (water hen)
Ardea Herodias (blue heron)	*Gallinula chloropus* (gallareta)
Anas Americana (chalcuán duck)*	*Himantopus mexicanus* (monjita or Mexican nun)
Anas discors (blue wing duck)	*Laurus pipixcan*
Anas cyanoptera	*Limnodromus scolopaceus* (Costurero pico largo)
Anas clypeata (big mouth duck)*	*Oxyura jamaicensis* (Tepalcate duck)*
Anas Cuta (golondrino duck)*	*Pelecanus arythrorhynchos* (white pelican)
Anas platyrhynchos diazi (Mexican duck)	*Plalacrocorax alcyon* (cormorant)
Avocetas recurvirostra americana,	*Phalaropus tricolor** (Falaropo pico-largo)
Bufeo jamaicensis (red tailed sparrow hawk)	*Podiceps nigricollis* (diver)
Casmerodius albus (long necked heron)	*Recurvirostra americana* (avoceta)
Cerly alcyon	*Rynchops niger* (rayador)
Charadrius vociferus (chorlo tildío)	Sparrow hawks
Charadrius alexandrinus (chorlo nevado or corredores = runners).	*Sterna caspia* (golondrina de mar or sea sparrow)
Falco peregrino (peregrine falcon)	*Tlanus jamaicensis* (black shouldered Milano)

*Migratory bird

Source: http://www.sagan-gea.org/hojared_AGUA/paginas/27agua.html; http://www.whsrn.org/es/perfil-de-sitio/lago-texcoco

Figure 24.5 Views of some birds of the Texcoco Lakes: a) *Himantopus mexicanus* (Monjita), b) *Recurvirostra americana* (Avoceta), c) *Charadrius vociferous* (chichicuilote) and d) *Recurvirostra americana* (Avoceta).

24.4 OTHER ASSOCIATED PROGRAMS

Reforestation

Even though this programme was named reforestation, in effect it constituted afforestation. The idea was to turn the base of the lake into a wood using the reclaimed water. Over 25 million pine, eucalyptus, acacia and cedar trees were planted over 4000 ha. Only 25–33% of the trees planted survived along the banks of the western rivers. The rest of the area is poorly vegetated with halophyte grass *Distichlis spicata* and the local tree (*Tamarix*). The trees along the rivers have helped to reduce erosion from a rate of 16.3 ton/ha·yr in 1973 to 0.26 ton/ha·yr in 1998.

Water exchange with farmers

As the treated wastewater is not suitable for irrigation because of its salinity (Table 24.2), farmers living around the Texcoco area refused to exchange their fresh well water for reclaimed water.

Table 24.2 Treated wastewater characteristics at the exit of the Nabor Carrillo in mg/L unless indicated (adapted from Maya & Jiménez, 2002).

Parameter	Influent	Effluent
Conductivity μS/cm	2161	5348
pH	8.3	10.4
Total dissolved solids	980	2954
Total suspended solids	32	204
Cl^-	268	840
$SO_4^=$	159	216
Na^+	320	1090
K^+	54	162
$N\text{-}NO_3^-$	1.05	0.007
$N\text{-}NO_2^-$	1.2	0.003
$N\text{-}NH_4^+$	34.3	0.08
Total coliforms, MPN/100 mL MNP	$2.4 \times 10^5 \pm 5.5$ log	ND
Estreptoccocus faecalis MPN	$2.1 \times 10^5 \pm 4.6$ log	ND
Salmonella spp., MPN/100 mL	$8.1 \times 10^3 \pm 3.4$ log	ND
Vibrio cholerae, MPN/100 mL	$3.2 \times 10^2 \pm 2.6$ log	ND

Production of flora and fauna for commercial purposes

Four different programmes were undertaken: three to produce food by cultivating fish, deer and bacteria, and one to produce a new variety of horse, the Aztec horse. Ultimately, only the production of bacteria was successful. *Spirulina* cyanobacteria is composed of 70% protein and has a high content of vitamin B12. It was used by the Aztecs as a source of food, and is used today as a dietary supplement sold in health food shops.

Regional development

As with many hydrological projects in Mexico, the Texcoco Lake programme promoted regional development throughout 50 nearby communities. Communal stores, libraries, medical services, agroindustry training programmes and secondary and high schools were built and sometimes even run by the Texcoco Lake Project. This further encouraged population growth in the area.

24.5 INSTITUTIONAL FRAMEWORK

The Texcoco Lake Project is run by the federal government through the National Water Commission (CONAGUA), part of the Ministry of Environment and Natural Resources (CONAGUA, 2011). The managers of the Texcoco Lake are in charge of hydraulically administering the project and the restored areas, treating the wastewater and running a bird and soil monitoring programme. However, many of the social programmes have been halted.

24.6 PUBLIC ACCEPTANCE AND INVOLVEMENT

Guided visits are provided to the site by the management of the Texcoco Lake, mainly to students from primary and secondary schools. Since 1999 on the first Saturday of December, "Fauna Sylvester Day" is celebrated in which local people, NGOs, academia and the government participate to promote birdwatching (birding). Each year, a 15 km run is organized around the Nabor Carrillo Lake, attracting more than 3000 competitors. In addition there are model aeroplane races, sport fishing and boat competitions during November.

24.7 CONCLUSIONS AND LESSONS LEARNED

The reuse of water to recreate the Texcoco Lake has given children the opportunity to discover that Mexico City was founded on a lake and that reclaimed water allowed the restoration of part of this historic site. This is important in a city where most of its rivers have now become the main streets, something of which children are generally unaware. The use of reclaimed water made it possible to construct a site suitable for birdwatching in the middle of a city of 21 million inhabitants. In addition, the development of the project allowed Mexico City's engineers to explore the potential for water reuse and also the limits of its implementation. On the one hand, many new uses for reclaimed water were explored and the knowledge acquired has been invaluable for many new projects that have been and are currently being implemented in Mexico. On the other hand, it is now clear that there are limits to the manipulation of the environment, and that even though reclaimed water is very useful in improving soil characteristics, it is impossible to use it to turn a saline and impermeable soil that was once the base of a lake into a forest.

The following Table 24.3 summarizes the main drivers of this water reuse project as well as the benefits obtained and some of the challenges for the future.

Table 24.3 Summary of lessons learned for the Texcoco Lake Water Reuse Project.

Drivers/Opportunities	Benefits	Challenges
Urban Floods Urban dust storms Restoration of local biodiversity Better urban planning	Avoidance of damage to urban infrastructure Reduction of respiratory illness Creation of a recreational area Halting the growth of slums in risky urban areas	Providing suitable storage capacity in a soil with poor building capacity Controlling soil erosion through reforestation and not afforestation Restoring the original environmental conditions as much as possible Encouraging communities to move to safer places to live

REFERENCES AND FURTHER READING

CONAGUA (2011). National Water Commission, http://www.whsrn.org/es/perfil-de-sitio/lago-texcoco.

Cruickshank García G. (1998). The Hydroecological Restoration of the Texcoco Lake. National Water Commission, Mexico, D.F., Mexico [In Spanish].

Jiménez B. (2008) Water and wastewater management in Mexico city in integrated urban water management. In: *Arid and Semi-arid Regions Around the World*, L. Mays (ed.), Taylor and Francis Group Ltd, Boca Raton, FL, USA.
Jiménez B. and Birrichaga D. (2012). Water services in Mexico City: The need to return to the IWRM principles of Tenochtitlán (700 years of water history). In: *Evolution of Water Supply Throughout the Millennia*, A. Angelakis (ed.), IWA Publishing, London, UK, 521–552.
León-Portilla M. (1963). A view of the defeated. Indian reports from the Conquest. Universidad Nacional Autónoma de México. México. [In Spanish].
Maya C. and Jiménez B. (2002). Physicochemical and microbial characteristics of the Nabor Carrillo Lake. México. *XIII National Congress Nacional: Mexican Federation of Sanitary and Environmental Engineers*, **I**(I), 289–294. Guanajuato, Gto.. [In Spanish].
Rojas Rabiela T. (2004). The basin from the Central High plateau. *Arqueología Mexicana*, **XII**(68), 20–27, [In Spanish].
Sagan (2001). http://www.sagan-gea.org/hojared_AGUA/paginas/27agua.html, consultado en mayo 2011.

WEB-LINKS FOR FURTHER READING

http://aztecgroup.blogspot.com/p/aztec-environment.html, consulted one January 2012.
http://www.whsrn.org/site-profile/lago-texcoco, consulted one January 2012.

Views of the Texcoco lake.

PART VII

Increasing drinking water supplies

Foreword

By The Editors

Historical development

There is a long history of cities drawing drinking water from river sources which receive treated water discharges from upstream communities. Well known examples are the Thames River in England and the Rhine River in Germany. A 1980 report published by the US Environmental Protection Agency studied the occurrence of treated discharges in the source waters of US cities with populations greater than 25,000. It identified 1246 supplies serving 525 cities serving 80 million people. The study found widespread unplanned reuse. About one-third of the surveyed population received their drinking water supplies from sources that contained from 5% to 100% treated wastewater discharges during low flow periods.

The history of planned use of recycled water to increase drinking water supplies now spans more than sixty years:

- Since 1948 recycled water has supplemented inflows to Lake Tegel in Berlin. Drinking water supplies are drawn from infiltration wells in the banks of Lake Tegel.
- Since 1959, one of the largest projects of artificial recharge has been operating in Croissy sur Seine in the West of Paris, France, providing up to 300,000 m^3/d of treated surface water from the Seine river downstream of wastewater discharges.
- In 1962 Los Angeles County commenced a major groundwater recharge project at the Whittier Narrows spreading basin. This was followed by the groundwater recharge project in Orange County in 1975 (Chapter 25) and by other projects including El Paso in Texas in 1985 and West Basin in California in 1992 (see Chapter 2).
- Direct potable use of recycled water commenced in Windhoek in Namibia in 1968 and has now been operating successfully for more than 40 years (Chapter 29).
- Use of recycled water to supplement inflows to the Upper Occaquan reservoir in Virginia commenced in 1978 (Chapter 27).
- Singapore has recently commenced adding small quantities of purified recycled water from its NEWater recycling plants to supplement inflows to its main drinking water reservoir (see Chapter 3).

Case studies of increasing drinking water supplies by water recycling

Part 7 of this book "Milestones in Water Reuse: The Best Success Stories" presents a number of case studies showing how water managers have used high quality recycled water to meet increased drinking water demand.

Chapter 25. The Orange County Water District in California has been a world leader on groundwater recharge. Recycled water has been used to prevent saline intrusion (direct injection) and to maintain water level (by spreading basins and river beds) in aquifers that supply drinking water to Orange County. The recharge system has recently been expanded and now supplies enough water for 600,000 people.

Chapter 26 describes the first European project of managed aquifer recharge by high-purity recycled water after advance membrane treatment in the Torreele dune aquifer in Belgium. A sustainable groundwater management strategy has been

Table 1 Highlights and lessons learned from the selected case studies of potable water reuse.

Project/ Location	Start-up/ Capacity	Type of uses	Key Figures	Drivers and Opportunities	Benefits	Challenges	Keys to Success
GWRS, Orange County, California United States	Start up: 1975 Water Factory 21 Jan 2008 GWRS 265,000 m^3/d expansion to 378,000 m^3/d	Indirect potable reuse by aquifer recharge: direct injection as salt intrusion barrier and surface spreading (infiltration basins)	Total volume of recycled water: 96.7 Mm^3/yr Capital costs: $480 million (2008) and $142 million (2011) Operation costs: 0.35 $\$/m^3$ Recycled water pricing: 0.20 $\$/m^3$ (2011) Drinking water standards	Droughts and increasing potable water demands. Seawater intrusion. Increasing cost of imported water supply.	Safe, local, reliable water supply for 600,000 residents. High quality water. Sustainable, using less power than imported water. Reduces discharges to the ocean. Lowers dependence on imported water.	Public acceptance. High cost and need of funding. Size/capacity of such a complex system. Regulatory approval. Keeping existing Water Factory 21 operational during construction.	Locally-controlled, drought-proof and reliable supply of high-quality water in an environmentally sensitive and economical manner. State-of-the-art technology. Strong Board and community support.
Torreele, Belgium	July 2002 7000 m^3/d	Indirect potable reuse by aquifer recharge: surface spreading in a dune aquifer	Total volume of treated wastewater: 3.3 $Mm^3/10$ yr Total volume of recycled water: 19.4 $Mm^3/10$ yrs Capital costs: €7 million Operation costs: 0.64 $€/m^3$ Recycled water tariffs: part of overall drinking-water tariff	Unsecure water supply in summer. Declining aquifer water table. Need for ecological management.	Improved drinking-water quality. Sustainable production near point of consumption. Enhanced natural values in dune area.	Increase infiltration capacity and reduce overall costs on long term. RO concentrate should be treated/reused.	Intense preparation of the project with pilot tests using different treatments. Infiltration project was integrated in a natural management plan for the area and enabled sustainable groundwater management. Improved quality of drinking-water and secured water supply. No excessive increase of drinking-water price.
Occoquan, United States	Since 1978 Treatment capacity: 204,400 m^3/d	Indirect potable reuse by replenishment of a drinking water surface reservoir	Total volume of reclaimed water: 44.1 Mm^3/yr (8% of annual average reservoir inflow) Capital and operation costs: $61.5 million Occoquan Reservoir safe yield increased by 71% to date; future increase expected to exceed 100%	Need to insure long-term ability to meet regional water supply needs and improve source water quantity and quality. Federal construction grant.	Regulatory support for adopting a reuse solution on a watershed scale. Enhanced safe yield of the drinking water system. Reduced local capital cost of initial project. High quality reclaimed water has enhanced both watershed and reservoir water quality.	Developing local support for more expensive reuse infrastructure. Continued growth may require the development of additional sources. Future expansions will all be locally funded. Increased emphasis on controlling pollution from urban runoff.	Foresight to envision requirement for increased water supply yield decades in advance of need. State and local government collaboration on adopting a reuse solution. Independent monitoring to provide unbiased information to stakeholders. Continued demonstration of improvements to water quality over project life.

Western Corridor, Australia	2007 Treatment capacity : 236,000 m^3/d	Indirect potable reuse by replenishment of a drinking water surface reservoir, Industrial uses	Total volume of recycled water produced: 74.3 Mm^3/yr Capital cost: AU$2.5 billion Australian Drinking Water Guidelines	Severe drought. Increasing population. Insurance policy for variable climate.	Augmentation for drinking supplies. Substitute potable supply for industrial users. Demonstration of the production of very high quality recycling water.	Stringent recycled water quality requirements. Managing the recycling facilities at low production rates. Managing facilities through flood conditions.	Largest recycled water scheme in Australia. Meets extremely high standards of water quality. Provides high quality water for industry. Reduces nutrient load to the receiving environment.
Windhoek Namibia	Since 1968 Treatment capacity: 21,000 m^3/d	Direct "pipe to pipe" water reuse	Total volume of treated wastewater: 6.4 Mm^3/yr Total volume of recycled water: 5.8 Mm^3/yr Total annualised costs: 0.95 €/m^3, including 0.75 €/m^3 O&M costs Recycled water tariffs: consumption-related, from 0.75 €/m^3 to 2.3 €/m^3 Drinking water standards: blending max 35%	Severe droughts in the past and water stress. Population growth.	Increased water supply security. Save potable water produced by an advanced multiple barriers approach. Secured social and economic development.	Optimum management of the multiple barrier approach (source control, treatment and operational barriers) for safe potable reuse. Increasing salinity. Emerging issue of micro-pollutants. Increasing water demand.	Vision and great dedication of the potable reclamation pioneers. Excellent information policy and education practice. No reclaimed water related health problems experienced. Multiple barrier approach. Reliable operation and on-line process and water quality control. Public-private partnership.

achieved. The recharge offsets an expected reduction in available water due to climate change and achieves a balance between economic and ecological factors.

Chapter 27. The Upper Occoquan water reuse project in Virginia USA began operation in 1978. The reservoir supplies more than 1.3 million people in Northern Virginia. Recycled water was recognised as the best quality source water to supplement drinking water supplies in the Upper Occoquan reservoir and offset water quality impacts from urban development in the catchment. It is estimated that full development of the scheme will double the water supply yield of the Occoquan reservoir system.

Chapter 28. The Western Corridor Water Recycling Scheme in Australia is part of an insurance policy that aims to ensure that the Brisbane region will have adequate water supplies available to handle the effects of drought and climate change. The scheme provides purified recycled water for industrial purposes, including power station cooling, to reduce demand on drinking water sources. The scheme could be used to supplement inflows into the main drinking water reservoir if necessary in severe drought events.

Chapter 29. The Windhoek direct reuse scheme has provided purified recycled water to supplement drinking water supplies in Windhoek, the capital city of Namibia since 1968. In more than 40 years of operation, no health problems have been reported. The system capacity was increased to 21,000 m^3/d in 2002. Recycled water now supplies about 30% of drinking water needs.

Keys to success

The common themes running through these case studies are that potable water reuse has offset pressures from growing demands and water shortages due to drought and climate change by producing high-purity alternative and drought-proof resource to cover drinking water needs. There have been worthwhile economic and environmental benefits. Keys to success have included use of state-of-the-art multi-barrier treatment technology to produce purified recycled water meeting drinking water requirements, detailed preparation and pilot plant testing, detailed independent monitoring, strong government and regulatory support, development of community support through effective public education programs, a demonstrated need for new reliable drought-proof supplies and clear demonstration of the economic and environmental benefits.

The major highlights and lessons learned from the selected case studies are summarised in Table 1.

25 Key to success of groundwater recharge with recycled water in California

Robert B. Chalmers and Mehul Patel

CHAPTER HIGHLIGHTS

The Groundwater Replenishment System (GWRS) is the world's largest wastewater purification system for indirect potable reuse. Beginning in 1975 with the successful implementation of Water Factory 21, the Orange County Water District (OCWD) has been a world leader in indirect potable reuse. Along with their partner agency, the Orange County Sanitation District, OCWD implemented the GWRS to supply water to take highly treated wastewater that would have previously been discharged into the Pacific Ocean, further treating it using a proven three step advance purification process, then discharging into the groundwater aquifer to provide drinking water for the residents of Orange County, California. Part of the water is injected into a hydraulic barrier to protect the basin from seawater intrusion while the rest of the water is percolated into the aquifer to improve water quality and recharge the groundwater. The GWRS produces a locally-controlled, drought-proof and reliable water supply of high-quality water in an environmentally sensitive and economically manner.

KEYS TO SUCCESS

- Locally-controlled, drought-proof and reliable supply of high-quality water in an environmentally sensitive and economical manner
- State-of-the-art technology
- Strong Board and community support
- Produces enough water for 600,000 people

KEY FIGURES

- Start up: 1975 (Water Factory 21) and Jan 2008 (GWRS)
- Treatment capacity : 265,000 m^3/d with on-going expansion to 378,000 m^3/d
- Total volume of recycled water: 96.7 Mm^3/yr
- Capital costs: $480 million (2008) and $142 million (2011)
- Operation costs: 0.35 $\$/m^3$
- Recycled water pricing: 0.20 $\$/m^3$ (2011)
- Drinking water standards

25.1 INTRODUCTION

The Groundwater Replenishment System (GWRS) is the world's largest wastewater purification system for indirect potable reuse (IPR). The GWRS takes highly treated wastewater that would have previously been discharged into the Pacific Ocean and purifies it using a three-step advanced treatment process consisting of membrane filtration (MF), reverse osmosis (RO) and ultraviolet light with hydrogen peroxide (UV-A). The process produces high-quality water that exceeds all state and federal drinking water standards. Operational since January 2008, this state-of-the-art water purification project can produce up to 265,000 m^3/d of high-quality water. This is enough water to meet the needs of nearly 600,000 residents in north and central Orange County, California.

The GWRS provides a locally controlled, drought-proof and reliable supply of high-quality water in an environmentally sensitive and economical manner. The design and construction of the GWRS was jointly funded by the Orange County Water District (OCWD) and the Orange County Sanitation District (OCSD). These two public agencies have worked together for more than 30 years and are leading the way in water recycling.

Main drivers for water reuse

The concept of the GWRS had been under consideration since the early 1990s. The project is an outgrowth of the wastewater reclamation work pioneered by OCWD and OCSD, beginning with implementation of Water Factory 21 in the mid 1970s to provide recycled water to the Talbert Gap Seawater Intrusion Barrier. Unlike Water Factory 21, which only supplied water for injection into the barrier, the GWRS is a large-scale recycled water program to provide water not only for an expanded seawater intrusion barrier, but also to provide a new supplemental water supply for groundwater recharge. OCSD joined with OCWD in the project to address its needs for peak flow disposal relief and to expand water recycling in Orange County. Therefore, the GWRS is both a water supply and a wastewater management project that provides many benefits to Orange County and California through the beneficial reuse of highly treated wastewater.

Brief history of the project development

Located in the coastal plain of Southern California, Orange County is a bustling suburb about 48.3 km (30 miles) southeast of Los Angeles. With an annual rainfall of less than 305 mm (12 inches), the population has historically relied on local groundwater to supply domestic demands. The northern two-thirds of the county are replenished by percolation from stormwater runoff carried by the Santa Ana River (SAR). The Santa Ana River Basin extends from the coast to the San Bernardino Mountains approximately 161 km (100 miles) to the east.

As early as the 1920s, the groundwater basin was being lowered from over-pumping. Artesian springs, once numerous in the area, were drying up even as the agricultural population was being replaced by an increasingly urban population. Now home to more than 2 million people, the county has been unable to meet its potable water demands from local groundwater sources on its own for years. Additional sources of water were required, not only for Orange County, but for all of southern California.

To meet the growing demand, large regional water projects were constructed to import water from sources outside of the area, first from the Colorado River (1930s) and then from northern California (1972). Imported water is available from the Metropolitan Water District (MWD). Currently, approximately 60% of the water in north Orange County is supplied from groundwater, while the rest is from imported sources. Almost all of the drinking water in the south part of the county is obtained from imported sources.

With most of the imported water sources already identified and exploited, new sources were required to keep up with the still increasing demand. Additional imported water from northern California is limited by strict environmental regulations and the difficulties with passing water through the Sacramento River Delta region between San Francisco and Sacramento. Demands from neighbouring arid states, such as Nevada and Arizona, as well as Mexico, threaten to even further decrease California's allotment of Colorado River water. In this political atmosphere, groundwater is an even more critical resource.

In 1933, the California legislature created the Orange County Water District to protect the Orange County portion of the Santa Ana River groundwater basin. OCWD's mandate was to protect the unadjudicated groundwater basin from "depletion and irreparable damage." OCWD's boundary is shown in Figure 25.1. Over the years, OCWD has implemented numerous projects to percolate imported water into the basin to supplement natural recharge. In the mid 1970s, OCWD envisioned and constructed Water Factory 21 to protect the basin from seawater intrusion. Water Factory 21 won international acclaim for its advanced technology, including the world's largest operating reverse osmosis plant, to transform wastewater to pure drinking water. In 1991, the California Department of Health Services (CDPH) granted OCWD the first ever permit to inject 100% recycled wastewater, without blending, into the groundwater basin. OCWD also built the Green Acres Project (GAP) to supply reclaimed water for industry and landscape irrigation.

Unfortunately, in the late 1990s, Water Factory 21 was near the end of its useful life. The technology that was state-of-the art in 1975 had been replaced by newer, more cost-efficient technology. OCWD looked to replace the existing facility with a larger, more productive facility. The project to replace Water Factory 21 was called the Groundwater Replenishment System.

Project objectives and incentives

The water produced by the GWRS is a "new" source, in that it is using water to satisfy potable demand from water that was previously lost when discharged to the ocean. Table 25.1 summarizes the major goals and objectives of the GWRS project.

Water supply objectives

Reliable New Source – GWRS product water is a new, reliable source of water for Orange County. It is a new water source because wastewater that was previously discharged to the ocean where it was lost is now available to meet the county's demands. Since wastewater is a never-ending stream, even during the deepest drought, the source is not affected by the weather, regional politics, or changing environmental regulations that can influence imported sources from northern California or the Colorado River. Even when imported water supplies are curtailed because adequate surface water is unavailable, the GWRS can continue to operate at full capacity.

Lower Groundwater Salinity – The salts or total dissolved solids (TDS) in the Santa Ana River groundwater basin that underlies most of central and north Orange County have increased significantly over the last 50 years, due to the use of high TDS imported water entering the watershed as discharges from upstream wastewater plants. As the secondary drinking water standard set by the U.S. Environmental Protection Agency (EPA), a TDS of 500 milligrams per litre (mg/L) is the level where regulatory concerns become a factor. Figure 25.2 shows the area that measured TDS concentrations greater than 500 mg/L in 1954 as compared to 1998, where more than half of the basin had a TDS in excess of the 500 mg/L. Unless lower TDS water was recharged into the groundwater basin, the salinity level would continue to rise to a point where the groundwater would become unusable for potable water use without wellhead treatment. With the GWRS, low TDS recycled water (less than 100 mg/L) is percolated into the northeast part of the basin, where over time, the lower TDS water will eventually reduce the salinity level of the whole groundwater aquifer.

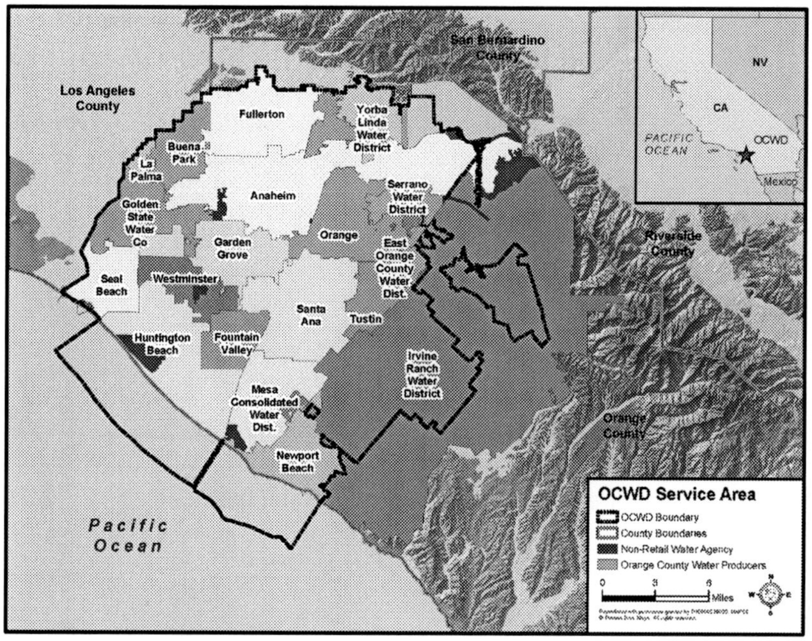

Figure 25.1 Orange County Water District boundary with local cities and retail water providers.

Seawater Intrusion Barrier Expansion – Water Factory 21 produced about 18,930 m^3/d (5 mgd) of recycled water for the Talbert seawater intrusion barrier located in Ellis Avenue. Ellis Avenue runs in an east/west alignment parallel to shoreline about 4.8 km (3 miles) inland from the ocean. Based on computer modelling, it was estimated that to maintain current pumping levels, more than 151,400 m^3/d (40 mgd) of water would be injected to keep the seawater out of the basin. The GWRS has the capacity to provide this higher flow to the barrier. It also increased the number of injection wells sites to a total of 36 by adding injection wells on the west and east ends of the barrier.

Table 25.1 Groundwater replenishment system goals and objectives.

Goals	Objectives
Water Supply	Provide a new, reliable source of water in times of drought and to meet future increasing demands. Reduce salinity of the groundwater. Expand the seawater intrusion barrier to protect groundwater quality. Maintain high groundwater utilization by replenishing the basin with a local source of water, reducing the region's dependence on imported water.
Wastewater Management	Divert peak effluent discharges to defer the need to construct a new ocean outfall. Reduce overall discharges to the outfall. Support water recycling to conform to the OCSD charter.

Reduced Dependence on Imported Water – In southern California, imported water was historically used for both potable water and recharge demands. New sources of imported water are not expected to grow significantly, since imported sources are considered to be fully exploited. With the GWRS, OCWD was able to use recycled water to supplement the imported water, reducing the demands for imported water by freeing up the potable water for direct domestic consumption. As shown in Figure 25.3, the GWRS, in conjunction with local conservation efforts, will account for about three-fourths of the additional quantity required by 2020 to meet future demands. Reducing the use of imported water to southern California also reduces the environmental impacts on the Sacramento River Delta and conforms to the state constitution, which supports water recycling.

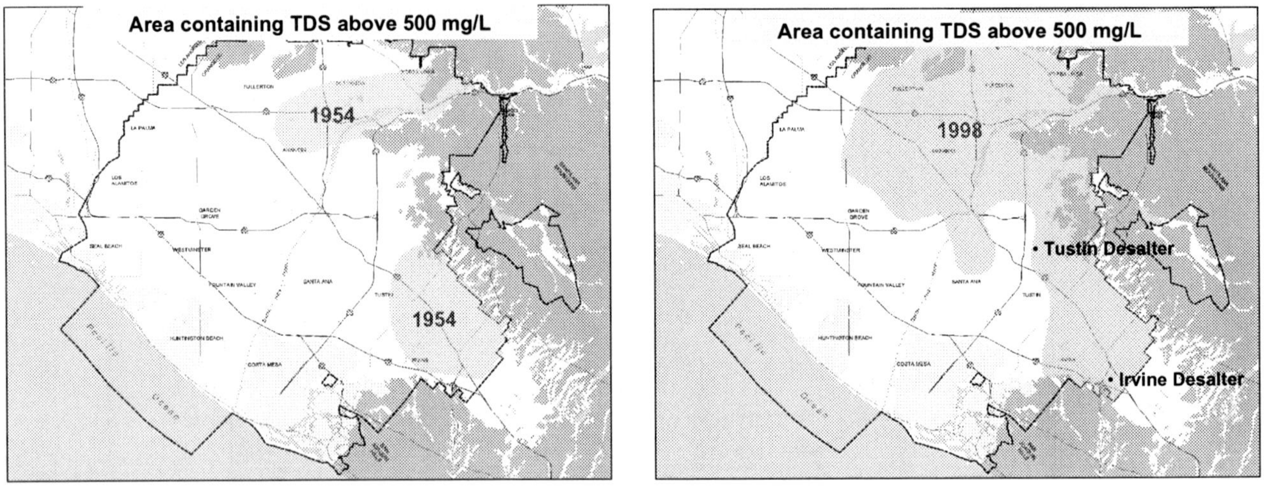

Figure 25.2 Salinity expansion within the Santa Ana River groundwater basin.

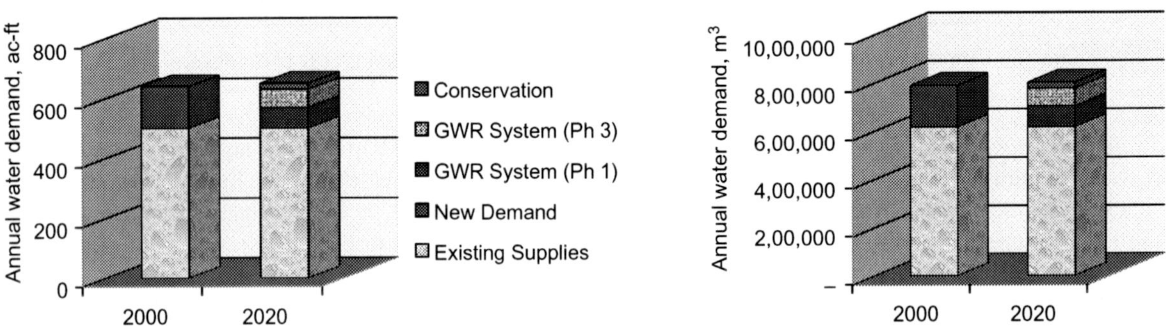

Figure 25.3 Sources to meet future demands.

Wastewater management objectives

Wastewater management objectives were developed to defer construction of a new outfall and decrease discharges to the ocean.

Divert Peak Flows to Defer Outfall Construction – Historically, population increases in Orange County increased the flows to OCSD. By the late 1990s, the peak flows had increased to a point where additional outfall capacity would soon be required. Due to the high cost and the regulatory uncertainty of constructing a new outfall, OCSD partnered with OCWD to use the GWRS to treat the peak flows and recycle the water instead of discharging that water to the ocean. This allowed OCSD to defer building a new outfall. The collaboration between the two agencies enabled OCSD to provide financing for about one-half of the capital costs for the GWRS from funds that were originally designated for outfall construction.

Reduce Overall Effluent Discharges – Overall effluent discharges to the ocean are reduced by recycling the wastewater that would normally be lost to the ocean through the outfall. The combination of reducing the ocean discharges and construction of additional secondary treatment capacity will allow OCSD to renew their ocean discharge waiver with fewer objections. The reduction in outfall discharges shows OCSD's commitment to being environmentally friendly as it lowers the potential for beach contamination.

Conforms to OCSD's Charter – OCSD's charter acknowledges that wastewater is a valuable resource. Participating with OCWD to use the GWRS to recycle symbolizes OCSD's commitment to beneficial reuse of the wastewater, continuing a relationship with OCWD that began with Water Factory 21 in the mid 1970s.

25.2 TECHNICAL CHALLENGES OF WATER QUALITY CONTROL

The GWRS consists of three major components:

1. Advanced water purification facility (AWPF)
2. The GWRS pipeline (a major pipeline connecting the AWPF to existing recharge basins)
3. Expansion of the existing seawater intrusion barrier

The locations of the major project components are shown on Figure 25.4, including the associated OCSD treatment facilities and the existing ocean outfall.

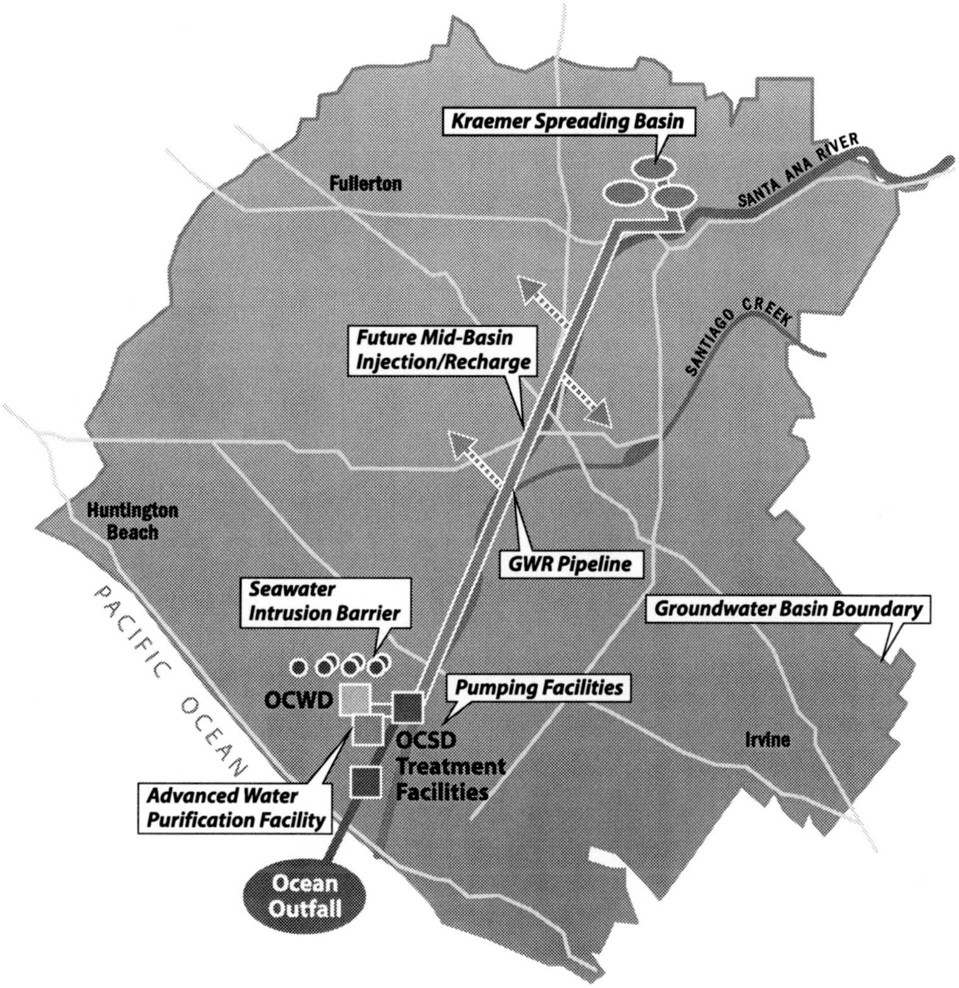

Figure 25.4 GWRS facilities map.

Implementation of the GWRS is being phased. Phase 1, completed in 2008, produces up to 89.3 million m^3/yr (Mm^3/yr) equivalent to 72,400 acre-feet per year (AFY) of recycled water (265,000 m^3/d or 70 mgd at 92% utilization). Phase 2 was bid in 2011 and will begin operation in 2013, increasing the capacity to 378,500 m^3/d (100 mgd). Ultimately, the GWRS will treat up to 492,000 m^3/d (130 mgd), or more than 165.3 Mm^3/yr (134,000 AFY) as anticipated demands for recycled water are realized and additional wastewater flows are available.

Treatment trains for water recycling

Advanced Water Purification Facility (AWPF)

The AWPF is the largest and most expensive component of the GWRS. The AWPF process train is often called the "California Model" of advanced recycled water treatment plants. The California Model uses membrane filtration (MF) as pre-treatment for reverse osmosis (RO), followed by disinfection and advanced oxidation using ultraviolet irradiation with peroxide (UV-A). MF pre-treatment is provided by fine screens, followed by either microfiltration or ultrafiltration systems. Low-pressure RO provides most of the treatment. UV-A is provided by an in-vessel, pressure UV system, which also destroys N-nitrosodimethylamine (NDMA) and other compounds of emerging concern. Figure 25.5 shows the process flow diagram for the GWRS. A summary of the most significant Phase 1 AWPF design criteria is shown in Table 25.2.

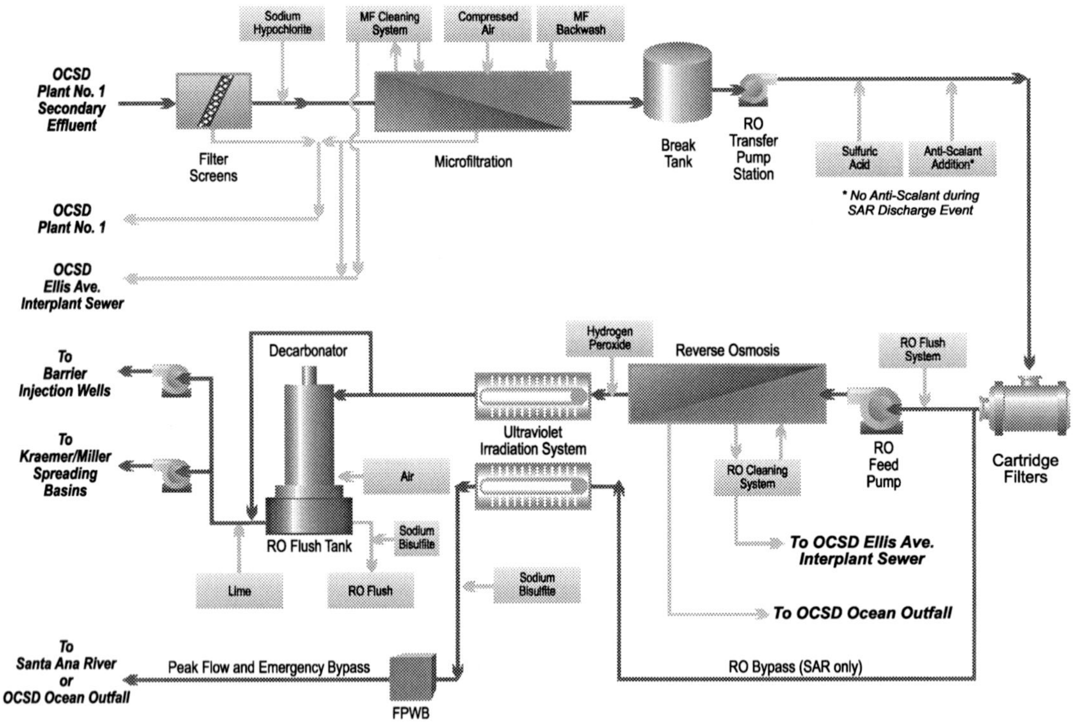

Figure 25.5 GWR process flow diagram.

Phase II will incorporate an influent equalization tank to even out fluctuations in influent flow caused by the diurnal flows into the wastewater pant. It will also expand the MF, RO and UV-A systems to increase the capacity of the plant to 378,500 m^3/d (100 mgd).

Water quality control and monitoring

The GWRS is fed from the OCSD's Plant No. 1. Plant No. 1 discharges two separate secondary effluent streams—one that has been treated in an activated sludge (AS) process and a second that has been processed through trickling filters (TF). The quality of the activated sludge effluent is better for reclamation purposes, and therefore is the primary source of feedwater to the GWRS. A goal of the blended feed water quality is to obtain maximum concentrations of 20 mg/L biochemical oxygen demand (BOD) and 20 mg/L total suspended solids (TSS), though typically the water is much better. Table 25.3 shows the average water quality for 2008.

The primary measure of the plant's performance is based on water quality, including total organic carbon (TOC), total nitrogen, TDS and NDMA. These parameters give an indication of the overall plant performance, especially in regard to the RO and AOP processes. Many of the water quality requirements are beyond those for primary and secondary drinking water standards. Critical water quality requirements defined in the operating permit include TOC (less than 0.5 mg/L in recycled water), total nitrogen (less than 5 mg/L), and NDMA (less than 10 micrograms per litre, 10 μg/L).

Table 25.2 Phase 1 Advanced Water Purification Facility design criteria.

System	Design criteria
Screening Facilities	Rotating 2-mm gravity screens to remove larger particles from clarified secondary effluents Gravity flow from OCSD Plant #1 to the MF.
Membrane Filtration	Capacity: 326,000 m^3/d (86 mgd) at a design flux rate of 34.7 L/m^2.h (20.4 gfd). 26 Siemens (US Filter) CMF-S microfiltration (cells) 3 trains of 8 cells each (plus 2 extra cells in 4th train) Cell: 19 racks per cell with 32 MF modules on each rack MF membrane pore size: 0.2 μm. Recovery rate: approximately 89% MF filtrate quality: turbidity less than 2 NTU and SDI less than 3 Required minimum cleaning cycle: 21 days. 22,700 m^3/d (6 mgd) of extra capacity was provided to supplement the T22 Green Acres Project
MF Break Tank/RO Transfer Pump Station (ROTPS)	Tank volume: 6057 m^3 (1.6 million gallons), separates the MF system from the RO system Storage time: approximately 16 minutes at Phase 3 flows. MF Break tank also supplies backwash water for the MF system. ROTPS: Five 113,600 m^3/d (30 mgd) vertical turbine pumps (1 spare) with 932 kW (1,250 Hp) motor
Reverse Osmosis	Capacity: 265,000 m^3/d (70 mgd) 5 trains with three 18,930 m^3/d (5 mgd) RO units each (1 spare unit) Array: X:Y:Z three stage array with 150 RO vessels each train, Membranes: >15,000 ESPA-2 RO thin film composite membrane elements Recovery rate: 80%–85%. Flux rate: 20.4 L/m^2.h (12 gfd), design RO Feed Pumps: dedicated SS vertical turbine variable speed pump for each RO unit and pressures ranging from 6.9 bar (100 psig) to 22.5 bar (325 psig) with 746 kW (1,000 Hp) motors
UV-A Disinfection/NDMA	Capacity: 265,000 m^3/d (70 mgd) Trojan UVPhox UV system Nine UV trains with one equivalent redundant train. Unit: three SS vessels each with 2 chambers per vessel Lamps: 72 lamps/chamber, 3456 active low pressure high output (LPHO) total Water Transmittance: >95% on RO permeate
Pumping Stations	Product Water Pump station (PWPS) Pumps: three 113,600 m^3/d (30 mgd) vertical turbine variable speed pumps, 1678 kW (2250 Hp) each Discharge: Kraemer Basin and mid-basin injection Barrier Pump Station (BPS). Pumps: four 56,780 m^3/d (15 mgd) vertical turbine variable speed pumps, 447 kW (600 Hp) each Discharge: Talbert Barrier injection wells
Chemical Storage and Feed Systems	Sodium hypochlorite (plant chloramine residual) Sulfuric acid (pH adjustment) Threshold inhibitor (mineral suspension) Citric acid (membrane cleaning) Sodium hydroxide (membrane cleaning) Hydrogen peroxide (contaminant oxidation) Lime (post treatment water stabilization) Sodium bisulfite (chloramine and peroxide residual neutralization)

The AWTF also removes compounds of emerging concern (CECs) that are not currently regulated in California, but may be considered for potential future regulations. The CECs most often found in wastewater include pharmaceuticals (acetaminophen, ibuprofen, caffeine, carbamazapine, gemfibrozil, primidone, sulfamethoxazole), flame retardants (TCEP), and pesticides (DEET, diuron, triclosan). None of these chemicals, nor any other CECs tested, are found in measurable concentrations after UV-A treatment.

Table 25.3 Average water quality of the GWRS advanced water treatment plant (2008 data).

Constituent	MCL[a]	Influent (mg/L)	MF Feed (mg/L)	RO Feed (mg/L)	ROP[b] (mg/L)	FPW[c] (mg/L)
Total Coliform (MPN/100 mL)		1.39 Million	45,200	<2	<2	<2
Total Suspended Solids, mg/L		6.41	6.17	2.80		
Total Dissolved Solids, mg/L	500–1000	920		963	19.9	81.0
Total Organic Carbon, mg/L	0.5/RWC	14.1	15.8	11.8	0.17	0.19
Turbidity (NTU)	<5	2.94		0.11		0.18
UV Transmittance % @ 254 nm			61.1		97.75	98.8
pH	6.5–8.5	7.69	7.74	7.14	6.12	8.22
Total Hardness, mg$CaCO_3$/L	290			290	<.1	22.9
Total Nitrogen, mgN/L	5	27.9		27.6		1.67
Ammonia, mgN/L		24.7		24.8	1.35	1.35
Total Phosphorus, mgP/L		0.79				<0.01
NDMA, ng/L	10	31.1			26.3	<2
1,4-Dioxane, µg/L	0.10	1.42				<1

[a]MCL: Maximum Concentration Level.
[b]ROP: the RO permeate flow stream.
[c]FPW: finished product water quality as it leaves the plant.

Main challenges for operation

OCWD has been operating the GWRS since January 2008 and has experienced several operational challenges. The large size of the AWPF (Figure 25.6), coupled with the complex nature of the equipment, caused some compatibility and monitoring problems that led to excessive alarms and premature RO membrane fouling. Operation of the post treatment lime stabilization system also proved to be a challenge to meet the water quality requirements while not fouling the injection wells.

Figure 25.6 View of GWR process buildings: microfiltration, reverse osmosis and lime treatment (Photo credit: Larny Mack).

Plant SCADA alarms

The plant supervisory control and data acquisition (SCADA) system, known as the Process Control System (PCS) for this facility (Figure 25.7a), uses an Emerson Delta V hybrid distributed control system (DCS) in conjunction with DeviceNet and Foundation Fieldbus digital technology for all plant instruments and motors. This type of hybrid system, while often used in large chemical and industrial facilities, is new to the water industry. It allows for a dramatic reduction in the amount of wiring and increases the amount of monitoring available within the plant. This system also allows for multiple plant instruments (i.e. temperature transmitters, pressure indicating transmitters, pH monitors, electrical conductivity analyzers, turbidimeters, etc.) to be grouped on a common fibre optic loop or segment. This grouping of equipment by segments allows for easier troubleshooting of any issues, but can also lead to frequent alarm notifications by the PCS, if one of the instruments or motors on a segment experiences a communication or input/output (IO) failure.

(a) (b)

Figure 25.7 Advanced water purification facility (AWPF): (a) SCADA system alarm monitoring and (b) staff checking the reverse osmosis vessels connections.

After spending significant operator effort trying to respond to alarms, an exhaustive investigation by OCWD plant operations staff determined that the sources of the nuisance alarms were caused by two issues: (1) loose wiring or incorrect wiring on communication segments and (2) incompatibility issues associated with firmware (software) imbedded in certain plant equipment. Because of the large size of the facility, there are large amounts of communication cables and electrical wiring where loose connections or faulty wiring could cause problems. Fortunately, the problems were isolated and easily repaired. Motors and instruments supplied on the project were all Foundation Fieldbus and DeviceNet certified and were to be compatible with the Delta V system. However, several certified vendors used on the project had only limited experience with Fieldbus and DeviceNet communication systems. The firmware supplied with the equipment was not always 100% compatible and required software "patches" to alleviate the problems. In some cases, this required a download of the new firmware to hundreds of pieces of equipment.

Process monitoring issues

The large size of the AWPF also led to difficulty in monitoring some parameters within individual process units. For example, the MF process contains 26 individual treatment basins or cells, with each cell independently operated. Given that fouling nature of the membranes within the cell is unique to each cell, the irregular fouling rate from one cell to another made it difficult to determine overall system trends. However, by monitoring performance trends from each individual cell, some abnormalities could be identified. During the first year of operation, the plant operators noticed that two of the 26 MF cells began to foul at a rate much greater than the other cells. It was found that a small valve on the aeration pipe was sticking, essentially cutting off air to the backwash cycle, allowing fouling particles to stay on the membranes. While the AWPF has monitored operation of the aeration system, there were no alarms on the individual isolation valves to each individual basin.

Third stage RO fouling by aluminium silicate also had to be resolved (Figure 25.7b). Since the RO feedwater contains relatively low levels of silica and aluminium (22 mg/L as silicate and 12 μg/L aluminium), the RO projection model did not predict these elements as key foulants. OCWD measures flow from each stage manually, since the RO system did not contain interstage flow meters or conductivity analyzers on each RO unit. After about 8 months of operation, it was observed that the third stage flows were decreasing at a far greater rate than flows in the first two stages. OCWD investigated cleaning regimes to remove this scale and found that cleaning of the membranes with ammonium bifluoride was effective. In addition, OCWD varied the feed pH to the RO system on a seasonal basis to help control scale, while also reducing sulfuric acid costs. In the hotter summer months, the feed pH is kept near 6.5 to 6.6, while in the colder winter months, the pH is held near 6.7 to 6.8. Finally, a new anti-scalant product specifically designed to limit silica-based scale was used to lengthen time between membrane clean-in-place procedures.

Injection well fouling

The AWPF post-treatment system uses hydrated lime (calcium hydroxide) to stabilize the low alkalinity RO product water prior to being sent to the seawater intrusion barrier and spreading ponds. This system consists of decarbonation with partial bypass followed by lime addition, lime storage silos, slurry mix tanks, slurry transfer pumps, saturators and a polymer addition system. The lime system makes a 7% slurry that is pumped to a saturator/clarifier, where the supernatant from the saturator is then dosed into the RO permeate water. The lime system helps produce a final product water (FPW) with a slightly positive Langelier Saturation Index (LSI) and an Aggressive Index (AI) of near 12. In addition, the plant operating permit requires the final product water pH to be between 6 and 9.

The saturator was initially operated as a solids contact clarifier; however, it was soon discovered that the downstream injection wells began to foul at an increased rate. Subsequent tests found the final product water silt density index (SDI) and modified fouling index (MFI) values to be well above 10. The SDI and MFI should be below 3 to ensure the water does not have a significant fouling potential. The saturator was changed to operate in a sludge blanket mode, allowing a layer of lime sludge to build up in the bottom of the saturator to increase the capture slowly settable lime solids. After this change, the rate of well fouling significantly decreased and the SDI/MFI numbers dramatically improved. Figure 25.8 shows the improvement of the final product water's SDI and MFI after the switch to sludge blanket operation of the saturator.

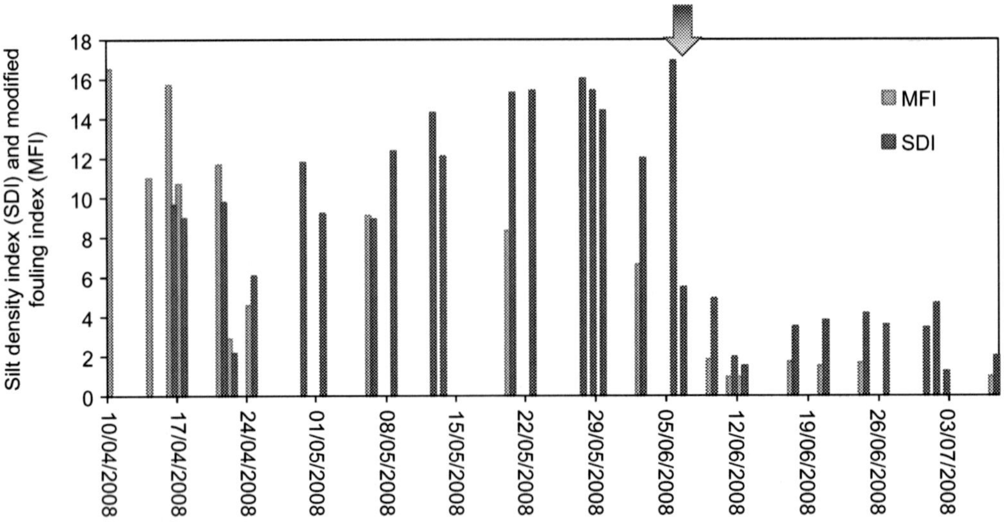

Figure 25.8 Improvement in SDI and MFI after change to sludge blanket operation.

25.3 WATER REUSE APPLICATIONS

Description of aquifer recharge facilities

The GWRS sends about one-half of the purified water to the seawater barrier along Ellis Avenue near the treatment plant, while the remaining water is pumped 20.9 km (13 miles) up the Santa Ana River to the Kraemer Basin for groundwater recharge.

Barrier facilities

Figure 25.9 shows the existing 28 OCWD injection wells and the additional eight GWRS well sites. The southeast wells are easily identified at the bottom right of the figure. The new Ellis Avenue wells are located on the left hand side of the map. A future line of 13 wells located south of the existing barrier will be needed to provide additional well capacity. This east-west alignment of these future injection wells will be located close to the aquifer mergence zone, which is an optimal injection location for the seawater barrier. These future injection wells will be installed when the older, existing Ellis Avenue wells reach the end of their useful life and can no longer be redeveloped efficiently to maintain the required injection flow rates.

The four well sites on the west end of the barrier have three injection wells located at each site. Each well injects water into a different level of the aquifer, ranging from 43 to 58 m (140 to 190 feet) below ground surface (bgs) for the shallowest aquifer and 189 to 210 m bgs (620 to 690 feet) for the deepest aquifer. Figure 25.8 also shows one of the multiple injection wells sites. The well sites on the east along the Santa Ana River only have a single well at depths of approximately 37 to 44 m bgs (120 to 145 feet).

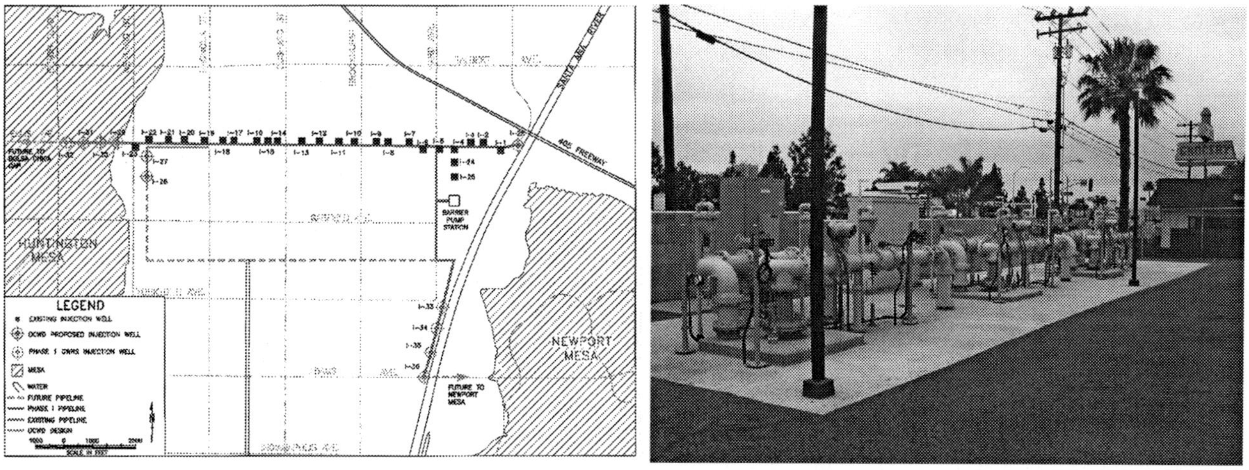

Figure 25.9 Talbert Barrier injection well facilities.

Kramer basin percolation pond

The Kramer Basin is located in the city of Anaheim, approximately 21 km (13 miles) from Fountain Valley, where the treatment plant is located. It is an old gravel pit that was originally used by OCWD to percolate Santa Ana River stormwater into the groundwater basin. The Kramer Basin was dedicated to the GWRS because it could operate year round without frequent cleaning, due to the high quality of the recharge water. Figure 25.10 shows the basin and the adjacent Miller flood control basin that can be used when the annual cleaning is performed.

Figure 25.10 Views of Kramer Basin recharge facilities and Miller basin, inlet structure and the district's cleaning machine.

Evolution of the volume of supplied recycled water

Table 25.4 shows the planned GWRS flows to the barrier or Kraemer Basin during the various seasons. Treated water flows from GWRS are directed either to the seawater barrier (barrier flow) or to the Kraemer Basin percolation ponds – recharge flow) in different amounts, depending on the prioritized needs of the barrier. This allows the GWRS (Figure 25.11) to operate at full capacity all year, with the majority of flow going to the barrier in the summertime, when basin pumping is high. Water flows to the percolation ponds for recharge during the winter, when recycled water needs in the barrier are reduced.

Table 25.4 GWRS annual flows by discharge location.

GWRS flows	Days	Barrier flow, m^3/d (mgd)	Total, Mm^3 (MG)	Days	Recharge flow m^3/d (mgd)	Total, Mm^3 (MG)	Total GWRS flow, Mm^3/yr	Total GWRS flow, MG/yr (AF/yr)
Winter w/ Recharge	50	56,780 (15)	2.8 (750)	50	208,200 (55)	10.4 (2750)	13.2	3500 (10,741)
Winter w/o Recharge	40	56,780 (15)	2.3 (600)	40	208,200 (55)	8.3 (2200)	10.6	2800 (8593)
Transition (Spring)	45	113,600 (30)	5.1 (1350)	45	151,400 (40)	6.8 (1800)	11.9	3150 (9667)
Summer	185	159,000 (42)	29.4 (7770)	185	106,000 (28)	19.6 (5180)	49.0	12,950 (39,742)
Transition (Fall)	45	113,600 (30)	5.1 (1350)	45	151,400 (40)	6.8 (1800)	11.9	3150 (9667)
Annual Total	365		44.7 (11,820)	365		52.0 (13,730)	96.7	25,550 (78,410)

Figure 25.11 Aerial view of the Advanced Water Purification Facility (Photo credit: Larny Mack).

The total annual volume of recycled water is 96.7 Mm^3/yr (78,410 AFY), 46% from which is supplied to the seawater barrier and 54% to the percolation ponds.

Relations and contracts with end-users

As a groundwater management agency created by the state of California, the OCWD's main function is to maintain the quality and sustainability of the underlying groundwater basin within its service area. Twenty-three drinking water provider entities within the OCWD service area pump groundwater as the main source of drinking water to their customers. The district's main end user contact and contractual obligation is to provide pricing and set quantities of groundwater that can be purchased each year. OCWD is governed by a 10-person Board of Directors publicly elected in a general election. Each year, the board votes on a replenishment assessment, which is the fee charged per volume of groundwater used by drinking water entities. Also, the board sets a basin production percentage or a percentage of groundwater that drinking water entities can use to supply their customer's needs. The basin production percentage is used by OCWD to manage production from the groundwater and protect it from overdraft. Finally, OCWD also administers a penalty known as a Basin Equity

Assessment (BEA) that serves as a deterrent to drinking water entities from exceeding the allotted groundwater usage established by the basin production percentage. OCWD's relationship with each of the drinking water entities ensures that the basin production percentage and replenishment assessment set each year are economically feasible and not burdensome to the ultimate end user customers. These entities meet regularly with OCWD staff and board members to discuss basin production percentage and replenishment assessment projections for upcoming year.

25.4 ECONOMICS OF WATER REUSE

Project funding and costs

The capital cost to build the Phase 1 GWRS project was US$481 million. Federal, state and local funding totaling US$92.8 million was secured for the project. The rest was obtained from loans or contributions from OCSD, OCWD's GWRS partner.

Construction costs

The GWRS project had over 11 major construction contracts, including the treatment plant, pipelines, barrier facilities and other appurtenances. The original GWRS project costs, including engineering, legal and administration costs are shown in Table 25.5.

Table 25.5 GWRS project capital costs.

Facility	Cost, US$
MF and UV Vendor Design Services	$860,000
AWPF	$298,330,000
GWRS Pipelines	$63,010,000
Barrier Facilities	$16,380,00
Miscellaneous Projects*	$20,490,000
Other	$14,950,000
Subtotal Capital Costs	$414,020,000
Engineering, Legal, Administration (ELA)	$66,890,000
Total Project Cost (2008)	$480,900,000
AWPF Expansion (2011)	$142,000,000

*Miscellaneous facilities include a 18,930 m^3/d (5 mgd) temporary treatment plant to continue operation of WF21 during GWRS AWTF construction, a Southern California Edison substation, and temporary project management trailer complex.

Operation costs

OCWD is responsible for GWRS program management and for funding day-to-day operations. The annual GWRS operating budget is approximately US$34 million. This amount includes electricity, chemicals, and labor and maintenance costs. Table 25.6 provides the net annual GWRS operating costs for 2009–2010. Annual cost is based on 81.3 Mm^3 (65,950 AF) produced. Operation and maintenance cost for 2010 was 0.35 $\$/m^3$ (432 $/AF).

Project funding

While OCWD is the owner of the GWRS project, OCWD and OCSD shared the construction costs for the Phase 1 implementation. OCSD paid for one-half of the construction cost, or approximately US$180 million. The project also receives an annual subsidy of US$7.5 million per year for 12 years from the MWD. MWD provides the subsidy because the GWRS project reduces demand on the state's imported water supplies by using the recycled water in the seawater barrier, instead of potable water. Grant and loan funding is summarized in Table 25.7.

The overall net annual cost of the purified water after the grants and the OCSD contribution is removed is shown in Table 25.8 and was 0.39 $\$/m^3$ (478 $/AF) in 2010. The cost of the alternative source, treated imported water, is approximately US$800 per AF which is equivalent to 0.65 $\$/m^3$.

Table 25.6 OCWD net annual GWRS operation costs (US$, 2009–2010).

Parameter	Treatment Cost, $/yr	Distribution Cost, $/yr	Total cost $/yr	Total cost $/m³	Total cost $/AF
Electricity	6,347,318	1,319,000	7,666,318	0.094	116
Chemicals	4,152,496	0	4,152,496	0.051	63
Labor	6,966,873	725,450	7,692,323	0.095	117
R&R Fund Contribution	3,904,866	910,134	4,815,000	0.059	73
Plant Maintenance	3,540,947	593,550	4,134,497	0.051	63
Operating Costs Subtotal	24,912,500	3,548,134	28,460,634	0.350	432
Debt Service	9,326,263	2,174,736	11,499,999	0.141	174
Operating and Capital Subtotal	34,238,763	5,721,870	39,960,633	0.491	606

Table 25.7 GWRS capital funding sources.

Funding or grant source	Type	Amount, US$
EPA	Grant	$500,000
U.S. Bureau of Reclamation Title XVI Program	Grant	$20,000,000
State Water Bond (Prop 13)	Grant	$37,000,000
State Water Resources Control Board (SWRCB)	Grant	$5,000,000
State Department of Water Resources	Grant	$30,000,000
California Energy Commission	Grant	$300,000
State Revolving Loan Fund	Loan	$145,000,000
OCSD (Phase 1 Construction only)	Partner Contribution	$180,000,000
Metropolitan Water District	Annual Subsidy	$7,500,000

Table 25.8 OCWD GWRS net costs (US$, 2009–2010).

Parameter	Treatment Cost, $/yr	Distribution Cost, $/yr	Total cost $/yr	Total cost $/m³	Total cost $/AF
OCWD's Operating and Capital Subtotal	34,238,763	5,721,870	39,960,633	0.491	606
MWD LRP Subsidy	−7,374,942	0	−7,374,942	−0.091	−112
Demand Response Credit	−547,823	−82,177	−630,000	−0.008	−10
OCSD Contribution to Maintenance (ended May 2010)	−416,667	0	−416,667	−0.005	−6
Operating Subsidies Subtotal	−8,339,432	−82,177	−8,421,609	−0.104	−128
OCWD Net Cost	25,899,331	5,639,693	31,539,024	0.388	478

Pricing strategy of recycled water

OCWD receives revenue from assessments paid by the groundwater basin pumpers, a percentage of the local property taxes and investments. OCWD is not a wholesale or retail water provider. OCWD produces the purified recycled water and uses it to recharge and protect the groundwater basin from seawater intrusion. OCWD does not charge by the amount of water that is recharged. Instead, funding is derived from a replenishment assessment (RA) charged when the retail water providers extract groundwater from the basin. The RA covers OCWD operations for administration, engineering and operations for the district staff and facilities. The RA fee charged per volume of groundwater pumped by the local retail water districts is currently 0.20 $\$/m^3$ (US$249 per AF) for 2011. The RA is set yearly by the OCWD board after negotiations between the pumpers and district.

Benefits of water recycling

Both water supply and wastewater management needs are addressed by the GWRS. The GWRS is a new, locally controlled source of water that can meet the future increasing demands from the region's groundwater producers. It is reliable, drought proof, and reduces the area's dependence on imported water transferred through the Sacramento-San Joaquin Delta. By replacing high TDS imported water from the Colorado River with low TDS recycled water, over time, the water produced by the GWRS will lower the salinity of groundwater. While difficult to quantify, the lower TDS water reduces consumer costs for cleaning and extends the life of equipment that use water, such as water heaters, washing machines, dishwashers and plumbing fixtures. Though a lot of power is required to purify the water (83,042 MWh/yr or 1.1 kWh/m^3), the GWRS project is sustainable, using less energy than would be required to import an equivalent amount of water from the State Water Project (2.59 kWh/m^3) or the Colorado River (1.62 kWh/m^3).

The project benefits OCSD as well as OCWD, since peak effluent discharges are diverted to the GWRS, deferring the need to construct a new ocean outfall. This saved OCSD approximately US$180 million, which was then used to help fund the GWRS. Since ocean discharges are decreased, there is less impact to the ocean environment, gaining the support of the environmental community.

25.5 HUMAN DIMENSION OF WATER REUSE

Public education and communication strategy

In the past, public education programs often consisted of telling the public after the project was built. After some high-profile projects in southern California were cancelled due to adverse public reaction, OCWD saw the need to develop a public outreach program that proactively communicated their message to interested stakeholders. Otherwise, there was a risk that the public would learn others' positions, instead of the agency's. Stakeholders included regulators, elected officials, environmentalists, private citizens, activists, the media, educators, customers, friends and opponents. OCWD understood that they were a part of the community and that positive relationships were essential to gaining public support.

OCWD also recognized that communicating with the public was more complex than in the past. With thousands of messages being sent every day from television, radio, magazines, cell phones, faxes, the internet and social media, getting the right message heard was not easy. Each individual has different characteristics that shape what they hear and what they will accept or reject. "Techno jargon" often confused or harmed the message, making it harder to overcome prejudices and messages that did not agree with one's general outlook on life.

The public outreach program began in 1997, even before the beginning of the design. Based on focus groups and telephone survey research, OCWD developed a message that the water was new, safe, reliable, and better than alternative sources, and that the time to implement it was now. They asserted that the proposed GWRS project would solve many of Orange County's water problems. To spread the word, OCWD invested in an outreach program that consisted of an effective four-step process: (1) research, (2) plan, (3) communicate and (4) evaluate.

Research provided information and insight regarding the major reasons the public supported or opposed the project, obstacles to be overcome, and messages that worked the best at reaching the community. OCWD found that safety was the most important issue and that support would increase if people were briefed on the technology that purified the water to near distilled quality. Use of simple terminology worked better and some words, such a purified, resonated much better than reclaimed, recycled or treated water. Community leaders indicated they would support the project, because it provided a new source of water that had better quality than many of the alternatives.

Based on the research, OCWD developed a clear message that focused on threats to the water supply and proven technology, and identified GWRS as the solution. The key GWRS public outreach messages are:

1. Future supplies are threatened by pollution and population growth.
2. Diversity and reliability of the water supply is good.
3. Efficiency and sustainability over the long term is important.

4. The quality of purified water is the highest available with multiple safeguards, proven technologies and constant safety monitoring.
5. Know your constituency, be transparent and build trust within the community. Support from legislators, business leaders, health experts and educators helps build that trust.
6. Messages should be kept short to be effective on short sound bites or video segments.

To communicate the message, OCWD held face-to-face meetings with every leadership group in Orange County, including political, business, education and health leaders. With hundreds of face-to-face meetings, the community outreach aspect of the plan was the most effective means of distributing the message and gaining public support. OCWD was also proactive with the media, not hiding what they were proposing to do. Newsletters, brochures, videos, websites, new releases and other forms of educational materials were used to distribute the message to the public. As the message began to be heard in the community, OCWD evaluated the effectiveness of the message and the communication tools. OCWD added a direct mail campaign and cable television spots to reach even more people. They also increased their effort to gain support of legislators, environmentalists and the minority community.

The role of decision makers

Elected officials

OCWD is the owner and operator of the GWRS; therefore, the OCWD Board of Directors must approve all major decisions. As a partner, the OSCD Board of Directors also reiterates their support at major milestones. The GWRS Steering Committee performs regular oversight of the project and is comprised of three members from the OCWD board and three members from the OCSD board. Public meetings are held on the second Monday of every other month to review and approve operational, financial, planning and joint agency issues.

Regulators

The GWRS produces high-quality water that exceeds all state and federal drinking water standards. The California Department of Public Health (CDPH) reviewed the project to ensure that the public's health was protected. The California Regional Water Quality Control Board (RWQCB) for the Santa Ana River Basin also reviewed GWRS for the protection of the groundwater quality and environmental compliance, and approved and permitted the project.

Approval for injection along the coast at Talbert Barrier was received on January 3, 2008, and the GWRS began sending water to the barrier on January 18, 2008. Approval for percolation via Kraemer and Miller Basin was received on January 17, 2008 and began the same day. The operating permit establishes criteria for GWRS treatment, including TOC limits, minimum travel time and diluent water blending requirements.

Independent advisory panel

The operating permit requires that an Independent Advisory Panel provide a periodic, ongoing scientific peer review of the GWRS. The panel of experts is appointed and administered by the National Water Research Institute (NWRI). Panel members are experts in water treatment technology, public health, environmental engineering, hydrogeology, microbiology, chemistry and toxicology.

Public acceptance and involvement

Over the approximate 4-year period of the outreach program, no organized opposition to the project surfaced. The citizens of Orange County put a high value on a reliable water supply and understood that water is a scarce resource and should not be wasted. Creating purified water from an advanced membrane treatment facility increased feelings of reliability and safety. Most of the OCWD presentations were made by OCWD staff, not consultants, with an Independent Advisory Panel providing an outside technical review, as required in California. Because the project reduced discharges to the ocean, the project had strong support from the environmental community.

The project has amassed more than 400 supporters, consisting of elected officials, environmental groups, civic groups, businesses, religious organizations, educational entities, water agencies, and health and medical officials. OCWD ensured that letters of support were obtained during planning, design and construction of the GWRS project, including support from all elected officials in the OCWD service area. As part of the public outreach program to gain public acceptance for the project, several hundred plant tours have been given by staff. The public outreach program has been credited with the fact there was no organized opposition to the project, even though similar projects in southern California had failed, due to negative public perception.

25.6 CONCLUSIONS AND LESSONS LEARNED

Beginning with Water Factory 21 in 1975 and continuing through the current expansion, the Orange County Water District, along with their partner, the Orange County Sanitation District, continue to be world leaders in sustainability and implementing recycled water projects, using an otherwise wasted resource that would be discharged to the ocean.

As the world's largest wastewater purification system, the GWRS produces a new, reliable source of water for indirect potable reuse. Since it began operation in January 2008 through August 2012, this state-of-the-art water facility has produced more than 340 Mm^3 (92.5 trillion gallons) of purified water. Its 265,000 m^3/d (70 mgd) provides enough water to meet the needs of nearly 600,000 residents in north and central Orange County, California. The GWRS is a complex project using MF, RO and UV-A to produce a product water that meets all regulatory requirements, has a low TDS, a low total nitrogen concentration and removes all of the CECs that are regularly tested. After treatment, the purified water is used to prevent seawater intrusion and recharge the groundwater basin. OCWD implemented a very successful and proactive public outreach program to gain governmental and community support. By the time the GWRS began operation, there was no significant organized opposition to the project.

The following Table 25.9 summarizes the main drivers, benefits and challenges for the Orange County Groundwater Replenishment System.

Table 25.9 Summary of lessons learned.

Drivers/Opportunities	Benefits	Challenges
Drought	Safe, local, reliable water supply for 600,000 residents	Public acceptance
Increasing potable water demands	High quality water	High cost and funding
Seawater intrusion	Sustainable, using less power than imported water	Size/capacity of such a complex system
Wastewater as a resource	Reduces discharges to the ocean	Regulatory approval
Increasing cost of imported water supply	Lowers dependence on imported water	Keeping existing Water Factory 21 operational during construction

REFERENCES AND FURTHER READING

Chalmers R. B., Patel M., Dunivin W. and Cutler D. (2008). Orange county water district's groundwater replenishment system is now producing water, Proceedings of the Water Environment Federation Technology Conference (WEFTE), Chicago, Illinois, AWWA Publishing.

Chalmers R. B. and Patel M. (2003). Meeting the challenge of providing a reliable water supply for the future, the Groundwater Replenishment System. Proceedings of the AWWA Annual Conference and Exposition, Washington, DC. AWWA Publishing, Denver, CO, USA.

Groundwater Replenishment System: Facts and Figures. (2011). Orange County Water District, Fountain Valley, CA.

Patel M. and Dunivin. (2010). Why operating membrane plants can drive you nuts, Proceedings of the AMTA Membrane Technology Conference, San Diego, WEF Publishing, Denver, CO, USA.

Wildermuth R. (2002), *Groundwater Replenishment System – A Public Education Project. AWWA Water Sources Conference Proceedings.* AWWA Publishing, Denver, CO, USA.

WEB-LINKS FOR FURTHER READING

http://www.ocwd.com/
http://www.gwrsystem.com/
http://cdmsmith.com/

Views of the RO building and purified water.

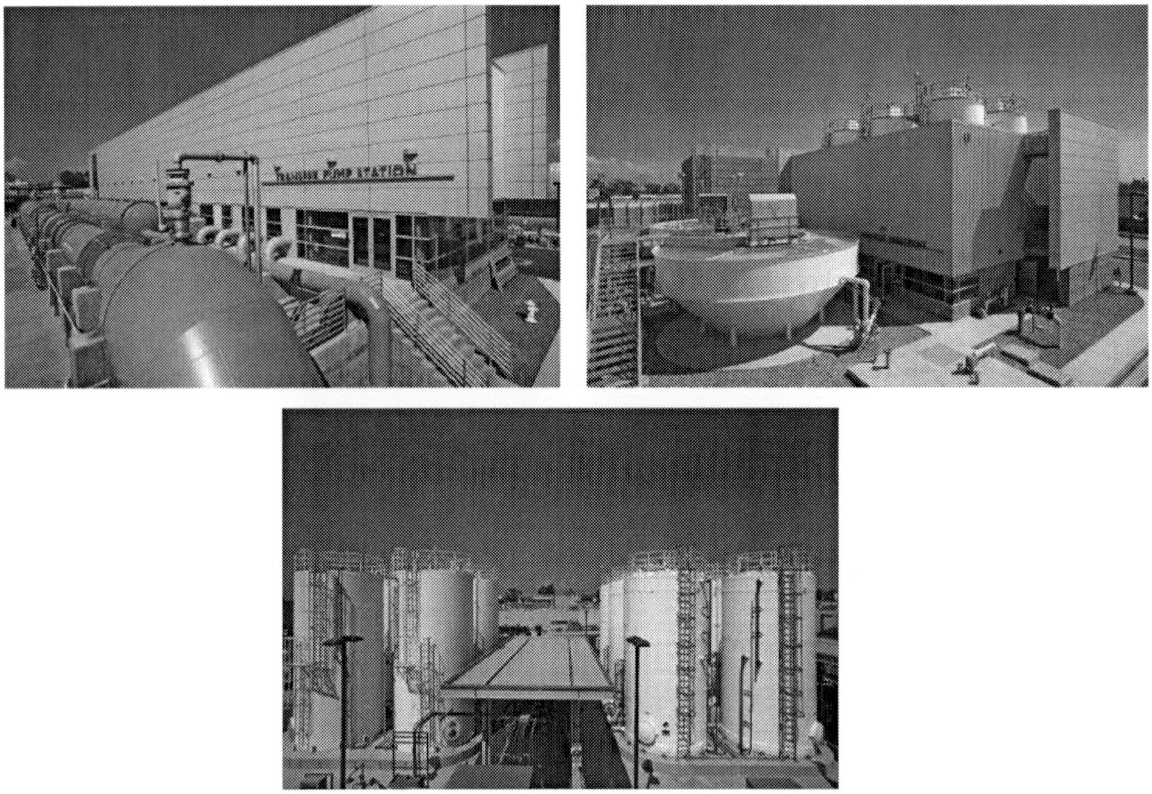

Views of the Advanced Water Purification Facility.

26 Torreele: Indirect potable water reuse through dune aquifer recharge

Emmanuel Van Houtte and Johan Verbauwhede

CHAPTER HIGHLIGHTS

The water reuse via managed aquifer recharge in the dunes of St-André resulted in a sustainable groundwater management and as such reconciled economy and ecology in this area where tourism is an important economic actor. Furthermore it is a pro-active measure to anticipate the expected effects of future climate change in this coastal area.

KEYS TO SUCCESS

- Intense preparation of the project with pilot tests using different techniques and membranes
- Infiltration project was integrated in a natural management plan for the area and enabled sustainable groundwater management
- Improved quality of drinking-water and secured supply
- No excessive increase of drinking-water price

KEY FIGURES

- Start up: July 2002
- Treatment capacity: max 7000 m^3/d
- Total annual volume of treated wastewater: max 3.3 Mm^3/yr (2005)
- Total volume of recycled water: 19.4 Mm^3/yr in 10 years of operation
- Capital costs: 7 M€
- Operation costs: 0.64 €/m^3 (data 2011)
- Recycled water tariffs: part of overall drinking-water tariff
- Benefits: high quality due to use of reverse osmosis, increased production capacity near point of use, sustainable groundwater management

26.1 INTRODUCTION

Since the 1950s, water demand in the western area of Belgium's Flemish coast expanded from 1 Mm^3 in 1950 to 5.8 Mm^3 in 1990. The three dune water catchments, where fresh groundwater is pumped from the unconfined aquifer by the Intermunicipal Water Company of the Furnes Region (IWVA), could not increase production to meet the new demand, as this would increase the risk of salt intrusion, which could negatively affect aquifer quality.

Main drivers for water reuse

Despite purchasing drinking-water from neighbouring regions since the 1970s, in summer periods water shortages were common in the area. The dune water catchments reached their maximum capacity and at the same time the ecological interest in the dunes was growing (Van Houtte & Vanlerberghe, 1998). At the beginning of the 1990s, alternative exploitation methods were studied to remediate the decreasing water table levels and to guarantee the current and future water extraction possibilities, resulting in the project for the artificial recharge of the unconfined dune aquifer. As no other year-round water sources are available in this area, the IWVA chose wastewater effluent from the nearby wastewater treatment plant of Wulpen as the source for the production of high-quality infiltration water.

Brief history of the project development

The start of the project was preceded by ten years of research. First, the Flemish Institute of Nature Conservation performed an ecological study of the water catchment of St-André resulting in a delineation of the infiltration area, the indication of ecological conditions (Kuijken *et al.* 1993) and an ecological management plan for this area. Two infiltration tests, using groundwater from sewage works, gave valuable information about the hydro-geological parameters of the dune aquifer in St-André (Lebbe *et al.* 1995). The impact of artificial recharge in St-André was simulated using MODFLOW (Van Houtte & Vanlerberghe, 1998).

Three major elements were considered when selecting the process to produce infiltration water (Van Houtte *et al.* 1998):

- The drinking-water should be supplied without any disinfection;
- The high ecological values of the dunes required high quality infiltration water;
- The land availability for buildings was restricted.

In view of the relative high nutrient and salt content of the wastewater effluent, a dual membrane treatment process was selected: first, a pre-treatment with either microfiltration (MF) or ultrafiltration (UF) membranes, followed by an additional desalination of the produced filtrate using reverse osmosis membranes (RO). Both processes needed less space than conventional plants and used a filtrate free of microorganisms.

Based on examples in the USA (Leslie *et al.* 1996), from 1997 until 1999 pilot tests were performed on the effluent from the wastewater treatment plant (WWTP) at Wulpen, operated by Aquafin (AQF). Different MF and UF systems were used for the pre-treatment (Van Houtte *et al.* 1998 and 2000) and two types of RO membranes were tested for the desalination (Van Houtte & Vanlerberghe, 2001).

Project objectives and incentives

The Managed Aquifer Recharge (MAR) scheme of Torreele/St-André has been developed to replenish the drinking water aquifer at St. André, levelling out seasonal variations of water availability and preventing seawater intrusion (Van Houtte *et al.* 2011) to achieve sustainable management of the groundwater resources. This approach also included environmental rehabilitation of the area and a bigger interest in the recreational and educational aspects of the dunes. This resulted in a "recreation centre" with full-time staff, redevelopment of the recreational structure and more active public involvement.

Drinking water produced through the dune aquifer is distributed to a resident population of 60,000 inhabitants. Because of tourism, daily consumption in summer can be as high as 2.5 times that of a normal day in winter.

26.2 TECHNICAL CHALLENGES OF WATER QUALITY CONTROL

Treatment train for water recycling

The scheme is based on the multiple barrier principle. It is composed of an activated sludge plant (WWTP Wulpen), operated by Aquafin, an advanced water reclamation scheme (Torreele facility), a groundwater recharge unit and a groundwater treatment plant (St-André), all operated by IWVA.

The *WWTP Wulpen*, built in 1987 for wastewater treatment in order to improve the quality of beach water, was extended in 1994 to comply with the new requirements for nutrient removal. The sewerage network has a combined usage and also conveys stormwater. Mechanical treatment consists of 2 step screens, a sand trap and rectangular primary clarifiers. Biological treatment is conventional low loaded activated sludge process with pre-denitrification. Phosphorus is removed by simultaneous chemical precipitation. Sludge-water separation is achieved by secondary settlement tanks. WWTP Wulpen meets all the EC Urban Waste Water Treatment Directive's limits for sensitive areas.

The Torreele facility (Figure 26.1) includes the following treatment steps:

- Pre-treatment by microscreens with 1 mm slots, followed by pre-chlorination to control bio-growth
- Five UF units (ZeeWeed® ZW500C), design flux 36 $L/h{\cdot}m^2$
- Cartridge filter (pore size of 15 μm)
- Two stage RO membranes (DOW 30LE-440, average recovery of 77%, design flux 20 $L/h{\cdot}m^2$) with injection of scale inhibitor, sulphuric acid, periodically $NaHSO_3$ to neutralize free chlorine and NaOH is dosed to RO filtrate to increase the pH above 6.5
- UV disinfection system as backup disinfection unit
- Storage tank

The purified water is transported by a pipeline over a distance of about 2,5 km to the recharge/extraction site of St-André in Koksijde.

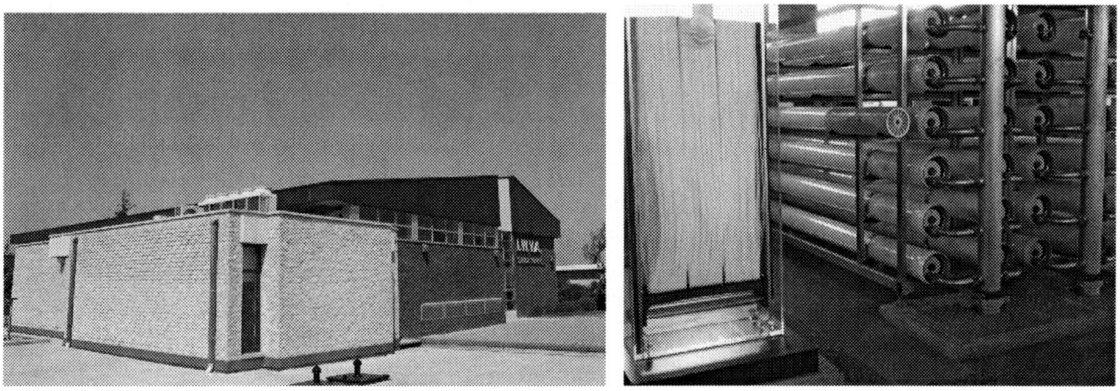

Figure 26.1 General view of the Torreele recycling facility and the membranes of ultrafiltration and reverse osmosis.

Water quality control and monitoring

A general overview of the water quality is shown in Table 26.1. Water reuse processes intended for drinking-water production, both direct and indirect, require intensive quality monitoring. The quality of both UF and RO filtrate was as expected. UF was capable to produce water free of suspended solids. With the exception of somatic coliphages, that were present in 100% of UF permeate samples (range 10–2.6 × 10^2 PFU/100 mL),the analyzed indicator microorganisms (Clostridium spores, *E. coli*, Enterococci, Giardia and Cryptosporidium) were never detected after membrane filtration (Levantesi *et al.* 2010). Hence UF proved to be a good pre-treatment for RO.

Table 26.1 Overview of water quality at Torreele in 2011.

Parameter	UF filtrate	RO filtrate	Infiltration water
Conductivity, µS/cm	1126 (454–1385)	19.5 (8–39)	24 (8–66)
pH	8.02 (7.58–8.23)	5.69 (5.35–6.07)	6.67 (6.25–6.92)
Total Organic Carbon, mg/L	9.2 (6.0–11.9)	0.3 (0–1.2)	0.3 (0.1–0.6)
Total hardness, mg/L as $CaCO_3$	27.5 (11–38)	<0.5	<0.5
Chlorides, mg/L	213 (70–300)	2.7 (2–9.1)	2.4 (1–3.3)
Fluorides, mg/L	0.06 (0.00–0.15)	0.00 (0.00–0.05)	<0.2
Sulphate, mg/L	64 (29–85)	0.2 (0–2)	0.1 (0–0.3)
Nitrate, mg NO_3/L	19 (6–53)	1.5 (0.3–5.8)	1.1 (0–5)
Ammonia, mg NH_4/L			0.18 (0.02–0.51)
Phosphate, mg PO_4/L	4.2 (1.0–9.3)	0.00 (0.00–0.07)	0.00 (0.00–0.03)
Total trihalomethanes, µg/L			2.8 (1.0–6.6)
Alumina, µg/L	35 (3–69)	0.91 (0–18.4)	2.6 (0–6)
Chromium, µg/L	0.05 (0–1.2)	0.15 (0–1.6)	<2.5
Copper, µg/L	1.1 (0–2.5)	0.3 (0–1.7)	<5
Iron, µg/L	3.5 (21.3–52.8)	0.15 (0–6)	1 (0–11)
Lead, µg/L	0.3 (0–11)	<5	<5
Manganese, µg/L	45 (21–87)	0.05 (0–0.3)	<10
Mercury, µg/L		<0.2	<0.2
Nickel, µg/L	1.04 (0–5.2)	0.5 (0–3.5)	<3
Sodium, µg/L	141 (38–200)	3 (0–6)	4.8 (1.9–6.2)
Zinc, µg/L	14.3 (0–31)	<20	<20
Total Coliforms, MPN/100 mL	0	0	0
E.coli, MPN/100 mL	0	0	0
HPC 22°C, CFU/mL	22 (2–50)	<1 (0–3)	<1 (0–3)

In recent years more attention goes to the presence of organic micro-pollutants in drinking-water, for example, pharmaceutical active compounds (PhACs) and endocrine disrupting compounds (EDCs). As the health effects related to the consumption of drinking water containing a cocktail of these substances is still unknown, their removal during the treatment is desirable (Verliefde, 2008).

In Torreele, RO is the major and ultimate single barrier against both microbial and chemical contamination of drinking water and the aquifer. The presence of *Giardia* sp p. and *Cryptosporidium* spp. oocysts, helminth eggs, *Salmonella* spp., *Campylobacter* spp. and pathogenic bacteria was monitored in three sampling campaigns at Torreele (February, July and October 2007) The advanced tertiary treatment technology produced infiltration water of microbiological drinking water quality (*E. coli* and enterococci absent in 100 mL of water, CEE98/83 EC) in which no pathogens detected (Levantesi, 2010).

Besides, RO can be considered as an adequate technology for the removal of trace contaminants (e.g. pesticides). Micro-pollutants were almost completely rejected by RO. Only NDMA, NMOR and bisphenol A were still detectable. All compounds were removed below the LOQ during MAR and this additional removal is attributed to subsoil processes (Kazner *et al.* 2011).

Main challenges for operation

The biggest challenges for the Torreele facility are the changing conditions, both based on the meteorological and seasonal variations. The operating conditions are varied according to these. The follow up of the plant is based not only on online measurements, but also on daily monitoring by the operator.

Submerged UF, using outside-in filtration and air, not only proved to be a good pre-treatment prior to RO but was also capable to cope with the expected variations of inlet water quality. The turbidity of the UF filtrate is constantly below 0,1 NTU, and most of the time below 0,05 NTU. Turbidity proved to be a good indicator when UF treatment failed and is used as a first quality control enabling the shut down of the facility when threshold values are exceeded.

Biofouling and scaling prevention are constant concerns for membrane-based water reclamation systems, as is the reduction of chemicals and energy consumption. Aeration on the UF system has been reduced from an initial rate of 50% to 30%. The average recovery (RO) was increased, without dosing more chemicals, as recovery was related to conductivity: when the electrical conductivity is lower, this means less risk for scaling and a higher recovery is allowed. This is called "recovery control". Additionally, especially during the colder periods chloramination is no longer constantly saving chemical consumption (Van Houtte & Verbauwhede, 2008).

The concentrates are discharged with the volume of effluent that is not treated. However IWVA investigated natural systems (test with reed bed from 2003 until 2009 and test with willows since 2007, Figure 26.2) to limit the nutrient content (Van Houtte & Verbauwhede, 2008; Van Houtte *et al.* 2012), as well as electrodialysis to minimize the concentrate volume (Zhang *et al.* 2010). More research is planned for the near future. However, the experience using natural systems, first using reed beds and now willows, prove that this process has the potential to treat concentrates. This will not only mitigate the effects of discharging this water into the environment, but will, by the production of biomass, contribute to the climate problem. The willows, by taking up the nutrient, organics and other elements from the concentrate, will harvest the energy out of this concentrate. This energy, in the form of biomass, can be used for heating or power production in a CO_2 neutral way (Van Houtte *et al.* 2012).

Figure 26.2 View of the willows for the treatment of reverse osmosis brines.

26.3 WATER REUSE APPLICATION

Description of the water reuse applications

From the start of infiltration until 2011, 18.5 Mm^3 of water was recharged in the unconfined sandy aquifer (Figure 26.3), while 25.1 Mm^3 was extracted using 112 wells with filter elements between 8 and 12 m depth. Analysis of residence time has indicated that 50% of the water recharged is reaching the extraction wells in less than 60 days. Because of the varying distance between the infiltration pond and the extraction wells (33 m to 153 m), it takes almost 5 years for the recharged water to reach the most distant ones. These calculations are confirmed by the results of a tracer test (Vandenbohede *et al.* 2008).

Figure 26.3 View of the recharge basin of St. André.

Recontamination occurs in the infiltration pond due to the presence of birds. Helminth eggs and Giardia cysts were detected in 33% of the samples (Levantesi *et al.* 2010). Monthly measurements since June 2008 show that *E. coli*, enterococci and Clostridium spores are present in 66% of the samples but the number never exceeded 2.7×10^3 cfu/100 ml. Bacterial indicators and pathogens, present in the infiltration water, were never detected in the ground water after the MAR as underground passage ensured elimination (Levantesi *et al.* 2010).

The recovered water is conveyed to the potable water production facility at St. André. The facility consists of an aeration step, rapid sand filtration, a reservoir and UV disinfection prior to distribution. Dosing of chlorine is possible as a preventive action to prevent re-growth and recontamination in the distribution network.

Thanks to the water reuse scheme of Torreele/St-André the groundwater extraction permits in two dune areas could be reduced from 3.7 Mm^3/yr in 2002 to 2.2 Mm^3/yr in 2010, while the (limited) pumping in the third area was totally stopped in 2002.

Evolution of the volume of supplied recycled water

The drinking water demand in the area decreased from 5.5 Mm^3/yr in 2002 to just over 4.8 Mm^3/yr in 2011. Education of the public on the proper use of drinking water; increased prices due to higher taxes for discharge of the used water and decreased leakage of the distribution network all contributed to this. It is difficult to make a prognosis on how the evolution will be in the next few years but the reduced use of drinking water, together with the fact that a dynamical equilibrium in the unconfined aquifer of the infiltration area seems to be reached, has meant that less infiltration water has been needed in recent years.

Relations and contracts with end-users

There are no contracts with industrial users, but discussions recently started for the direct delivery of RO filtrate to some industrial sites. However as the volumes are small, the bottleneck is the cost for installing and operating a new distribution network.

26.4 ECONOMICS OF WATER REUSE

Project funding and costs

The IWVA had enough capital to fund the project. The organisation decided to choose a 10-year maintenance contract. The total investment cost amounted €7 million.

The first set of RO membranes lasted for 6 to 7 years and the first set of UF membranes for 8 years. Membrane replacement is part of the maintenance contract.

The production of infiltration water decreased, partly due to decreased water demand and partly due to decreased infiltration rates, and as a consequence, the operation and investment cost has increased over recent years. In 2005 (2.17 Mm^3 produced) it cost 0.46 €/m^3 to produce infiltration water and this includes € 0.15/m^3 for both operational and investments costs (Van Houtte & Verbauwhede, 2008). In 2011, for a production just under 1.8 Mm^3, the cost amounted 0.62 €/m^3. The maintenance cost (chemicals, energy, personnel, lab) amounted 0.20 €/m^3 and the part of the maintenance contract and investment cost increased to 0.16 and 0.18 €/m^3 respectively. To this cost 0.02 €/m^3 should be added for infiltration but the total of 0.64 €/m^3 is still substantially lower than the average cost of purchasing drinking water from neighbouring areas which amounted 0.79 €/m^3 in 2011.

The cost of the recycled water is recovered as a part of the overall cost of the drinking water.

Benefits of water recycling

The water reuse/infiltration project allowed sustainable management of an area with high ecological value. This could prove the key factor for the future of drinking-water production in these dunes. As the water table rises it is also a preventive and pro-active measure against the expected effects of climate change. Preserving the production close to the point of consumption is very important in a tourist area where water consumption not only varies on seasonal basis (winter/summer) but also on daily basis subject to the weather conditions, especially during vacation periods when sudden high peaks in drinking water demand are frequent.

Having most of the production onsite means that the IWVA is less vulnerable to external risk factors.

26.5 HUMAN DIMENSION OF WATER REUSE

The role of decision makers

It took many years to obtain the permits needed to develop this project, especially as infiltration and membrane treatment were new concepts for the regulators. Future regulations should be more flexible to ensure that new developments can be implemented more easily if they prove to be of interest both to the environment and the economy.

Public acceptance and involvement

The whole project had a long history before it actually started with the local media regularly reporting on the preliminary plans and tests. Since the 1970's environmentalists have opposed groundwater extraction from the dune aquifer with hot summers often seeing the area affected by drinking-water distribution problems. As the water reuse scheme, involving groundwater recharge, proposed a solution both to the drinking-water shortage and to the environmental objections, the project was accepted by stakeholders and the large majority of the public. Plant visits and public forums were organized by IWVA to present the water recycling facility and its performance.

Since the project was implemented its results have been presented to the public. This is done through informative boards in IWVA's visitor's centre. In vacation periods guided walks to the infiltration area, which is usually closed to the public, are organized to inform the local community. It not only allows them to see the area but they can also discuss the project with IWVA members.

The quality of the drinking water distributed in a large part of the local area has improved thanks to the water reuse scheme. The hardness is substantially lower and the colour is better due to decreased organic and salt content in the water. In a benchmark study performed in 2000, the public rated the hardness and taste of the drinking-water poorly. The results of the internal inquiry that the IWVA is organizing since March 2006 show that the public is generally satisfied with taste, odour and colour. The rate of satisfaction concerning the hardness of the water is better in the area where water is distributed from the St-André area, thus coming from the water reuse/groundwater recharge scheme.

Despite the total investment of €7 million, drinking-water price only increased from 1.17 to 1.42 €/m^3.

26.6 CONCLUSIONS AND LESSONS LEARNED

The Torreele case shows that combined membrane treatment (UF and RO) enables the treatment of wastewater effluent in an effective and reliable way with reverse osmosis being the major and ultimate barrier against both microbial and chemical contamination. The produced recycled water is of excellent quality for artificial recharge of the dune aquifer of St-André and it enables sustainable groundwater management in an area with high ecological interest.

The RO membranes performed very well for over 6 years and chloramination did not damage the membranes. UF membrane replacement started in the 8th year of operation. These periods can be considered as very satisfactory.

As the production of infiltration water went down over the years, due to lower demand and reduced infiltration capacity, the cost to produce and infiltrate increased to 0.64 €/m^3 in 2011 but this is still lower compared to the average cost of 0.79 €/m^3 to purchase drinking-water from neighbouring companies.

The following Table 26.2 summarizes the main drivers of water reuse in Torreele as well as the benefits observed and some of the challenges to be overcome in the future.

Table 26.2 Summary of lessons learned.

Drivers/Opportunities	Benefits	Challenges
Unsecure supply in summer Declining water table Need for ecological management	Improved drinking-water quality Sustainable production near point of consumption Enhanced natural values in dune area	Increase infiltration capacity and reduce overall costs on longer term Concentrate should be treated/reused

REFERENCES AND FURTHER READING

Kazner C., Joss A., Ternes T., Van Houtte E. and Wintgens T. (2011). Membrane based treatment trains for managed aquifer recharge. In: *Advances in Water Reclamation Technologies for Safe Managed Aquifer Recharge*, C. Kazner, T. Wintgens and P. Dillon (eds), IWA Publishing, London, UK.

Krauss M., Longree P., Van Houtte E., Cauwenberghs J. and Hollender J. (2010). Assessing the fate of nitrosamine precursors in wastewater treatment by physico-chemical fractionation. *Environmental Science and Technology*, **44**(20), 7871–7877.

Kuijken E., Provoost S. and en Leten M. (1993). *Oppervlakte-infiltratie in de Doornpanne*. Instituut voor Natuurbehoud, Hasselt, 93.69, 86 p.

Leslie G. L., Dunivin W. R., Gabillet P., Conklin S. R., Mills W. R. and Sudak R. G. (1996). Pilot testing of microfiltration and ultrafiltration upstream of reverse osmosis during reclamation of municipal wastewater. Proceedings of the American Desalting Associations Biennal Conference, Monterey, California, August 1996, pp. 29–40.

Levantesi C., La Mantia R., Masciopinto C., Böckelmann U., Ayuso-Gabella M., Salgot M., Tandoi V., Van Houtte E., Wintgens T. and Grohmann E. (2010). Quantification of pathogenic microorganisms and microbial indicators in three wastewater reclamation and managed aquifer recharge facilities in Europe. *Science of the Total Environment*, **408**, 4923–4930.

Vandenbohede A., Van Houtte E. and Lebbe L. (2008). Groundwater flow in the vicinity of two artificial recharge ponds in the Belgian coastal dunes. *Hydrology Journal*, **16**(8), 1669–1681.

Van Houtte E. and Vanlerberghe F. (1998). Sustainable groundwater management by the integration of effluent and surface water to artificially recharge the phreatic aquifer in the dune belt of the western Flemish coastal plain. IAH International Groundwater Conference, Groundwater: Sustainable Solutions, Melbourne, Australia, pp. 93–99.

Van Houtte E., Verbauwhede J., Vanlerberghe F., Demunter S. and Cabooter J. (1998). Treating different types of raw water with micro- and ultrafiltration for further desalination using reverse osmosis. Proc. 'Membranes in drinking and industrial water production' Amsterdam, The Netherlands, *Desalination*, **117**(1), 49–60.

Van Houtte E., Verbauwhede J., Vanlerberghe F. and Cabooter J. (2000). Comparison between different out-to-in filtration MF/UF membranes. Proc. Int. Conf. on Membrane Technology in Water and Wastewater treatment, Lancaster 2000, pp. 190–197.

Van Houtte E. and Vanlerberghe F. (2001). Preventing biofouling on RO membranes for water reuse – Results of different tests. AWWA Membrane Technology Conference, San Antonio, USA.

Van Houtte E. and Verbauwhede J. (2005). Artificial recharge of treated wastewater effluent enables sustainable groundwater management of a dune aquifer in Flanders, Belgium. In: Unesco (2006), Recharge systems for protecting and enhancing groundwater resources. Proc. 5th International Symposium on Management of Aquifer Recharge ISMAR5, Berlin, Germany, 11–16 June 2005. IHP-VI, Series on Groundwater No. 13: pp. 236–243.

Van Houtte E. and Verbauwhede J. (2008). Operational experience with indirect potable reuse at the Flemish coast. *Desalination*, **218**, 198–207.

Van Houtte E., Cauwenbergh J., Weemaes M. and Thoeye C. (2011). Indirect potable reuse via managed aquifer recharge in the Torreele/St-André project. In: *Advances in Water Reclamation Technologies for Safe Managed Aquifer Recharge*, C. Kazner, T. Wintgens and P. Dillon (eds) , IWA Publishing, London, UK, pp 33–44.

Van Houtte E., Berquin S., Pinoy L. and Verbauwhede J. (2012). Experiment with willows to treat RO concentrate at Torreele' s water re-use facility in Flanders, Belgium. Proc. AMTA/AWWA Membrane Technology Conference, Glendale (AZ), USA.

Verliefde A. (2009). Rejection of Trace Organic Pollutants by High Pressure Membranes. PhD thesis at Delft University.

Zhang Y., Ghyselbrecht K., Vanherpe R., Meesschaert B., Pinoy L. and Van der Bruggen B. (2011). RO Concentrate Minimization by Electrodialysis: Techno-Economic Analysis and Environmental Concerns.

WEB-LINKS FOR FURTHER READING

www.iwva.be

View of the recharge basin.

27 34 Years of experience with potable water reuse in the occoquan reservoir

Robert W. Angelotti and Thomas J. Grizzard

CHAPTER HIGHLIGHTS

The Upper Occoquan Service Authority (UOSA) water reuse project began operation in mid-1978 with the objective to improve water quality problems observed in the existing surface water reservoir being used as the raw drinking water supply. Since the beginning of the project, the recycled water was recognised as the best quality source water in the Occoquan system, and the plant had several expansions. The drinking quality recycled water is produced by means of a multi-barrier conventional treatment, including biological nitrogen removal, lime clarification, multimedia filtration, granular activated carbon adsorption and chlorination/ dechlorination. More than 34 years of successful implementation has instilled confidence that the pioneered and visionary plan, originally conceived so many years ago, is still working.

KEYS TO SUCCESS

- Foresight to envision requirement for increased water supply yield decades in advance of need
- State and local government collaboration on adopting a reuse solution
- Independent monitoring to provide unbiased information to stakeholders
- Continued demonstration of improvements to water quality over project life

KEY FIGURES

- Treatment capacity: 204,400 m^3/d (54 MGD)
- Total volume of reclaimed water: 44.1 Mm^3/yr (11.68 bg)
- Reclaimed water is 8% of annual average reservoir inflow
- Capital and operation costs: $61.5 million
- Water reuse standards described in the Occoquan Policy
- Occoquan Reservoir safe yield increased by 71% to date; future increase expected to exceed 100%

27.1 INTRODUCTION

The Occoquan Reservoir is a critical component of the water supply for approximately 1.5 million residents of Northern Virginia, a highly urbanized region located just west of USA national capital, Washington, D.C. Reclaimed water represents a significant supplement to the potable water supply yield from the reservoir. The project has been in service for over three decades, and has been an unqualified success.

The story of potable water reuse in the Occoquan Reservoir began in 1950 when a low head dam was built on the Occoquan River near the Town of Occoquan to serve the water supply needs of the City of Alexandria. Responding to growing water supply needs in the U.S. Route 1 corridor, the low head facility was replaced in 1957 by a high dam, which created the Occoquan Reservoir. In 1967, ownership of the reservoir and associated treatment works passed to the Fairfax County Water Authority, which (doing business as Fairfax Water) continues to operate the system. A 2011 view of the Occoquan Reservoir as viewed from the high dam is shown in Figure 27.1.

At the time it was impounded the reservoir had a storage volume of 37 million m^3, Mm^3 (9.8 billion gallons), and a computed safe water supply yield of approximately 189,300 m^3/d (50 MGD).

Until the 1960s, most of the Occoquan Watershed had changed little since the American Civil War. However, the decade of the 1960s marked the beginning of a rapid transformation from a largely rural to a predominantly urban/suburban region. Northern Virginia experienced unprecedented growth which was stimulated by the opening of a new interstate highway (Route I-66) and the westward expansion of the urban core of Washington, D.C. By the mid-point of the decade, the resulting urbanization, including residential and commercial development, of the watershed began to adversely affect water quality of the Occoquan Reservoir (Robbins, 1993). The schematic of the geography of the Northern Virginia area is shown in Figure 27.1 (right) in the proximity of the Occoquan Watershed and Reservoir to Washington, D.C. and the UOSA service area (which will be referenced later).

Figure 27.1 Occoquan Reservoir (courtesy of Roger W. Snyder) and local geography for the Occoquan Watershed.

Main drivers for water reuse

The decline in water quality experienced in the 1960s led to extensive studies to determine the causes, and to develop plans to restore and protect the key water supply for much of Northern Virginia. The previously noted urban development had resulted in an unplanned and unintended indirect potable reuse scenario where eleven small wastewater treatment plants were discharging conventionally-treated effluents upstream of the Occoquan Reservoir. Together with pollutants from urban and agricultural runoff, poorly treated wastewater effluent threatened the continued use of the Occoquan Reservoir for public water supply. Typical problems being observed were frequent and intense blue-green algal blooms, depletion of dissolved oxygen, deep water sulphide concentrations, fish kills, taste and odour issues with finished water, increased treatment costs from algae mats, increased dissolved organics in the raw water and detectable levels of active pathogenic viruses (Robbins, 1979). Figure 27.2 shows an example of a summer cyanobacter bloom in the early 1970s.

Figure 27.2 Poor water quality of the Occoquan Reservoir occurring during the 1960s (Photograph courtesy of OWML).

Following a study by Metcalf and Eddy in the late 1960's, several alternative solutions for the long term management of the reservoir were considered. The conventional solution proposed was to export wastewater from the watershed, with treatment and ultimate disposal to take place elsewhere. After careful consideration, however, this alternative was rejected because it simply moved the pollution problem to another location, was politically untenable, and most importantly, did nothing to preserve the water for reservoir recharge during times of drought. A second alternative included a potable reuse scenario, in which the wastewater would be treated to the (then) state-of-the-art and pumped directly to the water treatment plant for further treatment and distribution. Although this option would have preserved the water for use within the region, it received little public support and was opposed by public health officials. The third option included an indirect potable reuse scenario, wherein watershed wastewaters would be reclaimed to the highest standards that available technology could practically achieve, and be subsequently released back to the Occoquan Reservoir to supplement the raw water supply. All of the alternatives included proposals to limit urban growth and population in order to restrain non-point pollutant loads from exceeding the reservoir's assimilative capacity (Robbins, 1993).

After careful review, in 1971, the Virginia State Water Control Board, with counsel and recommendation from the Virginia Department of Health, adopted a bold and innovative solution to protect the Occoquan Reservoir as a drinking water supply. The *Policy for Wastewater Treatment and Water Quality Management in the Occoquan Watershed*, more commonly known as the Occoquan Policy, mandated (with some refinements) the third option as a newly conceived framework for water reuse within the Occoquan Watershed. Although never before attempted at such a scale, the decision set in motion the first planned and intentional use of reclaimed water for the purpose of supplementing a potable surface water supply in the U.S. The Occoquan Policy established reclaimed water treatment standards that had never been achieved by a wastewater treatment plant before that time.

With the hindsight of over three decades of operation, the project may be seen as one of the pioneering efforts in potable reuse, and an unqualified success. The decision to move forward with such a bold solution serves as a continuing testament to the foresight and vision of those early decision makers. The clear demonstration of capability has since played a pivotal role in the advancement of potable reuse throughout the world.

Brief history of the project development

The Occoquan Policy mandated the creation of a regional State authority, the Upper Occoquan Service Authority (UOSA), to provide collection and reclamation of wastewater, and the Occoquan Watershed Monitoring Program (OWMP), to continuously monitor the watershed and reservoir with a view to providing independent water quality assessments and advice on protective measures for the reservoir. By the 1970's, Fairfax Water, was responsible for potable water production and distribution for much of Northern Virginia. The Virginia State Water Control Board and Virginia Department of Health were also highly involved in the technical development of the solution. In the early 1970's UOSA began the design of a regional wastewater collection and reclamation system, and the OWMP began monitoring streams and the reservoir to establish baseline conditions. In mid-1978, UOSA began operation of its water reclamation plant, initially producing about 18,900 m^3/d (5 MGD) of reclaimed water. The water quality of the Occoquan Reservoir was observed to exhibit dramatic improvements within the first few months of UOSA plant operation.

The UOSA plant was originally constructed with a reclaimed water capacity of 56,800 m^3/d (15 MGD). However, it was only permitted to deliver 41,260 m^3/d (10.9 MGD) of reclaimed water to the Occoquan Reservoir. Plant production was limited to curtail population growth within the watershed due to a limited water quality data base and because the impact of population growth on reservoir water quality was not well known. The 41,260 m^3/d (10.9 MGD) production limit remained in place until UOSA plant operation clearly demonstrated reservoir water quality restoration and more available reservoir assimilative capacity. The permit limit was increased to match the 56,800 m^3/d (15 MGD) water production capacity within a few years after UOSA plant operations commenced. In the early 1980s, it was clear that expansion of plant capacity would be needed to handle continued growth within the watershed. However, the Occoquan Policy initially allowed expansions to occur only in 28,390 m^3/d (7.5 MGD) increments. By 1985 UOSA, embarked on its first incremental plant capacity expansion to 85,200 m^3/d (22.5 MGD). While this expansion was still under construction, it was clear that the incremental expansion would soon be inadequate for the burgeoning population (Robbins, 1993).

Projections for continued and significant population growth within the area indicated that the UOSA plant required a larger expansion to 204,400 m^3/d (54 MGD). A significant water quality evaluation was commissioned in 1988 to better identify the assimilative capacity of the Occoquan Reservoir and to predict how the reservoir water quality might respond to the increased pollutant loads from reclaimed and urban storm water influences. Results of that evaluation were encouraging and some conclusions were a surprise. The modelling results indicated that the high quality reclaimed water from the UOSA plant was actually necessary to further improve water quality of the Occoquan Reservoir and the proposed expansion would be beneficial to the water supply role of the reservoir. The high level of point source load removal from the reclamation plant was found to be necessary to offset the increased pollutant loads from non-point sources and urban runoff from the developing watershed. Surprisingly, one recommendation was to actually import wastewater into the

Occoquan Watershed for treatment at the UOSA plant so that there would be more high quality water to dilute pollutants coming from storm water (Chen, 1988).

It was of note to find that in only a decade of operations, the original recommendation to export wastewater from the watershed to improve water quality had been replaced with a new recommendation to import wastewater into the watershed to not only supplement yield, but to enhance reservoir water quality. Continuing water quality studies supported the emerging view of reclaimed water as an important resource, and were supported by a decade of excellent performance with a high level of reliability from the UOSA water reclamation facility.

In the early 1980s, Fairfax Water commissioned a second major potable water treatment facility. The James J. Corbalis plant was named for the first Engineer-Director of the Authority, and was designed to use the free-flowing Potomac River as a source. In addition to adding considerable flexibility and robustness to the water supply system, the Potomac source had an additional, and ultimately beneficial, impact on the reuse system. The Corbalis Plant was designed to be the primary source of drinking water for much of the sewered portions of Fairfax County served by UOSA. As a result, the fraction of wastewater delivered to UOSA, but not having an Occoquan Watershed origin, has increased over time. This has had the benefit of further supplementing the water supply yield, but it has also served to maintain a more "open" reuse system with the regular introduction of a significant amount of water not derived from a reclaimed source.

The first UOSA expansion to 85,200 m^3/d (22.5 MGD) was completed in 1989. The next expansion to 204,400 m^3/d (54 MGD) occurred in two phases, the first completed in the mid-1990s which brought the reclaimed water production rate to 121,100 m^3/d (32 MGD) and the second phase was completed in 2004. The UOSA plant now reclaims about 113,600 m^3/d (30 MGD) of water on an annual average basis and the plant has the capacity to reclaim as much as 204,400 m^3/d (54 MGD) of water.

An aerial view of the UOSA water reclamation plant is shown in Figure 27.3. A plant flow of around 246,000 m^3/d (65 MGD) is associated with the build out condition within the service area of the reclamation plant. However, recent economic conditions have slowed population growth from the unprecedented rates previously observed.

Figure 27.3 Aerial photo of a portion of the 204,400 m^3/d (54 MGD) UOSA water reclamation plant known as the Millard H. Robbins, Jr. Water Reclamation Facility.

Project objectives, framework and incentives

While water quality improvement was the primary driver for implementing planned and intentional potable water reuse in the Occoquan system, supplementing the raw water supply was always an underlying objective. Although the mid-Atlantic region of the U.S. is not considered a dry or arid area, the population density results in stressed water supply, and limited per capita water availability. This situation becomes even more pronounced during periodic extended drought conditions. Given the current conditions, and also considering uncertainty from future climate change, a significant incentive exists for potable reuse not only from a water quality perspective but also from a need to augment available quantities of raw water supply. Figure 27.4 illustrates how the predicted safe yield of the Occoquan Reservoir will be increased by future expansions of reclaimed water production at the UOSA plant. When the UOSA service area reaches a build out condition, it is estimated that the safe yield of the Occoquan Reservoir may be effectively doubled compared to the safe yield from natural reservoir recharge. In most cases with constructed run-of-the-river water supply impoundments, safe yield may be expected to decline over time as the impoundment fills with sediment. The dramatic increase in safe yield observed for the Occoquan system is unusual and a remarkable outcome.

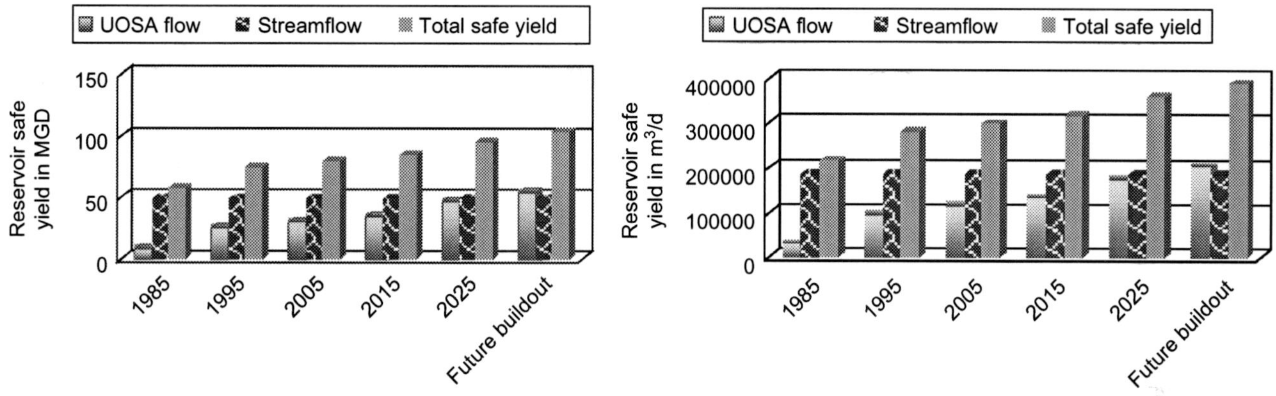

Figure 27.4 Safe yield of Occoquan Reservoir shown significantly increasing over time.

Fairfax Water acknowledges that UOSA provides reliable, high quality water to the Occoquan Reservoir, and the utility relies on reclaimed water as a key component of its water supply plan. The UOSA contribution is particularly important during times of drought. The UOSA product water makes up approximately 8% of the annual average inflow to the reservoir, but during periods of drought, that fraction may rise above 90%.

Because the reclaimed water flow is so important to the safe yield of the raw supply and to reservoir water quality, a framework of public service has been developed that requires very close collaboration among four main institutional areas:

1. UOSA, the water reclamation agent,
2. Fairfax Water, the potable water supply, treatment and distribution agent,
3. OWMP, the independent reservoir water quality monitoring agent, and
4. Virginia Department of Environmental Quality (VDEQ) and Virginia Department of Health (VDH) the regulatory agent(s).

These four institutional areas represent the foundational pillars that support successful implementation of potable reuse in Northern Virginia. While each entity has its own institutionally-defined role, it is the global mission to protect and enhance the Occoquan Reservoir as a raw water supply that weaves a common bond of cooperation and coordination among them.

The prospect of indirect potable water reuse within an urban area containing significant industrial and commercial activity comes with significant institutional challenges. The industrial contribution to the UOSA raw wastewater is significant. On occasion, flow from a single semiconductor manufacturer can be as much as 15% of the influent flow to the treatment plant. Rigorous attention must be paid to industrial and commercial contributions when water is being reclaimed for potable purposes.

Diligent industrial contaminant source controls are also required. Regional stakeholders take part in assessing the reclaimed and reservoir water quality and quantity impacts from significant industries proposing to locate within the UOSA service area (Angelotti, 2008). Industrial use of reclaimed water, particularly in consumptive uses such as process cooling, is an emerging institutional challenge for the region. Such uses have a regional impact, because of the potential effects on reservoir safe yield.

Favourable economics have been important to the success of the Occoquan story. The majority of funding for the original UOSA treatment plant was provided as a federal grant under the Clean Water Act in the early 1970s. In addition, UOSA does not charge Fairfax Water for the reclaimed water delivered to the Occoquan Reservoir, although Fairfax Water customers benefit from the increased drought tolerance and improved quality of the raw water supply.

27.2 TECHNICAL CHALLENGES OF WATER QUALITY CONTROL

Treatment train for water recycling

Converting raw wastewater into high quality product water suitable to supplementing drinking water supply is achieved by consideration of the multiple barrier approach as common practice. In the Occoquan System, elements of the multiple barrier concept include: wastewater source control and collection, water reclamation, local storage, natural environmental buffers and storage, raw water treatment, and potable water distribution processes. Aggressive evaluation and monitoring of industrial contributions to wastewater and changes in wastewater character during collection and conveyance, coupled with an understanding of how specific changes may impact potable reuse are needed for success.

The UOSA system contains both conventional and advanced water reclamation processes and also has a wide range of reliability features built into design and operation. Large retention basins located at pump stations and at the plant provide storage for emergencies, maintenance and high flow conditions. All major mechanical and electrical components have at least one backup unit on ready standby in case of a failure. Electric power is fed to the plant from two independent grid sources and an on-site power generation station is available to power the entire plant. Pump stations have similar power redundancies. All processes of the plant and collection system are monitored by a computerized supervisory control system and most of the pump station and plant site processes are automatically controlled by the computer network (Asano, 2007).

The plant flow diagram is shown in Figure 27.5. The major treatment processes include:

- Conventional preliminary and primary treatment
- Biological nitrogen removal (BNR) advanced secondary treatment
- High pH lime clarification for phosphorus and metal removal and virus inactivation
- Two stage recarbonation with intermediate settling for metal and hardness removal
- Multimedia filtration for turbidity, suspended solids, and protozoan cyst removal
- Deep bed granular activated carbon adsorption for further filtration and trace organic removal
- Free chlorine disinfection for added pathogen destruction
- Dechlorination (Angelotti, 1995, 2005)

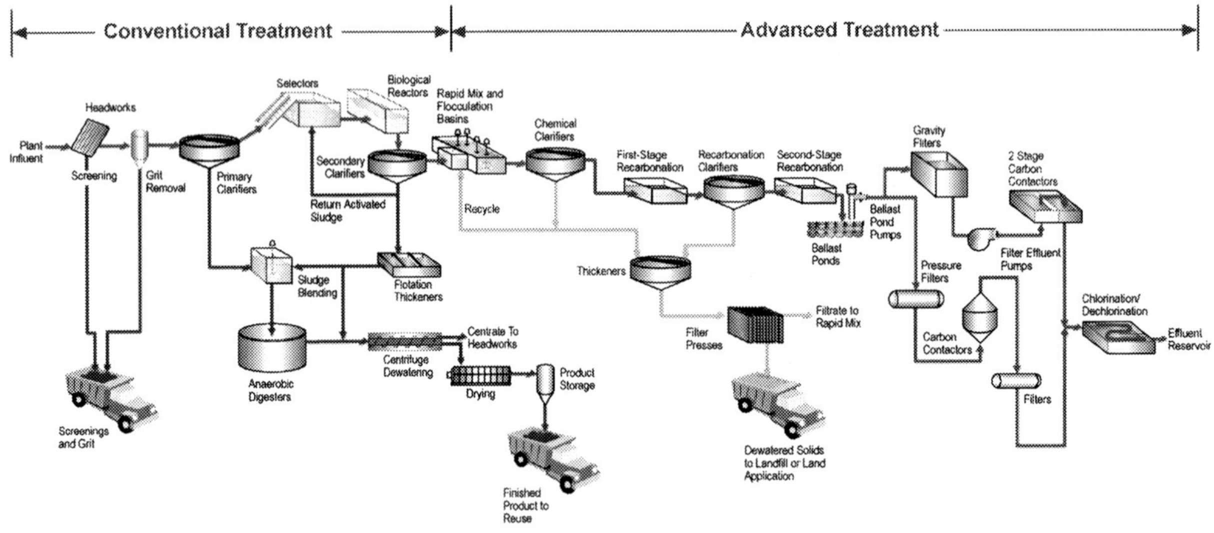

Figure 27.5 Schematic of the UOSA water reclamation process.

The water produced at the UOSA plant flows into a 681,400 m^3 (180 million gallon) impoundment on the UOSA site before discharging to an un-named tributary to Bull Run. The on-site reservoir provides the first of the natural barriers of the system, and provides significant dilution volume to mediate any short term deviations from normal constituent concentrations in the product water. The lake also allows for further environmental stabilization, maturation and some mixing with natural surface water. The impoundment discharge enters Bull Run and thence flows downstream to the headwaters of the Occoquan

Reservoir. The stream and reservoir environments provide opportunities for mixing, dilution, natural treatment and assimilation to occur.

Water withdrawn from the Occoquan Reservoir is treated in an advanced potable drinking water plant. The major treatment processes for the plant include:

- Conventional raw water screening and pumping
- Conventional chemical addition, flocculation and clarification
- Ozonation
- Deep-bed GAC biofiltration
- Chloramination

The potable water is pumped into the distribution system for use by the local population in Northern Virginia. Additional barriers, such as maintaining short water age, disinfectant residuals, periodic flushing of water mains, for example, are used in the potable distribution system to maintain finished water quality delivered to end use customers.

Water quality control and monitoring

The UOSA plant product water must meet performance standards as permitted by the Virginia Department of Environmental Quality. The current product water performance standards are shown in Table 27.1.

Table 27.1 Standard Requirements for UOSA Product Water Quality.

	Performance Standards			
Parameter	**Monthly Average Concentration**	**Weekly Average Concentration**	**Units of Measure**	**Typical Range of Performance**
Chemical Oxygen Demand (COD)	10	25	mg/L	6–9
pH	6–9*	N/A	pH units	7.1–7.7
Total Suspended Solids (TSS)	1	2.5	mg/L	0.3–0.6
Turbidity	0.5	1.25	NTU	0.1–0.2
Total Kjeldahl Nitrogen (TKN)	1	2.5	mg/L	0.3–0.5
MBAS	0.10	0.25	mg/L	0.01–0.05
E. coli (Geometric Mean)	<2	N/A	n/100mL	Non-detected
Total Residual Chlorine (after dechlorination)	0.008	0.010	mg/L	Non-detected
Total Phosphorus	0.10	0.25	mg/L	0.04–0.09

*Not monthly average. Any daily reading must fall within the range of 6–9 pH units.

Total coliforms are also measured in the product water and compared to EPA Guidelines for indirect potable reuse, but are very rarely detected in the reclaimed water. In addition to the routine laboratory analyses performed daily to monitor water quality, a number of continuous analyzers are used to monitor critical processes. Examples include ammonia and nitrate analyzers at BNR bioreactors, turbidimeters on individual filters to verify turbidity levels commensurate with good protozoan cyst removals and chlorine residual analyzers and flow meters at the chlorine contact basins to verify that contact time and chlorine concentrations are at levels above those known to kill or inactivate other pathogens (U.S. EPA, 2004).

Other monitoring is conducted to assure the appropriate water quality for potable reuse. Annual acute and chronic bioassays are conducted to verify that the product water has no observed effect on target organisms. Annual testing is performed to confirm that reclaimed water meets the latest requirements of the U.S. EPA National Primary and Secondary Drinking Water Standards. In addition, annual evaluations are performed to determine if pharmaceuticals, personal care products or other undesirable synthetic organic compounds are present at levels of concern in the reclaimed water. Hundreds of organic compounds are assayed each year at extremely low levels of detection in the UOSA product water, Bull Run and the Occoquan River.

A significant (and independent) effort is undertaken to monitor both point and nonpoint sources of pollution within the Occoquan Watershed. Given the levels of watershed development that have taken place in the years following start-up of the UOSA reclamation plant, it has become apparent that nonpoint sources and urban runoff have become an increasingly important component of the remaining nutrient, organic matter, and sediment loads reaching the Occoquan Reservoir. In order to accurately assess these sources, the OWMP operates and maintains a monitoring network of nine stream monitoring stations, 14 recording rain gages, and eight reservoir monitoring stations at appropriate locations. The OWMP also provides an independent review of UOSA facility performance and acts as a source of information and advice for all watershed stakeholders. The data derived from these activities have been of value to local governments in the development of ordinances for managing distributed sources of pollution in the watershed. Much of the impetus for controlling storm water, at least in the years prior to federal regulation, has derived from the desire to insure that poor quality watershed stream flows were not impacting the investment in high performance water reclamation. The OWMP is also responsible for the development and maintenance of a large, complexly-linked watershed and reservoir computer model which can be used to predict water quality changes resulting from increased urbanization or significant changes in land use patterns, changes in the character or quantity of UOSA flows, persistent droughts and a host of other scenario simulations. Figure 27.6 shows the physical geometry of the Occoquan Reservoir and the location of the OWML reservoir monitoring stations.

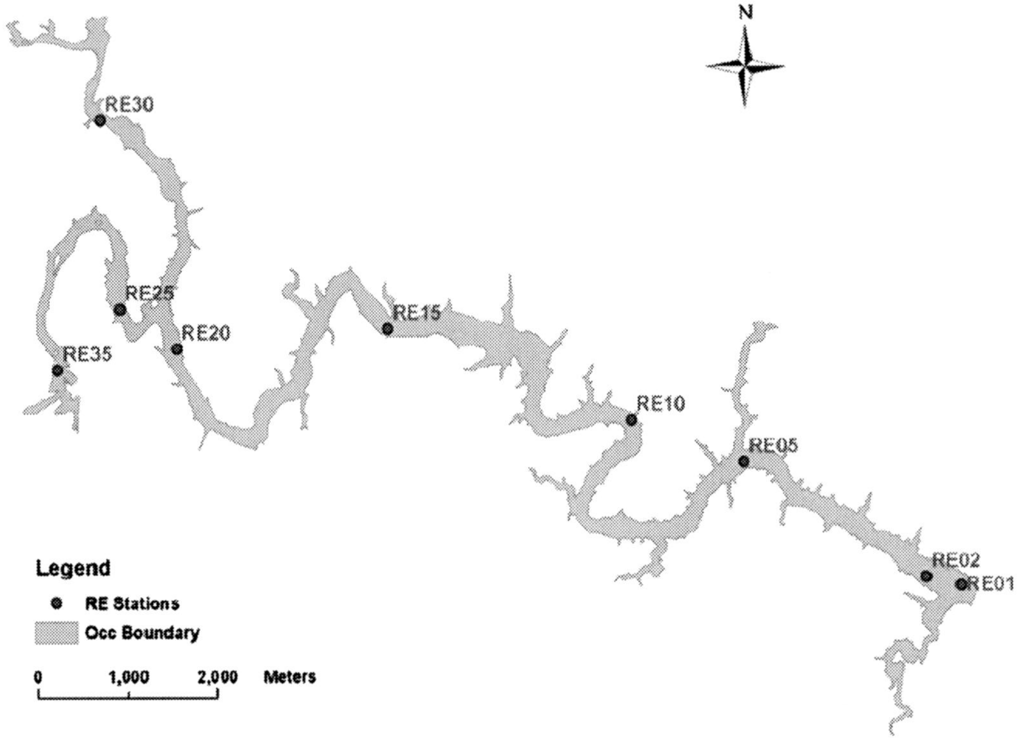

Figure 27.6 Physical geometry of the Occoquan Reservoir and reservoir monitoring stations.

Fairfax Water operates three water quality monitoring buoys on the Occoquan Reservoir beginning just upstream of the high dam and extending 8 km (5 miles) upstream. The buoys have instruments that monitor multiple water quality parameters, just below the water surface. Twenty trace organic compounds are monitored by Fairfax Water in the Occoquan Reservoir raw water supply (Rimer, 2011). OWMP staff also routinely conduct sampling on the reservoir throughout the year, with a view to maintaining a complete understanding of water quality in the longitudinal and vertical dimensions. The monitoring plan include a range of physical parameters, nutrients, algal indicators, sediment, and approximately 50 synthetic organic compounds. Although the reservoir remains eutrophic, the water reclamation plant performance and storm water management activities in the watershed have combined to maintain a stable raw water quality condition for over 30 years, in spite of a roughly 8-fold increase in watershed population.

Main challenges for operation

There are a number of significant operational challenges for the Occoquan System, but a principal one revolves around managing nitrogen. Nitrate is a regulated contaminant for drinking water with the maximum contaminant level (MCL) currently set at a concentration of 10 mgN/L. During the period of thermal stratification of the Occoquan Reservoir, UOSA operates the biological process to maximize the conversion of reduced wastewater nitrogen to nitrate, because it has been found that the nitrate plays an important role in managing water quality in the Occoquan Reservoir (OWMP, 2005). The UOSA discharge mixes with the natural stream flow of Bull Run and reaches the Occoquan Reservoir at a temperature substantially lower than the surface temperatures of the impoundment during thermal stratification. The resulting density difference causes the Bull Run flow, carrying the UOSA discharge, to plunge into the deeper waters of the reservoir. Because the reservoir is eutrophic, the summer stratification is always accompanied by development of an anoxic hypolimnion. Under such conditions, nitrate originating from the UOSA discharge serves as an alternate terminal electron acceptor (TEA) for organisms degrading organic matter in the reservoir.

As long as nitrate is available as an TEA, the dominant biological processes do not shift to anaerobic pathways, and the deep waters of the reservoir are poised at oxidation-reduction values that prevent the reduction of iron at the sediment-water interface. The observed result has been a dramatic decrease in the cycling of sediment-bound iron, manganese, phosphorus, and nitrogen into the water column. The additional effect on nitrogen is that the nitrate delivered to the hypolimnion is efficiently converted to nitrogen gas by denitrification, thereby removing it as a local algal nutrient, and also reducing its transport downstream to the Potomac River and the Chesapeake Bay. The decision to provide nitrogen to the reservoir has also resulted in the maintenance of relatively high ratios of nitrogen to phosphorus (N:P) in the system. This has had the observed effect of shifting algal dominance from undesirable cyanobacter towards green algae and diatoms. Overall algal production has been reduced through the control of phosphorus in the reclaimed water and through diffuse source management in the watershed.

During cold weather drought conditions, when the reservoir water column is well-oxygenated, the capacity of the reservoir to remove nitrogen is substantially reduced. For that reason, UOSA is typically operated to remove nitrogen during the cooler months, thereby reducing the risk of high concentrations of nitrate reaching the drinking water intake. Because UOSA is required to operate nitrogen removal facilities when the concentration at the drinking water intake reaches 5 mg/L as N, extraordinary cooperation and information sharing is needed between plant operations, OWMP, and Fairfax Water. UOSA must manipulate its biological nitrogen removal (BNR) process to provide sufficient nitrate in the summer, and to remove nitrate to a large degree during other months of the year. This is no mean feat given that the BNR process is very temperature dependent and process kinetics are slow in the cold weather conditions when nitrogen removal is most important.

UOSA plant operations are also challenged during periods of high precipitation or snowmelt. The collector sewers are owned and operated by UOSA's member jurisdictions, and are subject to significant infiltration and inflow. This condition creates high flows for the sewer system and the reclamation plant. If the hydraulic or storage and treatment capacity were to be exceeded, there is the potential for untreated or partially treated wastewater to be released to the environment, enter surface streams and be transported to the Occoquan Reservoir. Because of the sensitivity to sanitary sewer overflows upstream of the reservoir, UOSA and its participating jurisdictions continuously evaluate, identify and correct the source of leaks, or alternatively, size wastewater storage, conveyance and treatment systems to handle extraneous storm induced flows. Both approaches are expensive and complicated propositions. Finding and correcting leaks is particularly challenging because a significant portion of the leaks reside on private property and are inaccessible to UOSA or its participating members. Large wastewater retention basins are expensive and storing wastewater in large basins can create odour and vector nuisances for neighborhoods.

There are other operational challenges with reservoir management that affect drinking water treatment. When the reservoir stratifies during the summer season and there is not enough oxygen or nitrate in the bottom waters of the reservoir, the release of organic matter, ammonia, iron and manganese may be accelerated. This situation principally results from UOSA being unable to deliver sufficient nitrate during stratified conditions, or as a result of large storms that effectively flush the reservoir, thereby reducing the *in situ* concentrations of nitrate. Both situations have been observed to result in insufficient nitrate or oxygen in the bottom of the reservoir during stratification. The water quality problems manifest themselves as increased concentrations of dissolved organic matter, sulphide, iron and manganese in the raw water abstracted for drinking water production. Under such conditions, Fairfax Water typically increases the feed of potassium permanganate to oxidize the undesirable constituents. If not effectively removed, increased dissolved organics can lead to higher levels of disinfection by-products (DBPs) and may interfere with the permanganate oxidation and precipitation process. Increased concentrations of the inorganic constituents in finished water can cause increased customer complaints for taste, odour, colour or fixture staining problems with the finished potable water.

Because significant nutrient loads still enter the reservoir from non-point sources, algae blooms within the reservoir still occasionally cause problems for potable water treatment. Higher than normal concentrations of algae and certain algal species in the raw water can lead to taste and odour problems with the finished water. Fairfax Water has the ability to add copper sulphate to the reservoir as an algaecide when algal blooms reach undesirable levels.

27.3 WATER REUSE APPLICATIONS

The reuse application within the Occoquan system can be more specifically described as intentionally planned, indirect potable water reuse for the purpose of augmenting a surface water supply. In this scenario there is no separate reclaimed water storage and distribution system. Figure 27.7is a schematic showing how the wastewater collection, water reclamation, surface water, reservoir and potable water systems interact. It should be noted that wastewater arriving at the UOSA WRF is derived from three drinking water sources: the Potomac River, Lake Manassas, and the Occoquan Reservoir.

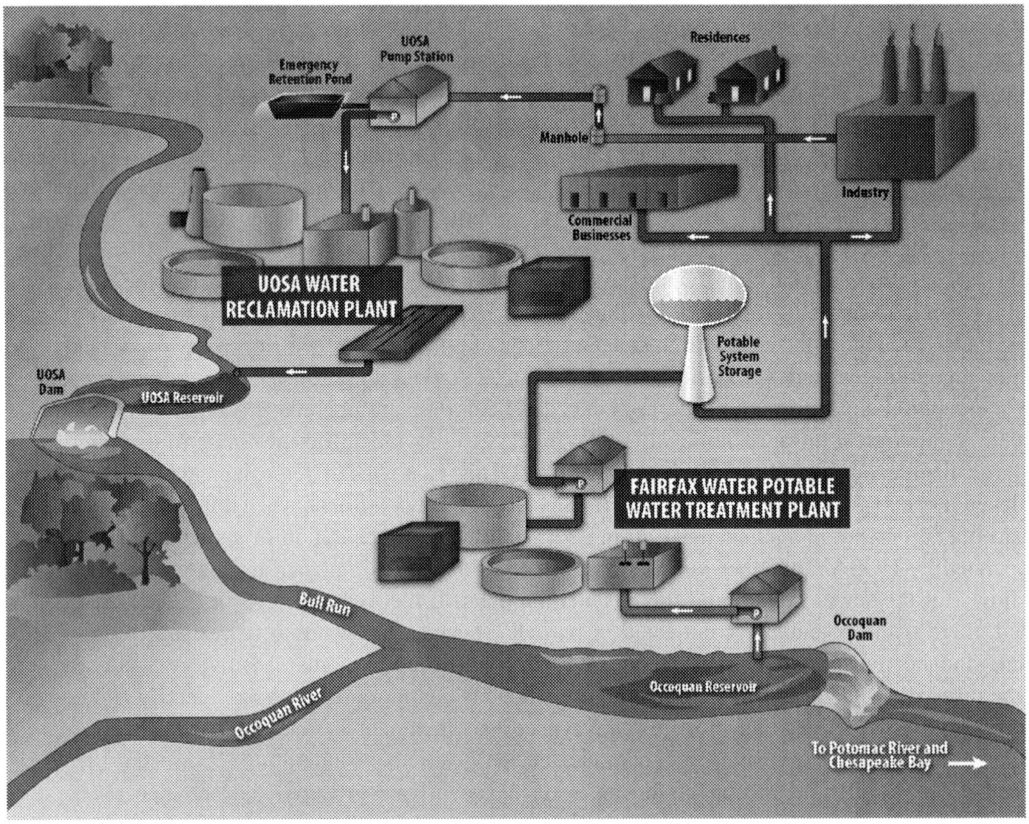

Figure 27.7 Potable reuse scenario being used for more than 34 years in Northern Virginia, USA (Courtesy of CDM).

In addition to augmenting the supply of potable water, the UOSA plant incorporates other water reuse measures. Reclaimed water is or has been successfully used onsite for chilled water, process water, fire protection, truck washing, dust suppression, construction, decorative water features, landscape impoundment and irrigation. Daily usage for these non-potable water reuses range from 3785 to 9840 m^3/d (1 to 3 MGD) depending on the season and reclamation process needs (Angelotti, 2010).

Evolution of the volume of supplied recycled water

The volume of recycled water within the Occoquan system has grown steadily over the past 50 years. This evolution has been described in prior sections. Post-World War II development in Northern Virginia led to unplanned indirect potable reuse of wastewaters from eleven small wastewater treatment plants. By 1970, these treatment plants had a combined treatment capacity of 10,980 m^3/d (2.9 MGD). The UOSA reclamation plant was constructed and began operation in 1978, which initiated the first planned indirect potable reuse operations in the region. By the mid-1980s, the volume of recycled water ranged from 18,930 to 37,850 m^3/d (5 to 10 MGD). The volume of recycled water has grown steadily over the years and reclaimed water produced at the UOSA plant has increased by an order of magnitude from those existing prior to the enactment of the Occoquan Policy. The reclaimed water produced from the UOSA plant presently ranges from 106,000 to 151,400 m^3/d (28 to 40 MGD). Currently, the UOSA facility has the capacity to reclaim 204,400 m^3/d (54 MGD), which is then released downstream to augment the Occoquan Reservoir.

Relations and contracts with end-users

UOSA has four customers and directly bills them for the cost of services. UOSA's customers are Fairfax County, Prince William County, City of Manassas and the City of Manassas Park. These four customers contract with UOSA for wastewater collection and water reclamation services. The relationship between these municipal customers and UOSA is described in a Service Agreement. UOSA is controlled by a Board of Directors. The Board is comprised of two representatives from each customer jurisdiction.

The Department of Civil and Environmental Engineering of the Virginia Tech College of Engineering is contracted to operate the OWMP, and to provide long term watershed monitoring services. Funding for the monitoring effort is secured through an agreement between the local governments of the watershed and the principal water purveyor. Half of the funding is derived from Fairfax Water, with the remaining amount distributed amongst the six jurisdictions of the watershed based on a formula that included watershed area, development, and wastewater allocation. The four UOSA jurisdictions are participants in the monitoring program, as well as Fauquier and Loudoun Counties.

Fairfax Water operates treatment works on the Potomac River and the Occoquan Reservoir with a total production capacity of 1,306,000 m^3/d (345 MGD). The production and distribution system is highly interconnected, and provides great flexibility in the production, treatment, storage, and delivery of finished water to the service area. The Section for Cooperative Water Supply Operations (CO-OP) of the Interstate Commission on the Potomac River Basin (ICPRB) helps to maintain adequate water supply for three major water suppliers in the Washington, DC, area, including Fairfax Water, by developing operating rules for both the Potomac River and the Occoquan Reservoir. Fairfax Water, a public non-profit authority, delivers water to both retail and wholesale customers in six different jurisdictions.

27.4 ECONOMICS OF WATER REUSE

Project funding and costs

The original 56,780 m^3/d (15 MGD) capacity UOSA reclamation plant was constructed with EPA grant funds that paid for approximately 75% of the construction cost of the plant. UOSA generally issues municipal revenue bonds to fund plant expansions and other major capital improvements. UOSA also takes advantage of other grant funding and low interest loans through the Virginia Water Quality Improvement Fund and Virginia Clean Water Revolving Loan Fund.

Smaller sized repairs, renovations and necessary improvements to facilities are budgeted annually through UOSA's $4 million Reserve Maintenance Fund. UOSA's other annual Operations and Maintenance budget funds the cost to operate and maintain its facilities. UOSA's Fiscal Year 2011 (FY-11) O&M Budget was $27.4 million and is described in Table 27.2. UOSA's annual funding requirement to pay the debt service on outstanding revenue bonds and loans was $30.1 million in FY-11. So, the combined annual budget for UOSA in FY-11 was approximately $61.5 million. UOSA produced nearly 41,6 Mm^3 (11 billion gallons) of water in FY-11. This brings the total unit cost of reclaimed water to around 1.48 $\$/m^3$ (5.60 $/kgals) produced.

Pricing strategy of recycled water

UOSA is responsible for wastewater collection and treatment within its service area and for producing the recycled water. UOSA is a wholesale service provider and charges its four jurisdictional customers for the direct cost of those services. The four jurisdictions are billed for their share of UOSA's Operations & Maintenance costs based on the volume of wastewater delivered to the plant for reclamation. Reserve Maintenance billings are prorated to member jurisdictions based upon the jurisdiction's allocation of treatment capacity in the reclamation plant. Debt service billings are determined based on jurisdictional need for the capital improvements funded. For example, plant expansion bills are prorated proportionally to the expanded flow capacity needed by each jurisdiction compared to the total capacity expansion.

Each jurisdiction charges individual residential, commercial and industrial user accounts according to each jurisdiction's own wastewater collection and treatment rate structures. The jurisdictions recover the cost of UOSA's billings through their own rates and billings to individual accounts or with other revenue sources.

Table 27.2 Major cost categories of UOSA's Operation and Maintenance Budget for Fiscal Year 2011 (UOSA, 2011).

UOSA O&M Fiscal Year Budget		
Budget Category	**FY-11**	**Brief Description**
Personnel Services	$15,963,500	Base salaries, health & life insurances, retirement plans, and so on
Electrical Power	$3,841,600	Power for treatment plant and pump stations
Process Chemicals	$2,071,400	Various process chemicals and filter media
Facilities Operations	$1,234,800	Fuels, telephone services, laboratory supplies
Facilities Maintenance	$1,430,000	Maintenance materials costing less the $5,000.
Contract Services	$1,845,600	Misc. contracted services for inspections, testing, equipment maintenance, and so on
Administration	$402,600	Auditing, general legal fees, financial services, and so on
Insurance	$429,300	Liability, workers' compensation, automobile, and so on
Miscellaneous	$228,500	Regulatory activity, safety clothing/equipment, and so on
Total O&M	$27,447,300	

Fairfax Water maintains over 233,000 mostly residential potable water accounts and bills potable water customers for volume used based on metered flows per thousand gallons delivered. These billings comprise about 40% of total water sales. Approximately 60% of total water sales are wholesaled in bulk to Loudoun and Prince William Counties and the City of Alexandria. Fairfax Water produces, on average, 605,700 m^3/d (160 MGD) (FCWA, 2011).

Benefits of water recycling

Potable water reuse in the Occoquan basin provides a number of monetary and non-monetary advantages over other possible alternatives. For example, reclaimed water could certainly be used for non-potable purposes, thus taking some beneficial advantage of the reclaimed water. However, such a non-potable reuse scenario would require a separate water distribution network of pump stations, piping, valves and storage facilities. That non-potable scenario might also require construction of a separate water reclamation facility that is distinct and geographically separate from the water reclamation plant or potable water treatment plant. Indirect potable reuse as implemented in the Occoquan only requires the plant and piping network infrastructure that municipalities are normally accustomed to. It comes with the distinct advantage of using the existing potable water distribution system to implement water reuse. There is no need to plan, design, construct, operate or maintain a separate reclaimed water piping network. So, in this respect it is very efficient with respect to infrastructure needs (Angelotti, 2010).

The Occoquan situation comes with another significant benefit which is avoided costs associated with the development of an alternate water supply equivalent to the magnitude of water contributed by UOSA to the reservoir. UOSA contributes about 113,600 m^3/d (30 MGD) of drought resistant supply to the downstream water supply. Presently, and increasingly in the future, water from UOSA is an important component of the Occoquan Reservoir drinking water supply.

One non-monetary benefit is improved environmental stewardship. Better water quality directly improves environmental conditions for downstream ecologies. Similarly, by keeping the reclaimed water in the basin and allowing it to flow to the reservoir via surface streams, rather than exporting it or reusing it consumptively, improves the minimum dry weather stream flows and has less impact on stream ecology during drought conditions.

Other potable reuse projects around the world have chosen to use reverse osmosis membranes as an additional barrier in the multiple barrier concept. This additional barrier may be more applicable near the coast where reject water is more easily disposed of. It may also be favoured for groundwater recharge with limited environmental buffers, in applications with little subsequent potable water treatment, in areas where total dissolved solids removal is needed or where public perception about the risk of unknowns has required it. The multiple barrier approach applied at the inland Occoquan project, which uses conventional filtration and granular activated carbon for water reclamation has been successful for more than three decades. Investigation suggests that process selection for water reclamation like the one at the UOSA facility comes with significantly lower capital, operations and maintenance costs when compared to a potable reuse reclamation project that uses reverse osmosis membranes. The environmental impact, as measured by carbon footprint, is also significantly less for a reclamation plant like the one at UOSA.

A triple bottom line evaluation should be considered when choosing technology for potable reuse projects. The UOSA facility appears to fall within the well balanced position when using a triple bottom line decision making approach (Schimmoller, 2011).

27.5 HUMAN DIMENSION OF WATER REUSE

Public education and communication strategy

When water reclamation was first proposed a number of hearings were conducted to explain what was to be implemented and to allow the public a venue to express their views. UOSA has always engaged in an active program to provide tours to local students, from grade school through college, during which potable reuse is thoroughly explained. These tours have been going on for more than 30 years, educating to a limited degree the local population on the importance of UOSA's mission. In addition, UOSA maintains a public accessible web site wherein UOSA's role in potable water reuse is clearly expressed. UOSA's success has not required dedicated public relations staff or a formal public outreach and communication program.

Periodically water related issues within the region result in the formation of technical advisory groups, citizen action committees and task forces. These may be composed of agency stakeholders, city or county government officials, community representatives, water experts and interested citizens (Ruetten, 2004). Examples of issues tackled by such groups include: down zoning of land around the reservoir to protect water quality, sitting of a major semiconductor industry within the UOSA service area and consumptive use of reclaimed water by a proposed power plant. These collaborative efforts with interested and affected parties are used to gather input before important decisions are made that might impact water quality or its availability to users.

The role of decision makers

Today, the concept of indirect potable reuse is well communicated to regulators and public official stakeholders within the region. Interested parties within local municipalities are well aware that a significant portion of the water supply is comprised of reclaimed water. Both Fairfax Water and UOSA are run by a Board of Directors. The Board members are representatives for their community constituents and make decisions in the best interest of the communities they serve. It is not uncommon for UOSA to collaborate closely with representatives of local governments about issues relating to water quality. The community and the independent water quality monitoring entity, OWMP, both openly acknowledge that the reclaimed water produced by UOSA is the most reliable and highest quality water entering the Occoquan Reservoir. The OWMP has a technical advisory panel that is comprised of members from the U.S. EPA, Virginia Department of Environmental Quality, Virginia Department of Health and an expert from an accredited and well renowned academic institution within the State (Virginia Polytechnic Institute and State University otherwise known as Virginia Tech). This provides even greater confidence and credence for potable reuse in the region.

Public acceptance and involvement

Because unplanned indirect potable reuse was already acknowledged as the root of the water quality problem and the new proposition involved solving that problem by implementing a planned potable reuse project; the perception of significantly adding value over the current situation was easily grasped and understood by the public (Ruetten, 2004). As a result there was less public opposition to this project than experienced historically with a number of other potable reuse projects. The greatest opposition to the concept was related to cost impacts to ratepayers. The cost of wastewater collection and treatment services dramatically increased in order to save the water supply. These cost increases were similar to those occurring throughout the country at a time when infrastructure improvements were being implemented for compliance with the Clean Water Act. This created some controversy, particularly in economically disadvantaged neighbourhoods. Soon after UOSA went into operation and a dramatic water quality improvement was observed, credibility was quickly established for the idea and there hasn't been any public opposition since that time.

27.6 CONCLUSIONS AND LESSONS LEARNED

Perhaps the greatest key to the success of this project is that it was implemented specifically to improve water quality problems observed in the existing surface water reservoir being used as the raw drinking water supply. The project was initiated by the Commonwealth of Virginia, via state regulation (the Occoquan Policy) which was developed by the Virginia Water Control Board and the Virginia Department of Health. Early water quality problems in the Occoquan Reservoir were clearly articulated and the best solution for the region was presented to stakeholders and interested citizens. Although water quality was the major driver, it was clearly recognized that treated wastewater flows returned to the reservoir would be a significant and valuable resource in the future (WEF & AWWA, 2008).

This project is unique in that it has a separate watershed management program (OWMP) along with its associated water quality monitoring laboratory (OWML) that provides oversight, independent accountability and recommendations to the water reclamation agent (UOSA), the potable water treatment and distribution entity (Fairfax Water) and state regulatory agencies. This was critical in establishing a credible voice of endorsement and recommendation for the plan.

Cooperative collaboration between the major institutional entities which work toward a common goal of protecting and improving the water quality of the reservoir demonstrates leadership for water related matters to the community (Ruetten, 2004).

More than 34 years of successful implementation has instilled confidence that the pioneered and visionary plan, originally conceived so many years ago, is still working.

The following Table 27.3 summarizes the main drivers and benefits obtained and some of the challenges for the future.

Table 27.3 Summary of lessons learned.

Drivers/Opportunities	Benefits	Challenges
Watershed development resulted in impacts on the drinking water source from wastewater and urban runoff	Regulatory support for adopting a reuse solution on a watershed scale (e.g. the Occoquan Policy)	Developing local support for more expensive reuse infrastructure
Need to insure long-term ability to meet regional water supply needs	The reuse solution has enhanced the safe yield of the drinking water system	Continued growth may require the development of additional sources
Federal construction grant funding was available for initial project development	Reduced local capital cost of initial project	Future expansions will all be locally funded
To improve source water quantity and quality	High quality reclaimed water has enhanced both watershed and reservoir water quality	Continued watershed development will place increased emphasis on controlling pollution from urban runoff

REFERENCES AND FURTHER READING

Angelotti R. W. (1995). Applying advanced treatment technology to achieve strict discharge limits, *Proceedings of Education Seminar: Tertiary Treatment Technology to Meet More Stringent Water Quality Standards*, Virginia Water Environment Association, Richmond, VA.

Angelotti R. W. (2010). Sustainable water practices surrounding indirect potable reuse within the occoquan basin, Proceedings 2010 Sustainable Water Management Conference and Exposition, American Water Works Association WateReuse Association and Water Research Foundation, Albuquerque, NM.

Angelotti R. W., Gallagher T. M., Brooks M. A. and Kulik W. (2005) Use of granular activated carbon as a treatment technology for implementing indirect potable reuse. Proceedings of the 20th Annual WateReuse Symposium, Denver, CO., WateReuse Association, Alexandria, VA.

Angelotti R. W., Maheiu E. and Kulik W. (2008). Strategies to minimize impacts of a microchip manufacturer on a Potable Water Reuse System, Proceedings 23rd WateReuse Symposium, Dallas, TX.

Asano T. (2007). *Water Reuse: Issues, Technologies, and Applications*. McGraw–Hill, New York, NY, pp. 1309–1329.

Chen C. W. and Gomez L. (1988). *Final Report: Water Quality Evaluation of Occoquan Reservoir*. Systech Engineering, Inc., Lafayette, CA.

FCWA (2011). 'Fairfax Water,' Fairfax County Water Authority, accessed at http://www.fcwa.org/

Rimer A. and Miller G. (2011). *Seasonal Storage of Reclaimed Water*. Report for WRF-09-05. WateReuse Research Foundation and U.S. Bureau of Reclamation, Alexandria, VA, pp. 71–82.

Robbins M. H., Jr. (1993). Supplementing a drinking water supply with reclaimed water. Proceedings AWWA Annual Conference and Exposition, San Antonio, TX.

Robbins M. H., Jr. and Gunn G. A. (1979). Water reclamation for reuse in Northern Virginia. Proceedings Water Reuse Symposium, AWWA, Denver, CO.

Ruetten J. (2004). Best Practices for Developing Indirect Potable Reuse Projects: Phase 1 Report. WateReuse Foundation, Alexandria, VA, pp. 43–47.

Schimmoller L. and Angelotti R. W. (2011). Indirect potable reuse without reverse osmosis: using GAC-based treatment to improve sustainability and reduce cost. Proceedings of the 2011 Potable Reuse Conference, Hollywood, FL, WateReuse Association, Alexandria, VA.

UOSA (2011). Upper Occoquan Service Authority, accessed at http://www.uosa.org/

U.S. EPA and U.S. AID (2004). Guidelines for Water Reuse, EPA/625/R-04/108. U.S. Environmental Protection Agency and U.S. Agency for International Development, Washington, DC.

Virginia Tech (2005). An Assessment of the Water Quality Effects of Nitrate in Reclaimed Water Delivered to the Occoquan Reservoir. Prepared by Occoquan Watershed Monitoring Laboratory, Department of Civil and Environmental Engineering, College of Engineering, Virginia Polytechnic Institute and State University (VPI), Manassas, VA.

WEF and AWWA (2008). *Using Reclaimed Water to Augment Potable Water Resources*, 2nd ed., Water Environment Federation, Alexandria, VA.

Views of the UOSA water reclamation facility.

28 Western Corridor Recycled Water Scheme

Troy Walker

CHAPTER HIGHLIGHTS

The Western Corridor Recycled Water Scheme is the largest recycled water scheme in Australia, developed in response to the most severe drought Australia had experienced in a century.

The scheme is a part of a long-term insurance policy to ensure sufficient resources are available for the region of South East Queensland, regardless of a constantly changing climate. The scheme is instrumental in conserving water by providing purified recycled water for industrial purposes and to relieve pressure on the region's potable water supplies.

The scheme has been validated for health safety and certified to the ISO 22000 food safety standard. It is operated to meet the stringent requirements of Australian Drinking Water Guidelines and even more stringent recycled water regulations in preparation for augmentation to a drinking water supply if drought conditions return to the region.

KEYS TO SUCCESS

- Largest recycled water scheme in Australia.
- Meets extremely high standards of water quality.
- Provides high quality water for industry.
- Reduces nutrient load to the receiving environment.

KEY FIGURES

- Start up 2007.
- Treatment capacity: 236,000 m^3/d.
- Total volume of recycled water produced: 74.3 Mm^3/yr.
- Capital cost - $2.5 billion Australian dollars.
- Australian Drinking Water Guidelines: high purity suitable for potable water Augmentation.

28.1 INTRODUCTION

> *"I love a sunburnt country,*
> *A land of sweeping plains,*
> *Of ragged mountain ranges,*
> *Of droughts and flooding rains."*

My Country - Dorothea Mackellar

Written over a century ago, this iconic poem is well known to most Australians, and is recited in schools around the country. The last line of the stanza encapsulates the challenge of managing water supplies in a country with a climate that is variable and difficult to predict.

In 2006, Australia was in the grip of the worst drought on record. All of the major capital cities were under water supply stress, with water restrictions commonplace. It was in this context, that the Western Corridor Recycled Water Scheme was conceived.

Main drivers for water re-use

Of Australia's major cities, Brisbane, the capital of the north eastern state of Queensland, was experiencing the worst water shortage in over 100 years. By 2006, water supplies were approaching 15% capacity, and combined with the country's highest rate of population growth, there was a real risk that the city may run out of water.

What in Australia became known as the Millenium Drought of 2000–2007 brought particular hardship to urban and rural communities in South East Queensland. It's impact here was more severe, as the rest of Australia had experienced above average rainfall in 1998–1999 and 2000–2001, while South East Queensland experienced drier conditions. An El Nino year was experienced in 2002–2003 which brought some of the highest aggregate daytime temperatures to Australia. These high temperatures impacted the water situation by increasing evaporation rates.

Adding to the pressure for South East Queensland is population growth. About two thirds of Queensland's population resides in the South East, which is the commercial, industrial and administrative centre for the state, as well as being the primary area for tourism. The population was approximately three million at the end of 2009, with the population predicted to grow to nearly four and a half million by 2031.

By 2006 reservoir levels in South East Queensland were at just 17% of capacity (Figure 28.1). Water inflows into these storages were only at 4% of the normal yearly average. In response to this situation, the Queensland Government immediately introduced tough water restrictions. This prohibited most outdoor uses of water, such as watering gardens on a daily basis and washing cars. There was also a campaign to urge residents to cut shower times form the average of seven minutes to four minutes, with the government distributing timers to be placed in the shower! Backed by a large communication effort and ongoing rebate program which incentivised householders to install rainwater tanks, water efficient shower heads and toilets, the target of water consumption per capita of 140 L/cap · d was achieved compared to the previous level of 300 L/cap · d.

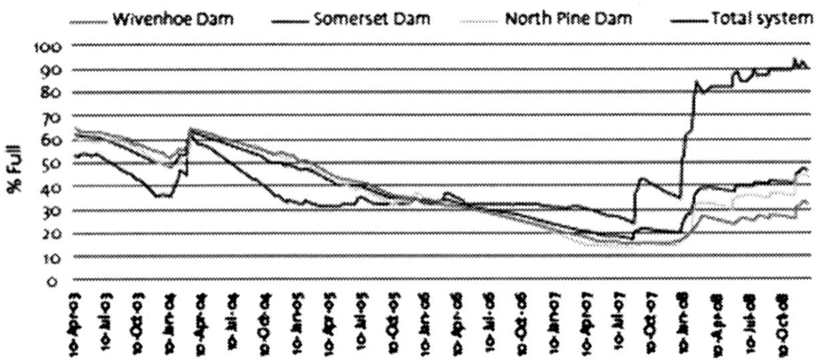

Figure 28.1 Evolution of reservoir (dam) levels in the South East Queensland.

The Queensland Government also announced the South East Queensland Water Strategy to create a safe and reliable long term water supply for this region and its future needs. The strategy included the development of the South East Queensland Water Grid (Figure 28.2). This grid, a network of pipelines connecting dams and other new water sources throughout the region, would provide water security and enable water to be moved from areas of water surplus to areas facing a shortage. The grid included the construction of the Gold Coast Desalination Plant and a major water recycling project named the Western Corridor Recycled Water Scheme.

Brief history of the scheme development

The Western Corridor Recycled Water Scheme (The Scheme), named after a developing region to the west of Brisbane which contains much of the project infrastructure, was constructed with a capacity to produce up to 236,000 m^3/d of highly purified recycled water (PRW). This large water recycling project consists of three Advanced Water Treatment Plants (AWTPs) which recycle water from six of Brisbane's major wastewater treatment plants. The AWTPs take biologically treated secondary effluent and produce a high quality water using a combination of microfiltration (MF), reverse osmosis (RO), advanced oxidation (AOX, using a combination of ultraviolet radiation and hydrogen peroxide dosing) and final chlorine disinfection. The capacities of the three plants are shown in Table 28.1.

To deliver the water for these uses, and to collect it for treatment from the wastewater plants, the Scheme has more than 200 km of pipeline network constructed in both urban and rural environments.

The Scheme's construction commenced in 2007 and with water supply reservoirs at historic low levels, it needed to be built as fast as possible. To manage the capacity of construction, all three plants (Bundamba, Luggage Point and Gibson Island AWTPs), and the interconnecting pipeline network, were designed and built by construction alliances consisting of local and international companies. The entire scheme is owned by Seqwater, the bulk water supply authority for South East Queensland (encompassing Brisbane) with operation and maintenance of the scheme managed on Seqwater's behalf by Veolia Water Australia.

Scheme construction began in 2006, with the Bundamba AWTP producing first water in August 2007. Final construction of the Scheme was completed in 2009 with the Gibson Island and Luggage Point AWTPs.

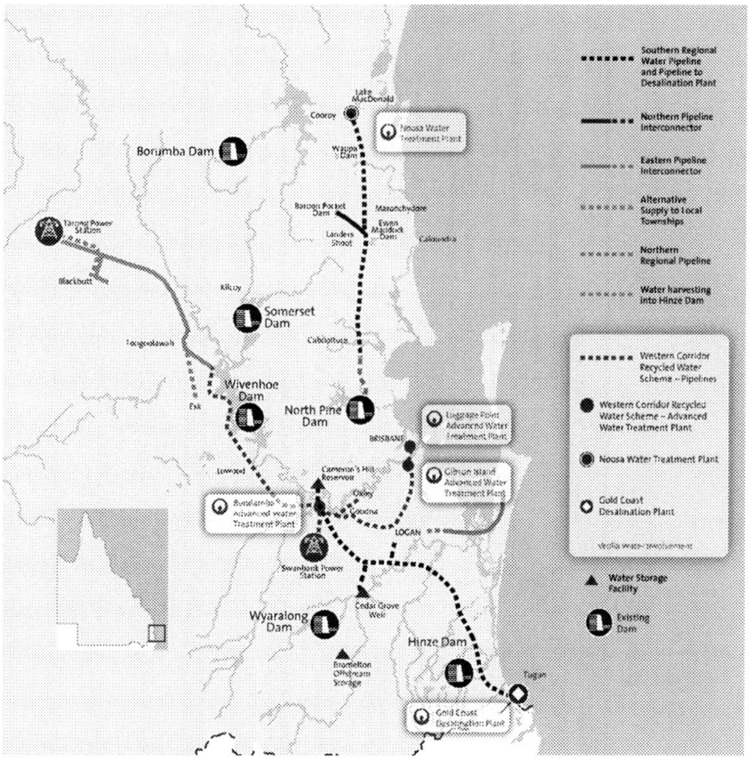

Figure 28.2 South East Queensland's Water Grid.

Table 28.1 Western Corridor AWTP Plant Capacities.

AWTP	Bundamba	Luggage Point	Gibson Island	Scheme total
Capacity, m^3/d	66,000	70,000	100,000*	236,000

* Note that the Gibson Island plant has current installed membrane capacity of 50,000 m^3/d. Membrane units are installed to achieve the full capacity of 100,000 m^3/d if membranes are procured.

Scheme objectives, incentives and water reuse applications

The Western Corridor Recycled Water Scheme is one element of the South East Queensland Water Grid, an integrated system that secures and efficiently manages the region's water supplies. The grid comprises an infrastructure network of treatment facilities and two way pipes that move water from new and existing sources across the region (Figure 28.1). It provides South East Queensland with more water sources – both climate dependent (dams and rainfall) and climate resilient (desalination and purified recycled water).

The purified recycled water produced from the Western Corridor Scheme provides an alternative water supply for power stations, to relieve reliance on drinking water sources, and is also available to supplement drinking water supplies if combined dam levels drop below 40% in the future. Two coal fired power stations currently use the recycled water for cooling and process applications, with additional industrial customers in the process of joining the scheme.

28.2 TECHNICAL CHALLENGES OF WATER QUALITY CONTROL

Treatment trains for water recycling

Secondary effluent is delivered to each AWTP, where the water is treated by a multiple barrier process (Figure 28.3). The water is initially dosed with liquid chlorine and ammonia to provide disinfection for downstream membrane systems. This dose forms chloramine, which minimises biological fouling of membranes but does not result in significant membrane oxidation damage.

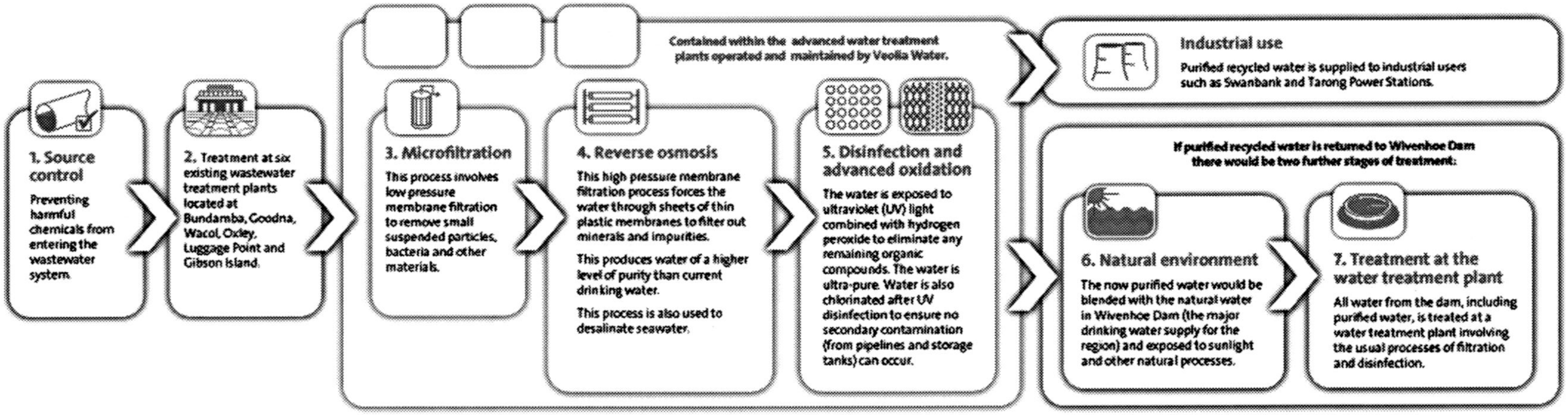

1. Source control
Preventing harmful chemicals from entering the wastewater system.
2. Treatment at six existing wastewater treatment plants located at Bundamba, Goodna, Wacol, Oxley, Luggage Point and Gibson Island.
Contained within the advanced water treatment plants operated and maintained by Veolia Water.
3. Microfiltration
This process involves low pressure membrane filtration to remove small suspended particles, bacteria and other materials.
4. Reverse osmosis
This high pressure membrane filtration process forces the water through sheets of thin plastic membranes to filter out minerals and impurities.
This produces water of a higher level of purity than current drinking water.
This process is also used to desalinate seawater.
5. Disinfection and advanced oxidation
The water is exposed to ultraviolet (UV) light combined with hydrogen peroxide to eliminate any remaining organic compounds. The water is ultra-pure. Water is also chlorinated after UV disinfection to ensure no secondary contamination (from pipelines and storage tanks) can occur.
Industrial use
Purified recycled water is supplied to industrial users such as Swanbank and Tarong Power Stations.
If purified recycled water is returned to Wivenhoe Dam there would be two further stages of treatment:
6. Natural environment
The now purified water would be blended with the natural water in Wivenhoe Dam (the major drinking water supply for the region) and exposed to sunlight and other natural processes.
7. Treatment at the water treatment plant
All water from the dam, including purified water, is treated at a water treatment plant involving the usual processes of filtration and disinfection.

Figure 28.3 The Seven Barrier Treatment Process.

Phosphorus is then removed from the water with coagulation and flocculation, using ferric chloride as the chemical. Phosphorus is removed to meet the tight environmental requirements on concentrate streams that are returned to the environment. The long term 50%ile (percentile) average phosphorus target is less than 7 mg/L, however performance well exceeds this requirement with an 50%ile result for 2011 of 0.35 mg/L.

The water is treated at all plants with membrane filtration, which provides a very high level of treatment to minimise fouling of the downstream reverse osmosis systems. The Gibson Island and Bundamba AWTPs utilise Siemens Memcor L20V ultrafiltration modules with 0.04 μm pore size (Figure 28.4a). The Luggage Point AWTP utilises a Pall Microza 0.2 μm microfiltration system. The membrane integrity of these systems is monitored automatically on line using a pressure decay integrity test, which demonstrates a 99.99% removal capability for micro-organisms such as *Cryptosporidium*, bacteria and *Giardia*.

(a) (b)

Figure 28.4 Views of the Bundamba AWTP: ultrafiltration membranes (a) and the large diameter 18" reverse osmosis system (b).

The next process barrier, reverse osmosis (RO), removes the majority of dissolved organic and inorganic compounds from the purified water. Each system operates a three stage, 85% recovery system using low pressure, brackish water, spiral wound thin film composite membranes. At the Bundamba AWTP, large 18" diameter membranes were installed (Figure 28.4b).

After the RO process, an additional treatment barrier of advanced oxidation is added which consists of a combination of ultraviolet radiation and hydrogen peroxide dosage (Figure 28.5a). This process is in place to ensure the removal of some specific trace organic chemicals, in particular NDMA (N-nitrosodimethylamine), which if present in the water can pass through the RO process (albeit at part per trillion levels).

(a) (b)

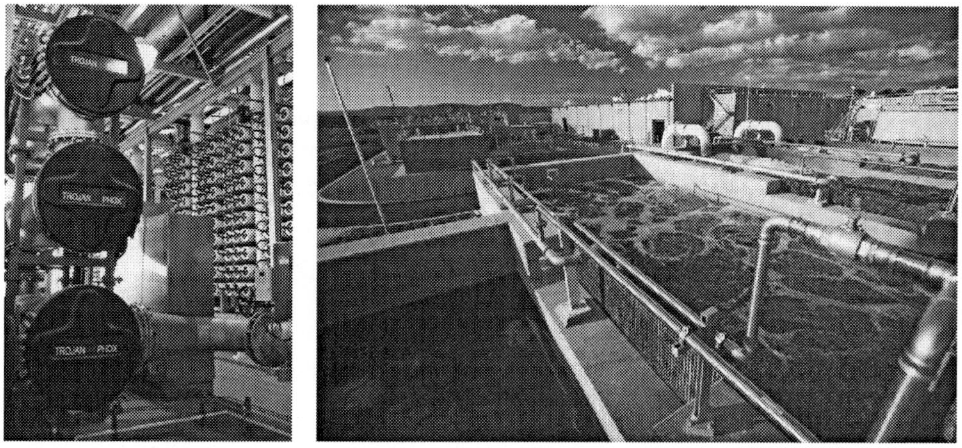

Figure 28.5 Views of the Bundamba AWTP: the UV advanced oxidation (a) and biological nutrient removal process for reverse osmosis concentrate (b).

Prior to delivery to customers, the water is stabilised with a combination of lime and carbon dioxide to ensure the water does not corrode cement lined pipework. Finally, the water is chlorinated with liquid chlorine to provide a disinfection residual in the delivery pipeline.

In addition to producing a high quality recycled water, the AWTPs also provide an additional environmental benefit by removing nutrients (nitrogen and phosphorus) to very low levels, in addition to that already provided by the upstream wastewater treatment plants. Phosphorus is removed at all plants by means of coagulation/flocculation and lamella settling. The Bundamba AWTP is unique with a biological treatment process (Figure 28.5b) for the reverse osmosis concentrate stream to further remove nitrogen from its waste stream prior to discharge to the upper reaches of the Brisbane River. This is an innovative process unique in the world.

Water quality control and monitoring

As the Western Corridor Scheme has been designed to augment drinking water supplies if required in the future, the protection of public health is paramount. The purified recycled water meets the requirements of the Australian Drinking Water Guidelines, as well as a separate Public Health Regulation enacted by the Queensland Government, which specifies over 400 individual water quality parameters. In order to ensure public health safety and secure public confidence in the high-quality recycled water, there is a rigorous program of water testing in place, with more than 100,000 test results gathered per year.

In addition to regular testing of purified recycled water, the source water for the recycling plants (secondary effluent from conventional wastewater treatment plants) is also regularly tested to ensure that water quality is suitable for treatment, and also to identify any emerging water quality hazards that may have entered the system.

The water is analysed for a range of contaminants (Water Quality Report, 2010), including:

- Herbicides
- Pesticides
- Endocrine Disrupting compounds
- Micro-organisms
- Inorganic chemicals
- Pharmaceuticals
- Radionuclides

A water quality report detailing 48,000 results, taken between December 2008 and June 2010 is available at the web (Water Quality Report, 2010). An abbreviated set of results, typically provided to industrial customers, is shown in Table 28.2.

Table 28.2 Major recycled water quality characteristics.

		Recycled Water	
Parameter	**Secondary Effluent**	**Measured**	**Required Value**
Total Organic Carbon (mg/L)	8.8–12.2	<100 ppb	<1.0
Turbidity (NTU)	1.8–6.7	0.3 (0.1–1.8)	<5
Total Dissolved Solids (mg/L)	506–1390	137 (90–190)	<250
Total Nitrogen (mg/L)	2.0–5.8	0.5 (0.3–1.2)	<1.5
Total Phosphorus (mg/L)	0.8–4.2	0.01 (0.01–0.02)	<0.15
E.Coli (CFU/100 mL)	–	<1	<1
NDMA, (ng/L)	–	<5	<10

*Range of average values across all 6 wastewater treatment plants.

Each AWTP is operated under a systematic preventative risk management approach developed using the principles of Hazard Analysis and Critical Control Point (HACCP) and the Scheme's operation is accredited to the food safety standard, ISO 22000. A critical control point is defined as a process step essential to prevent, eliminate or reduce a water quality hazard to an acceptable level. Across the Western Corridor Scheme, critical control points for monitoring treatment barriers have been identified to ensure the water quality requirements are consistently met at that barrier. Each of these control points is validated to ensure it can demonstrate the integrity of that barrier to both microbiological and chemical hazards. The barriers selected for the Scheme are shown in Table 28.3.

Table 28.3 Critical control points for water quality hazard analysis.

Critical Control Point	Process	Parameter monitored
CCP1	Wastewater treatment plant (Biological nutrient removal)	Online ammonia nitrogen.
CCP2	Microfiltration/ultrafiltration	Pressure decay integrity test.
CCP3	Reverse osmosis	Online permeate conductivity and permeate dissolved sulphate analysis.
CCP4	Advanced oxidation	Present Power Ratio
CCP5	Final chlorine disinfection	Concentration x time (CT)

Across these critical barriers, a removal capability for microorganisms is granted. In each case, the system can be operationally monitored to achieve, as a minimum, the log removal of microorganisms as shown in Table 28.4 (one log removal being equivalent to 90% removal, two log 99% etc.).

Table 28.4 Credit for log removal of microorganisms by critical treatment processes.

	Microbial removal, Log Reduction Value (LRV)			
Critical treatment process	Viruses	Bacteria	Protozoa	Helminths
Wastewater treatment	2	2	1	1
Microfiltration	0	3.5	3.5	3.5
Reverse osmosis	2	2	2	2
UV disinfection/AOX	4	4	4	2
Chlorine disinfection	2	2	0	0
Total log reduction	10	13.5	10.5	8.5
Required Log Removal for augmenting drinking water supplies	9.5	8	8	8

Major challenges for operation

In January 2011, South East Queensland was hit by the worst flooding in a century. These caused significant damage in the Brisbane area, including some damage to the Western Corridor Scheme's infrastructure. In particular, the Bundamba AWTP saw damage to a transfer pump station, which temporarily prevented the transfer of water along the main western pipeline to the Tarong Power Station (Figure 28.6).

Damage to the reverse osmosis concentrate disposal line, also at Bundamba, prevented operation of the plant for several months. During this time, the plant was placed into a medium term shutdown condition with membranes stored in preservative chemical solutions. In order to deliver water to the nearby Swanbank Power Station, temporary pumping arrangements were made to deliver water from the Gibson Island and Luggage Point AWTPs to the power station (the only supply to the Swanbank Power Station in normal conditions comes from Bundamba AWTP).

Under these conditions of higher rainfall, pumping of water from the Scheme to drinking water reservoirs has so far not been required, and thus Scheme production requirements have been lower than the available production capacity. While AWTPs are capable of operating at lower production rates, this has created some challenges in operating the delivery pipeline at low rates (and hence long detention times) increasing the risk of loss of disinfection residual, and in some cases, higher levels of some disinfection by-products. Careful attention has also been paid to the management of some storage chemicals on site where storage duration can impact product quality (e.g. sodium hypochlorite).

As reservoir levels have remained at high levels through to 2012, water is not predicted to be required for reservoir augmentation for some years. As a result, scheme production has been well below design capacity and a decision was taken by the Queensland Government to place the Gibson Island plant into a long term shutdown. This has involved the long term preservation of membranes and general maintenance planning to ensure the asset can be returned to service should drought conditions return.

Figure 28.6 Bundamba AWTP during the 2011 Brisbane flood.

Water quality testing and monitoring requirements for the scheme remain very stringent, as the AWTPs are still operated to the same high standard. Approximately 30,000 tests are conducted at each AWTP annually, which results in a relatively higher analysis cost than for many other water treatment facilities. In order to demonstrate compliance with stringent standards, many test must be conducted down to extremely low levels of detection. This has driven innovation in the improved development of analytical techniques at local government testing laboratories.

28.3 WATER REUSE APPLICATIONS

The Western Corridor Scheme was constructed in two separate stages. Stage 1 consisted of the construction of the Bundamba AWTP and pipelines to coal fired power stations where it is used to supplement both process and cooling water systems. This focused on reducing pressure on the region's drinking water supplies by providing 25 million m^3 (Mm^3) of water between August 2007 and September 2009.

Stage 2 of the Scheme, completed in mid-2009, increased the production capacity with the construction of Gibson Island and Luggage Point AWTPs. It was envisaged that this additional volume of recycled water could be supplied to industrial and agricultural users, and used to augment potable water supplies, if combined dam levels fell below 40%.

Forty eight hours of rain in May 2009 poured down the equivalent of half the volume of Sydney Harbour, and brought the drought to an end. This change can be seen clearly in Figure 28.1. At the time, the reservoir was running at 60% capacity, having recovered partially from the previous month at 50%, and the downpour topped it to 73% within a matter of days. The wet weather continued throughout 2010 and heavy rainfall filled the region's reservoirs to full capacity. With reservoirs now close to full, potable augmentation of recycled water will not be required for the next few years, and as a result demand for recycled water is well below the scheme's capacity.

Evolution of the volume of supplied recycled water

As discussed above, current production rates are well below capacity due to a lower demand. As shown in Figure 28.7, in 2010 the monthly production was between 1.0 and 3.6 Mm^3/month, compared to the maximum monthly capacity of 7.17 Mm^3/month. About 50% of the recycled water was delivered to power plants (from 0.56 to 1.61 Mm^3/month).

In 2011, the Gibson Island AWTP was placed into a long term shutdown condition, with equipment shut-down into reserve to save on operational costs. Equipment will be maintained such that the plant can be re-commissioned if drought conditions return.

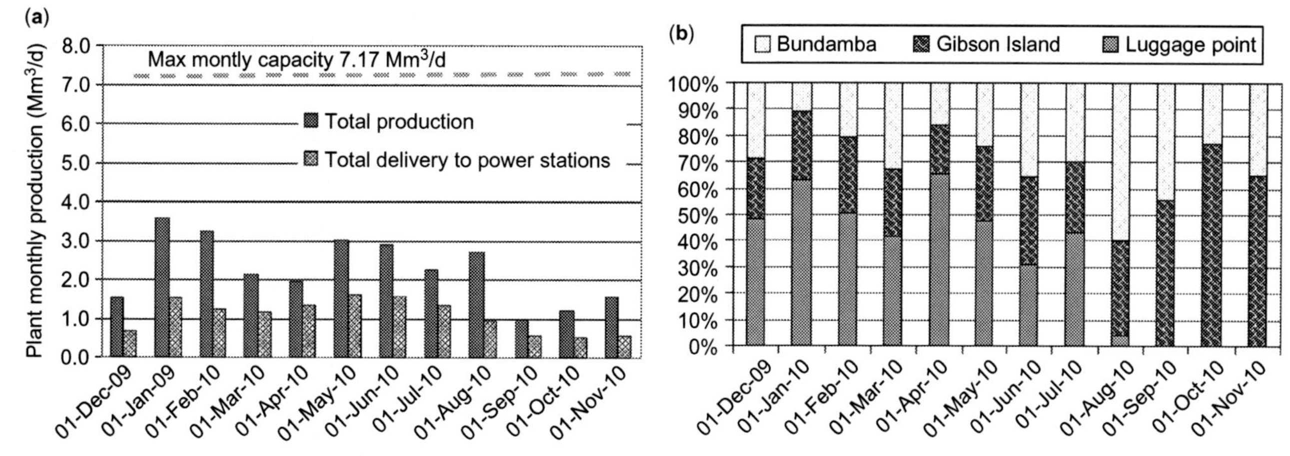

Figure 28.7 Recycled water production of the Western Corridor Scheme in 2010: total monthly production and delivery to industry (a) and specific production in percentage of the three recycling plants (b).

28.4 ECONOMICS OF WATER REUSE

Scheme funding and costs

The capital cost of the Scheme was approximately AU$2.5 billion (Australian dollars), with AU$408 million provided by the Australian Commonwealth Government, and the remainder by the State Government of Queensland.

The operational costs of the Western Corridor Scheme include operations of the AWTPs along with the management and maintenance of both raw water and treated water pipelines. Expenses include fixed costs for labour, repair and maintenance with variable costs for power and chemicals. Labour accounts for 48% of the operating costs, followed by energy 20% and chemicals 19%, and repair and maintenance 10% (Figure 28.8a). Water quality monitoring requires 3% of the operation costs which is significant for such large advanced treatment scheme.

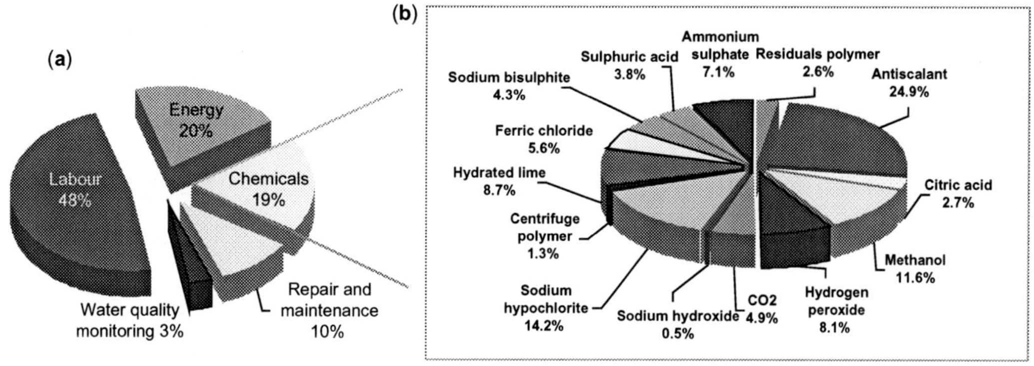

Figure 28.8 Distribution of operation costs including treatment and distribution (a) and chemicals' consumption (b).

Figure 28.8b provides also a further breakdown of chemical costs for Bundamba AWTP. Significant contributions are from:

- Antiscalant (24.9%), necessary to prevent the scaling of some soluble salts on the reverse osmosis membranes, is a significant contributor due to its high cost per litre.
- Sodium hypochlorite (14.2%), is used to create chloramines to prevent biofouling on membrane processes, used for ultrafiltration membrane cleaning and also dosed in the final treated water for final disinfection. While a less expensive chemical per litre, this contribution reflects a high usage.
- Methanol (11.8%), is used exclusively at Bundamba AWTP to provide a carbon source for denitrification, part of the reverse osmosis concentrate treatment process.

The energy consumption of a typical AWTP for the period of 2010 was 1.14 kWh/m^3 on average at full plant capacity. Adding in pumping energy to deliver water to power station customers requires 1.73 kWh/m^3. A breakdown of energy consumption for different unit processes within the AWTP is shown in Figure 28.9. Reverse osmosis accounts for 35% of the energy demand, followed by recycled water pumping, 22.7%.

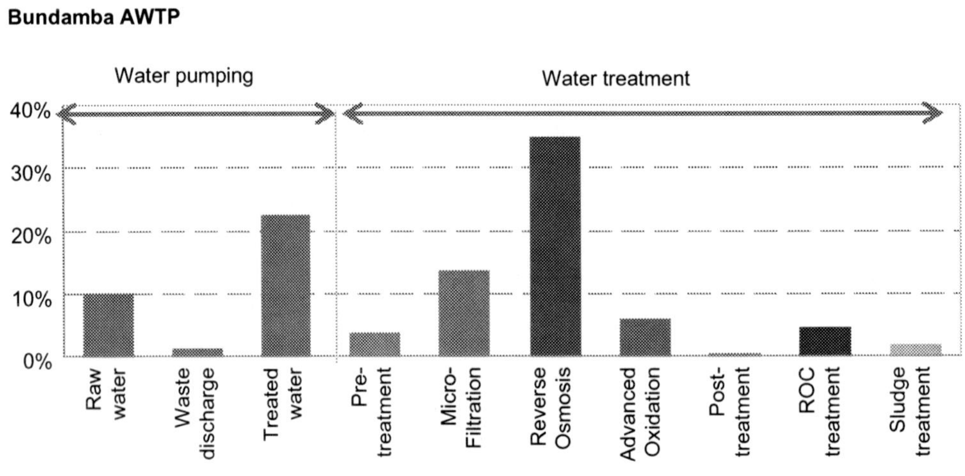

Figure 28.9 Breakdown of energy consumption at the Bundamba AWTP.

Benefits of water recycling

A key benefit of the Scheme is that it takes pressure off the region's water supply by providing an alternative source of suitable water quality for power stations and increasingly other industrial users around the Brisbane area. The Scheme minimises the risk to Brisbane facing a water supply shortage in the future, as it is ready to provide very high quality water, if required, to augment potable water supplies indirectly by delivery to reservoirs.

All three AWTPs provide an additional removal process for nutrients prior to discharge of return streams to the environments. Phosphorus is a pollutant present in the incoming secondary effluent which is further removed at the AWTPs (Figure 28.10). This removal of phosphorus is beneficial to the receiving water environments of the Brisbane River and Moreton Bay. Innovative nitrogen removal processes at the Bundamba AWTP also reduces the overall load of this to the Brisbane River. With environmental targets of 7 mg/L 50%ile phosphorus and 19 mg/L 50%ile total nitrogen, actual performance for 2011 was 0.35 mg/L 50%ile phosphorus and 7.6 mg/L 50%ile total nitrogen at Bundamba AWTP.

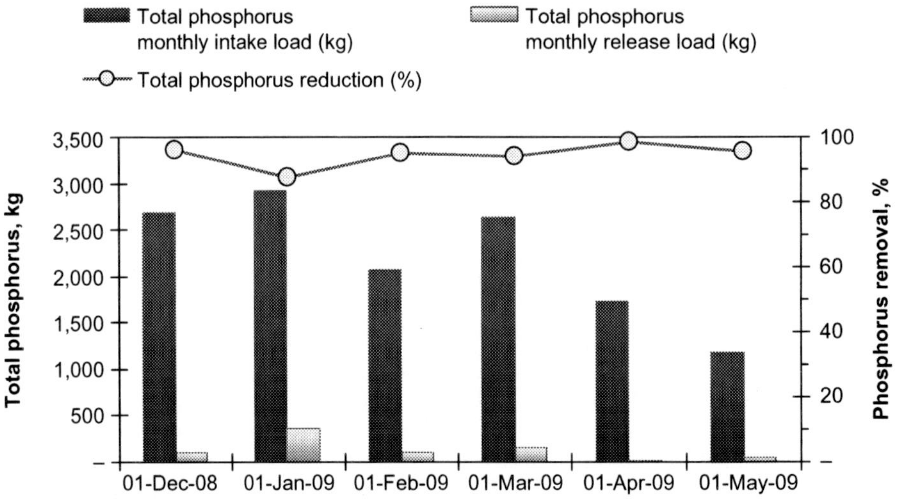

Figure 28.10 Phosphorus removal at Bundamba AWTP.

The Western Corridor Scheme has generated a significant amount of activity in the research and development arena, for example, 11 research and development projects with the University of Queensland and other institutions in the region and a AU$2.25 million water recycling research program for five years at the University of Queensland's Advanced Water Management Centre. The Australian Water Recycling Centre of Excellence headquartered in Brisbane was also established by the Federal Government in 2010 and is hosted by Seqwater to enhance the management and use of water recycling nationally and internationally through industry and research partnerships. It has a research budget of AU$20 million over five years funded by the Australian Government's Water for the Future initiative.

Four main research themes identified are 1) technology, efficiency and integration, 2) risk management and validation, 3) social, institutional and economic challenges and 4) sustainability in water recycling.

28.5 HUMAN DIMENSION OF WATER REUSE

Public education and communication strategy

The Western Corridor Recycled Water Scheme receives plenty of visitors from the community every year, organised by both the owner, Seqwater, and the operators, Veolia Water. Since 2007, more than 2500 people have visited the plants (Figure 28.11). This is not only an opportunity for the public to learn about the recycling technology, but also to understand the importance of the plant to the community and the South East Queensland region.

Figure 28.11 Western Corridor – view of a plant visit.

The Queensland Government put in a significant effort in the promotion of water recycling, particularly as the scheme was under construction. This was as a part of their education program around the drought. This included a significant amount of press advertising, television and radio advertisements.

Both during construction, and into the operations of the scheme, open days and site tours have been conducted on the sites. An interactive virtual tour of the Bundamba AWTP is available on line, and is also available at dedicated computers installed as kiosks at the Bundamba plant.

Both Veolia Water and Seqwater are also sponsors of the South East Queensland Healthy Waterways Awards which are held annually by the South East Queensland Healthy Waterways Partnership. This group was formed to improve management of catchments and the health of waterways in South East Queensland and the awards celebrate achievements in this area.

The role of decision makers

Operation of the scheme requires several important interfaces with regulators. As part of the South East Queensland Water Grid, production targets and water quality compliance is managed through the Office of the Water Supply Regulator, a part of the State of Queensland's Department of Environment and Resource Management (DERM). This office provides instructions to the scheme operators as to how much water to provide to customers. Water quality reporting is also managed through this office.

Environmental regulation is also managed via DERM, with several environmental licences for operations required. Good relationships with local representatives of DERM is a very important part of operations.

28.6 CONCLUSIONS AND LESSONS LEARNED

The Western Corridor Recycled Water Scheme forms an integral part of South East Queensland's Water Grid, and takes a key role in providing an alternative source of water in case of drought. The scheme was built in record time, and currently delivers water to two coal fired power stations to substitute, and thus remove pressure for drinking water supplies. By removing nutrients from the feed water, the scheme also works to improve environmental outcomes to local receiving water bodies of the Brisbane River and Moreton Bay.

Operation according to the principles of HACCP (Hazard Analysis and Critical Control) ensures the scheme consistently produces an exceptionally high quality of water, meeting both the Australian Drinking Water Guidelines and also local state quality regulations. Ongoing water quality risk assessments, bolstered by an extensive R&D program in recycled water will provide additional surety into the future.

While the scheme is not currently required at full capacity, it is operated to a level of reliability and quality suitable to augment potable supplies if required in the future.

The following Table 28.5 summarizes the main drivers and benefits of this water reuse project, as well as the major challenges for the future.

Table 28.5 Summary of lessons learned.

Drivers/Opportunities	Benefits	Challenges
Severe Drought Increasing Population Insurance Policy for Variable Climate	Augmentation for drinking supplies Substitute potable supply for industrial users Demonstration of the production of very high quality recycling water	Stringent quality requirements Managing at low production rates Managing through flood conditions

REFERENCES AND FURTHER READING

Final Progress Report Western Corridor Recycled Water Project Stage 2 (2009). Western Corridor Recycled Water Pty Ltd, Brisbane, http://www.environment.gov.au/water/publications/urban/ pubs/wsa-qld-08-final-report.pdf.

Poussade Y., Vince F. and Robillot C. (2010). Energy consumption and greenhouse gases emissions from the use of alternative water sources in South East Queensland, Proc. IWA Water and Energy Conference 2010, Copenhagen, Denmark.

Seqwater Reservoir Levels http://www.seqwater.com.au/public/dam-levels.

Water Quality Report (2010). Watersecure (a water quality report detailing 48,000 results, taken between December 2008 and June 2010), http://www.watersecure.com.au/pub/latest-news/water-quality-report-december-2008-to-june-2010.

29 More than 40 years of direct potable reuse experience in Windhoek

Josef Lahnsteiner, Piet du Pisani, Jürgen Menge and John Esterhuizen

CHAPTER HIGHLIGHTS

The Windhoek direct potable reclamation practise is still unique - worldwide. The advanced multi-barrier treatment process is producing purified water of a quality that consistently meets all the required drinking water standards. During the more than 40 years of operation, no health problems have been reported and the health safety was verified by epidemiological studies. It is all the more remarkable that this success story has been achieved in a country with limited technical and financial resources.

KEYS TO SUCCESS

- Vision and great dedication of the potable reclamation pioneers
- Excellent information policy and education practice
- No reclaimed water related health problems experienced
- Multiple barrier approach
- Reliable operation and on-line process and water quality control
- Public-private partnership

KEY FIGURES

- Treatment capacity: 21,000 m^3/d
- Total annual volume of treated wastewater: 6.4 Mm^3/yr
- Total volume of recycled water: 5.8 Mm^3/yr
- Total annualised costs: 0.95 €/m^3, including 0.75 €/m^3 operating and maintenance costs
- Recycled water tariffs: consumption-related, from 0.75 €/m^3 to 2.3 €/m^3
- Drinking water standards (blending with maximum 35% reclaimed water)

29.1 INTRODUCTION

The Republic of Namibia is located in the south-western part of Africa (Figure 29.1). It is flanked by two deserts, the Namib to the west and the Kalahari to the southeast, and is the driest country in sub-Saharan Africa. The country has a surface area of 825,000 km^2 and a population of 2.2 million. Thus the population density (2.7 inhabitants/km^2) is one of the lowest worldwide. Windhoek, the capital and economic centre of Namibia, is a fast growing city. It is located in the Central Highlands about 1600 m above sea level. The annual rainfall is roughly 360 mm, while the potential surface evaporation rate is approx. 3400 mm/yr. Roughly 700 km separates the city from the nearest perennial river, the Okavango, to the north-east, while the Atlantic Ocean, which borders the country in the west (coastline = 2650 km), is approx. 300 km away.

The population of Windhoek is about 350,000 inhabitants with a growth rate of around 5% per year. The city's water supply is based on the use of surface water (dams fed by ephemeral rivers) and groundwater (borehole water). However, as the region is one of the most arid in the world, regular water shortages result from repeated periods of erratic rainfall. Therefore, the water supply with ground and surface water (von Bach Dam, Swakoppoort Dam and Omatako Dam situated at a distance to Windhoek from 60 to 170 km) cannot be guaranteed. In view of this situation, in 1994 the Windhoek City Council approved an integrated Water Demand Management Program including policy matters, legislation, education, as well as technical and financial measures (Van der Merwe, 1994, 2000).

Today, Windhoek's total water demand amounts to 25 million m^3 per year (Mm^3/yr), that is an average demand of 68,500 m^3/d and depending on the season, a range of variation from 40,000 to 80,000 m^3/d. There are three main sources of potable water consisting of water from the aforementioned dams (65%), groundwater (7% from 50 municipal production boreholes) and reclaimed water from the New Goreangab Water Reclamation Plant, NGWRP (28%). Figure 29.2 shows the Von Bach Dam Water Treatment Plant (WTP) and a view of borehole drilling). Seawater desalination is not feasible due to Windhoek's high altitude and the relatively long distance to the ocean.

One option for the augmentation of the water supply to Windhoek is the managed aquifer recharge (MAR; water banking). A recent study (Tredoux, 2009) was aimed at developing water quality guidelines with specific management options for ensuring sustainable recharge. Borehole injection was considered as the preferred method. The sources are excess reclaimed (NGWRP) and Von Bach Dam water. However, the quality of both sources is not ideal for aquifer recharge. In the reclaimed water, the salinity is excessive (871 mg/L as a median value and 972 mg/L as 95% percentile) and in the Von Bach Dam water, the DOC is too high (3.6 mg/L as a median value and 7.2 mg/L 95%ile). Due to these facts, MAR on a full scale had to be postponed. Thus, the quality improvement of the aforementioned sources by (partial) desalination (reverse osmosis) at NGWRP and by advanced DOC removal (ozone and BAC) at the Von Bach Dam WTP is under discussion. Another option is to remove DOC by activated carbon adsorption. This process has been employed for polishing (DOC removal) in pilot scale injection tests.

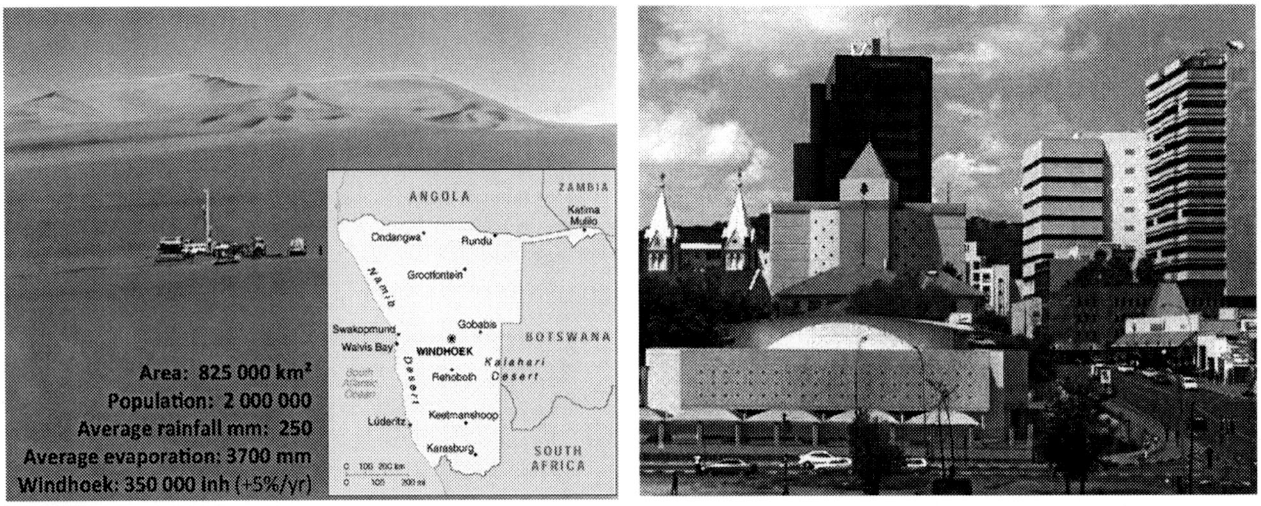

Figure 29.1 Location of Namibia and view of the Windhoek city center.

Figure 29.2 View of the Von Bach Dam WTP and Borehole Drilling in the Auas Mountains.

Main drivers for water reuse

In the 1960s, regular droughts in Namibia and a continuous severe shortage of potable water in Windhoek necessitated an investigation by the city authorities into alternative sources of raw water. It was therefore decided to reclaim water from the Gammams Water Care Works (Domestic Wastewater Treatment Plant) and from the polluted Goreangab Dam. It must also be stressed that the City of Windhoek had no other choice, as potable reclamation was the only affordable option. Other alternatives such as the transport of water from distant sources (e.g. Okavango River) or seawater desalination were far too expensive.

Brief history of the project development

Groundwater was the sole source until the 1930s when the first dam, the Avis Dam, was built. After decades of severe water stress, during the mid-1960s the Council of the Windhoek Municipality took the decision to implement water reclamation from secondary domestic effluent for potable reuse. The basis for this decision was provided by successful pilot tests at the Gammans Water Care Works. This led to the building of the Old Goreangab Water Reclamation Plant (OGWRP) in 1968. The OGWRP, which was modified (upgraded) several times between 1968 and 1995, produced drinking water utilizing the aforementioned source and Goreangab dam water. After nearly 30 years of successful potable reclamation, by the second half of the 1990s, this plant no longer met the latest technological standards or the increased demand for potable reuse. It was therefore decided to build a new and larger reclamation plant (the New Goreangab Water Reclamation Plant) adjacent to the old one. Since the start-up of the new facility in 2002, OGWRP has only been used for urban reclamation (reuse for golf course irrigation, etc.)

Project objectives

The objective of Windhoek's integrated water resource management was and is to secure supply by water saving, water reclamation, water banking (managed groundwater recharge) and water pollution control. The basic objective of the New Goreangab project was the realisation of a reclamation plant, which would meet the newest technological standards and cover increased potable water demand.

29.2 TECHNICAL CHALLENGES OF WATER QUALITY CONTROL

Domestic secondary effluent is used for potable reclamation,, as previously mentioned. In order to attain the highest safety levels possible for this unique practice, a multiple barrier approach is employed. There are three types of barriers: *non-treatment, treatment* and *operational barriers*.

An essential *non-treatment barrier* is the strict separation of domestic and industrial wastewater, that is only domestic sewage is used for potable reclamation. Industrial wastewater (approximately 1 Mm^3/yr), which is discharged mainly from a brewery, a tannery and a meat processing companies, is treated separately in sequentially operated anaerobic and aerobic ponds at another location. The reclaimed water is subsequently reused for garden turf cultivation. In a current project, a central treatment plant with an MBR as its core technology has been tendered on a BOOT basis. The process was piloted in 2011 and commissioning of the full-scale plant is expected in 2013. Another crucial *non-treatment barrier* is the comprehensive monitoring of the sewage treatment plant inlet and outlet, as well as the extensive monitoring of the drinking water quality. The blending of the reclaimed water with other potable sources (treated Von Bach Dam water and borehole water, maximum 35% reclaimed water) is also mentioned as a further important *non-treatment barrier*. Only blended water is distributed to the consumers.

The *treatment barriers*, which consist of purification units in constant operation are explained in detail in the next section.

Operational barriers represent additional treatment options or operational measures that can be used on demand. An additional treatment option is powdered activated carbon, which can be dosed if the adsorption capacity of the GAC is too low or the organic load of the reclamation plant inlet is too high. An example of an operational measure is switching to the recycle mode when the water quality fails to meet the absolute values set for the different process units.

Downstream treatment of domestic wastewater

The domestic wastewater is extensively pre-treated in the Gammans Sewage Treatment Plant named Gammans Water Care Works (Figure 29.3) for potable reclamation in the NGWRP. The aforementioned is a biological nutrient removal plant (UCT process), which employs primary treatment (fine screen, coarse screen, grit and grease removal, primary sedimentation) and secondary treatment (nitrogen and biological phosphorous removal) with both an activated sludge process and trickling filters in parallel operation.

The official name "Gammams Water Care Works", which has very positive connotations, represents an example of the city's excellent water marketing and communications practice. Further examples are described in the next sections.

In addition to the treatment in the Gammans Water Care Works, the secondary effluent (COD approximatly 50 mg/L) is polished in maturation ponds with a retention time of approximately three days. Figure 29.4a shows a section of the maturation pond, which apart from its polishing function, serves as a natural habita,t for example for water birds and snakes (such as zebra snakes, a spitting cobra species).

The outlet of the maturation ponds serves as raw water for the New Goreangab Water Reclamation Plant. The major quality parameters (95th percentile) of this effluent are shown in Table 29.1.

Figure 29.3 View of conventional biological wastewater treatment at the Gammans Water Care Works.

Figure 29.4 Maturation pond section (a) and a view of the polluted Goreangab Dam (b).

Table 29.1 Quality of the maturation pond effluent.

Parameter	Unit	95th tile	Parameter	Unit	95th tile
COD	mg/L	43	TDS	mg/L	344.5
Color	mg/L	80	Alkalinity	mg/L	153.1
DOC	mg/L	13.6	Aluminum	mg/L	1.7
Turbidity	NTU	102.2	Ammonia	mg N/L	1.7
UV_{254}	abs/cm	0.4	Chloride	mg/L	54.9
Heterotrophic plate count	per 1 mL	399,100	Fluorine	mg/L	0.8
Total coliforms	per 1 mL	464,000	Iron	mg/L	5.3
Fecal coliforms	per 1 mL	37,400	Manganese	mg/L	1.6
Chlorophyll (a)	µg/L	46.86	Nitrate and Nitrite	mg/L	0.1
Girardia	per 100 L	276.00			
Cryptosporidium	per 100 L	200.00			

Initially, the raw water fed to the New Goreangab Water Reclamation Plant consisted of 50% secondary (maturation pond) effluent and 50% Goreangab Dam water, which at the time of plant start-up was of a quality comparable to that of the maturation pond effluent. Currently, the raw water fed to the plant is composed entirely of maturation pond effluent, because the quality of the Goreangab Dam water has deteriorated due to pollution, including subsequent algae blooms (Figure 29.4b), and therefore, this source can no longer be utilized. This water is currently only abstracted for treatment in the Old Goreangab Water Reclamation Plant, OGWRP (Figure 29.5a) and used for landscape irrigation, sport fields and golf course (Figure 29.5b). The OGWRP treatment process currently consists of flocculation, dissolved air flotation, rapid sand filtration, and chlorine disinfection. It was refurbished and re-inaugurated in October 2010.

(a)

(b)

Figure 29.5 View of the Old Goreangab Water Reclamation Plant (a) and the golf course of the Windhoek Country Club (b).

Treatment train for water recycling

This globally unique direct potable reclamation plant (Figure 29.6a) transforms secondary domestic effluent into high-quality drinking water by means of an advanced multi-barrier system. It produces a maximum of 21,000 m^3/d of drinking water that is constantly controlled to ensure its suitability to be safe for human consumption. The plant was started up in mid-2002 and officially inaugurated on December 2, 2002 by the President of Namibia, Dr. Sam Nujoma. Figure 29.6b shows the plaque, which was unveiled during this historic event.

(a)

(b)

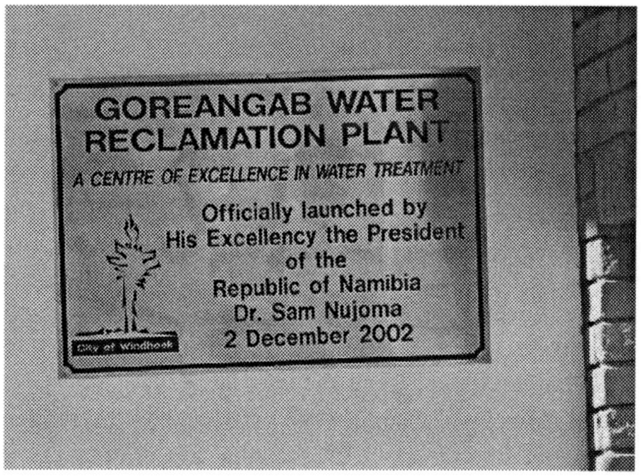

Figure 29.6 The New Goreangab Water Reclamation Plant.

Extensive pilot testing was conducted prior to the selection of the final treatment process shown in Figure 29.7 (Du Pisani, 2006; Van der Merwe, 2008). The treatment train for water recycling includes the following treatment barriers: powdered activated carbon (PAC) dosing (optional), pre-ozonation, enhanced coagulation and flocculation, dissolved air flotation (DAF), dual media filtration, main ozonation, biological activated carbon (BAC) filtration, granular activated carbon (GAC) adsorption, ultrafiltration (UF), disinfection with chlorine and stabilization with caustic soda (NaOH).

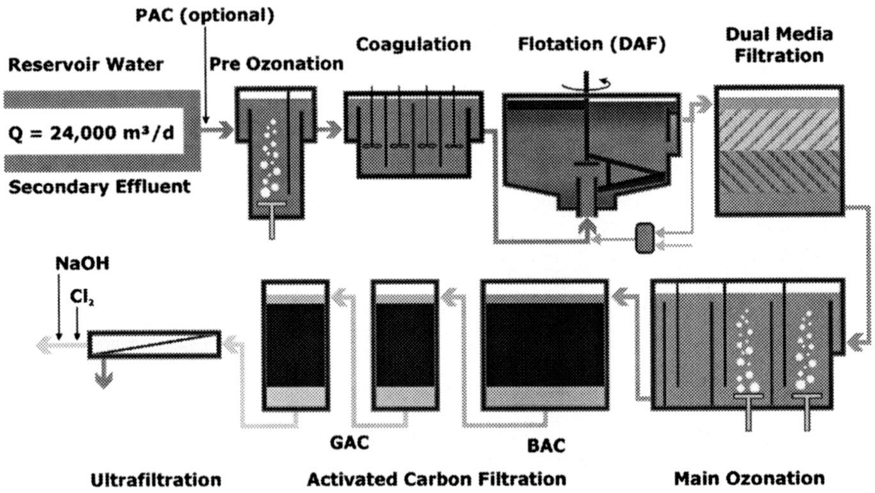

Figure 29.7 The New Goreangab Water Reclamation Plant Process Flow Diagram.

Pre-ozonation is the first of a series of barriers in the reclamation plant. It provides pre-disinfection, that is ozone eradicates part of the bacteria and viruses present in the raw water (maturation pond effluent). Furthermore, it partially splits long-chain, natural organic matter into smaller molecules, which are more easily removed during enhanced coagulation and subsequent dissolved air flotation. A further positive effect of pre-ozonation is micro-flocculation, which transforms parts of the dissolved impurities into micro-particles. Ozone is a highly reactive gas, which is generated at the reclamation plant.

$$\text{Ozonegeneration:}\quad O_2 + \text{energy} \longrightarrow O + O$$
$$O + O_2 \longrightarrow O_3(\text{ozone})$$

Coagulation and flocculation is accomplished by both the dosing of a coagulant/flocculant (ferric chloride) and energy input. The first step in this process is coagulation, during which the positively charged ferric ions adhere to suspended solids and dissolved impurities such as proteins and polysaccharides. These are thus destabilized and transformed into solid particles. The second step is flocculation in which the dosed ferric ions and the water react to create ferric hydroxide, which flocculates particles. A polymer can be added to assist flocculation.

$$Fe^{3+}Cl_3^- + 3H_2OFe(OH)_3 + 3H^+]] >$$

Dissolved air flotation removes impurities such as suspended solids and organics. This is achieved by introducing small air bubbles, which lift the flocs produced in the upstream coagulation/flocculation stage onto the surface. This floating sludge is then removed. Some suspended solids still remain in the clarified water and need to be removed in further steps. Prior to the next barrier (dual media filtration), NaOH and potassium permanganate ($KMnO_4$) are added to the partially cleaned water, in order to facilitate oxidation and the precipitation of dissolved iron (Fe) and manganese (Mn) in the dual media filter. Iron and manganese removal is important, as these metals cause colour and taste, which are aesthetically unacceptable to consumers. Furthermore, these metals are detrimental to the downstream membranes and lead to irreversible fouling when they precipitate on the membranes.

The main task of *dual media filtration* is the removal of turbidity including impurities such as algae and protozoa. It consists of two different filter media and is operated in a downflow mode. The water is first filtered by coarse-grained hydro-anthracite, which removes larger particles, and then by fine-grained sand, which removes fine particles. Dual media filters are physical barriers and become clogged with particles after several hours of operation. It is therefore necessary to clean the filters periodically by means of backwashing with an air scour and clean process water. The resultant heavily polluted backwash water is released into a nearby wastewater treatment plant. Sand filters also provide a barrier against protozoan cysts and oocysts, and for this reason, the backwash water and the first filtrate are disposed of and not recycled into the process.

The water is again treated with ozone (*main ozonation*), in order to inactivate persistent residual bacteria, virus and protozoa, as well as to oxidize organics and micro-pollutants. These include endocrine disrupting compounds, which as tests indicate are largely degraded and removed by pre- and main ozonation. Natural organic matter such as humic acids is cracked and thus transformed into biodegradable compounds, which are subsequently removed biologically by the next barrier (BAC). After ozonation, small amounts of H_2O_2 are added to destroy the residual ozone in the water and to protect the useful bacteria fixed in the porous structure of the BAC.

Biological activated carbon filtration - the filter medium not only consists of standard activated carbon, but also bacteria that are fixed in the porous surface of the carbon grains. This biomass partially consumes the cracked (biodegradable) humic acids (a fraction of DOC) whereby the disinfection by-product potential, that is Tri-Halo-Methane (THM) formation potential, is reduced.

Granular activated carbon adsorption - residual high and low molecular weight organics such as humic acids and micro-pollutants are adsorbed onto the porous surface of the carbon grains. This further reduces the disinfection by-product potential and the micro-pollutants concentrations to a "non-detectable level".

Membrane filtration is the most sophisticated filtration process (Figure 29.8). Membranes represent very fine filters through which water is either pressed or sucked. Any content, which is larger than the microscopic pores, is separated out. *Ultrafiltration* is a membrane process using extremely fine pores with a size of 0.045 μm. All particles larger than this cut-off are removed while dissolved ions and molecules with low molecular weights are allowed to pass. Therefore, ultrafiltration forms a complete barrier against turbidity and micro-biological contaminants such as bacteria and protozoa and a partial barrier against organic material (rejection of high molecular DOC). The UF system consists of six dedicated UF racks. Each rack has a maximum capacity of 200 m^3/h. A total of 336 pressure driven, hollow fiber membrane modules are employed–each module having a membrane area of 36 m^2–to give a membrane area of 12,096 m^2. The hollow fiber membranes are cleaned periodically by back-flushing with chemicals and water. During backwash the flow direction inside the membranes is reversed and all of the non-absorbed fouling is flushed from the system. The entire back-flush water is mixed with raw water and then recycled.

Figure 29.8 Views from the ultrafiltration process and the staff monitoring the direct potable reclamation process.

Safety disinfection with chlorine - The raw water (maturation pond effluent) has now passed through a highly sophisticated multi-barrier system, which guarantees the output of high quality potable water. However, for safety reasons and as a final barrier in the reclamation plant, the water is disinfected with chlorine (Cl_2) prior to being pumped into the Windhoek water distribution system. The water is also stabilized with caustic soda for pH correction or stabilization to prevent any erosion of the city's pipelines and household appliances. This water is pumped to the New Western pump station where it is blended with other potable water sources.

Water quality control and monitoring

Water quality monitoring includes three major elements (van der Merwe, 2008). The first involves the on-line monitoring of key parameters (pH, turbidity, conductivity, DOC, UV_{254}, dissolved oxygen, particle count, ozone concentration and free chlorine) at different process units.

The second consists of process unit monitoring with automatic samplers (24-hour composite samples) at critical control points. The parameters of this part can be divided into four groups:

1. Physical and organoleptic parameters (turbidity, color, pH, conductivity, TDS, hardness, corrosiveness),
2. Macroelements (potassium, sodium, chloride, sulfate, fluoride, iron, manganese, silica),
3. Organics and nutrients (DOC, COD, UV_{254}, THM, nitrate, nitrite, ammonia, TKN, orthophosphate,
4. Microbiological parameters (heterotrophic plate count, total coliform, fecal coliform, *Escherichia coli*, fecal streptococci, *Clostridium perfringens*, coliphages, *Giardia*, *Cryptosporidium* and chlorophyll a).

The third element comprises health-relevant monitoring programs that are still in research phase, which examine emerging parameters of new health concerns related to micropollutants (endocrines disrupters, pharmaceutical compounds, etc.), organic matter (NOM and EfOM) and viruses.

As summarized in Table 29.2, the New Goreangab Water Reclamation Plant (NGWRP) has been continuously producing good water quality that largely meets the required final water specification.

If the major parameters of turbidity, DOC, THM and UV_{254} in the water from the NGWRP and the Von Bach Dam are compared, the reclaimed water shows superior quality to that from the dam (Table 29.3). In this context, the pointed remark that blending with "natural" dam water compromises the good quality of the reclaimed water (the original source is domestic sewage) could be appropriate, but due to the increasing salinity of the reclaimed water, this parameter is vastly improved by blending.

Table 29.2 Final water quality criteria and operational results of the Goreangab Water Reclamation Plant.

			Actual Operational Results	
Parameter	**Units**	**Final Water Specification**	**50%tile**	**95%tile**
Physical and Organic				
Chemical Oxygen Demand	mg/L	10–15	6.6	11
Colour	mg/L Pt	8–10	0.5	0.5
Dissolvec Organic Carbon	mg/L	3*	1.7	2.8
Total Dissolved Solids	mg/L	1000 max or 200 above incoming	838	938
Turbidity	NTU	0.1–0.2	0.05	0.10
UV_{254}	abs/cm	0.00–0.06	0.015	0.027
Inorganic				
Aluminium	Al mg/L	0.15	0.005	0.05
Ammonia	N mg/L	0.1	0.05	0.18
Iron	Fe mg/L	0.05–0.10	0.01	0.03
Manganese	Mn mg/L	0.01–0.025	0.005	0.015
Microbiological				
Heterotrophic Plate Count	per 1 mL	80–100	0	4
Total Coliforms	per 100 mL	0	0	0
Faecal Coliforms	per 100 mL	0	0	0
Chlorophyll a	µg/L	1	0.27	2.58
Giardia	per 100 L	0 count/100 L or 5 log removal	0	0
Cryptosporidium	per 100 L	0 count/100 L or 5 log removal	0	0
Disinfection by products				
Trihalomethanes	µg/l	20–40	35	57

*A target was set for 3 mg/L DOC (with maximum of 5), based on the premise that in the final blended water not more than 1 mg/L DOC originating from sewage water should be present. A maximum blending ratio of 35% to 65% water from other sources was specified

Table 29.3 Comparison of water quality of reclaimed water and treated dam water.

		Treatment Plants	
Parameter	Unit	NGWRP median	Bach Dam WTP median
Turbidity	NTU	0.05	0.6
DOC	mg/L	1.7	3.6
THM	μg/L	35	73
UV_{254}	abs/cm	0.015	0.05

In the frame of the EU-funded research project TECHNEAU, the high quality of the NGWRP was confirmed during three sampling and monitoring campaigns. The performance of both overall performance and that of selected unit processes was evaluated for the removal of specific emerging and conventional contaminants, as well as the evaluation and demonstration of the state-of-the-art monitoring systems for drinking water treatment developed during this project. In addition to data collection and analysis, a risk assessment was carried out to highlight potential hazards in the treatment train. Finally, several types of hormonal activity were detected in the raw water, which are at least in part due to the natural and synthetic hormones present in the raw water. All types of activity were reduced by the treatment process to levels below the detection limit of the bioassays. The results of the Assimilable Organic Carbon (AOC) analysis from the third sampling campaign again showed good correlation with CoW Biodegradeable Dissolved Organic Carbon (BDOC) measurements. Both were removed to below detection limits in the final water. Even the microcystins present in the raw Goreangab Dam water were completely removed.

Main operational challenges

The main operating challenge is to keep all the process units running perfectly in order to produce excellent reclaimed water quality. A second challenge, with regard to the further improvement of the quality of the reclaimed water, relates to the evaluation of new scientific knowledge and the consideration of the relevant findings in the monitoring programs, applied R&D projects, and subsequently, in the treatment process.

A specific problem is that the salinity of the raw water at the inlet of the reclamation plant had been increasing, which makes blending with other potable sources more challenging due to the limits contained in drinking water standards. The plant operator, the Windhoek Goreangab Operating Company Ltd. (WINGOC) together with the City of Windhoek has therefore embarked on a project to pilot test desalination to reduce the salt content in the final water.

In order to deal with these challenges in optimum fashion and to include as much international specialist process and operating know-how as possible, the financing European Investment Bank requested the involvement of internationally recognised water sector players. Therefore a 20-year operation and maintenance contract was concluded between the City of Windhoek and WINGOC which has the three major international water treatment contractors, Berlinwasser International, VA TECH WABAG and Veolia Water as shareholders.

29.3 WATER REUSE APPLICATIONS

The major application, which is still unique worldwide, is direct potable reuse (Figure 29.9). The average citizen drinks water from the tap on a daily basis. As previously mentioned, the drinking water supplied to the consumers is a blend of reclaimed water (maximum 35%) and treated surface water. This represents one of the *non-treatment barriers* which has been implemented as an additional margin of safety because the raw water is domestic sewage and the reclamation plant is therefore operating in an area where no one else has experience. With emerging pollutants, which might not be known or catered for by the plant, blending was adopted as a fundamental principle. The blending ratio is based on an exposure to DOC from human origin of 1 mg/L and it is accepted that the natural water contains zero and the reclaimed water a maximum allowable of 3 mg/L. A blending ratio of 2:1 then accepts a DOC (from human origin) of 1mg/L in the blend.

Apart from potable reuse, urban water reuse (golf course irrigation, watering of parks, etc.) provides a substantial contribution to the coverage of the city's water demands. It can be further concluded that augmentation with both potable and irrigation water guarantees the supply security of water-stressed Windhoek. This is evidenced by the fact that Windhoek is a green city (Figure 29.10) in the heart of an arid country.

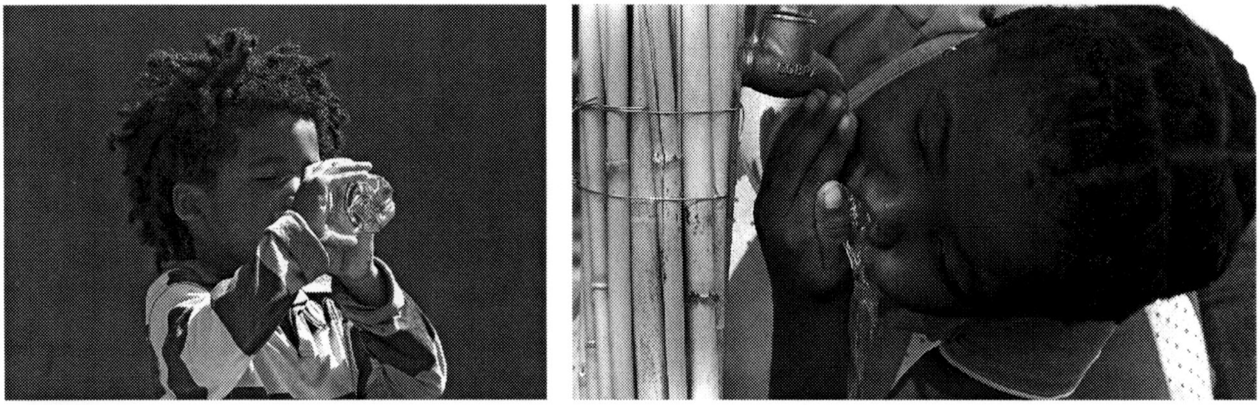

Figure 29.9 Direct potable reuse.

Figure 29.10 Windhoek, a green city.

The annual production in the last two years totalled 5.7 Mm^3 (2009) and 5.8 Mm^3 (2010) respectively, which corresponded to an average daily production of approximately 16,000 m^3/d (maximum capacity of 21,000 m^3/d). Due to increasing potable water demand, it is intended to fully use this capacity in the near future.

29.4 ECONOMICS OF WATER REUSE

Project funding and costs

The project was financed by the European Investment Bank (55%), the Kreditanstalt fuer Wiederaufbau (40%) and the City of Windhoek (5%). The investment costs for the reclamation plant were approximately €12.5 million including electrical and mechanical equipment of €8.3 million and civil works of €4.2 million. The total operating costs amount to 0.95 €/m^3 (capital costs 0.2 €/m^3, operational costs 0.75 €/m^3), which is far cheaper than the other options for importing water to Windhoek (e.g. transport from the Okavango River).

Pricing strategy of recycled water

Consumption-related, progressive pricing for the potable water (blend of reclaimed and treated dam water) plays an important role in achieving water saving targets. As a typical example, water charges in 2011 were as follows:

- 0 to 0.2 m^3/d–0.75 €/m^3,
- 0.201 to 1.2 m^3/d–1.26 €/m^3
- >1.2 m^3/d–2.3 €/m^3.

Depending on the consumer type, the price for non-potable reclaimed water in the OGWRP (semi-purified water reused mainly for landscape irrigation) ranges between 0.18 and 0.71 €/m^3 (Municipal Council of Windhoek, 2011). The price for municipal consumers is 0.24 €/m^3 while commercial consumers have to pay the maximum price of 0.71 €/m^3.

Benefits of water recycling

Other water supply options such as the use of distant sources (Okavango River 14 €/m^3 or the Tsumeb Karst Aquifer supply 4 €/m^3) would be much more expensive than direct reclamation. The major economic benefit is simply that potable water supply security is guaranteed by reclamation, that is without reuse there would be a lack of water, as approximately 30% of the potable water demand is covered by reclaimed water. Without reclamation (potable and urban) the city's development (population growth, tourism, industrial and commercial development, etc.) would have been substantially inhibited and the city would not be as green as it is.

The current supply system of the national bulk supplier NamWater, has a 95th percentile assured yield of 17 Mm^3/yr and a 98th percentile assured yield of 6 Mm^3/yr, as compared to the current demand of 25 Mm^3/yr for the City of Windhoek. Potable water reuse is therefore an indispensible part of the potable supply to Windhoek.

29.5 HUMAN DIMENSION OF WATER REUSE

Public education and communication strategy

In order to raise the level of water awareness, since 1995 the city of Windhoek has arranged suitable educational programs in schools (as part of the curriculum), on radio and television, as well as in the print media. Furthermore, open days and visits to the Gammans Water Care Works, the New Goreangab Water Reclamation Plant and the Gammans laboratory have been organized for the public, including citizens, students, national and international water experts, politicians.

Water matters have a generally high priority and are the object of political, media and public interest. This is best exemplified by the aforementioned inauguration of the NGWRP by the President of Namibia, Dr. Sam Nujoma, in December 2002. The related event was shown on TV and widely reported by the media. The high significance of water was also demonstrated at the Shanghai World Expo in 2010, where the Namibian presentation focused on two topics. Firstly, the natural beauty of the country (Namib desert, the Atlantic Ocean coast, wildlife, etc.) and secondly, water issues, for example with an excellent video about potable reuse showing the entire multiple barrier approach and including the New Goreangab Water Reclamation Plant as the main feature.

The information approach has been very open since the beginning of potable reclamation in 1968. Today, articles on water matters are published regularly in the local newspapers. The major topics are water saving, water treatment and water quality. Drinking water quality problems (taste and odour) are mainly caused by algae blooms in the surface water (von Bach reservoir). The national bulk water company (NamWater) supplies this source and therefore the City of Windhoek is closely involved with NamWater and the Ministry of Agriculture, Water and Forestry in a tripartite body called WARSCO (Windhoek aquifer recharge steering committee) in order to resolve the quality issues of the surface impoundments. However, as it is critically important to safeguard the public acceptance of potable reuse, all water complaints are taken seriously. When a complaint is received, the city informs the consumers of the reasons, takes a sample from the consumer's tap and provides a water quality report. In addition, the city informs the public via a press release if and when surface water supplies experience quality problems. Due to this proactive approach no instances have occurred where water quality problems have been associated with water reclamation.

The role of decision-makers

Due to severe water stress there has been intense political involvement for at least the last 60 years. As early as the 1950s, Windhoek's politicians had to propose additional water augmentation options including water reuse and during the early 1960s the Council of the Windhoek Municipality decided to conduct pilot tests at the Gammans Wastewater Treatment Plant. Based on the positive results from these tests, the City Council took the decision to implement potable reuse. In 1994, the City Council approved an integrated Water Demand Management Program including policy matters, legislation, education, as well as technical and financial measures. In 1997, Windhoek City Council also resolved to build the New Goreangab Reclamation Plant and to increase the level of reclaimed water to a maximum of 35%.

The availability of water is critical to the growth and expansion of the city. As the city has experienced massive population growth, from 140,000 people in 1990 to 350,000 people in 2010, this has to be managed in terms of water availability. Although the National Vision 2030 sees Namibia as an industrial nation by 2030, the City of Windhoek has resolved not to promote the establishment of water intensive industry. This has resulted in industrial investment that is slower than required, but it is crucial for the sustainability of the city.

Public acceptance

In 2012, people are not generally prepared to accept reclaimed drinking water from domestic secondary effluent and this requires public and psychological barriers to be broken down. This is practically impossible if potable reuse is not the only viable option. In Windhoek, which is still the only municipality employing direct potable reuse, the main reasons for public acceptance are the lack of affordable alternatives and the fact that since the beginning of potable reuse in 1968, no health problems have been experienced or verified in epidemiological studies with regard to reclaimed water. Additional reasons are the candid information policy implemented from the beginning in 1968 and the persistent and intelligent marketing contained in the aforementioned education programs. Over the years, acceptance has been further strengthened by the excellent water quality policy (multiple barrier approach including comprehensive monitoring programs) and from 2002 onwards, by the trust in the advanced technology of the New Goreangab Water Reclamation Plant. In fact, there have been practically no consumer complaints related to reclaimed water.

A public perception survey (with questionnaires) is planned in order to obtain an accurate and up-to-date picture of consumer acceptance of potable reuse. To promote the survey, an "Open Day" at the NGWRP is to be organized jointly with the monthly Municipal newsletter *Aloe*.

29.6 CONCLUSIONS

The Windhoek experience shows that treated domestic sewage can be successfully used for potable reclamation. The advanced process employed produces reclaimed water of a quality that constantly meets all the required drinking water standards. Approximately 30% of the potable water supply consists of reclaimed water. Therefore, this source is an essential part of the integrated water resource management (IWRM) and has contributed greatly to the social and economic development of the city. Further important elements of IWRM are urban reclamation (dual pipe system) and water demand management. An important water augmentation option for the future is managed aquifer recharge (water banking) with excess reclaimed and treated surface water. The required testing and planning work has already been carried out.

The main reasons for public acceptance of potable reclamation are the lack of other affordable choices and the fact that since the beginning of potable reuse, no reclaimed water related health problems have been experienced. Further important factors have been the open information policy employed from the start of the project in 1968, the excellent public education practice and the consumer confidence in both the quality management and the advanced water treatment technology employed.

It is all the more remarkable that this success story has been achieved in a country with limited technical and financial resources. In other words, it is quite exceptional that a developing country like Namibia leads the way in potable water reuse. A standing that represents a tribute to the vision and great dedication of the potable reclamation pioneers, as well as the on-going commitment of Windhoek's water professionals.

The following Table 29.4 summarizes the main drivers of the water reuse in Windhoek, as well as the benefits obtained and some of the challenges for the future.

Table 29.4 Summary of lessons learned.

Drivers/Opportunities	Benefits	Challenges
Severe droughts in the past/water stress Population growth	Increased water supply security Save potable water produced by an advanced multiple barrier approach Secured social and economic development	Optimum management of the multiple barrier approach (non-treatment, treatment and operational barriers) for safe potable reuse Increasing salinity Emerging micro-pollutants Increasing water demand

REFERENCES AND FURTHER READING

Du Pisani P. L. (2006). Direct reclamation of potable water at Windhoek's Goreangab reclamation plant. *Desalination*, **188**, 79–88.

Municipal Council of Windhoek (2011). General Amendment of Tariffs. *Government Gazette*, 15 July 2011, n°4756, pp 116–117.

Tredoux G. and van der Merwe (2009). Arificial recharge of the Windhoek aquifer, Namibia: water quality considerations. *Boletin Geologico y Minero*, **120**(2), 269–278.

Van der Merwe B. F. (1994) Water demand management in Windhoek, Namibia. Internal Report, City Engineer's Department, Windhoek Municipality, Windhoek, Namibia.

Van der Merwe B. F. (2000). Integrated water resource management in Windhoek, Namibia. *Water Supply*, **18**(1), 376–380.

Van der Merwe B. F., du Pisani P., Menge J. and Koenig E. (2008). Water reuse in Windhoek, Namibia: 40 years and still the only case of direct water reuse for human consumption. In: *Water Reuse – An International Survey of Current Practise, Issues and Needs*, B. Jimenez and T. Asano, IWA Publishing, London, UK, pp. 434–454.

WEB-LINKS FOR FURTHER READING

Web site of the Windhoek Goreangab Operating Company (WINGOC), http://www.wingoc.na
Web site of the City of Windhoek, http://www.windhoekcc.org.na
Web site of the EU funded project TECHNEAU, http://www.techneau.org

Views of the New Goreangab Plant in Windhoek.

Views of process units of the New Goreangab Plant in Windhoek.

Index

Plants, 7, 12–14, 16–17, 24–25, 39, 41, 56–57, 59, 61–62, 64–66, 68, 70–72, 76, 78–80, 82–87, 89–90, 93, 98–99, 102–104, 107–109, 112, 123–124, 127–128, 130, 142, 151–153, 162, 164, 175, 179, 181–185, 188, 192, 200, 202, 207, 210, 214, 217, 219, 224–228, 231–233, 243, 248, 251, 262, 274–275, 278–279, 293, 298, 302, 313, 316, 324, 332, 340, 343–344, 346–347, 349, 359

R